DIFFERENTIAL GEOMETRY
AND
PHYSICS

Proceedings of the 23rd International Conference of
Differential Geometric Methods in Theoretical Physics

NANKAI TRACTS IN MATHEMATICS

Published

Vol. 1 Scissors Congruences, Group Homology and Characteristic Classes
by J. L. Dupont

Vol. 2 The Index Theorem and the Heat Equation Method
by Y. L. Yu

Vol. 3 Least Action Principle of Crystal Formation of Dense Packing Type and Kepler's Conjecture
by W. Y. Hsiang

Vol. 4 Lectures on Chern–Weil Theory and Witten Deformations
by W. P. Zhang

Vol. 5 Contemporary Trends in Algebraic Geometry and Algebraic Topology
edited by Shiing-Shen Chern, Lei Fu & Richard Hain

Vol. 6 Riemann-Finsler Geometry
by Shiing-Shen Chern & Zhongmin Shen

Vol. 7 Iterated Integrals and Cycles on Algebraic Manifolds
by Bruno Harris

Vol. 8 Minimal Submanifolds and Related Topics
by Yuanlong Xin

Nankai Tracts in Mathematics – Vol. 10

DIFFERENTIAL GEOMETRY AND PHYSICS

Proceedings of the 23rd International Conference of Differential Geometric Methods in Theoretical Physics

Tianjin, China 20 – 26 August 2005

Editors

Mo-Lin Ge & **Weiping Zhang**

Chern Institute of Mathematics, Tianjin, China

World Scientific

NEW JERSEY • LONDON • SINGAPORE • BEIJING • SHANGHAI • HONG KONG • TAIPEI • CHENNAI

Published by

World Scientific Publishing Co. Pte. Ltd.

5 Toh Tuck Link, Singapore 596224

USA office: 27 Warren Street, Suite 401-402, Hackensack, NJ 07601

UK office: 57 Shelton Street, Covent Garden, London WC2H 9HE

British Library Cataloguing-in-Publication Data
A catalogue record for this book is available from the British Library.

Nankai Tracts in Mathematics — Vol. 10
DIFFERENTIAL GEOMETRY AND PHYSICS
Proceedings of the 23rd International Conference of Differential Geometric Methods in Theoretical Physics

ISBN 981-270-377-2

Printed in Singapore by World Scientific Printers (S) Pte Ltd

Dedicate to the memory of Professor Shiing-Shen Chern

FOREWORD

The Nankai Mathematical Institute, whose grand new premises were inaugurated on the occasion of the 23rd conference on Differential Geometric Methods in Theoretical Physics, is the creation of the great Chinese mathematician Shing-Shen Chern. Unfortunately he did not live long enough to attend the conference, but his spirit was present throughout.

Chern recognized many years ago the need for China to have its own centre for advanced mathematical research, a centre modelled on the Institute for Advanced Study at Princeton where Chern first went and on the Berkeley Institute (MSRI) which he later helped to establish. By his personal example and tireless efforts the Nankai Institute came into being and is well placed to play a leading role in the new China of the 21st century.

The 2005 conference will no doubt be just the first of many subsequent meetings at Nankai which will strengthen the international links between Chinese mathematicians and their colleagues in other countries.

I first met Chern in 1956, when I was a fresh Ph.D. on my first visit to the United States. He was very friendly and helpful and our association continued over subsequent years. When I was President of the London Mathematical Society in 1976, he came to London as the AMS bicentennial lecturer and brought me a Chinese poem, in beautiful calligraphy, which he had composed on the flight. Later he encouraged me to visit Nankai and meet some of his younger Chinese colleagues.

He remained active till the very end and his friends were all very pleased when he was awarded the first Shaw Prize in Mathematics, in recognition of his pioneering role in modern differential geometry.

Michael Atiyah

FOREWORD

This year, 2005, is the hundredth anniversary of Einstein's Annus Mirabilis. We recall his repeated emphasis on the need to geometrize the foundation of physics. It is thus especially appropriate this year to hold an International Conference on Differential Geometry Methods in Theoretical Physics. As a person associated with Nankai for many years, and as an early student and admirer of Professor S.S. Chern, I am particularly happy that this year's Conference site is his Nankai Institute of Mathematics.

Professor Chern had eagerly anticipated his participation at this Conference. He is no longer with us, but his work and his spirit will be with this, and indeed with all future International Conference on Differential Geometry Methods in Theoretical Physics.

Chen Ning Yang

PREFACE

The XXIII International Conference on Differential Geometric Methods in Theoretical Physics (XXIII DGMTP) was organized by Nankai Institute of Mathematics from August 20th to 26th, 2005. It was Professor S.S. Chern and Professor W. Nahm who proposed the XXIII DGMTP on the occasion of the 60th anniversary of Professor S.S. Chern's paper "Characteristic classes of Hermitian manifolds". Unfortunately, Professor S.S. Chern passed away in December 2004. So this Conference is in memory of Professor Chern dedicated by more than one hundred mathematicians and physicists actively working in the field, in particular differential geometry, topology, gauge theories, statistical mechanics, mathematical physics, and so on.

The XXIII DGMTP was held in the new building of Nankai Institute of Mathematics. It was completed one month before the Conference and named Shiing-Shen Building in memory of Professor S.S. Chern who founded the Institute in 1985.

The members of the International Advisory Committee include Professors Michael Atiyah, Jean-Michel Bismut, Shiing-Shen Chern, Alain Connes, Simon Donaldson, Ludvig Faddeev, Chaohao Gu, Vaughan F.R. Jones, Yuri. I. Manin, Edward Witten and Chen Ning Yang. We are greatly grateful to them for the very kind suggestions. We thank all of plenary and parallel session's speakers, not only for their bringing the newest developments in the frontier of the field, but also for their kind cooperation in many ways. We highly obliged to all of the session organizers including Professors Victor Batyrev, Jean-Pierre Bourguignon, Louis H. Kauffman, Xiao-Song Lin, Werner Nahm, Antti Niemi, Andrew Strominger, Fa Yueh Wu, Yong-Shi Wu and Xin Zhou for their most excellent jobs.

We are indebted to the Ministry of Education of China who mainly supported the Conference.

We sincerely thank Sir Michael Atiyah and Professor C.N. Yang who are close friends of Professor S.S. Chern for their kind contributions of the special preface and memory article.

Last but not the least we thank World Scientific Publishing Co. for their generous support for the publication.

Mo-Lin Ge
Weiping Zhang

CONTENTS

XXIII
of Differential Geometric Methods in
Physics
2005 Nankai Institute of Mathematics

Plenary Lectures

Yangian and Applications

Cheng-Ming Bai, Mo-Lin Ge

Theoretical Physics Division
Chern Institute of Mathematics
Nankai University
Tianjin 300071, P.R. China

Kang Xue, Hong-Biao Zhang

Department of Physics
Northeast Normal University
Changchun 130024, P.R. China

In this paper, the Yangian relations are tremendously simplified for Yangians associated to $SU(2)$, $SU(3)$, $SO(5)$ and $SO(6)$ based on RTT relations that much benefit the realization of Yangian in physics. The physical meaning and some applications of Yangian have been shown.

1. Introduction

Yangian was presented by Drinfel'd ([1-3]) twenty years ago. It receives more attention for the following reasons. It is related to the rational solution of Yang-Baxter equation and the RTT relation. It is a simple extension of Lie algebras and the representation theory of $Y(SU(2))$ has been given. Some physical models, say, two component nonlinear Schrodinger equation, Haldane-Shastry model and 1-dimensional Hubbard chain do have Yangian symmetry. Yangian may be viewed as the consequence of a "bi-spin" system. How to understand the physical meaning of Yangian is an interesting topic. In this paper, there is nothing with mathematics. Rather, we try to use the language of quantum mechanics and Lie algebraic knowledge to show the effects of Yangian.

2. Yangian and RTT Relations

Let $\mathcal{G}$ be a complex simple Lie algebra. The Yangian algebra $Y(\mathcal{G})$ associated to $\mathcal{G}$ was given as follows ([1-3]). For a given set of Lie algebraic generators I_μ of $\mathcal{G}$ the new generators J_ν were introduced to satisfy

$$[I_\lambda, I_\mu] = C_{\lambda\mu\nu} I_\nu, \quad C_{\lambda\mu\nu} \text{ are structural constants;} \tag{2.0.1}$$

$$[I_\lambda, J_\mu] = C_{\lambda\mu\nu} J_\nu; \tag{2.0.2}$$

and, for $\mathcal{G} \neq sl(2)$:

$$[J_\lambda, [J_\mu, I_\nu]] - [I_\lambda, [J_\mu, J_\nu]] = a_{\lambda\mu\nu\alpha\beta\gamma}\{I_\alpha, I_\beta, I_\gamma\}, \tag{2.0.3}$$

where

$$a_{\lambda\mu\nu\alpha\beta\gamma} = \frac{1}{4!} C_{\lambda\alpha\sigma} C_{\mu\beta\tau} C_{\nu\gamma\rho} C_{\sigma\tau\rho}, \tag{2.0.4}$$

$$\{x_1, x_2, x_3\} = \sum_{i\neq j\neq k} x_i x_j x_k, \quad \text{(symmetric summation);} \tag{2.0.5}$$

or for $\mathcal{G} = sl(2)$:

$$[[J_\lambda, J_\mu], [I_\sigma, J_\tau]] + [[J_\sigma, J_\tau], [I_\lambda, J_\mu]]$$

$$= (a_{\lambda\mu\nu\alpha\beta\gamma} C_{\sigma\tau\nu} + a_{\sigma\tau\nu\alpha\beta\gamma} C_{\lambda\mu\nu})\{I_\alpha, I_\beta, J_\gamma\}. \tag{2.0.6}$$

When $C_{\lambda\mu\nu} = i\varepsilon_{\lambda\mu\nu}(\lambda, \mu, \nu = 1, 2, 3)$, equation (2.0.3) is identically satisfied from the Jacobian identities. Besides the commutation relations there are co-products as follows.

$$\Delta(I_\lambda) = I\lambda \otimes 1 + 1 \otimes I_\lambda; \tag{2.0.7}$$

$$\Delta(J_\lambda) = J_\lambda \otimes 1 + 1 \otimes J_\lambda + \frac{1}{2} C_{\lambda\mu\nu} I_\mu \otimes I_\nu. \tag{2.0.8}$$

Further, the Yangian can be derived through RTT relations where R is a rational solution of Yang-Baxter equation (YBE) ([1-12]).

After lengthy calculations, we found the independent relations for $Y(SU(2))$, $Y(SU(3))$, $Y(SO(5))$ and $Y(SO(6))$ by expanding the RTT relations and also checked through equations (2.0.1)-(2.0.3) and (2.0.6) by substituting the structural constants ([13-17]), where RTT relation (Faddeev, Reshetikhin, Takhtajan — RFT [18]) satisfies

$$\check{R}(u-v)(T(u) \otimes 1)(1 \otimes T(v)) = (1 \otimes T(v))(T(u) \otimes 1)\check{R}(u-v). \tag{2.0.9}$$

2.1. $Y(SU(2))$

Let P_{12} be the permutation. Setting

$$\check{R}_{12}(u) = PR_{12}(u) = uP_{12} + I; \tag{2.1.1}$$

$$\begin{aligned} T(u) &= I + \sum_{n=1}^{\infty} u^{-n} \begin{bmatrix} T_{11}^{(n)} & T_{12}^{(n)} \\ T_{21}^{(n)} & T_{22}^{(n)} \end{bmatrix} \\ &= I + \sum_{n=1}^{\infty} u^{-n} \begin{bmatrix} \frac{1}{2}(T_0^{(n)} + T_3^{(n)}), & T_+^{(n)} \\ T_-^{(n)}, & \frac{1}{2}(T_0^{(n)} - T_3^{(n)}) \end{bmatrix}, \end{aligned} \tag{2.1.2}$$

and substituting the $T(u)$ into RTT relation it turns out that only

$$I_\pm = T_\pm^{(1)}, I_3 = \frac{1}{2} T_3^{(1)}; \tag{2.1.3}$$

$$J_\pm = T_\pm^{(2)}, J_3 = \frac{1}{2} T_3^{(2)} \tag{2.1.4}$$

are independent ones. The quantum determinant

$$\det T(u) = T_{11}(u)T_{22}(u-1) - T_{12}(u)T_{21}(u-1) = C_0 + \sum_{n=1}^{\infty} u^{-n} C_n \tag{2.1.5}$$

gives

$$C_0 = 1, \quad C_1 = T_0^{(1)} = trT^{(1)}, \tag{2.1.6}$$

$$C_2 = T_0^{(2)} - \mathbf{I}^2 + T_0^{(1)}(1 + \frac{1}{2} T_0^{(1)}), \quad \cdots, . \tag{2.1.7}$$

The independent commutation relations of $Y(SU(2))$ are:

$$[I_\lambda, I_\mu] = i\epsilon_{\lambda\mu\nu} I_\nu \qquad (\lambda, \mu, \nu = 1, 2, 3); \tag{2.1.8}$$

$$[I_\lambda, J_\mu] = i\epsilon_{\lambda\mu\nu} J_\nu; \tag{2.1.9}$$

and ($A_\pm = A_1 \pm iA_2$)

$$[J_3, [J_+, J_-]] = (J_- J_+ - I_- J_+) I_3 \tag{2.1.10}$$

that can be checked to generate all of relations of equations (2.0.1), (2.0.2) and (2.0.6) with the help of Jacobi identities.

The co-product is given through (RFT) as

$$\Delta T_{ab} = \sum_c T_{ac} \otimes T_{cb}. \tag{2.1.11}$$

The simplest realization of $Y(SU(2))$ is

$$\mathbf{I} = \sum_{i=1}^{N} \mathbf{I}_i \;\; (i: \text{lattice indices}), \tag{2.1.12}$$

$$\mathbf{J} = \sum_{i=1}^{N} \mu_i \mathbf{I}_i + \sum_{i<j}^{N} W_{ij} \mathbf{I}_i \times \mathbf{I}_j, \tag{2.1.13}$$

where

$$W_{ij} = \begin{cases} 1 & i < j \\ 0 & i = j \\ -1 & i > j \end{cases} \quad (\text{for any representation of } SU(2)) \tag{2.1.14}$$

or

$$W_{jk} = i \cot \frac{(j-k)\pi}{N} \quad (\text{only for spin } \frac{1}{2}, \text{ Haldane} - \text{Shastry model } [19-21]), \tag{2.1.15}$$

and μ_i arbitrary constants. Noting that μ_i plays important role for the representation theory of $Y(SU(2))$ given by Chari and Pressley ([22-24]).

The big difference between representations of Lie algebra and Yangian is in that in Yangian there appear free parameters μ_i depending on models.

Another example for single particle is finite W-algebra ([25-26]). Denoting by $\mathbf{L}$ and $\mathbf{B}$ angular momentum and Lorentz boost, respectively, as well as D the dilatation operator, the set of $\mathbf{L}$ and $\mathbf{J}$ satisfies $Y(SU(2))$ where ([13],[25])

$$\mathbf{I} = \mathbf{L} \tag{2.1.16}$$

$$\mathbf{J} = \mathbf{I} \times \mathbf{B} - i(D-1)\mathbf{B} \tag{2.1.17}$$

and

$$[J_\alpha, J_\beta] = i\epsilon_{\alpha\beta\gamma}(2\mathbf{I}^2 - c_2' - 4)\mathbf{I}_\gamma, \;\; c_2' \text{ casimir of } SO(4,2). \tag{2.1.18}$$

There are the following models whose Hamiltonians do commute with $Y(SU(2))$.

• Two component nonlinear Schrodinger equation (Murakami and Wadati [27])

$$i\psi_t = -\psi_{xx} + 2c|\psi|^2\psi, \tag{2.1.19}$$

$$\mathbf{I} = \int dx \psi_\alpha^+(x)(\frac{\sigma}{2})_{\alpha\beta}\psi_\beta(x); \tag{2.1.20}$$

$$\mathbf{J} = -i\int dx\psi_\alpha^+(x)(\frac{\sigma}{2})_{\alpha\beta}\psi_\beta(x) - \frac{ic}{2}\int dxdy\varepsilon(y-x)(\frac{\sigma}{2})_{\beta\lambda}\psi_\beta^+(x)\psi_\alpha^+(y)\psi_\alpha(x)\psi_\lambda(y). \tag{2.1.21}$$

- One-dimensional Hubbard model (for $N \to \infty$, [28])

$$H = -\sum_{i=1}^{N}(a_i^+ a_{i+1} + a_{i+1}^+ a_i + b_i^+ b_{i+1} + b_{i+1}^+ b_i) - U\sum_{i=1}^{N}(a_i^+ a_i - \frac{1}{2})(a_i^+ a_i - \frac{1}{2}); \tag{2.1.22}$$

$$\begin{aligned} J_\pm &= J_1 \pm iJ_2, \\ J_+ &= \sum_{i,j}\theta_{i,j}a_i^+ b_j - U\sum_{i\neq j}\varepsilon_{i,j}I_i^+ I_j^3, \\ J_- &= \sum_{i,j}\theta_{i,j}b_i^+ a_j + U\sum_{i\neq j}\varepsilon_{i,j}I_i^- I_j^3, \\ J_3 &= \frac{1}{2}[\sum_{i,j}\theta_{i,j}(a_i^+ a_j - b_i^+ b_j) + U\sum_{i<j}\varepsilon_{i,j}I_i^+ I_j^-, \end{aligned} \tag{2.1.23}$$

where

$$\theta_{i,j} = \delta_{i,j-1} - \delta_{i,j+1}, \quad \varepsilon_{i,j} = \begin{cases} 1 & i<j, \\ 0 & i=j, \\ -1 & i>j. \end{cases} \tag{2.1.24}$$

Essler, Korepin and Schoutens found the complete solutions ([29-30]) and excitation spectrum ([31]) of 1-D Hubbard model chain.

- Haldane-Shastry model ([19-21]) whose Hamiltonian is given by a family. The first member is

$$H_2 = \sum_{i,j}{}'(\frac{Z_i Z_j}{Z_{ij}Z_{ji}})(P_{ij} - 1), \tag{2.1.25}$$

where and henceforth the ' stands for $i \neq j$ in the summation and $P_{ij} = 2(\mathbf{S}_i \cdot \mathbf{S}_j + \frac{1}{4})$, $Z_k = \exp^{i\pi\frac{k}{N}}$, $Z_{ij} = Z_i - Z_j$. The next reads

$$H_3 = \sum_{i,j,k}{}'(\frac{Z_i Z_j Z_k}{Z_{ij}Z_{jk}Z_{ki}})(P_{ijk} - 1), \tag{2.1.26}$$

and

$$H_4 = \sum_{i,j,k,l}{}'(\frac{Z_i Z_j Z_k Z_l}{Z_{ij}Z_{jk}Z_{kl}Z_{li}})(P_{ijkl} - 1) + H_4', \tag{2.1.27}$$

$$H_4' = -\frac{1}{3}H_2 - 2\sum_{i,j}{}' (\frac{Z_i Z_j}{Z_{ij}Z_{ji}})^2 (P_{ij} - 1), \qquad (2.1.28)$$

where

$$\begin{aligned} P_{ijk} &= P_{ij}P_{jk} + P_{jk}P_{ki} + P_{ki}P_{ij}, \\ P_{ijkl} &= P_{ij}P_{jk}P_{kl} + (\text{cyclic for } i, j, k \text{ and } l). \end{aligned} \qquad (2.1.29)$$

The eigenvalues of H_2 and H_3 have been solved in Ref. [21] and numerical calculations were made for H_4. The H_2 and H_3 were shown to be obtained in terms of quantum determinant ([32]).

- Hydrogen atom (with and without monopole, [33])

$$H = \frac{\pi^2}{2\mu} + \frac{1}{2\mu}\frac{q^2}{r^2} - \frac{\kappa}{r}, \quad \pi = p - zeA \qquad (2.1.30)$$

where μ is mass, $q = zeg$, $\kappa = ze^2$ and g being monopole charge.

- Super Yang-Mills Theory ($N = 4$): $Y(SO(6))$ ([34])

$$H = 2\sum_{\alpha}\sum_{j} h(j)P^j_{\alpha\alpha+1}, \quad h(j) = \sum_{k=1}^{j}\frac{1}{k}, \quad h(0) = 1. \qquad (2.1.31)$$

where P^j is projector for the weight j of $SU(2)$ and α stands for "lattice" index.

2.2. $Y(SU(3))$

For the Yangian associated to $SU(3)$, there are the following independent relations

$$[I_\lambda, I_\mu] = if_{\lambda\mu\nu}I_\nu, \quad [I_\lambda, J_\mu] = if_{\lambda\mu\nu}J_\nu \quad (\lambda, \mu, \nu = 1, \cdots, 8). \qquad (2.2.1)$$

Define

$$I^{(1)}_\pm = I_1 \pm iI_2, \; U^{(1)}_\pm = I_6 \pm iI_7, \; V^{(1)}_\pm = I_4 \mp iI_5, \; \frac{\sqrt{3}}{2}I^{(1)}_8 = I_8 \qquad (2.2.2)$$

and J_μ represents the corresponding operator for $I^{(2)}_\pm, U^{(2)}_\pm, V^{(2)}_\pm$ and $I^{(2)}_8, I^{(2)}_3$. After lengthy calculation one finds that based on RTT relation there is only one independent relation for $Y(SU(3))$ additional to equation (2.2.1):

$$[I^{(2)}_8, I^{(2)}_3] - \frac{1}{3!}(\{I^{(1)}_+, U^{(1)}_+, V^{(1)}_+\} - \{I^{(1)}_-, U^{(1)}_-, V^{(1)}_-\}) \qquad (2.2.3)$$

where $\{\cdots\}$ stands for the symmetric summation. The conclusion can be verified through both the Drinfel'd formula ($C_{\lambda\mu\nu} = if_{\lambda\mu\nu}$) and RTT relations with replacing P_{12} in $SU(2)$ by

$$P_{12} = \frac{1}{3}I + \frac{1}{2}\sum_{\mu} \lambda_\mu \lambda_\mu, \tag{2.2.4}$$

where λ_μ are the Gell-Mann matrices. Setting

$$T(u) = \sum_{n=0}^{\infty} u^{-n} T^{(n)}, \tag{2.2.5}$$

$$T^{(n)} = \begin{bmatrix} \frac{1}{3}T_0^{(n)} + T_3^{(n)} + \frac{1}{\sqrt{3}}T_8^{(n)} & T_1^{(n)} - iT_2^{(n)} & T_4^{(n)} - iT_5^{(n)} \\ T_1^{(n)} + iT_2^{(n)} & \frac{1}{3}T_0^{(n)} - T_3^{(n)} + \frac{1}{\sqrt{3}}T_8^{(n)} & T_6^{(n)} - iT_7^{(n)} \\ T_4^{(n)} + iT_5^{(n)} & T_6^{(n)} + iT_7^{(n)} & \frac{1}{3}T_0^{(n)} - \frac{2}{\sqrt{3}}T_8^{(n)} \end{bmatrix}, \tag{2.2.6}$$

and substituting them into RTT relation we find equations (2.2.1)-(2.2.3) are independent relations together with the co-product, for example,

$$\begin{aligned} \Delta I_{\pm}^{(2)} &= I_{\pm}^{(2)} \otimes 1 + 1 \otimes I_{\pm}^{(2)} \pm 2(I_3^{(1)} \otimes I_{\pm}^{(1)} - I_{\pm}^{(1)} \otimes I_3^{(1)}) \\ &+ \frac{1}{2}(V_{\mp}^{(1)} \otimes U_{\mp}^{(1)} - U_{\mp}^{(1)} \otimes V_{\mp}^{(1)}) \end{aligned} \tag{2.2.7}$$

and others.

The quantum determinant of $T(u)$ which is 3 by 3 matrix for the fundamental representation of $gl(3)$ takes the form

$$\begin{aligned} \tilde{\det}_3 T(u) &= T_{11}(u)\{T_{22}(u-1)T_{33}(u-2) - T_{23}(u)T_{32}(u-2)\} \\ &\quad - T_{12}(u)\{T_{21}(u-1)T_{33}(u-2) - T_{23}(u-1)T_{31}(u-2)\} \\ &\quad + T_{13}(u)\{T_{21}(u-1)T_{32}(u-2) - T_{22}(u-1)T_{31}(u-2)\} \\ &= \sum_p (-1)^p T_{1p_1}(u) T_{2p_2}(u-1) T_{3p_3}(u-2) \end{aligned} \tag{2.2.8}$$

where p stands for all the possible arrangements of (p_1, p_2, p_3). In comparison with the quantum determinant

$$\tilde{\det}_2 T(u) = \sum_{k,l,m=0}^{\infty} \frac{(l-m-1)!}{(m-1)!l!} u^{-(m+l+k)} (T_{11}^{(k)} T_{22}^{(m)} - T_{12}^{(k)} T_{21}^{(m)}), \tag{2.2.9}$$

now we have

$$\begin{aligned}\tilde{\det}_3 T(u) =& \sum_{k,l,m,p,q=0}^{\infty} \frac{(l+m-1)!}{(m-1)!l!}\frac{2^q(p+q-1)!}{(p-1)!q!} u^{-(m+l+k+p+q)} \\ & \{T_{11}^{(k)}(T_{22}^{(m)}T_{33}^{(p)} - T_{23}^{(m)}T_{32}^{(p)}) - T_{12}^{(k)}(T_{21}^{(m)}T_{33}^{(p)} - T_{23}^{(m)}T_{31}^{(p)}) \\ & +T_{13}^{(k)}(T_{21}^{(m)}T_{32}^{(p)} - T_{22}^{(m)}T_{31}^{(p)})\} \\ =& \sum_{n=0}^{\infty} u^{-n} C_n, \end{aligned} \tag{2.2.10}$$

i.e.,

$$C_0 = 1, C_1 = T_0^{(1)}, C_2 = T_0^{(2)} + T_0^{(1)} + 2(T_0^{(1)})^2 - \mathbf{I}^2, \tag{2.2.11}$$

$$\mathbf{I}^2 = \sum_{\lambda=1}^{\infty} \mathbf{I}_\lambda^2. \tag{2.2.12}$$

When we constrain $\tilde{\det} T(u) = 1$ it leads to $Y(SU(2))$ and $Y(SU(3))$ that are formed by the set $\{I_\lambda, J_\lambda\}$, $\lambda = 1,2,3$ and $\lambda = 1,2,\cdots,8$ for $SU(2)$ and $SU(3)$, respectively.

An example of realization of $Y(SU(3))$ is the generalization of Haldane-Shastry model ([19-21]) for the fundamental representation of generators of $SU(3)$:

$$I_\mu = \sum_i F_i^\mu, \tag{2.2.13}$$

$$J_\mu = \sum_i \mu_i F_i^\mu + \lambda f_{\mu\lambda\nu} \sum_{i\neq j} W_{ij} F_i^\nu F_j^\lambda, \tag{2.2.14}$$

where W_{ij} satisfies the same relation as in Haldane-Shastry model given in section 2.1 and F^μ are the Gell-Mann matrices.

2.3. $Y(SO(5))$ and $Y(SO(6))$

For $SO(N)$ it holds

$$[L_{ij}, L_{kl}] = iC_{ij,kl}^{st} L_{st}, \tag{2.3.1}$$

where

$$C_{ij,kl}^{st} = \delta_{ik}\delta_{js}\delta_{lt} - \delta_{il}\delta_{js}\delta_{kt} - \delta_{jk}\delta_{is}\delta_{lt} + \delta_{jl}\delta_{is}\delta_{kt}. \tag{2.3.2}$$

The rational solutions of YBE for $SO(N)$ were firstly given by Zamolodchikov's ([35]). They are also re-derived by taking the rational limit of the trigonometric R-Matrix:

$$\breve{R}(u) = f(u)[u^2P + u(A - I - \frac{3}{2}P)\xi + \frac{3}{2}I\xi^2], \qquad (2.3.3)$$

where u stands for spectral parameter and ξ the other free parameter ([36-37]). The elements of $\breve{R}(u)$ are $(a,b,c,d = -2,-1,0,1,2)$

$$[\breve{R}(u)]^{ab}_{cd} = u^2\delta_{ab}\delta_{bc} + u(\delta_{a-b}\delta_{c-d} - \delta_{ac}\delta_{bd} - \frac{3}{2}\delta_{ad}\delta_{bc})\xi + \frac{3}{2}\delta_{ac}\delta_{bd}\xi^2. \quad (2.3.4)$$

For $SO(5)$, we introduce

$$T^{(1)} = \xi \begin{bmatrix} E_3 - \frac{3}{2} & U_+ & E_+ & V_+ & 0 \\ U_- & F_3 - \frac{3}{2} & F_+ & 0 & -V_+ \\ E_- & F_- & -\frac{3}{2} & -F_+ & -E_+ \\ V_- & 0 & -F_- & -F_3 - \frac{3}{2} & -U_+ \\ 0 & -V_- & -E_- & -U_- & -E_3 - \frac{3}{2} \end{bmatrix}, \qquad (2.3.5)$$

where

$$\begin{array}{l} E_3 = E_{22} - E_{-2,-2},\ F_3 = E_{11} - E_{-1-1},\ U_+ = E_{21} - E_{-1-2}, \\ V_+ = E_{2-1} - E_{1-2},\ E_+ = E_{20} - E_{0,-2},\ F_+ = E_{10} - E_{0-1}, \\ U_- = E_{12} - E_{-2-1},\ V_- = E_{-12} - E_{-2}\ \ E_- = E_{02} - E_{-20}, \\ F_- = E_{01} - E_{-10}; \end{array} \qquad (2.3.6)$$

$$T^{(2)}_{ab} = \frac{3}{2}\xi^2 E^{(2)}_{ab} \qquad (a,b = -2,-1,0,1,2). \qquad (2.3.7)$$

Substituting $T^{(n)}$ (only $n = 1,2$ are needed to be considered) into RTT relation, there appears 35 relations for J_μ besides the Jacobi identities. However , a lengthy computation shows that besides

$$\begin{array}{l} [I_\alpha, I_\beta] = C^\gamma_{\alpha\beta}I_\gamma \\ [I_\alpha, J_\beta] = C^\gamma_{\alpha\beta}J_\gamma \end{array} \qquad (\alpha = i,j), \qquad (2.3.8)$$

there is only one independent relation

$$[E^{(2)}_3, F^{(2)}_3] = \frac{1}{4!}(\{U_-,E_+,F_-\} - \{U_+,E_-,F_+\} - \{V_+,E_-,F_-\} + \{V_-,E_+,F_+\}), \qquad (2.3.9)$$

where again { } stands for the symmetric summation.

A realization of $Y(SO(5))$is given as follows. Set

$$I_{ab}(x) = \frac{1}{2}\psi^+_\alpha(x)(I^{ab})_{\alpha\beta}\psi_\beta(x) \qquad (a,b = -2,-1,0,1,2), \qquad (2.3.10)$$

$$\{\psi_\alpha^+(x), \psi_\beta(y)\}_+ = \delta(x-y)\delta_{\alpha\beta}. \tag{2.3.11}$$

Then

$$I_{ab} = \sum_x I_{ab}(x), \tag{2.3.12}$$

$$J_{ab} = \sum_{x,y,c\neq a,b} \epsilon(x-y) I_{ac}(x) I_{cb}(y) \tag{2.3.13}$$

satisfies the commuting relations for $Y(SO(5))$. The following Hamiltonian of ladder model not only commutes with I_{ab}, i.e., it possesses $SO(5)$ symmetry, but also commutes with J_{ab}.

$$H = H_1 + \sum_x H_2(x) + \sum_x H_3(x); \tag{2.3.14}$$

$$H_1 = 2t_1 \sum_{<x,y>} [c_\sigma^+(x)c_\sigma(y) + d_\sigma^+(x)d_\sigma(y) + H.C.]; \tag{2.3.15}$$

$$\begin{aligned} H_2(x) &= U(n_{c\uparrow} - \frac{1}{2})(n_{c\downarrow} - \frac{1}{2}) + (c \to d) + V(n_c - 1)(n_d - 1) + J\mathbf{S}_c \cdot \mathbf{S}_d \\ &= \frac{J}{4}\sum_{a<b} I_{ab}^2 + (\frac{1}{8}J + \frac{1}{2}U)(\psi_\alpha^+\psi_\alpha - 2); \end{aligned} \tag{2.3.16}$$

$$H_3(x) = -2t_3(c_\sigma^+(x)d_\sigma(x) + H.C.). \tag{2.3.17}$$

Because locally $SO(6) \simeq SU(4)$ we introduce (15 generators)

$$T_{ab}^{(1)} = I_{ab}, \quad T_{ab}^{(2)} = I_{ab}^{(2)} (a, b = 1, 2, \ldots, 6.) \tag{2.3.18}$$

and the $\check{R}(u)$-matrix reads

$$\check{R}(u) = f(u)[u^2 P + u\xi(A - 2P - I) + 2\xi^2 I]. \tag{2.3.19}$$

The RTT relation gives 4+4+441+315+225 more relations. After careful calculations one finds ([15-16]) that there are the following independent relations for J_{ab} themselves:

$$[I_{12}^{(2)}, I_{34}^{(2)}] = \frac{i}{24}(\{I_{23}, I_{16}, I_{46}\} + \{I_{23}, I_{15}, I_{45}\} + \{I_{14}, I_{25}, I_{35}\} + \{I_{14}, I_{26}, I_{36}\} - \{I_{13}, I_{26}, I_{46}\} - \{I_{13}, I_{25}, I_{45}\} - \{I_{24}, I_{15}, I_{35}\} - \{I_{24}, I_{16}, I_{36}\}); \tag{2.3.20}$$

$$[I_{12}^{(2)}, I_{56}^{(2)}] = \frac{i}{24}(\{I_{15}, I_{23}, I_{36}\} + \{I_{15}, I_{24}, I_{46}\} + \{I_{26}, I_{13}, I_{35}\} + \{I_{26}, I_{14}, I_{45}\} - \{I_{25}, I_{13}, I_{36}\} - \{I_{25}, I_{14}, I_{46}\} - \{I_{16}, I_{23}, I_{35}\} - \{I_{16}, I_{24}, I_{45}\}); \tag{2.3.21}$$

$$[I_{34}^{(2)}, I_{56}^{(2)}] = \frac{i}{24}(\{I_{45}^{(1)}, I_{13}^{(1)}, I_{16}^{(1)}\} + \{I_{45}^{(1)}, I_{23}^{(1)}, I_{26}^{(1)}\} + \{I_{36}^{(1)}, I_{14}^{(1)}, I_{16}^{(1)}\} + \{I_{36}^{(1)}, I_{24}^{(1)}, I_{26}^{(1)}\} - \{I_{35}^{(1)}, I_{14}^{(1)}, I_{16}^{(1)}\} - \{I_{35}^{(1)}, I_{24}^{(1)}, I_{26}^{(1)}\} - \{I_{46}^{(1)}, I_{13}^{(1)}, I_{16}^{(1)}\} - \{I_{46}^{(1)}, I_{23}^{(1)}, I_{26}^{(1)}\}). \tag{2.3.22}$$

3. Applications of Yangian

The first example was given by Belavin ([38]) in deriving the spectrum of nonlinear σ model. Here we only show briefly some interpretations of Yangian through the particular realizations of Yangian.

3.1. *Reduction of $Y(SU(2))$*

The simplest realization of $Y(SU(2))$ is made of two-spin system with $\mathbf{S_1}$ and $\mathbf{S_2}$ (any dimensional representations of $SU(2)$):

$$\mathbf{J}' - \frac{\mathbf{1}}{\mu+\nu}\mathbf{J} - \frac{\mathbf{1}}{\mu+\nu}(\mu\mathbf{S_1} \times \mathbf{1} + \nu\mathbf{S_2} \times \mathbf{1} + 2\lambda\mathbf{S_1} \times \mathbf{S_2}), \tag{3.1.1}$$

that contains the (antisymmetric) tensor interaction between $\mathbf{S_1}$ and $\mathbf{S_2}$. For example, for Hydrogen atom $\mathbf{S_1} = \mathbf{L}$ and $\mathbf{S_2} = \mathbf{K}$ (Lung-Lenz vector).

For $S_1 = S_2 = 1/2$, when

$$\mu\nu = \lambda^2, \tag{3.1.2}$$

we prove that after the following similar transformation

$$\mathbf{Y} = A\mathbf{J}'A^{-1}, \quad A = \begin{bmatrix} 1 & 0 & 0 & 0 \\ 0 & \nu & i\lambda & 0 \\ 0 & i\lambda & \nu & 0 \\ 0 & 0 & 0 & 1 \end{bmatrix}, \tag{3.1.3}$$

the Yangian reduces to $SO(4)$: $(\rho = \nu + i\lambda = \sqrt{\nu^2 + \lambda^2}e^{i\theta})$

$$Y_1 = \begin{bmatrix} M_1 & 0 \\ 0 & L_1 \end{bmatrix}, \; M_1 = \frac{1}{2}\begin{bmatrix} 0 & \rho \\ \rho^{-1} & 0 \end{bmatrix}, \; L_1 = \frac{1}{2}\begin{bmatrix} 0 & \rho^{-1} \\ \rho & 0 \end{bmatrix},$$
$$Y_2 = \begin{bmatrix} M_2 & 0 \\ 0 & L_2 \end{bmatrix}, \; M_2 = \frac{1}{2}\begin{bmatrix} 0 & -i\rho \\ i\rho^{-1} & 0 \end{bmatrix}, \; L_2 = \frac{1}{2}\begin{bmatrix} 0 & -i\rho^{-1} \\ i\rho & 0 \end{bmatrix},$$
$$Y_3 = \begin{bmatrix} \frac{1}{2}\sigma_3 & 0 \\ 0 & \frac{1}{2}\sigma_3 \end{bmatrix}, \; M_3 = \frac{1}{2}\sigma_3. \tag{3.1.4}$$

and

$$\mathbf{Y}^2 = \frac{1}{2}(\frac{1}{2} + 1) = \frac{3}{4}. \tag{3.1.5}$$

Namely, under $\mu\nu = \lambda^2$, the $\mathbf{Y}$ reduces to $SO(4)$ by $M_\pm = M_1 \pm iM_2$, $M_+ = \rho\sigma_+$, $M_- = \rho^{-1}\sigma_-$. The scaled $M_\pm$ and M_3 still satisfy the $SU(2)$ relations:

$$[M_3, M_\pm] = \pm M_\pm, \;\; [M_+, M_-] = 2M_3 \tag{3.1.6}$$

and there are the similar relations for $\mathbf{L}$.

It should be emphasized that here the new "spin" $\mathbf{M}$ (and $\mathbf{L}$) is the consequence of two spin($\frac{1}{2}$) interaction. As usual for two 2-dimensional representations of $SU(2)$ (Lie algebra)

$$\underline{2} \otimes \underline{2} = \underline{3} \text{ (spin triplet)} \oplus \underline{1} \text{ (singlet)}. \tag{3.1.7}$$

However, here we meet a different decomposition:

$$\underline{2} \otimes \underline{2} = \underline{2}(\mathbf{M}) \oplus \underline{2}(\mathbf{L}). \tag{3.1.8}$$

The idea can be generalized to $SU(3)$'s fundamental representation

$$J_\lambda = uI_1^\lambda + vI_2^\lambda + \lambda f_{\lambda\mu\nu} \sum_{i<j} F_{1i}^\mu F_{2j}^\nu, \tag{3.1.9}$$

$$[F_{i\mu}, F_{j\nu}] = if_{\mu\nu\lambda}F_{i\lambda}\delta_{ij} \qquad (\lambda, \mu, \nu = 1, 2, \cdots, 8). \tag{3.1.10}$$

Under the condition

$$uv = \lambda^2, \qquad v + i\lambda = \rho, \tag{3.1.11}$$

and the similar transformation

$$Y_\mu = AJ_\mu A^{-1}/(u+v), \quad A = \begin{bmatrix} 1 & 0 & 0 & 0 & 0 & 0 & 0 & 0 & 0 \\ 0 & \nu & 0 & i\lambda & 0 & 0 & 0 & 0 & 0 \\ 0 & 0 & \nu & 0 & 0 & 0 & i\lambda & 0 & 0 \\ 0 & i\lambda & 0 & \nu & 0 & 0 & 0 & 0 & 0 \\ 0 & 0 & 0 & 0 & 1 & 0 & 0 & 0 & 0 \\ 0 & 0 & 0 & 0 & 0 & \nu & 0 & i\lambda & 0 \\ 0 & 0 & i\lambda & 0 & 0 & 0 & \nu & 0 & 0 \\ 0 & 0 & 0 & 0 & 0 & i\lambda & 0 & \nu & 0 \\ 0 & 0 & 0 & 0 & 0 & 0 & 0 & 0 & 1 \end{bmatrix}, \tag{3.1.12}$$

the Yangian then reduces to

$$Y(I_-) = \begin{bmatrix} \rho^{-1}I_- & 0 & 0 \\ 0 & \rho I_- & 0 \\ 0 & 0 & I_- \end{bmatrix}, \; Y(I_+) = \begin{bmatrix} \rho I_+ & 0 & 0 \\ 0 & \rho^{-1}I_- & 0 \\ 0 & 0 & I_3 \end{bmatrix},$$

$$Y(I_8) = \frac{\sqrt{3}}{3}\begin{bmatrix} \lambda_3 & 0 & 0 \\ 0 & \lambda_3 & 0 \\ 0 & 0 & \lambda_3 \end{bmatrix}, \; Y(I_3) = \frac{1}{2}\begin{bmatrix} \lambda_3 & 0 & 0 \\ 0 & \lambda_3 & 0 \\ 0 & 0 & \lambda_3 \end{bmatrix},$$

$$Y(U_+) = \begin{bmatrix} U_+ & 0 & 0 \\ 0 & \rho U_+ & 0 \\ 0 & 0 & \rho^{-1}U_+ \end{bmatrix}, \; Y(U_-) = \begin{bmatrix} U_- & 0 & 0 \\ 0 & \rho^{-1}U_- & 0 \\ 0 & 0 & \rho U_- \end{bmatrix},$$

$$Y(V_+) = \begin{bmatrix} \rho^{-1}V_- & 0 & 0 \\ 0 & V_- & 0 \\ 0 & 0 & \rho V_- \end{bmatrix}, \; Y(V_-) = \begin{bmatrix} \rho V_- & 0 & 0 \\ 0 & V_- & 0 \\ 0 & 0 & \rho^{-1}V_- \end{bmatrix}. \tag{3.1.13}$$

The usual decomposition through the Clebsch-Gordan coefficients for the representations of Lie algebra $SU(3)$ is $\underline{3} \otimes \underline{3} = \underline{6} \oplus \underline{3}$. However, here we have

$$\underline{3} \otimes \underline{3} = \underline{3} \oplus \underline{3} \oplus \underline{3}, \tag{3.1.14}$$

and

$$\sum_{\lambda=1}^{8} Y_\lambda^2 = \frac{1}{u+v}\sum_{\lambda=1}^{\infty} J_\lambda^2 = \frac{1}{3}. \tag{3.1.15}$$

It is easy to check that the rescaling factor ρ does not change the commutation relations for $SU(3)$ formed by $I_\pm$, $U_\pm$, $V_\pm$, I_3 and I_8. In general, we guess for the fundamental representation of $SU(n)$ we shall meet

$$\underline{n} \otimes \underline{n} = \underline{n} \oplus \underline{n} \oplus \underline{n} + \cdots + \underline{n} \quad (n \text{ times}). \tag{3.1.16}$$

Next we consider Yang-Mills gauge field for reduced $Y(SU(2))$. For a tensor wave function ($x \equiv \{x_1, x_2, x_3, x_0\}$),

$$\Psi(x) = \|\psi_{ij}(x)\| \quad (i,j = 1,2,3,4). \tag{3.1.17}$$

An isospin transformation yields

$$\Psi'(x) = U(x)\Psi(x), \quad U(x) = 1 - i\theta^a J_a, \tag{3.1.18}$$

where

$$J^a = uS_a \otimes \mathbf{1} + v\mathbf{1} \otimes S_a + 2\lambda\epsilon_{abc}S^b \otimes S^c, \tag{3.1.19}$$

or

$$[J_a]^{\alpha\beta}_{\gamma\delta} = u(S^a)_{\alpha\gamma}\delta_{\beta\delta} + v(S^a)_{\beta\delta}\delta_{\alpha\gamma} + i\alpha\varepsilon_{abc}(S^b)_{\alpha\gamma}(S^c)_{\beta\delta}. \tag{3.1.20}$$

Define

$$D_\mu = \partial_\mu + gA_\mu, \tag{3.1.21}$$

i.e.,

$$[D_\mu\psi]_{\alpha\beta} = \partial_\mu\psi_{\alpha\beta} + gA^a_\mu[Y_a]^{\alpha\beta}_{\gamma\delta}\psi_{\gamma\delta}(x), \quad A_\mu = A^a_\mu J_a. \tag{3.1.22}$$

The gauge-covariant derivative should preserve

$$\delta(D_\mu\psi) = 0, \tag{3.1.23}$$

i.e.,

$$(-i\partial_\mu\theta^a(x) + g\delta A^a_\mu)[Y_a]^{\alpha\beta}_{\gamma\delta} - ig\theta^a(x)A^b_\mu[J_b, J_a]^{\alpha\beta}_{\gamma\delta} = 0. \tag{3.1.24}$$

When $uv = \lambda^2$ and by rescaling

$$Y_a = (u+v)J_a, \tag{3.1.25}$$

we have

$$\delta A^a_\mu = \epsilon_{abc}\theta^b(x)A^c_\mu(x) + \frac{i}{g}\partial_\mu\theta^a(x), \tag{3.1.26}$$

and

$$F_{\mu\nu} = \frac{1}{g}[D_\mu, D_\nu] = F^a_{\mu\nu}Y_a, \tag{3.1.27}$$

$$F^a_{\mu\nu} = \partial_\mu A^a_\gamma - \partial_\nu A^a_\mu + ig\epsilon_{abc}A^b_\mu A^c_\gamma. \tag{3.1.28}$$

Here the tensor isospace has been separated to two irrelevant spaces, i.e., $\Psi = \begin{bmatrix} \Psi_1 & 0 \\ 0 & \Psi_2 \end{bmatrix}$ where Ψ_1 and Ψ_2 are 2×2 wavefunction.

3.2. *Illustrative examples: NMR of Breit-Rabi Hamiltonian and Yangian*

The Breit-Rabi Hamiltonian is given by

$$H = \mathbf{K} \cdot \mathbf{S} + \mu \mathbf{B} \cdot \mathbf{S}, \tag{3.2.1}$$

where $S = \frac{1}{2}$ and $B = \mathbf{B}(t)$ is magnetic field.

The Hamiltonian can easily be diagonalized for any background angular momentum (or spin) $\mathbf{K}$. The $\mathbf{S}$ stands for spin of electron and for simplicity $\mathbf{K} = \mathbf{S_1}(S_1 = 1/2)$ is an average background spin contributed by other source, say, control spin. Denoting by

$$H = H_0 + H_1(t), \;\; H_0 = \alpha \mathbf{S_1} \cdot \mathbf{S_2}, \;\; H_1(t) = \mu \mathbf{B}(t) \cdot \mathbf{S_2}. \tag{3.2.2}$$

Let us work in the interaction picture:

$$H_I = \mu \mathbf{B}(t) \cdot (e^{i\alpha \mathbf{S_1}\cdot\mathbf{S_2}} \mathbf{S_2} e^{-i\alpha \mathbf{S_1}\cdot\mathbf{S_2}}) = \mu \mathbf{B}(t) \cdot \mathbf{J}, \tag{3.2.3}$$

$$\mathbf{J} = \mu_1 \mathbf{S_1} + \mu_2 \mathbf{S_2} + 2\lambda(\mathbf{S_1} \times \mathbf{S_2}), \tag{3.2.4}$$

where $\mu_1 = \frac{1}{2}(1 - cos\alpha)$, $\mu_2 = \frac{1}{2}(1 + cos\alpha)$, $\lambda = \frac{1}{2} sin\alpha$. Obviously, here we have $\mu_1\mu_2 = \lambda^2$. It is not surprising that the $Y(SU(2))$ reduces to $SO(4)$ here because the transformation is fully Lie-algebraic operation. This is an exercise in quantum mechanics.

For generalization we regard μ_1 and μ_2 as independent parameters, i.e., drop the relation $\mu_1\mu_2 = \lambda^2$. Looking at

$$\mathbf{J} = \mu_1 \mathbf{S_1} + \mu_2 \mathbf{S_2} - \frac{1}{2}(\mu_1 + \mu_2)(\mathbf{S_1} + \mathbf{S_2}) + \gamma(\mathbf{S_1} + \mathbf{S_2}) + 2\lambda \mathbf{S_1} \times \mathbf{S_2}. \tag{3.2.5}$$

When $\gamma = \frac{1}{2}$, $\mu_2 - \mu_1 = cos\alpha$ and $\lambda = \frac{1}{2} sin\alpha$, it reduces to the form in the interacting picture. Putting

$$\mathbf{S_1} + \mathbf{S_2} = \mathbf{S}, \;\; 2\lambda = -\frac{h}{2}(h \text{ is not Plank constant}). \tag{3.2.6}$$

In accordance with the convention we have

$$\mathbf{J} = \gamma \mathbf{S} + \sum_{i=1}^{2} \mu_i \mathbf{S_i} + \frac{h}{2} \mathbf{S_1} \times \mathbf{S_2} - \frac{1}{2}(\mu_1 + \mu_2)\mathbf{S} = \gamma \mathbf{S} + \mathbf{Y}. \tag{3.2.7}$$

Since $\mathbf{J} \to \xi \mathbf{S} + \mathbf{J}$ still satisfies Yangian relations, it is natural to appear the term $\gamma \mathbf{S}$. The interacting Hamiltonian then reads

$$H_I(t) = -\gamma \mathbf{B}(t) \cdot \mathbf{S} - \mathbf{B}(t) \cdot \mathbf{Y}. \tag{3.2.8}$$

When $\mu_i = 0$, $h = 0$, it is the usual NMR for spin 1/2. To solve the equation, we use

$$i\frac{\partial\Psi(t)}{\partial t} = H_I(t)\Psi(t), \ \ |\Psi(t)\rangle = \sum_{\alpha=\pm,3;0} a_\alpha(t)|\chi_\alpha\rangle, \tag{3.2.9}$$

where $\{\chi_\pm, \chi_3\}$ is the spin triplet and χ_0 singlet. Setting

$$B_\pm(t) = B_1(t) \pm iB_2(t) = B_1 e^{\mp i\omega_0 t}, \ \text{ and } \ B_3 = \text{const.} \tag{3.2.10}$$

and rescaling by

$$a_\pm(t) = e^{\pm i\omega_0 t} b_{\pm(}t), \tag{3.2.11}$$

we get

$$\begin{aligned} i\frac{db_\pm(t)}{dt} &= -\gamma\{\frac{1}{\sqrt{2}}B_1 a_3(t) \mp (\omega_0\gamma^{-1} - B_3)b_\pm(t)\} \pm \frac{1}{2\sqrt{2}}\mu_- B_1 a_0(t), \\ i\frac{da_3(t)}{dt} &= -\frac{\gamma B_1}{\sqrt{2}}\{b_+(t) + b_-(t)\} - \frac{1}{2}\mu_- B_3 a_0(t), \\ i\frac{da_0(t)}{dt} &= -\frac{1}{2}\mu_+\{\frac{1}{\sqrt{2}}B_1[b_-(t) - b_+(t)]\} + B_3 a_3(t), \end{aligned} \tag{3.2.12}$$

where $\mu_\pm = (\mu_1 - \mu_2 \pm i\frac{h}{2})$, i.e.,

$$|\Phi(t)\rangle = \begin{bmatrix} b_1(t) \\ a_3(t) \\ b_-(t) \\ a_0(t) \end{bmatrix}, \mathcal{H}_I = \begin{bmatrix} \omega_0 - \gamma B_3 & -\gamma B_1 \frac{1}{\sqrt{2}} & 0 & \frac{1}{2\sqrt{2}}\mu_- B_1 \\ -\gamma B_1 \frac{1}{\sqrt{2}} & 0 & -\gamma B_1 \frac{1}{\sqrt{2}} & -\frac{1}{2}\mu_- B_3 \\ 0 & -\gamma B_1 \frac{1}{\sqrt{2}} & -(\omega_0 - \gamma B_3) & -\frac{1}{2\sqrt{2}}\mu_- B_1 \\ \frac{1}{2\sqrt{2}}\mu_+ B_1 & -\frac{1}{2}\mu_+ B_3 & -\frac{1}{2\sqrt{2}}\mu_+ B_1 & 0 \end{bmatrix}, \tag{3.2.13}$$

$$i\frac{d|\Phi(t)\rangle}{dt} = H_I|\Phi(t)\rangle. \tag{3.2.14}$$

Noting that $\mathcal{H}_\mathcal{I}$ is independent of time, we get

$$|\Phi(t)\rangle = e^{-iEt}|\Phi(t)\rangle. \tag{3.2.15}$$

Then

$$\det|H_I - E| = 0 \tag{3.2.16}$$

leads to

$$E^4 - [(\omega_1 - \gamma B_3)^2 + \gamma^2 B_1^2 + \frac{1}{4}\mu_+\mu_-(B_1^2 + B_3^2)]E^2 +$$

$$\frac{1}{4}\mu_+\mu_-[B_3^2(\omega_0 - \gamma B_3)^2 - 2\gamma B_3 B_1^2(\omega_0 - \gamma B_3) + \gamma^2 B_1^4] = 0. \tag{3.2.17}$$

There is a transition between the spin singlet and triplet in the NMR process, i.e., the Yangian transfers the quantum information through the evolution. The simplest case is $B_1 = 0$, then the eigenvalues are

$$E = \pm(\omega_0 - \gamma B_3), E = \pm\omega = \pm\frac{B_3}{2}\sqrt{(\mu_1 - \mu_2)^2 + \frac{h^2}{4}}. \tag{3.2.18}$$

It turns out that there is a vibration between $s = 0$ and $s = 1$.

$$< s^2 >= 0 \text{ at } t = \frac{\pi}{2\omega} \text{ (total spin } = 0), \tag{3.2.19}$$

$$< s^2 >= 2 \text{ at } t = \frac{\pi}{\omega} \text{ (total spin } = 1). \tag{3.2.20}$$

Under adiabatic approximation it can be proved that it appears Berry's phase. Obviously, only spin vector can make the stereo angle. The role of spin singlet here is a witness that shares energy of spin=1 state.

Actually, if

$$B_\pm(t) = B_0 \sin\theta e^{\mp i\omega_0 t}, \quad B_3 = B_0\cos\theta, \tag{3.2.21}$$

and

$$|\chi_{11}\rangle = |\uparrow\uparrow\rangle, \quad |\chi_{1-1}\rangle = |\downarrow\downarrow\rangle, \quad |\chi_{10}\rangle = \frac{1}{\sqrt{2}}(|\uparrow\downarrow\rangle + |\downarrow\uparrow\rangle),$$

$$|\chi_{00}\rangle = \frac{1}{\sqrt{2}}(|\uparrow\downarrow\rangle - |\downarrow\uparrow\rangle), \tag{3.2.22}$$

then let us consider the eigenvalues of

$$H = \alpha\mathbf{S}_1 \cdot \mathbf{S}_2 - \gamma B_0 S_3 - g B_0 J_3, \tag{3.2.23}$$

under adiabatic approximation which are

$$E_\pm = \frac{1}{2}(-\frac{\alpha}{2} \pm \sqrt{\alpha^2 + g^2 B_0^2 \mu_+\mu_-}), \tag{3.2.24}$$

and

$$f_1^{(\pm)} = [2(\alpha^2 + g^2B_0^2\mu_+\mu_-)]^{-1/2}[(\alpha^2 + g^2B_0^2\mu_+\mu_-)^{1/2} \pm \alpha]^{1/2}, \tag{3.2.25}$$

$$f_2^{(\pm)} = [2(\alpha^2 + g^2B_0^2\mu_+\mu_-)]^{-1/2}[\frac{\mu_+}{\mu_-}(\alpha^2 + g^2B_0^2\mu_+\mu_-)^{1/2} \mp \alpha]^{1/2}. \tag{3.2.26}$$

We obtain the eigenstates of H besides $|\chi_{1i}\rangle$ $(i = 1, 2)$

$$|\chi_\pm\rangle = f_1^{(\pm)}|\chi_{10}\rangle + f_2^{(\pm)}|\chi_{00}\rangle, \tag{3.2.27}$$

where

$$|\chi_{11}(t)\rangle = \cos^2\frac{\theta}{2}|\chi_{11}\rangle + \frac{1}{\sqrt{2}}\sin\theta e^{-i\omega_0 t}|\chi_{10}\rangle + \sin^2\frac{\theta}{2}e^{-2i\omega_0 t}|\chi_{1-1}\rangle,$$

$$|\chi_{1-1}(t)\rangle = \sin^2\frac{\theta}{2}e^{2i\omega_0 t}|\chi_{11}\rangle - \frac{1}{\sqrt{2}}\sin\theta e^{i\omega_0 t}|\chi_{10}\rangle + \cos^2\frac{\theta}{2}|\chi_{1-1}\rangle,$$

$$|\chi_{\pm}(t)\rangle = \frac{1}{\sqrt{2}}f_1^{(\pm)}\{-\sin\theta e^{i\omega_0 t}|\chi_{11}\rangle + \sqrt{2}\cos\theta|\chi_{10}\rangle + \sin\theta e^{-i\omega_0 t}|\chi_{1-1}\rangle\} + f_2^{(\pm)}|\chi_{00}\rangle. \quad (3.2.28)$$

We then obtain

$$\langle\chi_{11}(t)|\frac{\partial}{\partial t}|\chi_{11}(t)\rangle = -i\omega_0(1-\cos\theta),$$

$$\langle\chi_{1-1}(t)|\frac{\partial}{\partial t}|\chi_{11}(t)\rangle = i\omega_0(1-\cos\theta),$$

$$\langle\chi_{\pm}(t)|\frac{\partial}{\partial t}|\chi_{\pm}(t)\rangle = 0. \quad (3.2.29)$$

The Berry's phase is then

$$\gamma_{1\pm1} = \pm\Omega, \quad \Omega = 2\pi(1-\cos\theta), \quad (3.2.30)$$

whereas $\gamma_{10} = \gamma_{00} = 0$. The Yangian changes the eigenstates of H, but preserves the Berry's phase.

3.3. *Transition between S-wave and P-wave superconductivity*

We set for a pair of electrons:

$$S: \quad \text{spin singlet}, \quad L = 0; \quad (3.3.1)$$

$$P: \quad \text{spin triplet}, \quad L = 1. \quad (3.3.2)$$

Due to Balian-Werthamer ([39]), we have

$$\triangle(\mathbf{k}) = -\frac{1}{2}\sum_{\mathbf{k}'} V(\mathbf{k},\mathbf{k}')\frac{\triangle(\mathbf{k}')}{E(\mathbf{k}')}\tanh\frac{\beta}{2}E(\mathbf{k}'), \quad (3.3.3)$$

$$E(\mathbf{k}) = (\epsilon^2(k) + |\triangle(\mathbf{k})|^2)^{\frac{1}{2}}. \quad (3.3.4)$$

Therefore, still by Balian-Werthamer ([39]), we know

$$\triangle(\mathbf{k}) = \triangle(k)(\frac{4\pi}{3})^{\frac{1}{2}}\begin{bmatrix}\sqrt{2}Y_{1,1}(\hat{\mathbf{k}}) & Y_{1,0}(\hat{\mathbf{k}}) \\ Y_{1,0}(\hat{\mathbf{k}}) & \sqrt{2}Y_{1,-1}(\hat{\mathbf{k}})\end{bmatrix}^* = (-\sqrt{6})\triangle(k)(\frac{4\pi}{3})^{\frac{1}{2}}\Phi_{0,0}(\hat{\mathbf{k}}), \quad (3.3.5)$$

$$\Phi_{0,0}(\hat{\mathbf{k}}) = \frac{1}{\sqrt{3}}\{Y_{1,-1}(\hat{\mathbf{k}})\chi_{11} - Y_{1,0}(\hat{\mathbf{k}})\chi_{10} + Y_{1,1}(\hat{\mathbf{k}})\chi_{1-1}\} = \frac{1}{\sqrt{8}}\begin{bmatrix} \hat{\mathbf{k}}_- & -\hat{\mathbf{k}}_z \\ -\hat{\mathbf{k}}_z & -\hat{\mathbf{k}}_+ \end{bmatrix}, \tag{3.3.6}$$

where χ_{11},χ_{10} and χ_{1-1} stand for spin triplet:

$$\Phi_{0,0} \equiv \Phi_{J=0,m=0}. \tag{3.3.7}$$

The wave function of SC is

$$\phi_{0,0} = \frac{1}{\sqrt{2}}\begin{bmatrix} 0 & Y_{0,0} \\ -Y_{0,0} & 0 \end{bmatrix}. \tag{3.3.8}$$

Introducing

$$I_\mu = \sum_{i=1}^{2} S_\mu(i); \qquad (\mu = 1, 2, 3), \tag{3.3.9}$$

$$J_\mu = \sum_{i=1}^{2} \lambda_i S_\mu(i) - \frac{ihv}{4}\epsilon_{\mu\lambda\nu}(S^\lambda(1)S^\nu(2) - S^\lambda(2)S^\nu(1)), \tag{3.3.10}$$

and noting that $J_\mu \to J_\mu + fI_\mu$ does not change the Yangian relations, we choose for simplicity $f = -\frac{1}{2}(\lambda_1 + \lambda_2)$. Then we obtain for $G = \hat{\mathbf{k}} \cdot (\mathbf{J} + f\mathbf{I})$

$$G\phi_{0,0} = \hat{\mathbf{k}} \cdot (\mathbf{J} + f\mathbf{I})\phi_{0,0} = \frac{\sqrt{3}}{2}(\lambda_2 - \lambda_1 + \frac{hv}{2})\Phi_{0,0}, \tag{3.3.11}$$

$$G\Phi_{0,0} = \hat{\mathbf{k}} \cdot (\mathbf{J} + f\mathbf{I})\Phi_{0,0} = \frac{1}{2\sqrt{3}}(\lambda_2 - \lambda_1 - \frac{hv}{2})\phi_{0,0}. \tag{3.3.12}$$

The transition directionally depends on the parameters in $Y(SU(2))$. For instance,

$$SC \to PC: \; G\phi_{0,0} = \frac{\sqrt{3}}{2}\Phi_{0,0}, \; G\Phi_{0,0} = 0, \; \text{if } \lambda_1 - \lambda_2 = -\frac{hv}{2}, \tag{3.3.13}$$

and

$$PC \to SC: \; G\phi_{0,0} = 0, \; G\Phi_{0,0} = -\frac{hv}{2\sqrt{3}}\phi_{0,0}, \; \text{if } \lambda_1 - \lambda_2 = \frac{hv}{2}. \tag{3.3.14}$$

We call the type of the transition “directional transition” ([40]). The controlled parameters are in the Yangian operation. They represent the interaction coming from other controlled spin.

We have got used to apply electromagnetic field A_μ to make transitions between l and $l \pm 1$ states. Now there is Yangian formed by two spins that plays the role changing angular momentum states.

3.4. $Y(SU(3))$-directional transitions

Setting

$$F_\mu = \frac{1}{2}\lambda_\mu, \ [F_\lambda, F_\mu] = if_{\lambda\mu\nu}F_\nu, \tag{3.4.1}$$

$$I_\mu = \sum_i F_i^\nu, \tag{3.4.2}$$

$$J_\mu = \sum_i \mu_i F_i^\mu - ihf_{\mu\nu\lambda}\sum_{i\neq j} W_{ij}F_i^\nu F_j^\lambda, \ (W_{ij} = -W_{ji}), \tag{3.4.3}$$

$$[F_i^\lambda, F_j^\mu] = if_{\lambda\mu\nu}\delta_{ij}F_i^\nu, \tag{3.4.4}$$

where $\{F_\mu\}$ is the fundamental representation of $SU(3)$ and $(i, j, k = 1, 2, ..., 8)$

$$\triangle_{ijk} = W_{ij}W_{jk} + W_{jk}W_{ki} + W_{ki}W_{ij} = -1. \tag{3.4.5}$$

(Here, no summation over repeated indices, $i \neq j \neq k$). The reason that such a condition works only for 3-dimensional representation of $SU(3)$ is similar to Haldane's (long-ranged) realization of $Y(SU(2))$([19]). In $SU(2)$ long-ranged form, the property of Pauli matrices leads to $(\sigma^\pm)^2 = 0$. Instead, for $SU(3)$ the conditions of J_μ satisfying $Y(SU(3))$ read

$$\sum_{i\neq j} (1 - w_{ij}^2)(I_j^+V_i^+U_i^+ - U_i^-V_i^-I_j^- + I_i^+V_j^+U_i^+ - U_i^-V_j^-I_i^- \\ +I_j^+V_j^+U_i^+ - U_i^-V_j^-I_j^-) = 0, \tag{3.4.6}$$

and

$$\sum_i (I_i^+V_i^+U_i^+ - U_i^-V_i^-I_i^-) = 0, \tag{3.4.7}$$

that are satisfied for Gell-Mann matrices.

The simplest realization of $Y(SU(3))$ is then

$$W_{ij} = \begin{cases} 1 & i > j \\ 0 & i = j \\ -1 & i < j \end{cases} \quad (W_{ij} = -W_{ji}). \tag{3.4.8}$$

Recalling ($I_8 = \frac{\sqrt{3}}{2}Y$)

$$I^+ = \begin{bmatrix} 0&1&0 \\ 0&0&0 \\ 0&0&0 \end{bmatrix}, \quad U^+ = \begin{bmatrix} 0&0&0 \\ 0&0&1 \\ 0&0&0 \end{bmatrix}, \quad V^+ = \begin{bmatrix} 0&0&0 \\ 0&0&0 \\ 1&0&0 \end{bmatrix},$$

$$I^3 = \begin{bmatrix} 1&0&0 \\ 0&-1&0 \\ 0&0&0 \end{bmatrix}, \quad Y = \frac{1}{3}\begin{bmatrix} 1&0&0 \\ 0&1&0 \\ 0&0&-2 \end{bmatrix}. \tag{3.4.9}$$

We find

$$\begin{aligned}
J_\mu &= \{\bar{I}_\pm, \bar{U}_\pm, \bar{V}_\pm, \bar{I}_3, \bar{I}_8\}, \\
\bar{I}_\pm &= \sum_i \mu_i I_i^\pm \mp 2h \sum_{i \neq j} W_{ij}(I_i^\pm I_j^3 + \frac{1}{2}U_i^\mp V_j^\mp), \\
\bar{U}_\pm &= \sum_i \mu_i U_i^\pm \pm h \sum_{i \neq j} W_{ij}[U_i^\pm (I_j^3 - \frac{3}{2}Y_j) + I_i^\mp V_j^\mp], \\
\bar{V}_\pm &= \sum_i \mu_i V_i^\pm \pm h \sum_{i \neq j} W_{ij}[V_i^\pm (I_j^3 + \frac{3}{2}Y_j) + U_i^\mp I_j^\mp], \\
\bar{I}_3 &= \sum_i \mu_i I_i^3 + h \sum_{i \neq j} W_{ij}[I_i^+ I_j^- - \frac{1}{2}(U_i^+ U_j^- - V_i^+ V_j^-)], \\
\bar{I}_8 &= \sum_i \mu_i Y_i + h \sum_{i \neq j} W_{ij}(U_i^+ U_j^- - V_j^+ V_j^-),
\end{aligned} \tag{3.4.10}$$

where μ_i and h (not Planck constant) are arbitrary parameters. Notice again that the simplest choice of W_{ij} is given by equation (3.4.8).

When $i = 1, 2$, $Y(SU(2))$ makes transition between spin singlet and triplet. Now $Y(SU(3))$ transits $SU(3)$ singlet and Octet. For instance, setting

$$\begin{aligned}
&| \pi^- \rangle = |d\bar{u}\rangle, \;\; |\pi^0\rangle = \frac{1}{\sqrt{2}}(|u\bar{u}\rangle - |d\bar{d}\rangle), |K^-\rangle = |d\bar{u}\rangle, \;\; |K^0\rangle = |d\bar{s}\rangle, \\
&| \eta^0 \rangle = \frac{1}{\sqrt{(6)}}(-|u\bar{u}\rangle - |d\bar{d}\rangle + 2|s\bar{s}\rangle), | \eta^{0'} \rangle = \frac{1}{\sqrt{(3)}}(|u\bar{u}\rangle + |d\bar{d}\rangle + |s\bar{s}\rangle).
\end{aligned} \tag{3.4.11}$$

Special interest is the following. When

$$\mu_1 - \mu_2 = -3h, \;\; f = -\frac{1}{2}(\mu_1 - \mu_2), \tag{3.4.12}$$

by acting the Yangian operators on the Octet of $SU(3)$, we obtain (see Figure 3.1)

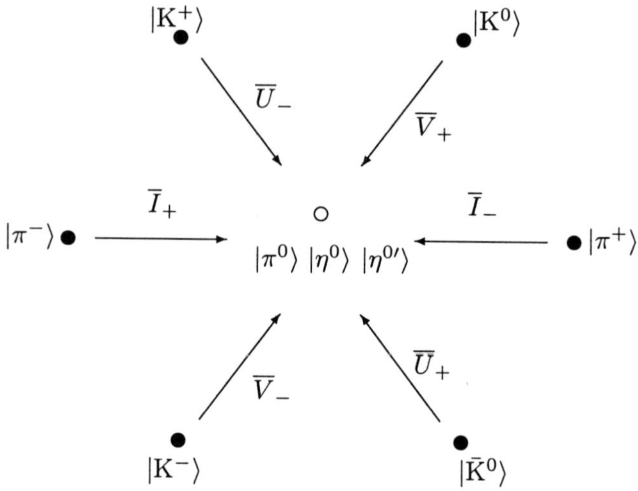

Fig. 3.1. Representation of $SU(3)$

$$
\begin{aligned}
&\overline{I}_-|\pi^+ >= \frac{1}{\sqrt{6}}(\mu_1-\mu_2)|\eta^0 > +\frac{1}{\sqrt{2}}(\mu_1+\mu_2)|\pi^0 > -\frac{1}{\sqrt{3}}(\mu_1-\mu_2+3h)|\eta^{0'} >,\\
&\overline{U}_+|\overline{K}^0 >= \frac{1}{\sqrt{6}}(\mu_1+2\mu_2)|\eta^0 > +\frac{1}{\sqrt{2}}\mu_1|\pi^0 > -\frac{1}{\sqrt{3}}(\mu_1-\mu_2+3h)|\eta^{0'} >,\\
&\overline{U}_-|K^0 >= \frac{1}{\sqrt{6}}(2\mu_1+\mu_2)|\eta^0 > +\frac{1}{\sqrt{2}}\mu_2|\pi^0 > +\frac{1}{\sqrt{3}}(\mu_1-\mu_2+3h)|\eta^{0'} >,\\
&\overline{V}_+|K^+ >= \frac{1}{\sqrt{6}}(2\mu_1+\mu_2)|\eta^0 > -\frac{1}{\sqrt{2}}\mu_2|\pi^0 > +\frac{1}{\sqrt{3}}(\mu_1-\mu_2+3h)|\eta^{0'} >,\\
&\overline{V}_-|K^- >= -\frac{1}{\sqrt{6}}(\mu_1+2\mu_2)|\eta^0 > +\frac{1}{\sqrt{2}}\mu_1|\pi^0 > +\frac{1}{\sqrt{3}}(\mu_1-\mu_2+3h)|\eta^{0'} >,\\
&\overline{I}_3|\pi^0 >= -\frac{1}{2\sqrt{3}}(\mu_1-\mu_2)|\eta^0 > +\frac{1}{\sqrt{6}}(\mu_1-\mu_2+3h)|\eta^{0'} >,\\
&\overline{I}_8|\eta^0 >= -\frac{1}{3}(\mu_1-\mu_2)|\eta^0 > -\frac{\sqrt{2}}{3}(\mu_1-\mu_2+3h)|\eta^{0'} >, \qquad (3.4.13)
\end{aligned}
$$

i.e.,

$$
\begin{aligned}
&(\overline{I}_\pm + fI_\pm)|\eta^{0'} >= \pm 2\sqrt{3}h|\pi^\pm >,\ (\overline{U}_+ + fU_+)|\eta^{0'} >= -2\sqrt{3}h|K^0 >,\\
&(\overline{U}_- + fU_-)|\eta^{0'} >= 2\sqrt{3}h|\overline{K}^0 >,\ (\overline{V}_\pm + fV_\pm)|\eta^{0'} >= -2\sqrt{3}h|K^\mp >,\\
&(\overline{I}_3 + fI_3)|\eta^{0'} >= -\sqrt{6}h|\pi^0 >,\ (\overline{I}_8 + fI_8)|\eta^{0'} >= 2\sqrt{2}h|\eta^0 >, \qquad (3.4.14)
\end{aligned}
$$

and

$$
\begin{aligned}
&(\overline{I}_{\pm} + fI_{\pm})|\pi^{\mp} >= \pm\sqrt{\frac{3}{2}}h|\eta^0 >, \\
&(\overline{U}_{+} + fU_{+})|K^0 >= -\frac{\sqrt{3}}{2\sqrt{2}}h(\sqrt{3}|\pi^0 > -|\eta^0 >), \\
&(\overline{U}_{-} + fU_{-})|K^0 >= \frac{\sqrt{3}}{2\sqrt{2}}h(\sqrt{3}|\pi^0 > -|\eta^0 >), \\
&(\overline{V}_{\pm} + fV_{\pm})|K^{\pm} >= -\frac{\sqrt{3}}{2\sqrt{2}}h(\sqrt{3}|\pi^0 > +|\eta^0 >), \\
&(\overline{I}_3 + fI_3)|\pi^0 >= \sqrt{\frac{3}{2}}h|\eta^0 >,\ (\overline{I}_8 + fI_8)|\eta^0 >= \sqrt{3}h|\eta^0 > . \qquad (3.4.15)
\end{aligned}
$$

The Yangian operators play the role to transit the Octet states to the singlet state of $SU(3)$.

Whereas, if

$$
\mu_1 - \mu_2 = 3h,\ f = -\frac{1}{2}(\mu_1 + \mu_2), \qquad (3.4.16)
$$

with the notations

$$
(\overline{A}^{(2)} + fA^{(1)})|\eta^{0'} >= 0,\ A = I_\alpha,\ (\alpha = \pm, 3, 8),\ U_{\pm},\ V_{\pm}, \qquad (3.4.17)
$$

we have

$$
\begin{aligned}
&(\overline{I}_{\pm} + fI_{\pm})|\pi^{\mp} > = \mp\sqrt{\frac{3}{2}}h|\eta^0 > \pm 2\sqrt{3}h|\eta^{0'} >, \\
&(\overline{U}_{+} + fU_{+})|\overline{K}^0 > = \frac{\sqrt{3}}{2\sqrt{2}}h(\sqrt{3}|\pi^0 > -|\eta^0 >) - 2\sqrt{3}h|\eta^{0'} >, \\
&(\overline{U}_{-} + fU_{-})|K^0 > = -\frac{\sqrt{3}}{2\sqrt{2}}h(\sqrt{3}|\pi^0 > -|\eta^0 >) + 2\sqrt{3}h|\eta^{0'} >, \\
&(\overline{V}_{\pm} + fV_{\pm})|K^{\pm} > = \frac{\sqrt{3}}{2\sqrt{2}}h(\sqrt{3}|\pi^0 > +|\eta^0 >) + 2\sqrt{3}h|\eta^{0'} >, \\
&(\overline{I}_3 + fI_3)|\pi^0 > = -\frac{\sqrt{3}}{2}h|\eta^0 > +\sqrt{6}h|\eta^{0'} >, \\
&(\overline{I}_8 + fI_8)|\eta^0 > = h|\eta^0 > -2\sqrt{2}h|\eta^{0'} > . \qquad (3.4.18)
\end{aligned}
$$

Obviously, in this case the Yangian operators make the transition from the Octet to a "combined" singlet state of $SU(3)$.

3.5. $\mathbf{J}^2$ *as a new quantum number*

Because $[\mathbf{I}^2, \mathbf{J}^2] = 0$, $[\mathbf{I}^2, I_z] = 0, [\mathbf{J}^2, I_z] = 0$, but $[\mathbf{J}^2, J_z] \neq 0$, we can take $\{\mathbf{I}^2, I_z, \mathbf{J}^2\}$ as a conserved set.

First we consider the case $\mathbf{S}_1 \bigotimes \mathbf{S}_2 \bigotimes \mathbf{S}_3$, where $S_1 = S_2 = S_3 = \frac{1}{2}$. We shall show that instead of 6-j coefficients and Young diagrams, $\mathbf{J}^2$ can be viewed as a "collective" quantum number that describes the "history" besides $\mathbf{S}^2$ ($\mathbf{S} = \mathbf{S}_1 + \mathbf{S}_2 + \mathbf{S}_3$) and S_z.

As representations of Lie algebra $SU(2)$, we have

$$(\frac{1}{2} \bigotimes \frac{1}{2}) \bigotimes \frac{1}{2} = (1 \bigoplus 0) \bigotimes \frac{1}{2} = \frac{3}{2} \bigoplus \frac{1}{2} \bigoplus \frac{1}{2}'. \tag{3.5.1}$$

Noting that $|\frac{1}{2}\rangle$ and $|\frac{1}{2}'\rangle$ are degenerate regarding the total spin $\frac{1}{2}$. The usual Lie algebraic base can be easily written as

$$\begin{aligned}
\phi_{\frac{3}{2},\frac{3}{2}} &= |\uparrow\uparrow\uparrow\rangle, \\
\phi_{\frac{3}{2},\frac{1}{2}} &= \frac{1}{\sqrt{3}}(|\uparrow\uparrow\downarrow\rangle + |\uparrow\downarrow\uparrow\rangle + |\downarrow\uparrow\uparrow\rangle), \\
\phi_{\frac{3}{2},-\frac{1}{2}} &= \frac{1}{\sqrt{3}}(|\uparrow\downarrow\downarrow\rangle + |\downarrow\uparrow\downarrow\rangle + |\downarrow\downarrow\uparrow\rangle), \\
\phi_{\frac{3}{2},-\frac{3}{2}} &= |\downarrow\downarrow\downarrow\rangle,
\end{aligned} \tag{3.5.2}$$

and the two degeneracy states with respect to $\mathbf{S}^2$ and S_z are given by:

$$\begin{aligned}
\phi'_{\frac{1}{2},\frac{1}{2}} &= \frac{1}{\sqrt{6}}(|\downarrow\uparrow\uparrow\rangle + |\uparrow\downarrow\uparrow\rangle - 2|\uparrow\uparrow\downarrow\rangle), \\
\phi'_{\frac{1}{2},-\frac{1}{2}} &= \frac{1}{\sqrt{6}}(|\uparrow\downarrow\downarrow\rangle + |\downarrow\uparrow\downarrow\rangle - 2|\downarrow\downarrow\uparrow\rangle), \\
\phi_{\frac{1}{2},\frac{1}{2}} &= \frac{1}{\sqrt{2}}(|\downarrow\uparrow\uparrow\rangle - \uparrow\downarrow\uparrow\rangle), \\
\phi_{\frac{1}{2},-\frac{1}{2}} &= \frac{1}{\sqrt{2}}(|\uparrow\downarrow\downarrow\rangle - |\downarrow\uparrow\downarrow\rangle).
\end{aligned} \tag{3.5.3}$$

To distinguish ϕ' from ϕ we introduce $\mathbf{J}$:

$$\mathbf{J} = \sum_{i=1}^{3} u_i \mathbf{S}_i + ih \sum_{i<j}^{3} (\mathbf{S}_i \times \mathbf{S}_j), \tag{3.5.4}$$

and calculate $\mathbf{J}^2$. It turns out that

$$\begin{aligned}
\mathbf{J}^2\phi_{\frac{3}{2},m} &= [\frac{3}{4}(u_1^2+u_2^2+u_3^2)+\frac{1}{2}(u_1u_2+u_2u_3+u_1u_3)-h^2]\Phi_{\frac{3}{2},m};\\
\mathbf{J}^2\phi'_{\frac{1}{2},m} &= [\frac{3}{4}(u_1^2+u_2^2+u_3^2)+\frac{1}{2}u_1u_2-u_2u_3-u_1u_3-\frac{7}{4}h^2]\Phi'_{\frac{1}{2},m}\\
&\quad -\frac{\sqrt{3}}{2}(u_1-u_2+h)(u_3+h)\Phi_{\frac{1}{2},m};\\
\mathbf{J}^2\phi_{\frac{1}{2},m} &= -\frac{\sqrt{3}}{2}(u_1-u_2-h)(u_3-h)\Phi'_{\frac{1}{2},m}+[\frac{3}{4}(u_1-u_2)^2\\
&\quad +\frac{3}{4}u_3^2-\frac{3}{4}h^2]\Phi_{\frac{1}{2},m}.
\end{aligned}\tag{3.5.5}$$

In order to make the matrix of $\mathbf{J}^2$ be symmetric (then it surely can be diagonalized), one should put

$$u_2 = u_1 + u_3. \tag{3.5.6}$$

The eigenvalues of $\mathbf{J}^2$ are given by

$$\begin{aligned}
\lambda_{\frac{3}{2}} &= 2u_1^2+2u_3^2+3u_1u_3-h^2,\\
\lambda_{\frac{1}{2}}^{\pm} &= u_1^2+u_3^2-\frac{5}{4}h^2\pm\frac{1}{2}[(2u_1^2-u_3^2-h^2)^2+3(u_3^2-h^2)^2]^{\frac{1}{2}}.
\end{aligned}\tag{3.5.7}$$

The eigenstates of $\mathbf{J}^2$ are the rotation of $\phi'_{\frac{1}{2},m}$ and $\Phi_{\frac{1}{2},m}$:

$$\begin{pmatrix}\alpha^+_{\frac{1}{2},m}\\ \alpha^-_{\frac{1}{2},m}\end{pmatrix} = \begin{pmatrix}\cos\frac{\varphi}{2} & -\sin\frac{\varphi}{2}\\ \sin\frac{\varphi}{2} & \cos\frac{\varphi}{2}\end{pmatrix}\begin{pmatrix}\phi'_{\frac{1}{2},m}\\ \phi_{\frac{1}{2},m}\end{pmatrix}, \quad \mathbf{J}^2\alpha^{\pm}_{\frac{1}{2}} = \lambda^{\pm}_{\frac{1}{2}}\alpha^{\pm}_{\frac{1}{2},m}, \tag{3.5.8}$$

where

$$\sin\varphi = \sqrt{3}(u_3^2-h^2)/\omega, \ \ \omega^2 = (2u_1^2-u_3^2-h^2)^2+3(u_3^2-h^2)^2. \tag{3.5.9}$$

It is worth noting that the conclusion is independent of the order, say, $(\frac{1}{2}\bigotimes\frac{1}{2})\bigotimes\frac{1}{2}$, $\frac{1}{2}\bigotimes(\frac{1}{2}\bigotimes\frac{1}{2})$ and the other way. The difference is only in the value of φ.

The above example can be generalized to $\mathbf{S}_1\bigotimes\mathbf{S}_2\bigotimes\mathbf{L}$ where $S_1 = S_2 = \frac{1}{2}$ and $\mathbf{L}^2 = l(l+1)$. As representations of Lie algebra $SU(2)$, we have

$$\begin{aligned}
(\frac{1}{2}\bigotimes\frac{1}{2})\bigotimes l = (1\bigoplus 0)\bigotimes l = l+1 \qquad & l \qquad l-1\\
& l
\end{aligned}\tag{3.5.10}$$

There are no degeneracy for $l \pm 1$, but two l states can be distinguished in terms of $\mathbf{J}^2$

$$\begin{aligned}
\mathbf{J}^2\Phi_{l+1,m} &= \{\frac{3}{4}(u_1^2+u_2^2)+l(l+1)u_3^2+\frac{1}{2}u_1u_2+l(u_2u_3+u_1u_3)\\
&\quad -h^2[l(l+1)+\frac{1}{4}]\}\Phi_{l+1,m},\\
\mathbf{J}^2\Phi_{l-1,m} &= \{\frac{3}{4}(u_1^2+u_2^2)+l(l+1)u_3^2+\frac{1}{2}u_1u_2-(l+1)u_1u_3-(l+1)u_2u_3\\
&\quad -h^2[l(l+1)+\frac{1}{4}]\}\Phi_{l-1,m},\\
\mathbf{J}^2\Phi_{l,m}^1 &= \{\frac{3}{4}(u_1^2+u_2^2)+l(l+1)u_3^2+\frac{1}{2}u_1u_2-u_2u_3-u_1u_3\\
&\quad -2h^2[l(l+1)\frac{1}{8}]\}\Phi_{l,m}^1-\sqrt{l(l+1)}(u_1-u_2+h)(u_3+h)\Phi_{l,m}^2,\\
\mathbf{J}^2\Phi_{l,m}^2 &= -\sqrt{l(l+1)}(u_1-u_2-h)(u_3-h)\Phi_{l,m}^1\\
&\quad +[\frac{3}{4}(u_1-u_2)^2+l(l+1)u_3^2-\frac{3}{4}]\Phi_{l,m}^2.
\end{aligned} \tag{3.5.11}$$

Again in order to guarantee the symmetric form of the matrix we put

$$u_2 = u_1 + u_3, \tag{3.5.12}$$

then the eigenvalues and eigenstates of $\mathbf{J}^2$ are given by

$$\lambda_l^{\pm} = u_1^2 + [l(l+1)+\frac{1}{4}]u_3^2 - h^2[l(l+1)+\frac{1}{2}] \pm \frac{1}{2}\sqrt{P}, \tag{3.5.13}$$

$$\begin{pmatrix}\alpha_{l,m}^+\\ \alpha_{l,m}^-\end{pmatrix} = \begin{pmatrix}\cos\frac{\varphi}{2} & -\sin\frac{\varphi}{2}\\ \sin\frac{\varphi}{2} & \cos\frac{\varphi}{2}\end{pmatrix}\begin{pmatrix}\Phi_{l,m}^1\\ \Phi_{l,m}^2\end{pmatrix}, \tag{3.5.14}$$

where

$$\omega^2 = P = [2u_1^2 - u_3^2 - h^2(2l(l+1)-\frac{1}{2})]^2 + 4l(l+1)(u_3^2-h^2)^2, \tag{3.5.15}$$

$$\sin\varphi = \frac{2\sqrt{l(l+1)}}{\omega}(u_3^2-h^2). \tag{3.5.16}$$

As a simple example, we consider the spin structure of rare gas

$$H = -a\mathbf{L}\cdot\mathbf{S}_1 - b\mathbf{S}_1\cdot\mathbf{S}_2, \quad (\lambda = \frac{b}{a}). \tag{3.5.17}$$

It describes the interaction of spin $\mathbf{S}_1$ of an electron exited from l-shell and the left hole $\mathbf{S}_2$.

$$H\Phi_{l+1,m} = -\frac{1}{2}(al + \frac{1}{2}b)\Phi_{l+1,m},$$
$$H\Phi_{l-1,m} = \frac{1}{2}[(l+1)a - \frac{1}{2}b]\Phi_{l-1,m},$$
$$H\begin{bmatrix}\Phi^{\pm}_{l,m}\\ \Phi^2_{l,m}\end{bmatrix} = \frac{1}{2}\begin{bmatrix}(a-\frac{1}{2}b) & a\sqrt{l(l+1)}\\ a\sqrt{l(l+1)} & \frac{3}{2}b\end{bmatrix}\begin{bmatrix}\Phi^1_{l,m}\\ \Phi^2_{l,m}\end{bmatrix}. \tag{3.5.18}$$

The eigenstates of H associated to l, m are

$$\begin{pmatrix}\alpha^+_{l,m}\\ \alpha^-_{l,m}\end{pmatrix} = \begin{pmatrix}\cos\frac{\varphi}{2} & -\sin\frac{\varphi}{2}\\ \sin\frac{\varphi}{2} & \cos\frac{\varphi}{2}\end{pmatrix}\begin{pmatrix}\Phi^1_{l,m}\\ \Phi^2_{l,m}\end{pmatrix}. \tag{3.5.19}$$

where

$$\sin\varphi = \frac{\sqrt{l(l+1)}}{\omega}, \quad \omega^2 = (\frac{1}{2} - \lambda)^2 + l(l+1), \lambda = \frac{b}{a}. \tag{3.5.20}$$

The eigenvalues are

$$\lambda_{l+1} = -\frac{1}{2}(la + \frac{b}{2}), \quad \lambda_{l-1} = \frac{1}{2}[(l+1)a - \frac{b}{2}];$$
$$\lambda^{\pm}_l = \frac{1}{4}(a+b) \pm \frac{1}{2}[l(l+1)a^2 + (\frac{a}{2} - b)^2]^{\frac{1}{2}}. \tag{3.5.21}$$

The rotation should be made in such a way that

$$[H, \mathbf{J}^2] = 0 \tag{3.5.22}$$

which is satisfied if the matrix $\mathbf{J}^2$ is symmetric, i.e.,

$$\gamma = \frac{\{2u_1^2 - 2h^2[l(l+1) + \frac{1}{4}]\}}{(u_3^2 - h^2)} = 2(1-\lambda). \tag{3.5.23}$$

Therefore, the parameter γ in $Y(SU(2))$ determines the rotation angle φ. It is reasonable to think that the appearance of "rotation" of degenerate states is closely related to the "quantum number" of $\mathbf{J}^2$. Transition between $\alpha^+_{l,m}$ and $\alpha^-_{l,m}$ ($l = 1$) can be made by J_3. Because there are two independent parameters u_1 and u_3 in $\mathbf{J}$, one can choose a suitable relation between u_3 and $\lambda = \frac{b}{a}$ such that

$$J_3\alpha^+ \sim \alpha^-, \tag{3.5.24}$$

i.e., the transition between two degenerate states in Lie-algebra is made trough J_3 operator, because of

$$[\mathbf{J}^2, J_3] \neq 0. \tag{3.5.25}$$

3.6. *Happer degeneracy*

In the experiment for $^{87}R_b$ molecular there appears new degeneracy ([41]) at the special $\pm B_0$ (magnetic field), i.e., the Zeeman effect disappears at $\pm B_0$. The model Hamiltonian reads ([42]) (x is scaled magnetic field)

$$H = \mathbf{K}\cdot\mathbf{S} + x(k+\frac{1}{2})S_z, \tag{3.6.1}$$

where $\mathbf{K}$ is angular momentum and $\mathbf{K}^2 = K(K+1)$. It only occurs for spin $S = 1$. It turns out that when $x = \pm 1$ there appears the curious degeneracy, that is, there is a set of eigenstates corresponding to

$$E = -\frac{1}{2}. \tag{3.6.2}$$

The conserved set is $\{\mathbf{K}^2, G_z = K_z + S_z\}$. For $\mathbf{G} = \mathbf{K} + \mathbf{S}$ we have $G = k \pm 1, k$. The eigenstates are specified in terms of three families: T, B and D. Only D-set possesses the degeneracy.

Happer gives, for example,the eigenstates for $x = \pm 1$ ([42]):

$$\begin{array}{ll} x = +1 & H\alpha_{Dm} = (-\frac{1}{2})\alpha_{Dm} \\ x = -1 & H\beta_{Dm} = (-\frac{1}{2})\beta_{Dm} \end{array}, \tag{3.6.3}$$

and shows that

$$\begin{aligned} \alpha_{Dm} = {} & [2(K+\frac{1}{2})(K+m+\frac{1}{2})]^{-\frac{1}{2}}\{-[\frac{(K-m+1)(K+m+1)}{2}]^{\frac{1}{2}}\alpha_1 \\ & +[(K+m)(K+m+1)]^{\frac{1}{2}}\alpha_2 + [\frac{(K-m)(K+m)}{2}]^{\frac{1}{2}}\alpha_3\}; \end{aligned} \tag{3.6.4}$$

$$\begin{aligned} \beta_{Dm} = {} & [2(K+\frac{1}{2})(K-m+\frac{1}{2})]^{-\frac{1}{2}}\{[\frac{(K-m)(K+m)}{2}]^{\frac{1}{2}}\alpha_1 \\ & +[(K-m)(K-m+1)]^{\frac{1}{2}}\alpha_2 - [\frac{(K-m+1)(K+m+1)}{2}]^{\frac{1}{2}}\alpha_3\}, \end{aligned} \tag{3.6.5}$$

where $\alpha_1 = e_1 \otimes e_{m-1}$, $\alpha_2 = e_0 \otimes e_m$ and $\alpha_3 = e_{-1} \otimes e_{m+1}$.

It is natural to ask what is the transition operator between α_{Dm} and β_{Dm}? The answer is Yangian operator. In fact, introducing

$$J_\pm = aS_+ + bK_- \pm (s_\pm K_z - s_z K_\pm), \tag{3.6.6}$$

we find that by choosing $a = -\frac{k+1}{2}, b = 0$, we have

$$\beta_{Dm} \xrightarrow{J_+} \lambda_1(m)\alpha_{Dm+1} \text{ and } \alpha_{Dm} \xrightarrow{J_-} \lambda_2(m)\beta_{Dm-1}; \tag{3.6.7}$$

and by choosing $a = \frac{k}{2}, b = 0$, we have

$$\beta_{Dm} \xrightarrow{J_-} \lambda_1'(m)\alpha_{Dm-1} \text{ and } \alpha_{Dm} \xrightarrow{J_+} \lambda_2'(m)\beta_{Dm+1}. \tag{3.6.8}$$

The Yangian makes the transition between the states with B and $-B$, which here is only for $S = 1$. The reason is that for $S = 1$ there are two independent coefficients in the combination of α_1, α_2 and α_3 and there are two free parameters in **J**. Hence the number of equations are equal to those of free parameters (a and b), so we can find a solution. The numerical computation shows that only $S = 1$ gives rise to the new degeneracy ([42]) that prefers the Yangian operation ([43]).

3.7. *New degeneracy of extended Breit-Rabi Hamiltonian*

As was shown in the Happer's model ($H = \mathbf{K} \cdot \mathbf{S} + x(k + \frac{1}{2})S_3$) there appeared new degeneracy for $S = 1$. It has been pointed out that the above degeneracy with respect to Zeeman effect cannot appear for spin=$\frac{1}{2}$. Actually, in this case it yields for $S = \frac{1}{2}$ ([42]),

$$E = -\frac{1}{4} - \omega_m S_3, \tag{3.7.1}$$

where

$$\omega_m^2 = [(1 + x^2)(k + \frac{1}{2}) + 2xm](k + \frac{1}{2}). \tag{3.7.2}$$

Therefore if the Happer's type of degeneracy can occurs, there should be $\omega_m = 0$ that means

$$x_0 = -\frac{m}{k} \pm i\sqrt{1 - \frac{m^2}{k^2}} \quad (k = K + \frac{1}{2}), \tag{3.7.3}$$

i.e., the magnetic field should be complex.

However, the situation will be completely different, if a third spin is involved. For simplicity we assume $S_1 = S_2 = S_3 = \frac{1}{2}$ in the Hamiltonian:

$$H = -(a\mathbf{S}_2 + b\mathbf{S}_3) \cdot \mathbf{S}_1 + x\sqrt{ab}S_1^z, \lambda = b/a, \tag{3.7.4}$$

then besides two non-degenerate states, there appears the degenerate family:

$$H\alpha^{\pm}_{D,\pm\frac{1}{2}} = -(\frac{a+b}{4})\alpha^{\pm}_{D,\pm\frac{1}{2}}, \quad \text{for } x = \pm 1, \tag{3.7.5}$$

where

$$\alpha^{\pm}_{D,+\frac{1}{2}} = -\sqrt{2}\lambda| \uparrow\uparrow\downarrow> \pm\sqrt{\lambda}| \uparrow\downarrow\uparrow +(1 \pm \sqrt{\lambda})| \downarrow\uparrow\uparrow>; \tag{3.7.6}$$

$$\alpha^{\pm}_{D,-\frac{1}{2}} = -\sqrt{2}\lambda|\downarrow\downarrow\uparrow> \mp\sqrt{\lambda}|\downarrow\uparrow\downarrow +(1\mp\sqrt{\lambda})|\uparrow\downarrow\downarrow> . \quad (3.7.7)$$

The expecting value of S_1^z are

$$<\alpha^{+}_{D,\pm\frac{1}{2}}|S_1^z|\alpha^{+}_{D,\pm\frac{1}{2}} >\sim \sqrt{\lambda}\,(x=1); \quad (3.7.8)$$

$$<\alpha^{-}_{D,\pm\frac{1}{2}}|S_1^z|\alpha^{-}_{D,\pm\frac{1}{2}} >\sim -\sqrt{\lambda}\,(x=-1), \quad (3.7.9)$$

namely, at the special magnetic field ($x = \pm 1$) the observed $< S_1^z >$ still opposite to each other for $x = \pm 1$, but without the usual Zeeman split.

The reason of the appearance of the new degeneracy is obvious. The two spins $\mathbf{S}_2$ and $\mathbf{S}_3$ here play the role of $S = 1$ in comparison with Happer model.

3.8. *Super Yang-Mills ($N = 4$)-Lipatov model and $Y(SO(6))$*

Beisert et al([44-45]), Dolan-Nappi-Witten (DNW,[34]) and other authors ([46-47]) proposed to take the quantum correction of the dilatation operator δD ($D \in SO(4,2)$ is a subalgebra of $PSU(2,2|4)$) as Hamiltonian for supper Yang-Mills ($N = 4$):

$$H = \sum_{\alpha} H_{\alpha\alpha+1}, \quad (3.8.1)$$

$$H_{\alpha\alpha+1} = 2\sum_{j} h(j)P^{j}_{\alpha\alpha+1},\quad h(j) = \sum_{k=1}^{j}\frac{1}{k}, h(0) = 1, \quad (3.8.2)$$

where P^j is projector for the weight j of $SU(2)$ and α stands for "lattice" index. DNW showed that ([34])

$$[H, Y(SO(6))] = 0. \quad (3.8.3)$$

It turns out that the Hamiltonian H is nothing but Lipatov model ([48]) which was related to the Yang-Baxter form by Lipatov ([49]), Faddeev and Korchemsky ([50]).

Based on Tarasov, Takhtajan and Faddeev([51]) the $\breve{R}$-matrix associated with any spin S reads

$$\breve{R}(u) = \frac{\Gamma(u-s)\Gamma(u+2s+1)}{\Gamma(u-\hat{J})\Gamma(u+\hat{J}+1)}, \quad (3.8.4)$$

where u is spectrum parameter and s the spin (arbitrary). The trigonometric Yang-Baxterization ([52]) gives

$$\breve{R}(u) = \sum_{j=0} \rho_j(x) P_j(q) \quad (x = e^{iu}), \tag{3.8.5}$$

where $P_j(q)$ is the q-deformed product with weight j. Taking the rational limit ([9],[36]) we have

$$\rho_j \Rightarrow \frac{\Gamma(u)\Gamma(u+1)}{\Gamma(u-j)\Gamma(u+j+1)}, \quad P_j(q) \Rightarrow P_j. \tag{3.8.6}$$

The Hamiltonian for the lattices α and $\alpha+1$

$$H_{\alpha\alpha+1} = I_1 \times I_2 \times \cdots \times I_{\alpha-1} \times \frac{d}{du}\breve{R}(u)|_{u=0}[\breve{R}(0)]^{-1} \times I_{\alpha+2} \times \cdots \tag{3.8.7}$$

is then

$$H = \sum_{\alpha} H_{\alpha\alpha+1} \tag{3.8.8}$$

where

$$\begin{aligned} H_{\alpha\alpha+1} &= \{-\psi(-\hat{J}_{\alpha\alpha+1}) - \psi(\hat{J}_{\alpha\alpha+1}+1) + \psi(1+2s) + \psi(1-2s) - \frac{1}{2s}\}|_{s=0} \\ &= \sum_j \{-\psi(-j) - \psi(j+1) + 2\psi(1) - \lim_{x\to 0}\frac{1}{x}\} P^j_{\alpha\alpha+1}. \end{aligned} \tag{3.8.9}$$

It describes the QCD correction to the parton model shown by Lipatov ([48-49]). The diagonalization of Lipatov model has probably been achieved by de Vega and Lipatov ([53-54]). Noting that the j indicates the block in the reducible block-diagonal form.

Using

$$\begin{aligned} &\psi(x+1) = \psi(x) + \frac{1}{x}, \\ &\psi(x+n) = \psi(x) + \sum_{k=0}^{n-1} \frac{1}{x+k}, \\ &\psi(1) = -c, \end{aligned} \tag{3.8.10}$$

and hence

$$\begin{aligned} &\psi(j+1) = \psi(1) + \sum_{k=1}^{j} \frac{1}{k} = \psi(1) + h(j) \\ &\psi(-j) = \psi(1) + h(j) - \lim_{x\to 0}\frac{1}{x}. \end{aligned} \tag{3.8.11}$$

We obtain

$$H_{\alpha,\alpha+1} = (-2)\sum_j h(j)P^j_{\alpha\alpha+1}. \tag{3.8.12}$$

Separating the finite part from the infinity the H is nothing but the δD derived in super Yang-Mills ($N = 4$) with the approximation. Of course, the derivation of δD based on super Yang-Mills ($N = 4$) explores much larger symmetry than Lipatov model. Therefore, DNW's result shows that the Lipatov's model possesses $Y(SO(6))$ symmetry.

To obtain $Y(SO(6))$ in terms of RTT relation we start from the rational solution of $\breve{R}$-matrix whose general form for $O(N)$ was firstly by Zamolodchikov and Zamolodchikov ([35]) and extended through rational limit of trigonometric Yang-Baxterization ([36]):

$$\breve{R} = u[u - \frac{1}{2}(N-2)a]P + \alpha u A_N + [-u\alpha + \frac{\alpha^2}{2}(N-2)]I, \tag{3.8.13}$$

where u is spectrum parameter and α a free parameter allowed by YBE. Here we adopt the convention of Jimbo:

$$P^{ab}_{cd} = \delta^a_d\delta^b_c, \quad (A_N)^{ab}_{cd} = \delta^{a,-b}\delta_{c,-d} \tag{3.8.14}$$

where

$$a, b, c, d = [-(\frac{N-1}{2}), -(\frac{N-1}{2}) + 1, \cdots, (\frac{N-1}{2})] \tag{3.8.15}$$

and $N = 2n+1$ for B_n and $N = 2n$ for C_n, D_n.

The R-matrix is given by

$$R = \breve{R}P = u(u-2\alpha)I + u(2u-\alpha)P + 2u\alpha A_N, \tag{3.8.16}$$

that coincides with Zamolodchikov's S-matrix (up to an over all factor considering the CDD poles) with $\alpha = 1$ and $u = \frac{\theta}{i\lambda}$. Actually, Zamolodchikov's S-matrix is universal, i.e., model independent.

$$\begin{aligned} S(\theta) = R(u) &= Q^{\pm}(u)u(u-2)[I + \frac{\sigma_3}{\sigma_2}P + \frac{\sigma_1}{\sigma_2}A_N] \\ &= Q^{\pm}(u)u(u-2)[I - \frac{1}{u}P + \frac{2}{u-2}A_N], \\ Q^{\pm}(u) &= \frac{\Gamma(\pm\frac{\lambda}{2\pi} - i\frac{\theta}{2\pi})\Gamma(\frac{1}{2} - i\frac{\theta}{2\pi})}{\Gamma(\frac{1}{2} \pm \frac{\lambda}{2\pi} - i\frac{\theta}{2\pi})\Gamma(-i\frac{\theta}{2\pi})} \end{aligned} \tag{3.8.17}$$

where $\lambda = \frac{2\pi}{N-2}$, $\theta = i\lambda u$. The spectrum parameter u is one-dimensional, but u can be taken to be the cut-off in 4-dimensional quantum field theory, for example

$$u \sim \ln \Lambda^2, \tag{3.8.18}$$

where Λ^2 is Lorentz invariant, i.e., scalar. This is the reason why asymptotic behavior of quantum field theory model may be related to Yang-Baxter system. The Bethe Ansatz for $S(\theta)$ with $SO(6)$ was discussed by Minahan and Zarembo ([46]).

For given $\check{R}(u)$ one can easily obtain Hamiltonian by

$$H = [\frac{\partial \check{R}(u)}{\partial u}\check{R}(u)]|_{u=0}, \tag{3.8.19}$$

for $O(N)$.

However, the essential connection between Lipatov model and $SO(6)$-RTT formulation is still missing.

4. Remarks

Although there has been certain progress of Yangian's application in physics, there are still open questions:

(1) How can the Yangian representations help to solve physical models, in particular, in strong correlation models?

(2) Direct evidences of Yangian in the real physics.

(3) What is the geometric meaning of Yangian?

Acknowledgement

We thank Professors F. Dyson and W. Happer for encouragement and enlighten discussions. This work is in part supported by NSF of China.

References

1. V. Drinfel'd, Sov. Math. Dokl. **32**(1985) 32.
2. V. Drinfel'd, Quantum group (PICM, Berkeley, 1986) 269.
3. V. Drinfel'd, Sov. Math. Dokl. **36** (1985) 212.
4. L.D. Faddeev, Sov. Sci. Rev. **C1** (1980) 107.
5. L.D. Faddeev, Les Houches, Session **39**, 1982.
6. L.D. Faddeev, Proc. of Les Houches Summer School, Session **LXIV** (1998) 149.
7. C.N. Yang, Phys. Rev. Lett. **19** (1967) 1312.
8. R. Baxter, Exactly Solved Methods in Statistical Mechanics, Academic, London, 1982.
9. M. Jimbo (ed), Yang-Baxter Equations in Integrable Systems, World Scientific, Singapore, 1990.
10. P.P. Kulish, I.K. Sklyanin, Lecture Notes in Physics, **151**(1982) 1.
11. V.E. Korepin, N.M. Bogoliubov, A.G. Izergin, Quantum inverse scattering method and correlation function, (Cambridge Univ. Press, Cambridge, 1993).

12. E.K. Sklyanin, *Quantum Inverse Scattering Methods*, Selected Topics, in M.L. Ge (ed.) Quantum Groups and Quantum Integrable Systems, World Scientific, Singapore, 63-88 (1991).
13. M.L. Ge, K. Xue and Y.M. Cho, Phys. Lett. A **260**(1999) 484.
14. M.L. Ge, K. Xue and Y.M. Cho, Phys. Lett. A **249**(1998) 358.
15. H.B. Zhang, M.L. Ge, K. Xue, J. Phys. A. **33** (2000) L345.
16. H.B. Zhang, M.L. Ge, K. Xue, J. Phys. A. **34** (2001) 919.
17. H.B. Zhang, M.L. Ge, K. Xue, J. Phys. A. **35** (2001) L7.
18. L.D. Faddeev, N.Yu. Reshetikhin, L.A. Takhtajan, Algebraic Analysis, **1** (1988) 178 (in Russian).
19. F.D.M. Haldane, Phys. Rev. Lett. **60** (1988) 635.
20. S.Shastry, Phys. Rev. Lett. **60** (1988) 639.
21. F.D.M. Haldane, Physics of the ideal semion gas: spinions and quantum symmetries of the integrable Haldnade-Shastry spin-chain, Proc. 16th Taniguch Symp. on Condensed Matter (Springer, Berlin, 1994).
22. V. Chari, A.Pressley, L'Enseignement mathematique **36** (1990) 267.
23. V. Chari, A.Pressley, J. Reine Angew. Math. **417** (1991) 87.
24. V. Chari, A.Pressley A guide to quantum groups (Cambridge Univ. Press, Cambridge, 1994).
25. E. Ragoucy, P. Sorba, Comm. Math. Phys. **203** (1999) 551.
26. M. Mintchev, E. Ragoucy, P. Sorba, Ph. Zaugg, J. Phys. A. **32** (1999) 5885.
27. S. Murakami, M. Wadati, J. Phys. Soc. Japan **62** (1993) 4203.
28. D.B. Uglov, V.E. Korepin, Phys. Lett. A **190** (1994) 238.
29. F. Essler, V.E. Korepin, K. Schoutens, Nucl. Phys. B **372** (1992) 559.
30. F. Essler, V.E. Korepin, K. Schoutens, Nucl. Phys. B **384** (1992) 431.
31. F. Essler, V.E. Korepin, Phys. Rev. Lett. **72** (1994)908.
32. Z.F. Wang, M.L. Ge, K, Xue, J. Phys. A. **30** (1997) 5023.
33. C.M. Bai, M.L. Ge and K. Xue, J. Stat. Phys. **102** (2001) 545.
34. L. Dolan, C.R. Nappi, E. Witten, arXiv: hep-th/0401243 (2004).
35. A.B. Zamolodchikov, Al.B. Zamolodchikov, Ann. Phys. **120** (1979) 253.
36. Y. Cheng, M.L. Ge, K. Xue, Comm. Math. Phys. **136** (1991) 195.
37. M.L. Ge, K. Xue, Phys. Lett. A. **120** (1991) 266.
38. A. Belavin, Phys. Lett. B **283** (1992) 67.
39. R. Balian, N.R. Werthamer, Phys. Rev. **131** (1963) 1553.
40. C.M. Bai, M.L. Ge and K. Xue, Directional Transitions in spin systems and representations of $Y(sl(2))$, Nankai preprint, APCTP-98-026.
41. C.J. Erickson, D. Levron, W. Happer, S. Kadlecek, B. Chann, L.W. Anderson, T.G. Walker, Phys. Rev. Lett. **85** (2000) 4237.
42. W. Happer, Degeneracies of the Hamiltonian $x(K+1/2)S_z+\mathbf{K}\cdot\mathbf{S}$, preprint, Princeton University, November, 2000.
43. C.M. Bai, M.L. Ge, K. Xue, Inter. J. Mod. Phys. B **16** (2002) 1876.
44. N. Beisert, arXiv: hep-th/0307015 (2003).
45. N. Beisert, M. Staudacher, arXiv: hep-th/0307042 (2003).
46. J.A. Minahan, K. Zarembo, JHEP **0303** (2003) 013.
47. V.A. Kazakov, A. Marshakov, J.A. Minahan, K. Zarembo, arXiv: hep-th/0402207 (2004).

48. L.N. Lipatov, Phys. Lett. B **309** (1993) 394.
49. L.N. Lipatov, JETP Lett. **59** (1994) 596.
50. L.D. Faddeev, G.P. Korchemsky, Phys. Lett. B **342** (1995) 311.
51. V.O. Tarasov, L.A. Takhtajan, L.D. Faddeev, Theor. Math. Phys. **57** (1983) 163.
52. M. Jimbo, Lett. Math. Phys. **10** (1985) 63.
53. H.J. de Vega, L.N. Lipatov, Phys. Rev. D **66** (2002) 074013.
54. H.J. de Vega, L.N. Lipatov, Phys. Lett. B **578** (2004) 335.

The hypoelliptic Laplacian and Chern-Gauss-Bonnet *

Jean-Michel Bismut

Département de Mathématique, Université Paris-Sud
Bâtiment 425, 91405 Orsay, France
E-mail: Jean-Michel.Bismut@math.u-psud.fr

This paper is dedicated to the memory of Professor S.S. Chern

We construct a new Hodge theory on the cotangent bundle of a Riemannian manifold X. The corresponding Laplacian is a second order hypoelliptic operator, which is self-adjoint with respect to a Hermitian form whose signature is (∞, ∞). This Hodge theory interpolates between the classical Hodge theory on X and the geodesic flow on T^*X. We also give results obtained with G. Lebeau on the analysis of the hypoelliptic Laplacian and on the hypoelliptic analytic torsion. Finally we explain the connections of this construction with Chern's proof of Chern-Gauss-Bonnet.

Introduction

The purpose of this paper is to describe a deformation of the classical Hodge theory of a compact Riemannian manifold X, whose corresponding Laplacian is a hypoelliptic operator on the cotangent bundle T^*X.

This construction came from the author's attempt to develop the Hodge theory of the loop space LX of X, and to construct the Witten deformation [W82] of the Hodge Laplacian of LX which would be associated to the energy functional E. Such a Witten deformation, if it existed, would interpolate between the Hodge Laplacian $\square^{LX}$ on LX and the Morse theory for E, whose critical points are the closed geodesics in X. There is indeed no Hodge theory on LX, one difficulty being the construction of a L^2 scalar product on the de Rham complex of LX. Still one can think of our construction as being the semiclassical limit of the non existing Hodge theory of LX.

---

*2000 *Mathematics Subject Classification.* 35H10, 58A14, 58J20, 58J52
Key words and Phases. Hypoelliptic equations, Hodge theory, Index theory and related fixed point theorems, Determinants and determinant bundles, analytic torsion

Needless to say, the construction of the hypoelliptic Laplacian can be done without any explicit reference to the loop space LX. Still many of the remarkable properties of this operator can be anticipated if one accepts the fact it is the 'shadow' of a Hodge theory to be on LX.

Another impetus came from the realization of the fact that many properties of the Witten deformation are related to an infinite dimensional version of the proof by Chern [C44] of Chern-Gauss-Bonnet. Indeed our strategy was to try finding what exotic Hodge theory corresponded to a formally defined supersymmetric path integral associated to the energy functional E on LX.

This paper is organized as follows. In section 1, we construct the adjoint of the de Rham operator d^{T^*X} with respect to an exotic bilinear form on the de Rham complex of T^*X.

In section 2, we give the Weitzenböck formula for the corresponding Laplacian, which turns out to be a hypoelliptic operator on T^*X.

In section 3, we show that the new Laplacian interpolates between classical Hodge theory and the geodesic flow.

In section 4, we give a self-adjointness property of the hypoelliptic Laplacian with respect to a Hermitian form of signature (∞, ∞).

In section 5, we summarize some of the results on the analysis of the new Laplacian obtained in [BL06] jointly with Lebeau.

In section 6, we state the main result we obtained in [BL06] saying that the Ray-Singer metric for the hypoelliptic Laplacian is the same as the Ray-Singer metric associated to the classical Laplacian.

Finally in section 7, we relate the above constructions to infinite dimensional versions of Chern-Gauss-Bonnet.

The construction of the hypoelliptic Laplacian was announced in [B04a, B04b, B04c]. It is detailed in [B05]. For a survey, we also refer to [B04d]. The analysis of the hypoelliptic Laplacian, and applications to analytic torsion are carried through in joint work with Lebeau [BL06].

1. A non standard Hodge theory

Let M be a smooth manifold. Let η be a nondegenerate bilinear form on TM. Let $\phi : TM \to T^*M$ be the morphism such that if $U, V \in TM$,

$$\eta(U,V) = \langle U, \phi V \rangle . \tag{1.1}$$

Let η^* the bilinear form on T^*M which corresponds to η by the morphism ϕ. Then η^* induces a nondegenerate bilinear form on $\Lambda^{\cdot}(T^*M)$. Let dv_M be a volume form on M. Let $\left(\Omega^{\cdot}(M), d^M\right)$ be the de Rham complex of smooth compactly supported differential forms on M. We equip $\Omega^{\cdot}(M)$ with the nondegenerate bilinear form,

$$\langle s, s' \rangle = \int_M \eta^* (s, s') \, dv_M. \tag{1.2}$$

Note that this bilinear form is in general neither symmetric nor antisymmetric.

Let $\overline{d}^M$ be the formal adjoint of d^M with respect to the bilinear form (1.2), so that if $s, s' \in \Omega^{\cdot}(M)$, then

$$\langle s, d^M s' \rangle = \left\langle \overline{d}^M s, s' \right\rangle. \tag{1.3}$$

Note that in general the formal adjoint of $\overline{d}^M$ in the sense of (1.3) is not equal to d^M.

Let X be a compact manifold of dimension n. Let $\pi : T^*X \to X$ be the cotangent bundle on X. Let $\theta = \langle p, dx \rangle$ be the canonical 1-form on T^*X. Let $\omega = d^{T^*X}\theta$ be the canonical symplectic form on T^*X. This is a nondegenerate bilinear form on TT^*X.

Let $\overline{d}^{T^*X}$ be the formal adjoint of d^{T^*X} with respect to the bilinear form $\langle \rangle$ on $\Omega^{\cdot}(T^*X)$, which is associated to ω and to the symplectic volume dv_{T^*X}.

It is easy to show that

$$\left[d^{T^*X}, \overline{d}^{T^*X} \right] = 0. \tag{1.4}$$

Observe that equation (1.4) is valid on any symplectic manifold. Indeed by using Darboux's theorem, equation (1.4) is just a reflection of the fact that $\omega(\xi, \xi) = 0$.

Equation (1.4) says that the Laplacian which is associated to the above bilinear form vanishes identically. Recall that our ultimate purpose is to produce a hypoelliptic Laplacian. The vanishing of our symplectic Laplacian simply indicates we have gone too far in the right direction.

Let us now explain in more detail the construction of the hypoelliptic Laplacian. Let g^{TX} be a metric on TX. We identify TX and T^*X by the metric g^{TX}. Let ∇^{TX} be the Levi-Civita connection on TX, and let R^{TX} be its curvature. The connection ∇^{TX} induces the splittings,

$$TT^*X = \pi^* (TX \oplus T^*X), \qquad T^*T^*X = \pi^* (T^*X \oplus TX). \tag{1.5}$$

From (1.5), we get the isomorphism,

$$\Lambda^{\cdot}(T^*T^*X) = \pi^* \left(\Lambda^{\cdot}(T^*X) \widehat{\otimes} \Lambda^{\cdot}(TX) \right). \tag{1.6}$$

We denote with a $\widehat{}$ the objects which refer to the second factor in the right-hand side in (1.6). Let $\nabla^{\Lambda^{\cdot}(T^*T^*X)}$ be the connection induced by ∇^{TX} on $\Lambda^{\cdot}(T^*T^*X)$.

Put

$$\phi = \begin{pmatrix} 1 & -1 \\ 1 & 0 \end{pmatrix}. \tag{1.7}$$

We identify ϕ with an automorphism of $TT^*X = TX \oplus T^*X$. The bilinear form η which is associated to ϕ as in (1.1) is given by

$$U, V \to \eta(U,V) = \langle \pi_* U, \pi_* V \rangle_{g^{TX}} + \omega(U,V). \tag{1.8}$$

Let $\langle\,\rangle_\phi$ be the associated nondegenerate bilinear form on $\Omega^{\cdot}(T^*X)$. Let $\overline{d}_\phi^{T^*X}$ the formal adjoint of d^{T^*X} with respect to η and to the symplectic volume form dv_{T^*X}.

Let $\mathcal{H} : T^*X \to \mathbf{R}$ be a smooth function. Let $Y^{\mathcal{H}}$ be the corresponding Hamiltonian vector field, so that

$$d^{T^*X}\mathcal{H} + i_{Y^{\mathcal{H}}}\omega = 0. \tag{1.9}$$

Set

$$\langle s, s' \rangle_{\phi,\mathcal{H}} = \int_{T^*X} \eta^* (s, s') \, e^{-2\mathcal{H}} dv_{T^*X}. \tag{1.10}$$

Put

$$d_{\mathcal{H}}^{T^*X} = e^{-\mathcal{H}} d^{T^*X} e^{\mathcal{H}}, \qquad \overline{d}_{\phi,\mathcal{H}}^{T^*X} = e^{\mathcal{H}} \overline{d}_\phi^{T^*X} e^{-\mathcal{H}}. \tag{1.11}$$

Then $\overline{d}_{\phi,2\mathcal{H}}^{T^*X}$ is the formal adjoint of d^{T^*X} with respect to $\langle\,\rangle_{\phi,\mathcal{H}}$, and $\overline{d}_{\phi,\mathcal{H}}^{T^*X}$ is the formal adjoint of $d_{\mathcal{H}}^{T^*X}$ with respect to $\langle\,\rangle_\phi$.

Set

$$A_{\phi,\mathcal{H}} = \frac{1}{2}\left(\overline{d}_{\phi,2\mathcal{H}}^{T^*X} + d^{T^*X}\right), \qquad \mathfrak{A}_{\phi,\mathcal{H}} = \frac{1}{2}\left(\overline{d}_{\phi,\mathcal{H}}^{T^*X} + d_{\mathcal{H}}^{T^*X}\right). \tag{1.12}$$

Clearly,

$$\mathfrak{A}_{\phi,\mathcal{H}} = e^{-\mathcal{H}} A_{\phi,\mathcal{H}} e^{\mathcal{H}}. \tag{1.13}$$

If Z is a vector field on T^*X, let L_Z be the corresponding Lie derivative operator acting on $\Omega^{\cdot}(T^*X)$.

More generally, let $\left(F, \nabla^F\right)$ be a complex flat vector bundle on X, and let g^F be a non necessarily flat Hermitian metric on F. Let

$\left(\Omega^{\cdot}\left(T^*X,\pi^*F\right),d^{T^*X}\right)$ be the de Rham complex of smooth compactly supported forms on T^*X with coefficients in F. The operator L_Z still acts naturally on $\Omega^{\cdot}\left(T^*X,\pi^*F\right)$. Set

$$\omega\left(\nabla^F,g^F\right)=\left(g^F\right)^{-1}\nabla^F g^F. \tag{1.14}$$

The 1-form $\omega\left(\nabla^F,g^F\right)$ takes values in self-adjoints endomorphisms of F.

Also there is an obvious extension of the bilinear form in (1.10) to a skew-linear form on $\Omega^{\cdot}\left(T^*X,\pi^*F\right)$, in which the metric g^F is incorporated in the obvious way. It is then possible to extend the above constructions, and still obtain operators like the ones in (1.11)-(1.13), which now act on $\Omega^{\cdot}\left(T^*X,\pi^*F\right)$. In the sequel, we will deal with this more general situation.

2. The Weitzenböck formula for the hypoelliptic Laplacian

Let $e_1,\ldots,e_n$ be an orthonormal basis of TX, let $e^1,\ldots,e^n$ be the corresponding dual basis of T^*X. Let $\widehat{e}_1,\ldots,\widehat{e}_n$ and $\widehat{e}^1,\ldots,\widehat{e}^n$ be other copies of these bases.

Then $e_1,\ldots,e_n,\widehat{e}^1,\ldots,\widehat{e}^n$ is a basis of TT^*X, and $e^1,\ldots,e^n,\widehat{e}_1,\ldots,\widehat{e}_n$ is the dual basis of T^*T^*X. Set

$$\widehat{\nabla^V\mathcal{H}}=\nabla_{\widehat{e}^i}\mathcal{H}\widehat{e}^i. \tag{2.1}$$

We give the Weitzenböck formula established in B05 .

Theorem 2.1. *The following identities hold,*

$$\begin{aligned}A^2_{\phi,\mathcal{H}}={}&\frac{1}{4}\left(-\Delta^V-\frac{1}{2}\left\langle R^{TX}\left(e_i,e_j\right)e_k,e_l\right\rangle e^ie^ji_{\widehat{e}^k}i_{\widehat{e}^l}+2L_{\widehat{\nabla^V\mathcal{H}}}\right)\\&-\frac{1}{2}\left(L_{Y^{\mathcal{H}}}+\frac{1}{2}e^ii_{\widehat{e}^j}\nabla^F_{e_i}\omega\left(\nabla^F,g^F\right)\left(e_j\right)+\frac{1}{2}\omega\left(\nabla^F,g^F\right)\left(e_i\right)\nabla_{\widehat{e}^i}\right),\end{aligned} \tag{2.2}$$

$$\begin{aligned}\mathfrak{A}^2_{\phi,\mathcal{H}}={}&\frac{1}{4}\Bigg(-\Delta^V-\frac{1}{2}\left\langle R^{TX}\left(e_i,e_j\right)e_k,e_l\right\rangle e^ie^ji_{\widehat{e}^k}i_{\widehat{e}^l}+\left|\nabla^V\mathcal{H}\right|^2\\&\qquad-\Delta^V\mathcal{H}+2\nabla_{\widehat{e}^i}\nabla_{\widehat{e}^j}\mathcal{H}\widehat{e}_ii_{\widehat{e}^j}+2\nabla_{\widehat{e}^i}\nabla_{e_j}\mathcal{H}e^ji_{\widehat{e}^i}\Bigg)\\&-\frac{1}{2}\Bigg(L_{Y^{\mathcal{H}}}+\frac{1}{2}\omega\left(\nabla^F,g^F\right)\left(Y^{\mathcal{H}}\right)+\frac{1}{2}e^ii_{\widehat{e}^j}\nabla^F_{e_i}\omega\left(\nabla^F,g^F\right)\left(e_j\right)\\&\qquad+\frac{1}{2}\omega\left(\nabla^F,g^F\right)\left(e_i\right)\nabla_{\widehat{e}^i}\Bigg).\end{aligned}$$

Given $c \in \mathbf{R}$, set

$$\mathcal{H} = \frac{|p|^2}{2}, \qquad \mathcal{H}^c = c\frac{|p|^2}{2}. \tag{2.3}$$

When $c \in \mathbf{R}^*$, put $c = \pm 1/b^2, b > 0$. We state a result which was established in B05 .

Theorem 2.2. *The following identity holds,*

$$L_{Y^{\mathcal{H}^c}} = \nabla_{Y^{\mathcal{H}^c}}^{\Lambda^{\cdot}(T^*T^*X)\otimes F} + c\widehat{e}_i i_{e_i} + c\left\langle R^{TX}(p,e_i)p,e_j\right\rangle e^i i_{\widehat{e}^j}. \tag{2.4}$$

Moreover,

$$A^2_{\phi,\mathcal{H}^c} = \frac{1}{4}\left(-\Delta^V + 2cL_{\widehat{p}} - \frac{1}{2}\left\langle R^{TX}(e_i,e_j)e_k,e_l\right\rangle e^i e^j i_{\widehat{e}^k} i_{\widehat{e}^l}\right)$$
$$-\frac{1}{2}\left(L_{Y^{\mathcal{H}^c}} + \frac{1}{2}e^i i_{\widehat{e}^j}\nabla^F_{e_i}\omega\left(\nabla^F,g^F\right)(e_j) + \frac{1}{2}\omega\left(\nabla^F,g^F\right)(e_i)\nabla_{\widehat{e}^i}\right), \tag{2.5}$$
$$\mathfrak{A}^2_{\phi,\mathcal{H}^c} = \frac{1}{4}\left(-\Delta^V + c^2|p|^2 + c\left(2\widehat{e}_i i_{\widehat{e}^i} - n\right) - \frac{1}{2}\left\langle R^{TX}(e_i,e_j)e_k,e_l\right\rangle e^i e^j i_{\widehat{e}^k} i_{\widehat{e}^l}\right)$$
$$-\frac{1}{2}\left(L_{Y^{\mathcal{H}^c}} + \frac{1}{2}\omega\left(\nabla^F,g^F\right)\left(Y^{\mathcal{H}^c}\right) + \frac{1}{2}e^i i_{\widehat{e}^j}\nabla^F_{e_i}\omega\left(\nabla^F,g^F\right)(e_j)\right.$$
$$\left. + \frac{1}{2}\omega\left(\nabla^F,g^F\right)(e_i)\nabla_{\widehat{e}^i}\right).$$

For $c \in \mathbf{R}^*$, the operators $\frac{\partial}{\partial u} - A^2_{\phi,\mathcal{H}^c}, \frac{\partial}{\partial u} - \mathfrak{A}^2_{\phi,\mathcal{H}^c}$ are hypoelliptic.

Proof. Observe here that the result of hypoellipticity follows from a well-known result by Hörmander H67. □

Observe that the operators $A^2_{\phi,\mathcal{H}^c}, \mathfrak{A}^2_{\phi,\mathcal{H}^c}$ are not elliptic and not self-adjoint.

3. An interpolation property

Let $r : T^*X \to T^*X$ be the map $(x,p) \to (x,-p)$. Set

$$\mathfrak{a}_\pm = \frac{1}{2}\left(-\Delta^V \pm 2L_{\widehat{p}} - \frac{1}{2}\left\langle R^{TX}(e_i,e_j)e_k,e_l\right\rangle e^i e^j i_{\widehat{e}^k} i_{\widehat{e}^l}\right), \tag{3.1}$$
$$\mathfrak{b}_\pm = -\left(\pm L_{Y^{\mathcal{H}}} + \frac{1}{2}e^i i_{\widehat{e}^j}\nabla^F_{e_i}\omega\left(\nabla^F,g^F\right)(e_j) + \frac{1}{2}\omega\left(\nabla^F,g^F\right)(e_i)\nabla_{\widehat{e}^i}\right).$$

Then $\mathfrak{a}_\pm$ commutes with r^*, and that $\mathfrak{b}_\pm$ anticommutes with r^*.

For $a \in \mathbf{R}$, let $r_a : T^*X \to T^*X$ be the dilation $(x,p) \to (x,ap)$, so that $r = r_{-1}$. For $c = \pm 1/b^2$, set

$$A_{\phi_b,\pm\mathcal{H}} = r_b^* A_{\phi,\mathcal{H}^c} r_b^{*-1}. \tag{3.2}$$

By (2.5), we get

$$2A^2_{\phi_b,\pm\mathcal{H}} = \frac{1}{b^2}\mathfrak{a}_\pm + \frac{1}{b}\mathfrak{b}_\pm. \tag{3.3}$$

Let $o(TX)$ be the orientation bundle of TX. Let Φ^{T^*X} be the Thom form on T^*X of Mathai-Quillen MQ86 which is associated to the metric g^{TX} and to the connection ∇^{TX}. The form Φ^{T^*X} is a closed form of degree n with coefficients in $o(TX)$, such that $\pi_*\Phi^{T^*X} = 1$. It is normalized in such a way that

$$\Phi^{T^*X} = \exp\left(-|p|^2 + \dots\right). \tag{3.4}$$

In (3.4), ... designates an explicit complicate expression involving curvature. As is suggested by (3.4), the form Φ^{T^*X} restricts to a Gaussian form along the fibre.

One verifies easily that the operators $\mathfrak{a}_\pm$ are semisimple. The kernel of $\mathfrak{a}_+$ is generated by the function 1, and the corresponding projector $Q_+^{T^*X}$ on this kernel is given by $\alpha \to \pi_*\left(\alpha \wedge \Phi^{T^*X}\right)$. The kernel of $\mathfrak{a}_-$ is generated by Φ^{T^*X}, and the corresponding projector $Q_-^{T^*X}$is given by $\alpha \to (\pi_*\alpha)\wedge\Phi^{T^*X}$.

Let d^X be the de Rham operator acting on $\Omega^\cdot(X,F)$ in the + case or on $\Omega^\cdot(X, F \otimes o(TX))$ in the − case, and let d^{X*} be its formal adjoint with respect to the standard L^2 Hermitian product. Let $\square^X = \left[d^X, d^{X*}\right]$ denote the corresponding Hodge Laplacian.

The following result is established in B05

Theorem 3.1. *The following identity holds,*

$$-Q_\pm^{T^*X}\mathfrak{b}_\pm\mathfrak{a}_\pm^{-1}\mathfrak{b}_\pm Q_\pm^{T^*X} = \frac{1}{2}\square^X. \tag{3.5}$$

Observe that a formula similar to (3.5) plays a key role in the paper by Bismut et Lebeau BL91, where the Hodge theory of a compact complex manifold is deformed into the Hodge theory of a submanifold. Identities (3.3) and (3.5) indicate that the matrix structure of the operator in (3.3) is essentially similar to the one in BL91.

Also observe that in degree 0, equation (3.5) is equivalent to

$$\int_{T^*X} \nabla_p\nabla_p e^{-|p|^2}\frac{dp}{\pi^{n/2}} = \frac{1}{2}\Delta^X, \tag{3.6}$$

which itself is equivalent to

$$\sum_1^n \nabla_{e_i}^2 = \Delta^X. \tag{3.7}$$

The contribution of $\mathfrak{a}_{\pm}^{-1}$to equation (3.6) is in fact equal to 1.

In BL06, equation (3.5) provides one of the key algebraic results from which one shows that in the proper sense, when $c \to \pm\infty$, the resolvent of a suitably conjugate version of the operator $2A^2_{\phi,\mathcal{H}^c}$ converges to the resolvent of $\frac{1}{2}\square^X$. The relevant conjugation is described in B05 and in BL06.

Suppose again that $F = \mathbf{R}$. Let $N^V = \sum_1^n \widehat{e}_i i_{\widehat{e}^i}$ be the vertical number operator, i.e. the operator which counts the vertical degree of forms in $\Omega^{\cdot}(T^*X, \pi^*F)$. We have the identity of B05 ,

$$\begin{aligned} r_{b^2}^* 2\mathfrak{A}^2_{\phi,\mathcal{H}^c} r_{b^2}^{*-1} = {} & \frac{1}{2}\left(-c^2\Delta^V + |p|^2 - cn\right) + cN^V \\ & - \frac{c^2}{4}\left\langle R^{TX}(e_i,e_j)e_k, e_l\right\rangle e^i e^j i_{\widehat{e}^k} i_{\widehat{e}^l} \mp L_{Y^{\mathcal{H}}}, \end{aligned} \tag{3.8}$$

so that as $b \to +\infty$,

$$r_{b^2}^* 2\mathfrak{A}^2_{\phi,\mathcal{H}^c} r_{b^2}^{*-1} \simeq \frac{1}{2}|p|^2 \mp L_{Y^{\mathcal{H}}}. \tag{3.9}$$

In the right-hand side of (3.9), there is essentially the Lie derivative operator $\mp L_{Y^{\mathcal{H}}}$.

This should convince the reader that as when $b \to +\infty$, the trace of the heat kernel $\exp\left(-tA^2_{\phi,\mathcal{H}^c}\right)$ should localize near the closed geodesics in X.

From the above, we find that up to scaling, $2A^2_{\phi,\mathcal{H}^c}$ interpolates in a proper sense between the Hodge Laplacian and the geodesic flow.

4. A self-adjointness property

The operator $A^2_{\phi,\mathcal{H}^c}$ is certainly not self-adjoint in the classical sense. However it is shown in B05 that it is self-adjoint with respect to a nondegenerate Hermitian form of signature (∞,∞), which we now describe.

Let g^{T^*X} be the metric on the fibres of T^*X which is dual to g^{TX}. Let $\mathfrak{g}^{TT^*X}$ be the Riemannian metric on T^*X whose matrix with respect to the splitting $TT^*X = \pi^*(TX \oplus T^*X)$ is given by

$$\mathfrak{g}^{TT^*X} = \begin{pmatrix} g^{TX} & 1|_{T^*X} \\ 1|_{TX} & 2g^{T^*X} \end{pmatrix}. \tag{4.1}$$

The volume form attached to $\mathfrak{g}^{TT^*X}$ is the symplectic volume form dv_{T^*X}.

Let F be the $\mathbf{g}^{TT^*X}$ isometric involution of TT^*X whose matrix with respect to the above splitting is given by

$$F = \begin{pmatrix} 1|_{TX} & 2g^{T^*X} \\ 0 & -1|_{T^*X} \end{pmatrix}. \tag{4.2}$$

Then F acts like $\widetilde{F}^{-1}$ on $\Lambda^{\cdot}(T^*T^*X)$.

Let $\langle\rangle_{\mathbf{g}^{\Omega^{\cdot}(T^*X,\pi^*F)}}$ be the Hermitian product on $\Omega^{\cdot}(T^*X,\pi^*F)$ associated to the metrics $\mathbf{g}^{TT^*X}, g^F$.

Let u be the isometric involution of $\Omega^{\cdot}(T^*X,\pi^*F)$,

$$us(x,p) = Fs(x,-p). \tag{4.3}$$

Let $\mathfrak{h}^{\Omega^{\cdot}(T^*X,\pi^*F)}$ be the Hermitian form on $\Omega^{\cdot}(T^*X,\pi^*F)$,

$$\langle s,s'\rangle_{\mathfrak{h}^{\Omega^{\cdot}(T^*X,\pi^*F)}} = \langle us,s'\rangle_{\mathbf{g}^{\Omega^{\cdot}(T^*X,\pi^*F)}}. \tag{4.4}$$

Note here that a Hermitian form has the same properties as a Hermitian product, except for positivity.

If $\mathcal{H}$ is a r-invariant smooth function on T^*X, if $s,s' \in \Omega^{\cdot}(T^*X,\pi^*F)$, set

$$\langle s,s'\rangle_{\mathfrak{h}_{\mathcal{H}}^{\Omega^{\cdot}(T^*X,\pi^*F)}} = \langle e^{-2\mathcal{H}}s,s'\rangle_{\mathfrak{h}^{\Omega^{\cdot}(T^*X,\pi^*F)}}. \tag{4.5}$$

Note that since $\mathcal{H}$ is r-invariant, $\mathfrak{h}_{\mathcal{H}}^{\Omega^{\cdot}(T^*X,\pi^*F)}$ is still a Hermitian form. The Hermitian forms in (4.4), (4.5) have signature (∞,∞).

Now we state a result established in [B05].

Theorem 4.1. *If $\mathcal{H}$ is r-invariant, then $A_{\phi,\mathcal{H}}$ (resp. $B_{\phi,\mathcal{H}}$) is $\mathfrak{h}_{\mathcal{H}}^{\Omega^{\cdot}(T^*X,\pi^*F)}$ self-adjoint (resp skew-adjoint), and $\mathfrak{A}_{\phi,\mathcal{H}}$ (resp. $\mathfrak{B}_{\phi,\mathcal{H}}$) is $\mathfrak{h}^{\Omega^{\cdot}(T^*X,\pi^*F)}$ self-adjoint (resp. skew-adjoint).*

Of course, Theorem 2.1 applies to the operators associated to $\mathcal{H} = \mathcal{H}^c$ which were considered in section 2. Its implications are discussed in [B05] and [BL06].

5. The analysis of the hypoelliptic Laplacian

Now we briefly describe some results on the analysis of the operator $\mathfrak{A}^2_{\phi,\mathcal{H}^c}$ which are established in [BL06]. One of the key results is that $\mathfrak{A}^2_{\phi,\mathcal{H}^c}$ has compact resolvent, that its spectrum is discrete, and that the corresponding characteristic subspaces are finite dimensional and included in the Schwartz space $\mathcal{S}^{\cdot}(T^*X,\pi^*F)$.

Of special interest from the point of view of Hodge theory is the characteristic subspace $\mathcal{S}^{\cdot}\left(T^{*}X,\pi^{*}F\right)_{0}$ attached to the eigenvalue 0. The spectral projection provides a natural supplementary subspace $\mathcal{S}^{\cdot}\left(T^{*}X,\pi^{*}F\right)_{*}$ to $\mathcal{S}^{\cdot}\left(T^{*}X,\pi^{*}F\right)_{0}$.

Let $\mathfrak{H}^{\cdot}(X,F)$ denote the ordinary cohomology of $T^{*}X$ when $c>0$, and the compactly supported cohomology of $T^{*}X$ for $c<0$. For $c>0$, $\mathfrak{H}^{\cdot}(X,F)=H^{\cdot}(X,F)$, and for $c<0$, $\mathfrak{H}^{\cdot}(X,F)=H^{\cdot-n}(X,F\otimes o(TX))$, this last identification being the Thom isomorphism.

In BL06, it is shown that the complex $\left(\mathcal{S}^{\cdot}\left(T^{*}X,\pi^{*}F\right)_{*},d^{T^{*}X}_{\mathcal{H}^{c}}\right)$ is acyclic, and that the cohomology of $\left(\mathcal{S}^{\cdot}\left(T^{*}X,\pi^{*}F\right)_{0},d^{T^{*}X}_{\mathcal{H}^{c}}\right)$ is just $\mathfrak{H}^{\cdot}(X,F)$.

We will say that $b>0$ is of Hodge type if all the classical consequences of Hodge theory hold for the hypoelliptic Laplacian $\mathfrak{A}^{2}_{\phi,\mathcal{H}^{c}}$, which means in particular that $d^{T^{*}X}_{\mathcal{H}^{c}}$ vanishes on $\mathcal{S}^{\cdot}\left(T^{*}X,\pi^{*}F\right)_{0}$.

In BL06, it is shown that for $b>0$ small enough, b is of Hodge type, and also that the set of $b>0$ which are not of Hodge type is discrete. The proof relies in particular on the fact that classical Hodge theory is... of Hodge type, and moreover that being of Hodge type is an open property.

Finally it is shown in BL06 that, as explained in section 3, the resolvent of a suitably conjugate version of $A^{2}_{\phi,\mathcal{H}^{c}}$ converges in the strongest possible sense to the resolvent of $\square^{X}/4$, and also that the corresponding heat kernels converge in a very strong sense.

6. The hypoelliptic Laplacian and analytic torsion

Set

$$\lambda(F)=\det H^{\cdot}(X,F). \tag{6.1}$$

Put

$$\begin{aligned}\lambda=&\lambda(F)\text{ if }c>0,\\&\left(\lambda\left(F\otimes o(TX)\right)\right)^{(-1)^{n}}\text{ if }c<0.\end{aligned} \tag{6.2}$$

The line λ can be equipped with the classical Ray-Singer metric $\|\;\|^{2}_{\lambda,0}$, which one obtains via the Ray-Singer analytic torsion for the Hodge Laplacian $\square^{X}$.

On the other hand, for $b>0$, one can define a generalized metric $\|\lambda\|^{2}_{\lambda,b}$ on λ, which is obtained via the analytic torsion or $A^{2}_{\phi,\mathcal{H}^{c}}$. Its construction also involves the Hermitian form $\mathfrak{h}^{\Omega^{\cdot}(T^{*}X,\pi^{*}F)}_{\mathcal{H}^{c}}$. Contrary to an usual metric, this generalized metric has a sign. When the sign is positive, it is a usual metric.

The main result established in [BL06] is as follows.

Theorem 6.1. *For $b > 0$, we have the identity,*

$$\| \|^2_{\lambda,b} = \| \|^2_{\lambda,0} . \tag{6.3}$$

The proof of Theorem 6.1 is difficult. Besides the functional analytic machine which is needed to handle the hypoelliptic Laplacian properly, one also needs to develop a local index theory for this operator. It is rather easy to show that the generalized metric $\| \|^2_{\lambda,b}$ does not depend on $b > 0$. Showing equality in (6.3) is much harder. One has to take full advantage of the convergence of resolvents which was described in sections 3 and 5.

In fact equality in (6.3) should not be taken for granted. Indeed the small time asymptotics of the heat kernels associated to elliptic or hypoelliptic operators are very different. On a priori grounds, one could expect a term measuring the transition from the hypoelliptic regime to the elliptic one. In fact such a term appears when one considers the equivariant version of the above metrics.

7. The hypoelliptic Laplacian and Chern-Gauss-Bonnet

Let $\left(E, g^E, \nabla^E\right)$ be a real Euclidean vector bundle of dimension n on a manifold M, which is equipped with a metric preserving connection. Let Φ^E be the Mathai-Quillen Thom form [MQ86] associated to $\left(g^E, \nabla^E\right)$. The Mathai-Quillen Thom form, which is a form of degree n, will be normalized in such a way that if p is the generic element of E,

$$\Phi^E = \exp\left(- |p|^2 /2 + \dots\right). \tag{7.1}$$

Note that the normalization in (7.1) is different from the one which is used in (3.4).

Let s be a smooth section of E on M. Then $s^*\Phi^E$ is a closed n-form on M, whose cohomology class does not depend on s. For $T > 0$, set

$$a_T = (Ts)^* \Phi^E. \tag{7.2}$$

Then a_T is a family of closed n-forms, which lie in the same cohomology class. The form a_0 is just the Chern form $e\left(E, \nabla^E\right) = \mathrm{Pf}\left[R^E/2\pi\right]$ which appears in Chern's version of Chern-Gauss-Bonnet [C44]. By (7.1), we get

$$a_T = \exp\left(-T^2 |s|^2 /2 + \dots\right). \tag{7.3}$$

Equation (7.3) indicates that when $T \to +\infty$, the current a_T localizes near the vanishing locus Y of s. If the section s is generic, then Y is a submanifold of M. One can establish that when $T \to +\infty$, a_T converges as a current to an explicit current localized on Y.

The strategy used by Chern [C44] to prove Chern-Gauss-Bonnet is closely related to the above argument. Indeed he constructs directly a transgressed version of the Thom form, from which the Gauss-Bonnet theorem follows by an argument essentially similar to the one outlined above.

Physicists have taught us that some version of the Chern-Gauss-Bonnet theorem still holds in infinite dimensions, thereby establishing a connection between an often mathematically ill-defined functional integral and its localisation on the zero set of the section of some infinite dimensional vector bundle, which is directly accessible to mathematical understanding.

We will illustrate this point in the context of the Witten deformation of classical Hodge theory, and later explain the relevance of Chern's point of view to the hypoelliptic Laplacian.

Indeed let X be a Riemannian manifold as above. We take here F to be just $\mathbf{R}$ equipped with its canonical metric. Let $f : X \to \mathbf{R}$ be a smooth function. In [W82], Witten proposed a deformation of Hodge theory associated to the function f. Given $T \in \mathbf{R}$, the idea is to replace the de Rham operator d^X by the twisted version $d^X_T = e^{-Tf} d^X e^{Tf}$, and to form the corresponding Laplacian $\square^X_T$.

Observe the following simple three points:

- For $T = 0$, $\square^X_T = \square^X$.
- The operator $\square^X_T$ is still a second order elliptic self-adjoint nonnegative operator.
- The Hodge theorem still holds for $\square^X_T$, i.e. $\ker \square^X_T$ still represents $H^{\cdot}(X, \mathbf{R})$.

Assume that f is a Morse function. In [W82], Witten showed that when $T \to +\infty$, most of the eigenvalues of $\square^X_T$ tend to $+\infty$, except a finite family of them which are either 0 or are exponentially small. Moreover the finite dimensional complex $\left(F^{\cdot}_T, d^X_T\right)$ of eigenforms associated to small eigenvalues localizes near the critical points of f, the forms of degree i localizing near the critical points of index i, from which the Morse inequalities immediately follow.

We will not focus here on the refinements suggested by Witten concerning the explicit description of the complex $\left(F^{\cdot}_T, d^X_T\right)$ in terms of the Morse-Smale complex associated to the gradient field $-\nabla f$. The main point we want to make is that $\square^X_T$ provides an interpolation between Hodge theory and Morse theory.

When f is Morse, the gradient field ∇f is a generic section of TX. The corresponding forms a_T as in (7.2) interpolate between the Chern form $\mathrm{Pf}\left[\frac{R^X}{2\pi}\right]$ and a signed sum of Dirac masses concentrated at the critical points of f.

We will briefly explain how the fact that $\square_T^X$ interpolates between classical Hodge theory and Morse theory can be interpreted as a consequence of the same localization principle on the loop space LX of X, which is the set of smooth maps $s \in S^1 \to x_s \in X$.

We start from observations of Atiyah and Witten A85. Note that LX is a Riemannian manifold, which inherits its Riemannian metric g^{TLX} from the metric g^{TX}. Also S^1 acts isometrically on LX , so that if $t \in S^1, x \in LX$, $k_t x_. = x_{.+t}$. The generator of this action is the Killing vector field $K(x) = \dot{x}$. The manifold X sits inside LX as the zero set of K.

The function f lifts to the S^1-invariant function F on LX,

$$F(x) = \int_{S^1} f(x_s)\, ds. \tag{7.4}$$

By the McKean-Singer formula McKS67, we find that if $\chi(X)$ is the Euler characteristic of X, then

$$\chi(X) = \mathrm{Tr_s}\left[\exp\left(-\square_T^X/2\right)\right]. \tag{7.5}$$

Using functional integration, and more specifically the theory of Brownian motion, we can rewrite the right-hand side of (7.5) in the form,

$$\mathrm{Tr_s}\left[\exp\left(-\square_T^X/2\right)\right] = \int_{L^0X} d\mu_T. \tag{7.6}$$

In (7.6), μ_T is a signed measure on L^0X, the set of continuous loops in X, which is S^1-invariant. The fact that μ_T is carried by L^0X and not by LX is a well-known pathology associate with functional integration.

By using arguments developed first by Atiyah and Witten in A85 and later pursued in B85; B86, one can transform the well defined integral in the right-hand side of (7.6) into an ill defined integral of a current on LX. More specifically, we rewrite (7.6) in the form,

$$\mathrm{Tr_s}\left[\exp\left(-\square_T^X/2\right)\right] = \int_{LX} \alpha \wedge (T\nabla F)^* \, \Phi^{TLX}. \tag{7.7}$$

Note that we have replaced L^0X by LX for notational expediency. Let us briefly describe the two forms which appear in the right-hand side of (7.7). First they are both closed with respect to the operator $d + i_K$, which is the equivariant version of the de Rham operator. Since $L_K = (d + i_K)^2$, these forms are also S^1-invariant. The vanishing under $d + i_K$ is called supersymmetry in the physics literature.

Let $E(x) = \frac{1}{2}\int_{S^1} |\dot{x}|^2 ds$ be the energy functional on LX. The form α takes the form

$$\alpha = \exp(-E + \omega). \tag{7.8}$$

The form ω is a closed 2-form which we will not describe more precisely.

The form Φ^{TLX} is the equivariant Thom form for TLX equipped with the metric g^{TLX}, the Levi-Civita connection ∇^{TLX} and the action of K. In view of (7.1), (7.7), (7.8), we get

$$\mathrm{Tr_s}\left[\exp\left(-\square^X_T/2\right)\right] = \int_{LX} \exp\left(-\frac{1}{2}\int_{S^1} |\dot{x}|^2 ds - \frac{T^2}{2}\int_{S^1} |\nabla f(x_s)|^2 ds + \dots\right). \tag{7.9}$$

The point about (7.9) is that for $T = 0$, we get a classical Brownian integral which is known to be connected with the Hodge Laplacian $\square^X/2$. For $T \to +\infty$, the integral (7.9) should localize on $\nabla f = 0$.

The above picture gives us a geometric understanding of the localization of the heat kernels on the diagonal near the critical points of f, of which the standard localization of the form a_T associated with ∇f appears as a semiclassical limit, when scaling the metric g^{TX} by a factor $1/t$ and making $t \to 0$.

Now LX carries many natural S^1 functionals like the energy E or more generally any functional

$$I(x) = \int_{S^1} L(x, \dot{x})\, ds, \tag{7.10}$$

where L is a classical Lagrangian. Of course when $L(x, \dot{x}) = \frac{1}{2}|\dot{x}|^2$, then $I = E$. The idea is then to replace F by E in (7.7). More precisely consider a path integral of the type

$$\int_{LX} \alpha \wedge (T\nabla E)^* \Phi^{TLX}. \tag{7.11}$$

One can ask whether there is a new Hodge theory which would extend (7.7) to an expression of the type (7.11).

This is exactly what the the hypoelliptic Laplacian $2A^2_{\phi,\mathcal{H}^c}$ does, with $c = \pm 1/b^2, T = b^2$. Indeed in this case equation (7.9) is replaced by

$$\mathrm{Tr_s}\left[\exp\left(-2A^2_{\phi,\mathcal{H}^c}\right)\right] = \int_{LX} \exp\left(-\frac{1}{2}\int_{S^1} |\dot{x}|^2 ds - \frac{T^2}{2}\int_{S^1} |\ddot{x}|^2 ds + \dots\right). \tag{7.12}$$

For $T = 0$, we should recover the classical Hodge theory for $\square^X/2$, and for $T \to +\infty$, the integral in (7.12) should localize on closed geodesics.

The results which were described in the previous sections come as close as possible to fulfil this dream.

References

A85. M. F. Atiyah. Circular symmetry and stationary-phase approximation. *Astérisque*, (131):43–59, 1985. Colloquium in honor of Laurent Schwartz, Vol. 1 (Palaiseau, 1983).

B85. J.-M. Bismut. Index theorem and equivariant cohomology on the loop space. *Comm. Math. Phys.*, 98(2):213–237, 1985.

B86. J.-M. Bismut. Localization formulas, superconnections, and the index theorem for families. *Comm. Math. Phys.*, 103(1):127–166, 1986.

B04a. J.-M. Bismut. Une déformation de la théorie de Hodge sur le fibré cotangent. *C. R. Math. Acad. Sci. Paris*, 338(6):471–476, 2004.

B04b. J.-M. Bismut. Le laplacien hypoelliptique sur le fibré cotangent. *C. R. Math. Acad. Sci. Paris*, 338(7):555–559, 2004.

B04c. J.-M. Bismut. Une déformation en famille du complexe de de Rham-Hodge. *C. R. Math. Acad. Sci. Paris*, 338(8):623–627, 2004.

B04d. J.-M. Bismut. Le Laplacien hypoelliptique. In *Séminaire: Équations aux Dérivées Partielles, 2003–2004*, Sémin. Équ. Dériv. Partielles, pages Exp. No. XXII, 15. École Polytech., Palaiseau, 2004.

B05. J.-M. Bismut. The hypoelliptic Laplacian on the cotangent bundle. *J. Amer. Math. Soc.*, 18(2):379–476 (electronic), 2005.

BL91. J.-M. Bismut and G. Lebeau. Complex immersions and Quillen metrics. *Inst. Hautes Études Sci. Publ. Math.*, (74):ii+298 pp. (1992), 1991.

BL06. J.-M. Bismut and G. Lebeau. The hypoelliptic Laplacian and Ray-Singer metrics. *to appear*, 2006.

C44. S.S. Chern. A simple intrinsic proof of the Gauss-Bonnet formula for closed Riemannian manifolds. *Ann. of Math. (2)*, 45:747–752, 1944.

H67. L. Hörmander. Hypoelliptic second order differential equations. *Acta Math.*, 119:147–171, 1967.

MQ86. V. Mathai and D. Quillen. Superconnections, Thom classes, and equivariant differential forms. *Topology*, 25(1):85–110, 1986.

McKS67. H. P. McKean, Jr. and I. M. Singer. Curvature and the eigenvalues of the Laplacian. *J. Differential Geometry*, 1(1):43–69, 1967.

W82. E. Witten. Supersymmetry and Morse theory. *J. Differential Geom.*, 17(4):661–692 (1983), 1982.

S. S. Chern and Chern-Simons Terms*

R. Jackiw

Center for Theoretical Physics,
Department of Physics,
Massachusetts Institute of Technology
Cambridge, Massachusetts 02139
E-mail: jackiw@lns.mit.edu

Some properties of Chern-Simons terms are presented and their physical utility is surveyed.

1. Meeting S.S. Chern

I first met Professor Chern in Durham, England a quarter century ago in the summer of 1979 at a symposium sponsored by the London Mathematical Society. The event brought together physicists and mathematicians because both discovered that after many years of separation we were again interested in common problems. This was a time when physicists realized that the axial anomaly involves the Chern-Pontryagin density, whose integral measures the topological properties of gauge fields; that the anomaly equation is a local version of the Atiyah-Singer index theorem, which in turn counts the number of zero modes in various linear elliptic equations, like the physicists' Euclidean Dirac equation [1].

I wanted to get Chern's reaction to the fact, noted by physicists, that $<^* FF>$, the axial anomaly as well as the 4-dimensional Chern-Pontryagin density, can be written as the 4-divergence of a 4-vector constructed from connections — a quantity physicists call the anomaly current. Whereupon he informed me of the Chern-Simons secondary characteristic class, which he had put forward some years earlier [2]. The sobriquet "secondary characteristic class" seems to demote that entity to a secondary class of importance. Nevertheless I was not discouraged, and with colleagues proposed using it, after renaming it simply and neutrally as the 3-dimensional Chern-

*Chern Memorial, Taijin China, August 2005

Simons term [3]. The envisioned physical application was to dynamics in 3-dimensional space-time, *i.e.*, on a plane. This suggestion was taken up by many physicists for analyzing a variety of physical, planar processes (Hall effect, high T_c super conductivity, motion in presence of cosmic and other vortices). Eventually physics returned the favor to mathematics, where the Chern-Simons term describes knot invariants.

Chern was happy that his secondary class found first class uses in physics. He thanked me for spreading the word among physicists and gave me an inscribed book, containing some of his relevant papers.

Here I shall explore some further properties of Chern-Simons terms.

2. Chern-Pontryagin/Simons Topological Entities

We begin by recalling that the Chern-Pontryagin densities appeared in physics when anomalous Feynman diagrams were computed. These diagrams carry vector indices, and formal arguments led us to expect that the evaluated expressions would be transverse in each index. But in fact the explicit expressions fail to be transverse, and the appropriate longitudinal part in the anomaly. In a 4-dimensional abelian gauge theory, the anomaly reads

$$\mathcal{A}_{(4)} = \frac{1}{2}\, {}^*F^{\mu\nu}\, F_{\mu\nu} = \frac{1}{4}\, \varepsilon^{\mu\nu\alpha\beta}\, F_{\mu\nu}\, F_{\alpha\beta}, \tag{2.1a}$$

where $F_{\mu\nu}$ is the gauge field strength (curvature). In the non-Abelian theory the expression is similar, except that the gauge fields carry a Lie algebra index a, that is summed.

$$\mathcal{A}_{(4)} = \frac{1}{2}\, {}^*F^{\mu\nu\ a}\, F_{\mu\nu}^{\ a} = \frac{1}{4}\, \varepsilon^{\mu\nu\alpha\beta}\, F_{\mu\nu}^{\ a}\, F_{\alpha\beta}^{\ a} \tag{2.1b}$$

An anomaly also exists in an Abelian 2-dimensional gauge theory; it is simply

$$\mathcal{A}_{(2)} = {}^*F = \frac{1}{2}\, \varepsilon^{\mu\nu}\, F_{\mu\nu}. \tag{2.2}$$

The anomalies are recognized to be densities, which upon integration over the appropriate manifold, produce the Chern-Pontryagin gauge field invariant. Note that (2.1) and (2.2) are generally covariant densities, giving world scalars upon integration — no metric tensor is needed. Because of this metric-independence, they are topological entities, independent of local, geometric properties of the manifold.

Physicist usually work on open, unbounded spaces, and the integrals are taken over these spaces. One may imagine that the integration is performed over a spherical ball, with very large radius. The ball is bounded by its spherical surface, which passes to infinity as the radius increases without limit. Since the Chern-Pontryagin entities are topological, one may expect that they can be determined by behavior at the large-distance boundary. For this to be the case, it should be possible to represent these scalar densities as divergences of vector densities, so that by Gauss' theorem their volume integral can be cast onto the boundary surface (at infinity).

Indeed this is possible, but the field strengths (curvatures) must be expressed in terms of potentials (connections). The Abelian formula involves the familiar curl,

$$F_{\mu\nu} = \partial_\mu A_\nu - \partial_\nu A_\mu \tag{2.3}$$

while the non-Abelian expression includes a non-linear term constructed with Lie algebra structure constraints f^{abc}.

$$F_{\mu\nu}^{\ a} = \partial_\mu A_\nu^a - \partial_\mu A_\nu^b + f^{abc} A_\mu^b A_\nu^c \tag{2.4}$$

Inserting (2.3), (2.4) in (2.1), (2.2) exhibits the desired result.

Abelian, 4-d:

$$\begin{aligned} \mathcal{A}_4 &= \frac{1}{2}\ {}^*F^{\mu\nu} F_{\mu\nu} = \partial_\mu C_4^\mu \\ C_4^\mu &= \varepsilon^{\mu\alpha\beta\gamma} A_\alpha \partial_\beta A_\gamma \end{aligned} \tag{2.5a}$$

non-Abelian, 4-d:

$$\mathcal{A}_4 = \frac{1}{2}\ {}^*F^{\mu\nu\, a} F_{\mu\nu}^{\ a} = \partial_\mu C_4^\mu \tag{2.5b}$$

$$C_4^\mu = \varepsilon^{\mu\alpha\beta\gamma} \left(A_\gamma^a \partial_\beta A_\gamma^a + \frac{1}{3} f^{abc} A_\alpha^a A_\beta^b A_\gamma^c\right) \tag{2.5c}$$

Abelian, 2-d:

$$\begin{aligned} \mathcal{A}_2 &= \frac{1}{2} \varepsilon^{\mu\nu} F_{\mu\nu} = \partial_\mu C_2^\mu \\ C_2^\mu &= \varepsilon^{\mu\nu} A_\nu \end{aligned} \tag{2.6}$$

The vectors C^μ whose divergence gives the anomalies A are called anomaly currents or Chern-Simons currents.

The above is recapitulated succinctly in form notation. The anomaly or the Chern-Pontryagin density is a 4-form in four dimensions and a 2-form in two dimensions. These forms are closed, and can be presented as exact forms; they are given by the exterior derivative of the Chern-Simons form, which is a 3-form in the former case and a 1-form in the latter.

While the Chern-Pontryagin and the Chern-Simons forms are defined on even dimensional manifolds, one may restrict the latter, in a natural way, to one lower, odd dimensional manifold. The restriction proceeds as follows. Observe from (2.5) and (2.6) that the Chern-Simons/anomaly currents involve a free index carried by the Levi-Civita epsilon tensor. Choose a definite coordinate for that index. Because of the total anti-symmetry of the Levi-Civita tensor, the remaining indices will not repeat the chosen, external index, and therefore neither will the quantities (gauge potentials, derivatives) comprising the Chern-Simons current. Furthermore, if all dependence of the potentials on the selected coordinate is suppressed, we are left with the so-called Chern-Simons terms, defined in odd-dimensional spaces.

$$\text{Abelian, 3-d: } CS(A) = \varepsilon^{ijk}\,(A_i\,\partial_j\,A_k) \tag{2.7a}$$

$$\text{non-Abelian, 3-d: } CS(A) = \varepsilon^{ijk}\,(A_i^a\,\partial_i\,A_k^a + \tfrac{1}{3}\,f^{abc}\,A_i^a\,A_j^b\,A_k^c) \tag{2.7b}$$

$$\text{Abelian, 1-d: } CS(A) = A_1 \tag{2.8}$$

Evidently the Chern-Simons terms can be integrated over 3-dimensional or 1-dimensional spaces, thereby producing world scalars without the intervention of a metric tensor. Thus we again encounter topological entities. Some of these integrals have been known in physics and mathematics for a long time, as encoding interesting properties of vector fields and gauge fields. For example, if in (2.7a) A_i is identified with the electromagnetic vector potential, and $\varepsilon^{ijk}\,\partial_j\,A_k$ with the magnetic field B^i, the integral defines the "magnetic helicity" $\int d^3r\,\mathbf{A}\cdot\mathbf{B}$, which measures linkage of magnetic flux lines. Alternatively, if A_i is the velocity vector of a fluid v_i, then $\varepsilon^{ijk}\,\partial_j\,v^k \equiv \omega^i$ is the vorticity and the integral of (2.7a) becomes $\int d^3r\,\mathbf{v}\cdot\boldsymbol{\omega}$; this is the "kinetic vorticity," which provides an obstruction to a canonical formulation of fluid mechanics [4]. When the non-Abelian Chern-Simons term is evaluated at a pure gauge connection $A = g^{-1}dg$ (in matrix notation), then the integrated Chern-Simons term involves $\int d^3r\ tr(g^{-1}dg)^3$, and evaluates the winding of the gauge function g. Moreover, it is known that $tr(g^{-1}dg)^3$ is a total derivative, so that the winding number integral is given by a surface term [3].

3. Chern-Simons Terms as Total Derivatives

The question, which this essay addresses, is whether the Chern-Simons terms can be expressed as total derivatives, so that their integrals over all space are given by contributions from the bounding surface.

The answer is clearly "yes" for the 1-dimensional Chern-Simons term, which according to (2.8) is just the single function A_1. This can always be presented as the derivative of another quantity — of another "secondary" potential θ,

$$A_4 = \partial_1 \theta \tag{3.1}$$

so that $\int_{-\infty}^{\infty} dx\, A_1(x) = \theta(\infty) - \theta(-\infty)$.

Also in the 3-dimensional, Abelian case one can write the Chern-Simons term (2.7a) as a total derivative, provided the vector A_i is presented in terms of further, "secondary" potentials.

$$A_i = \partial_i \theta + \alpha\, \partial_i \beta \tag{3.2a}$$

The representation (3.2a) is called the Clebsch parameterization of a 3-vector; it involves a "gauge" part $(\partial_i \theta)$ and two more scalars (α, β), called Monge potentials. Altogether three functions appear; thus there is sufficient generality to represent the arbitrary 3-vector A_i. An analytic procedure for finding the Clebsch parameterization for a given vector A_i has been know since the 19th century. On the other hand, when (3.2a) is written in form notation

$$A = d\theta + \alpha\, d\beta \tag{3.2b}$$

one recognizes this as an instance of Darboux' theorem.

With A_i parameterized in the Clebsch parameterization manner, as in (3.2), the Abelian Chern-Simons term indeed becomes a total derivative.

$$\varepsilon^{ijk} A_i\, \partial_i\, A_k = \partial_i\, (\theta \varepsilon^{ijk}\, \partial_j \alpha\, \partial_k\, \beta) \tag{3.3}$$

With $\mathbf{B}$ or $\boldsymbol{\omega}$ given by $\boldsymbol{\nabla}\alpha \times \boldsymbol{\nabla}\,\beta$, the magnetic helicity becomes

$$\int d^3r\ \mathbf{A} \cdot \mathbf{B} = \int d\,\mathbf{S} \cdot \theta \mathbf{B} \tag{3.4a}$$

and similarly for the kinetic vorticity.

$$\int d^3r\ \mathbf{v} \cdot \boldsymbol{\omega} = \int d\,\mathbf{S} \cdot \theta \boldsymbol{\omega} \tag{3.4b}$$

Thus the volume integral of the Abelian Chern-Simons term is found from the surface integral of the potentials in the Clebsch parameterization.

This result is important for the canonical (symplectic) formulation of Eulerian fluid mechanics. As remarked previously the kinetic vorticity provides an obstruction to a canonical formulation of that dynamical system.

To make progress, the obstruction must be removed. By using the Clebsch parameterization for the velocity places the kinetic vorticity at spatial inplicity, away from the finite regions of the 3-dimensional space, and a canonical formulation becomes possible. That is why the Clebsch parameterization is needed in fluid mechanincs [4].

How about the non-Abelian, 3-dimensional Chern-Simons term? We have already remarked that in the special case when A_i is a pure gauge $g^{-1}\partial_i g$, the Chern-Simons term (2.7b) is a total derivative. We shall now show that there exists a parameterization for arbitrary (not only pure gauge) non-Abelian vectors, such that their Chern-Simons term is a total derivative.

4. Mathematical Sidebar

Before proceeding, let us reformulate our problem, and also describe work of Bott and Chern who posed and solved a related but different problem.

We know that the Chern-Pontryagin entities are exterior derivatives of the Chern-Simons entities, as in (2.5) and (2.6).

$$\text{Chern-Pontryagin} = d(\text{Chern-Simons}) \tag{4.1}$$

We have set for ourselves the problem of further demonstrating that the Chern-Simons quantities also are exterior derivatives of further entities.

$$\text{Chern-Simons} \stackrel{?}{=} d(\Omega) \tag{4.2}$$

But this also entails that

$$\text{Chern-Pontryagin} = d(\text{Chern-Simons}) = dd\,\Omega = 0 \tag{4.3}$$

So our result can hold only when the anomaly, the Chern-Pontryagin density, is absent. Thus if we work in three dimensions with a Chern-Simons 3-form or in one dimension with the Chern-Simons 1-form, the Chern-Pontryagin 4-form and 2-form are absent — they cannot be constructed. The Chern-Simons forms are closed because they are maximal for the considered dimensionality, and it comes as no surprise that locally exact expressions for them can be constructed.

Nevertheless, we call attention to the fact that this situation for Chern-Simons forms is different from the situation with Chern-Pontryagin forms. The latter are closed without regard to dimensionality, whereas the former are closed for dimensional reasons.

Bott and Chern have derived a representation for the Chern-Simons term as a sum of (different) total derivative expressions, in the special case that the field strength (curvature) satisfies a further condition [5]. To contrast and compare with our investigation we now describe their result in the 4-dimensional case.

Bott and Chern work with the two complex coordinates that can be constructed in four dimensions.

$$(z, \bar{z}) \equiv \frac{1}{\sqrt{2}}(x_1 \pm i x_2), \qquad (w, \bar{w}) = \frac{1}{\sqrt{2}}(x_3 \pm i x_4) \tag{4.4}$$

They further require that the holomorphic and anti-holomorphic components of the (non-Abelian) curvature $F_{\mu\nu}$ vanish.

$$F_{zw} = 0 = F_{\bar{z}\bar{w}} \tag{4.5}$$

They show that then the Chern-Pontryagin density takes the form

$$\begin{aligned} \text{Chern-Pontryagin} &= d_-(\text{Chern-Simons}) \\ &= d_-\, d_+\, \Omega, \end{aligned} \tag{4.6}$$

which implies that

$$\text{Chern-Simons} = d_+\, \Omega + d_- \chi .$$

Here $d_\pm$ are the holomorphic and anti-holomorphic exterior derivatives

$$d_+ = dz \frac{\partial}{\partial z} + dw \frac{\partial}{\partial w}, \tag{4.7}$$

$$d_- = d\bar{z} \frac{\partial}{\partial \bar{z}} + d\bar{w} \frac{\partial}{\partial \bar{w}}. \tag{4.8}$$

Thus for restricted curvatures, as in (4.5), the Chern-Simons term is a sum of terms that are exact on the holomorphic and anti-holomorphic submanifolds.

In contrast to the Bott-Chern result, we consider the case with no restriction on the curvature, but vanishing Chern-Pontryagin density (because of dimensionality) and construct an exact (total derivative) expression for the Chern-Simons term.

5. The Result

Our result in the non-Abelian case is not found by an analytic method, as is done for the Abelian case via the Clebsch parameterization. Rather we develop a group theoretical argument [6]. To illustrate our method, we first apply it to the Abelian case in a rederivation of the Clebsch parameterization.

To deal with the U(1) (Abelian) Chern-Simons term, we begin with SU(2) and consider a pure gauge connection.

$$A = g^{-1}\, d\, g \equiv V^a \frac{\sigma^a}{2i}, \; g \,\epsilon\, SU(2). \tag{5.1}$$

The $\frac{\sigma^a}{2i}$ are the anti-Hermitian matrix generators of the Lie algebra (σ^a = Pauli matrices). It follows that $tr(g^{-1}dg)^3$, which is known to be a total derivative, is given by

$$tr(g^{-1}dg)^3 = -\frac{1}{4}\, \varepsilon_{abc}\, V^a V^b V^c = -\frac{3}{2} V^1 V^2 V^3 = \text{total derivative.} \tag{5.2}$$

Here ε_{abc} are the $SU(2)$ structure constants. Also, because the non-Abelian connection is a pure gauge, V^a obeys

$$dV^a = \frac{1}{2}\, \varepsilon_{abc} V^b V^c. \tag{5.3}$$

We define the Abelian connection, relevant to our $U(1)$ problem as

$$A = V^3 = tr\, i\, \sigma^3\, g^{-1} dg. \tag{5.4}$$

NOTE: A is not a pure gauge within $U(1)$. It now follows that the Abelian Chern-Simons term satisfies the following sequence of equalities.

$$CS(A) = AdA = V^3 d\, V^3 = -V^1 V^2 V^3 = \frac{2}{3}\, tr(g^{-1}dg)^3 \tag{5.5}$$

But the last term is known to be a total derivative, and this establishes that property for $CS(A)$.

Since g lies in $SU(2)$, it depends on three functions, and so does V^3. Thus there is sufficient generality to represent an arbitrary 3-dimensional Abelian vector A_i.

It is instructive to see how this works explicitly. Parameterizing $g \,\epsilon\, SU(2)$ as

$$g = e^{\frac{\sigma^3}{2i}\beta}\; e^{\frac{\sigma^2}{2i}\gamma}\; e^{\frac{\sigma^3}{2i}\theta} \tag{5.6}$$

we find

$$A = V^3 = tr\, i\, \sigma^3\, g^{-1} dg = d\,\theta + \cos\gamma\, d\,\beta$$

The Clebsch parameterization is regained!

The argument for the non-Abelian Chern-Simons term proceeds in an analogous, but generalized manner. We seek to parameterize a connection 1-form A^a, belonging to the Lie algebra of H, whose generators are T^a.

Consider a larger group G, with H as a subgroup. With $g \epsilon G$ construct a pure gauge connection for $G : g^{-1}dg$. The H-connection is then defined as

$$A^a \propto tr\, T^a\, g^{-1}dg \tag{5.7}$$

A^a is not a pure gauge in H. Since an arbitrary 3-dimensional gauge potential for H contains 3(dim H) components, we require $\dim G > 3(\dim H)$.

Our goal is to show that the H-Chern-Simons term constructed from A^a coincides with the G-Chern-Simons term constructed from $g^{-1}dg$. The latter involves $tr(g^{-1}dg)^3$ and is known to be a total derivative. The coincidence of two then establishes that the H-Chern-Simons term also is a total derivative.

The desired coincidence occurs when G/H is a symmetric space. In terms of the Lie algebra for H and G, this means that the generators T^a of $H, a = 1..., \dim H$, and the additional generators, S^M, $M = 1..., (\dim G - \dim H)$, which together with the T^a comprise the generators of G, must satisfy

$$[T^a, T^b] = f_{abc}\, T^c, \tag{5.8a}$$

$$[T^a, S^M] = h^{aMN}\, S^N, \tag{5.8b}$$

$$[S^M, S^N] \propto h^{aMN}\, T^a. \tag{5.8c}$$

Eqs. (5.8) record the Lie algebra of G, with (5.8a) being Lie algebra of H (structure constraints f_{abc}), with (5.8b) showing that the S^M provide a representation for that algebra, and with (5.8c) giving the closure of S-generators on the T-generators.

A straightforward, but tedious, sequence of manipulations then establishes the coincidence of G and H Chern-Simons terms. To see them carried out, see the published literature [6].

6. Conclusion

The 3-dimensional Chern-Simons term first entered physics to provide a gauge-invariant mass gap for a 3-dimensional gauge theory [3]. The 1-dimensional Chern-Simons term is related to the 2-dimensional Chern-Simons current $C_2^\mu = \varepsilon^{\mu\nu} A_\nu$. It has recently been realized that the gauge fields in the Schwinger model — 2-dimensional electrodynamics with massless fermions — can be presented solely in terms of 2-dimensional topological entities: the kinetic term $\sim F^{\mu\nu}F_{\mu\nu}$ is just the square of the 2-dimensional Chern-Pontryagin density; the interaction with the vector current J^μ, $J^\mu A_\mu$, is also given by $J^5_\mu C^\mu_2$ since the axial vector is dual to the vector in 2-dimensions: $J^5_\mu \varepsilon^{\mu\nu} = J^\nu$. Moreover, the divergence of the axial vector current is anomalous and again leads to the Chern-Pontryagin density. This viewpoint on the Schwinger model suggests that it can be lifted to any even-dimensional space-time with approximate higher dimensional Chern-Pontryagin densities and Chern-Simons currents coupled to anomalous axial vector currents. This would effect a Schwinger-model-like topological mass generation in even-dimensional space-time [7]. Thus it is clear that the Chern-Simons term continues to provide physicists with ideas for new physical mechanisms.

References

1. For a summary see *e.g.* R. Jackiw "Fifty Years of Yang-Mills Theory and Our Moments of Triumph" in *Fifty Years of Yang-Mills Theory*, G. 't Hooft ed. (World Scientific, Singapore 2005).
2. S. S. Chern, *Complex Manifolds Without Potential Theory*, 2nd edition (Springer Verlag, Berlin 1979).
3. S. Deser, R. Jackiw and S. Templeton, "Topologically Massive Gauge Theories," *Annals Phys.* **140**, 372 (1982); "Three-Dimensional Massive Gauge Theories," *Phys. Rev. Lett.* **48** 975 (1982).
4. See for example, R. Jackiw, V. P. Nair, S. Y. Pi and A. P. Polychronakos, "Perfect fluid theory and its extensions," *J. Phys. A* **37**, R327 (2004).
5. R. Bott and S. S. Chern, "Hermitian Vector Bundles and the Equidistribution of the Zeroes of their Holomorphic Section", *Acta. Math.* **114**, 71 (1965).
6. R. Jackiw, V. P. Nair and S. Y. Pi, "Chern-Simons reduction and non-Abelian fluid mechanics," *Phys. Rev. D* **62**, 085018 (2000).
7. G. Dvali, R. Jackiw and S. Y . Pi, "Topological Mass Generation in Four Dimensions," [arXiv:hep-th/0511175].

Localization and Conjectures from String Duality*

Kefeng Liu
Center of Mathematical Sciences, Zhejiang University, Hangzhou, China,
Department of Mathematics
University of California at Los Angeles
Los Angeles, CA 90095-1555, USA
Email: liu@math.ucla.edu, liu@cms.zju.edu.cn

We describe the applications of localization methods, in particular the functorial localization formula, in the proofs of several conjectures from string theory. Functorial localization formula pushes the computations on complicated moduli spaces to simple moduli spaces. It is a key technique in the proof of the general mirror formulas, the proof of the Hori-Vafa formulas for explicit expressions of basic hypergeometric series of homogeneous manifolds, the proof of the Mariño-Vafa formula, its generalizations to two partition analogue. We will also discuss our development of the mathematical theory of topological vertex and simple localization proofs of the ELSV formula and Witten conjecture.

1. Introduction

According to string theorists, String Theory, as the most promising candidate for the grand unification of all fundamental forces in the nature, should be the final theory of the world, and should be unique. But now there are five different looking string theories. As argued by physicists, these theories should be equivalent, in a way dual to each other. On the other hand all previous theories like the Yang-Mills and the Chern-Simons theory should be parts of string theory. In particular their partition functions should be equal or equivalent to each other in the sense that they are equal after certain transformation. To compute partition functions, physicists use localization technique, a modern version of residue theorem, on infinite dimensional spaces. More precisely they apply localization formally to path integrals which is not well-defined yet in mathematics. In many cases such computations reduce the path integrals to certain integrals of various Chern classes

*The author is supported by the NSF and NSFC.

on various finite dimensional moduli spaces, such as the moduli spaces of stable maps and the moduli spaces of vector bundles. The identifications of these partition functions among different theories have produced many surprisingly beautiful mathematical formulas like the famous mirror formula [29], as well as the Mariño-Vafa formula [44].

The mathematical proofs of these conjectural formulas from the string duality also depend on localization techniques on these various finite dimensional moduli spaces. The purpose of this note is to discuss our works on the subject. I will briefly discuss the proof of the mirror conjecture and its generalizations, the proof of the Hori-Vafa formula, the proof of the Marinõ-Vafa formula and its generalizations, the related topological vertex theory [1] [26], and simple localization proofs of the ELSV formula and the Witten conjecture [20]. More precisely we will use localization formulas in various form to compute the integrals of Chern classes on moduli spaces, and to prove those conjectures from string duality. For the proofs of these conjectures such as the mirror formula, the Mariño-Vafa formula and the theory of topological vertex, we note that many aspects of mathematics are involved, such as the Chern-Simons knot invariants, combinatorics of symmetric groups, representations of Kac-Moody algebras, Calabi-Yau manifolds, geometry and topology of moduli space of stable maps, etc. The spirit of our results is the duality among various string theories. In particular the duality between IIA and IIB string theory gives the mirror formulas, the duality between gauge theory, Chern-Simons theory and the Calabi-Yau geometry in string theory leads to the Mariño-Vafa conjecture and the theory of topological vertex.

Localization techniques have been very successful in proving many conjectures from physics, see my ICM 2002 lecture [41] for more examples. The reason may be that physical systems always have natural symmetry which can be used to do localizations. One of our major tools in the proofs of these conjectures is the functorial localization formula which is a variation of the classical localization formula, it transfers computations on complicated spaces to simple spaces, and connects computations of mathematicians and physicists.

In this note we will discuss the following results:

1. The proof of the mirror formulas and its generalizations which we call the mirror principle. The mirror principle implies all of the conjectural mirror formulas of counting rational curves for toric manifolds and their Calabi-Yau submanifolds from string theory. In this case we apply the

functorial localization formula to the map from the nonlinear moduli space to the linearized moduli space. This transfers the computations of integrals on complicated moduli space of stable maps to computations on rather simple spaces like projective spaces. From this the proof of the mirror formula and its generalizations become conceptually clean and simple.
In fact the functorial localization formula was first found and used in Lian-Liu-Yau's proof of the mirror conjecture.

2. The proof of the Hori-Vafa conjecture and its generalizations for Grassmannian and flag manifolds. This conjecture predicts an explicit formula for the basic hypergeometric series of a homogeneous manifold in terms of the basic series of a simpler manifold such as the product of projective spaces. In this case we use the functorial localization formula twice to transfer the computations on the complicated moduli spaces of stable maps to the computations on quot-schemes. The first is a map from moduli space of stable maps to product of projective spaces, and another one is a map from the quot-scheme into the same product of projective spaces. A key observation we had is that these two maps have the same image.
This approach was first sketched in [31], the details for Grassmannians were carried out in [28] and [3]. The most general case of flag manifolds was carried out in [35] and [4].
3. The proof of the Mariño-Vafa conjecture on Hodge integrals in [38]. This conjecture gives a closed formula for the generating series of a class of triple Hodge integrals for all genera and any number of marked points in terms of the Chern-Simons knot invariant of the unknot. This formula was conjectured by M. Mariño and C. Vafa in [44] based on the duality between large N Chern-Simons theory and string theory. Many Hodge integral identities, including the ELSV formula for Hurwitz numbers [8] and the λ_g conjecture [10], can be obtained by taking various limits of the Mariño-Vafa formula [39]. The Mariño-Vafa formula was first proved by applying the functorial localization formula to the branch morphism from the moduli space of relative stable maps to a projective space.
4. The proof of the generalization of the Mariño-Vafa formula to two partitions cases, and the theory of topological vertex. The mathematical theory of topological vertex was motivated by the physical theory as first developed by the Vafa group [1], who has been working on string duality for the past several years. Topological vertex theory is a high point of their work starting from their geometric engineering theory and Wit-

ten's conjecture that Chern-Simons theory is a string theory [50]. While the Marinõ-Vafa formula gives a close formula for the generating series of triple Hodge integrals on the moduli spaces of all genera and any number marked points, topological vertex [26] gives the most effective ways to compute the Gromov-Witten invariants of any open toric Calabi-Yau manifolds. Recently Pan Peng was able to use our results on topological vertex to give a complete proof of the Gopakumar-Vafa integrality conjecture for any open toric Calabi-Yau manifolds [48]. Kim also used our technique to derive new effective recursion formulas for Hodge integrals on the moduli spaces of stable curves [18].

5. We describe a very simple proof of the ELSV formula [8] following our proof of the Mariño-Vafa formula, by using the cut-and-join equation from localization and combinatorics. The proof of the ELSV formula is particularly easy by using functorial localization, it is reduced to the fact that the push-forward in equivariant cohomology of a constant between two equal dimensional varieties is still constant. We will also show how to directly derive the ELSV formula from the Mariño-Vafa formula by taking a scaling limit.
6. By using functorial localization formula we have the simple proofs of the Witten conjecture [20]. Our simple proof of the Witten conjecture in [19] is to study the asymptotic expansion of the simple cut-and-join equation for one Hodge integrals which is derived from functorial localization. This immediately gives a recursion formula which implies both the Virasoro constraints and the KdV relation satisfied by the generating series of the ψ integrals.

I will start with brief discussions about the proofs of the mirror conjecture and the Hori-Vafa formula for Grassmannians, then I will go to the proofs of the Marinõ-Vafa conjecture and its generalizations to two partitions and the topological vertex theory. After that we discuss the simple proofs of the ELSV formula and the Witten conjecture. This note is partly based on my plenary lecture at the International Conference of Differential Geometry Method in Theoretical Physics held in August 2005. It is an much more expanded version of a previous survey I wrote for the 2004 International Complex Geometry Conference held in the Eastern Normal University of China. This survey is intended for readers from physics and from other fields of mathematics. The materials on mirror conjectures and the Hori-Vafa formulas were taken from a previous survey of Chien-Hao Liu, Shing-Tung Yau and myself written for the Gelfand symposium. Our purpose to combine the discussions together is to give the reader a more complete picture about the applications of localization techniques in solving conjectures from string duality. I hope this note has accomplished this goal. I would like to thank the organizers of the conferences, especially Professor Chengming Bai, Professor Shengli Tan, Professor Weiping Zhang and Professor Zhijie Chen for their hospitality during my visits. I would also like to thank my collaborators for the past 10 years, Bong Lian, Shing-Tung Yau, Chien-Hao Liu, Melissa C.-C. Liu, Jian Zhou, Jun Li, Yon Seo Kim for the wonderful experience in solving these conjectures and to develop the theory together.

2. Localization

In this section we will explain the *Functorial Localization Formula.* We start with a review of the Atiyah-Bott localization formula. Recall that the definition of equivariant cohomology group for a manifold X with a torus T action:

$$H_T^*(X) = H^*(X \times_T ET)$$

where ET is the universal bundle of T, we will use $\mathbb{R}$ or $\mathbb{Q}$ as coefficients through this note.

Example We know $ES^1 = S^\infty$. If S^1 acts on $\mathbf{P}^n$ by

$$\lambda \cdot [Z_0, \ldots, Z_n] = [\lambda^{w_0} Z_0, \ldots, \lambda^{w_n} Z_n],$$

with $w_0, \cdots, w_n$ as weights, then

$$H_{S^1}^*(\mathbf{P}^n; \mathbb{Q}) \cong \mathbb{Q}[H, u]/\langle (H - w_0 u) \cdots (H - w_n u) \rangle$$

where u is the generator of $H^*(BS^1, \mathbb{Q})$. We have the following important *Atiyah-Bott Localization Formula:*

Theorem 2.1.

For $\omega \in H_T^*(X)$ an equivariant cohomology class, we have

$$\omega = \sum_E i_{E*} \left(\frac{i_E^* \omega}{e_T(E/X)} \right).$$

where E runs over all connected components of T fixed points set, i_E denotes the inclusion map, i_E^* i_{E*} denote the pull-back and push-forward in equivariant cohomology.

This formula is very effective in the computations of integrals on manifolds with torus T symmetry. The idea of localization is fundamental in many subjects of geometry. In fact Atiyah and Witten proposed to formally apply this localization formula to loop spaces and the natural S^1-action, from which one gets the Atiyah-Singer index formula. In fact the Chern characters can be interpreted as equivariant forms on loop space, and the $\hat{A}$-class is the inverse of the equivariant Euler class of the normal bundle of X in its loop space LX:

$$e_T(X/LX)^{-1} \sim \hat{A}(X),$$

which follows from the normalized infinite product formula

$$\left(\prod_{n\neq 0}(x+n)\right)^{-1} \sim \frac{x}{\sin x}.$$

I observed in [42] that the normalized product

$$\prod_{m,n}(x+m+n\tau) = 2q^{\frac{1}{8}}\sin(\pi x)\cdot\prod_{j=1}^{\infty}(1-q^j)(1-e^{2\pi i x}q^j)(1-e^{-2\pi i x}q^j),$$

where $q = e^{2\pi i\tau}$, also has deep geometric meaning. This formula is the Eisenstein formula. It can be viewed as a double loop space analogue of the Atiyah-Witten observation. This formula gives the basic Jacobi θ-function. As observed by in [42], formally this gives the $\hat{A}$-class of the loop space, and the Witten genus which is defined to be the index of the Dirac operator on the loop space:

$$e_T(X/LLX) \sim \hat{W}(X),$$

where LLX is the double loop space, the space of maps from $S^1 \times S^1$ into X. $\hat{W}(X)$ is the Witten class. See [42] for more detail.

The variation of the localization formula we will use in various situations is the following *Functorial Localization Formula*

Theorem 2.2. *Let X and Y be two manifolds with torus action. Let $f : X \to Y$ be an equivariant map. Given $F \subset Y$ a fixed component, let $E \subset f^{-1}(F)$ be those fixed components inside $f^{-1}(F)$. Let $f_0 = f|_E$, then for $\omega \in H_T^*(X)$ an equivariant cohomology class, we have the following identity on F:*

$$f_{0*}\left[\frac{i_E^*\omega}{e_T(E/X)}\right] = \frac{i_F^*(f_*\omega)}{e_T(F/Y)}.$$

This formula will be applied to various settings to prove various conjectures from physics. It first appeared in [29]. In many cases we will use a virtual version of this formula. It is used to push computations on complicated moduli spaces to simpler moduli spaces. A K-theory version of the functorial localization formula also holds [30], interesting applications are expected.

Remark Consider the diagram:

$$\begin{array}{ccc} H_T^*(X) & \xrightarrow{f_*} & H_T^*(Y) \\ \downarrow i_E{}^* & & \downarrow i_F{}^* \\ H_T^*(E) & \xrightarrow{f_{0*}} & H_T^*(F)\,. \end{array}$$

The functorial localization formula is like Riemann-Roch with the inverted equivariant Euler classes of the normal bundle as "weights", in a way similar to the Todd class for the Riemann-Roch formula. In fact if we formally apply this formula to the map between the loop spaces of X and Y, equivariant with respect to the rotation of the circle, we do formally get the differentiable Riemann-Roch formula. We believe this can be done rigorously by following Bismut's proof of the index formula which made rigorous of the above argument of Atiyah-Witten.

3. The Mirror Principle

There have been many discussions of mirror principle in the literature. Here we only give a brief account of the main ideas of the setup and proof of the mirror principle. We will use two most interesting examples to illustrate the algorithm. These two examples give proofs of the mirror formulas for toric manifolds as conjectured by string theorists.

The goal of mirror principle is to compute the characteristic numbers on moduli spaces of stable maps in terms of certain hypergeometric type series. This was motivated by mirror symmetry in string theory. The most interesting case is the counting of the numbers of curves which corresponds to the computations of Euler numbers. More generally we would like to compute the characteristic numbers and classes induced from the general Hirzebruch multiplicative classes such as the total Chern classes. The computations of integrals on moduli spaces of those classes pulled back through evaluation maps at the marked points and the general Gromov-Witten invariants can also be considered as part of mirror principle. Our hope is to develop a "black-box" method which makes easy the computations of the characteristic numbers and the Gromov-Witten invariants.

The general set-up of mirror principle is as follows. Let X be a projective manifold, $\overline{\mathcal{M}}_{g,k}(d,X)$ be the moduli space of stable maps of genus g and degree d with k marked points into X, modulo the obvious equivalence. The points in $\overline{\mathcal{M}}_{g,k}(d,X)$ are triples $(f;C;x_1,\cdots,x_k)$ where $f: C \to X$ is a degree d holomorphic map and $x_1,\cdots,x_k$ are k distinct smooth points on the genus g curve C. The homology class $f_*([C]) = d \in H_2(X,\mathbb{Z})$ is identified as integral index $d = (d_1,\cdots,d_n)$ by choosing a basis of $H_2(X,\mathbb{Z})$, dual to the Kähler classes.

In general the moduli space may be very singular, and may even have different dimension for different components. To define integrals on such singular spaces, we need the virtual fundamental cycle of Li-Tian [25], and also Behrend-Fantechi [5] which we denote by $[\overline{\mathcal{M}}_{g,k}(d,X)]^v$. This is a homology class of the expected dimension

$$2\left(c_1(TX)[d] + (\dim_{\mathbb{C}} X - 3)(1-g) + k\right)$$

on $\overline{\mathcal{M}}_{g,k}(d,X)$.

Let us consider the case $k = 0$ first. Note that the expected dimension of the virtual fundamental cycle is 0 if X is a Calabi-Yau 3-fold. This is the most interesting case for string theory.

The starting data of mirror principle are as follows. Let V be a concavex bundle on X which we defined as the direct sum of a positive and a negative bundle on X. Then V induces a sequence of vector bundles V_d^g on $\overline{\mathcal{M}}_{g,0}(d,X)$ whose fiber at $(f;C;x_1,\cdots,x_k)$ is given by $H^0(C,f^*V)\oplus H^1(C,f^*V)$. Let b be a multiplicative characteristic class. So far for all applications in string theory, b is the Euler class.

The problem of mirror principle is to compute

$$K_d^g = \int_{[\overline{\mathcal{M}}_{g,0}(d,X)]^v} b(V_d^g).$$

More precisely we want to compute the generating series

$$F(T,\lambda) = \sum_{d,\,g} K_d^g\,\lambda^g\ e^{d\cdot T}$$

in terms of certain hypergeometric type series. Here λ, $T = (T_1,\cdots,,T_n)$ are formal variables.

The most famous formula in the subject is the Candelas formula as conjectured by P. Candelas, X. de la Ossa, P. Green, and L. Parkes [6]. This formula changed the history of the subject. More precisely, Candelas formula considers the genus 0 curves, that is, we want to compute the so-called *A-model potential* of a Calabi-Yau 3-fold M given by

$$\mathcal{F}_0(T) = \sum_{d \in H_2(M;\mathbb{Z})} K_d^0 \, e^{d \cdot T},$$

where $T = (T_1, \ldots, T_n)$ are considered as the coordinates of the Kahler moduli of M, and K_d^0 is the genus zero, degree d invariant of M which gives the numbers of rational curves of all degree through the multiple cover formula [29]. The famous mirror conjecture asserts that there exists a mirror Calabi-Yau 3-fold M' with *B-model potential* $\mathcal{G}(T)$, which can be computed by period integrals, such that

$$\mathcal{F}(T) = \mathcal{G}(t),$$

where t accounts for coordinates of complex moduli of M'. The map $t \mapsto T$ is called the *mirror map.* In the toric case, the period integrals are explicit solutions to the GKZ-system, that is the Gelfand-Kapranov-Zelevinsky hypergeometric series. While the A-series are usually very difficult to compute, the B-series are very easy to get. This is the magic of the mirror formula. We will discuss the proof of the mirror principle which includes the proof of the mirror formula.

The key ingredients for the proof of the mirror principle consists of

(1) Linear and non-linear moduli spaces;
(2) Euler data and hypergeometric (HG) Euler data.

More precisely, the non-linear moduli is the moduli space $M_d^g(X)$ which is the stable map moduli of degree $(1, d)$ and genus g into $\mathbf{P}^1 \times X$. A point in $M_d^g(X)$ consists of a pair $(f, C) : f : C \to \mathbf{P}^1 \times X$ with C a genus g (nodal) curve, modulo obvious equivalence. The linearized moduli W_d for toric X were first introduced by Witten and used by Aspinwall-Morrison to do approximating computations.

Example Consider the projective space $\mathbf{P}^n$ with homogeneous coordinate $[z_0, \cdots, z_n]$. Then the linearized moduli W_d is defined as projective space with coordinates

$$[f_0(w_0, w_1), \cdots, f_n(w_0, w_1)]$$

where $f_j(w_0, w_1)$'s are homogeneous polynomials of degree d.

This is the simplest compactification of the moduli spaces of degree d maps from $\mathbf{P}^1$ into $\mathbf{P}^n$. The following lemma is important. See [32] for its proof. The $g = 0$ case was given in [11] and in [29].

Lemma 3.1. *There exists an explicit equivariant collapsing map*

$$\varphi: \; M_d^g(\mathbf{P}^n) \longrightarrow W_d.$$

For general projective manifold X, the nonlinear moduli $M_d^g(X)$ can be embedded into $M_d^g(\mathbf{P}^n)$. The nonlinear moduli $M_d^g(X)$ is very "singular" and complicated, but the linear moduli W_d is smooth and simple. The embedding induces a map of $M_d^g(X)$ to W_d. Functorial localization formula pushed the computations onto W_d. Usually mathematical computations should be done on the moduli of stable maps, while physicists tried to use the linearized moduli to approximate the computations. So functorial localization formula connects the computations of mathematicians and physicists. In some sense the mirror symmetry formula is more or less the comparison of computations on nonlinear and linearized moduli.

Mirror principle has been proved to hold for balloon manifolds. A projective manifold X is called balloon manifold if it admits a torus action with isolated fixed points, and if the following conditions hold. Let

$$H = (H_1, \cdots, H_k)$$

be a basis of equivariant Kahler classes such that

(1) the restrictions $H(p) \neq H(q)$ for any two fixed points $p \neq q$;
(2) the tangent bundle T_pX has linearly independent weights for any fixed point p.

This notion was introduced by Goresky-Kottwitz-MacPherson.

Theorem 3.1. *Mirror principle holds for balloon manifolds and for any concavex bundles.*

Remarks

1. All toric manifolds are balloon manifolds. For $g = 0$ we can identify the hypergeometric series explicitly. Higher genus cases need more work to identify such series.
2. For toric manifolds and $g = 0$, mirror principle implies all of the mirror conjectural formulas from string theory.
3. For Grassmannian manifolds, the explicit mirror formula is given by the Hori-Vafa formula to be discussed in Section 4.
4. The case of direct sum of positive line bundles on $\mathbf{P}^n$, including the Candelas formula, has two independent approaches by Givental, and by Lian-Liu-Yau.

Now we briefly discuss the proof of the mirror principle. The main idea is to apply the functorial localization formula to φ, the collapsing map and the pull-back class $\omega = \pi^* b(V_d^g)$, where $\pi: \; M_d^g(X) \to \overline{\mathcal{M}}_{g,0}(d, X)$ is the natural projection.

Such classes satisfy certain induction property. To be precise we introduce the notion of *Euler Data*, which naturally appears on the right hand side of the functorial localization formula, $Q_d = \varphi_!(\pi^* b(V_d^g))$ which is a sequence of polynomials in equivariant cohomology rings of the linearized moduli spaces with simple quadratic relations. We also considered their restrictions to X.

From functorial localization formula we prove that, by knowing the Euler data Q_d we can determine the K_d^g. On the other hand, there is another much simpler Euler data, the *HG Euler data* P_d, which coincides with Q_d on the "generic" part of the nonlinear moduli. We prove that the quadratic relations and the coincidence on generic part determine the Euler data uniquely up to certain degree. We also know that Q_d always have the right degree for $g = 0$. We then use mirror transformation to reduce the degrees of the HG Euler data P_d. From these we deduced the mirror principle.

Remarks

1. Both the denominator and the numerator in the HG series, the generating series of the HG Euler data, are equivariant Euler classes. Especially the denominator is exactly from the localization formula. This is easily seen from the functorial localization formula.
2. The quadratic relation of Euler data, which naturally comes from gluing and functorial localization on the A-model side, is closely related to special geometry, and is similar to the Bershadsky-Cecotti-Ooguri-Vafa's *holomorphic anomaly* equation on the B-model side. Such relation can determine the polynomial Euler data up to certain degree.
   It is an interesting task to use special geometry to understand the mirror principle computations, especially the mirror transformation as a coordinate change.
3. The Mariño-Vafa formula to be discussed later is needed to determine the hypergeometric Euler data for higher genus computations in mirror principle. The Mariño-Vafa formula comes from the duality between Chern-Simons theory and Gromov-Witten theory. This duality and the matrix model for Chern-Simons theory indicate that mirror principle may have matrix model description.

Let us use two examples to illustrate the algorithm of mirror principle.

Example Consider the Calabi-Yau quintic in $\mathbf{P}^4$. In this case

$$P_d = \prod_{m=0}^{5d} (5\kappa - m\alpha)$$

with α can be considered as the weight of the S^1 action on $\mathbf{P}^1$, and κ denotes the generator of the equivariant cohomology ring of W_d.

The starting data of the mirror principle in this case is $V = \mathcal{O}(5)$ on $X = \mathbf{P}^4$. The hypergeometric series, after taking $\alpha = -1$, is given by

$$HG[B](t) = e^{H\,t} \sum_{d=0}^{\infty} \frac{\prod_{m=0}^{5d}(5H+m)}{\prod_{m=1}^{d}(H+m)^5} e^{d\,t},$$

where H is the hyperplane class on $\mathbf{P}^4$ and t is a formal parameter.

We introduce the series

$$\mathcal{F}(T) = \frac{5}{6}T^3 + \sum_{d>0} K_d^0 \, e^{d\,T}.$$

The algorithm is as follows. Take the expansion in H:

$$HG[B](t) = H\{f_0(t) + f_1(t)H + f_2(t)H^2 + f_3(t)H^3\},$$

from which we have the famous *Candelas Formula*: With $T = f_1/f_0$,

$$\mathcal{F}(T) = \frac{5}{2}\left(\frac{f_1}{f_0}\frac{f_2}{f_0} - \frac{f_3}{f_0}\right).$$

Example Let X be a toric manifold and $g = 0$. Let $D_1, .., D_N$ be the T-invariant divisors in X. The starting data consist of $V = \oplus_i L_i$ with $c_1(L_i) \geq 0$ and $c_1(X) = c_1(V)$. Let us take $b(V) = e(V)$ the Euler class. We want to compute the A-series

$$A(T) = \sum K_d^0 \, e^{d\cdot T}.$$

The HG Euler series which is the generating series of the HG Euler data can be easily written down as

$$B(t) = e^{-H\cdot t} \sum_d \prod_i \prod_{k=0}^{\langle c_1(L_i), d\rangle} (c_1(L_i) - k) \frac{\prod_{\langle D_a, d\rangle < 0} \prod_{k=0}^{-\langle D_a, d\rangle - 1} (D_a + k)}{\prod_{\langle D_a, d\rangle \geq 0} \prod_{k=1}^{\langle D_a, d\rangle} (D_a - k)} e^{d\cdot t}.$$

Then mirror principle implies that there are explicitly computable functions $f(t), g(t)$, which define the mirror map, such that

$$\int_X \left(e^f B(t) - e^{-H\cdot T} e(V)\right) = 2A(T) - \sum T_i \frac{\partial A(T)}{\partial T_i}$$

where $T = t + g(t)$. From this equation we can easily solve for $A(T)$.

In general we want to compute:

$$K_{d,k}^g = \int_{[\mathcal{M}_{g,k}(d,X)]^v} \prod_{j=1}^k ev_j^* \omega_j \cdot b(V_d^g)$$

where $\omega_j \in H^*(X)$ and ev_j denotes the evaluation map at the j-th marked point. We form a generating series with t, λ and ν formal variables,

$$F(t, \lambda, \nu) = \sum_{d,g,k} K_{d,k}^g e^{dt} \lambda^{2g} \nu^k.$$

The ultimate mirror principle we want to prove is to compute this series in terms of certain explicit HG series. It is easy to show that those classes in the integrand can still be combined to induce Euler data. Actually the Euler data really encode the geometric structure of the stable map moduli.

We only use one example to illustrate the higher genus mirror principle.

Example Consider open toric Calabi-Yau manifold, say $\mathcal{O}(-3) \to \mathbf{P}^2$. Here $V = \mathcal{O}(-3)$. Let

$$Q_d = \sum_{g \geq 0} \varphi_!(\pi^* e_T(V_d^g)) \, \lambda^{2g}.$$

Then it can be shown that the corresponding HG Euler data is given explicitly by

$$P_d \, J(\kappa, \alpha, \lambda) J(\kappa - d\alpha, -\alpha, \lambda),$$

where P_d is exactly the genus 0 HG Euler data and J is generating series of Hodge integrals with summation over all genera. J may be considered as the degree 0 Euler data. In fact we may say that the computations of Euler data include computations of all Gromov-Witten invariants, and even more. Zhou has obtained some closed formulas. We have proved that the mirror principle holds in such general setting. The remaining task is to determine the explicit HG Euler data. But the recently developed topological vertex theory has given complete closed formulas for all open toric Calabi-Yau manifolds in terms of the Chern-Simons invariants. See the discussion in Section 7 for details.

Finally we mention some recent works. First we have constructed refined linearized moduli space for higher genus, the *A-twisted moduli stack* $\mathcal{AM}_g(X)$ of genus g curves associated to a smooth toric variety X, induced from the gauged linear sigma model studied by Witten.

This new moduli space is constructed as follows. A morphism from a curve of genus g into X corresponds to an equivalence class of triples $(L_\rho, u_\rho, c_m)_{\rho,m}$, where each L_ρ is a line bundle pulled back from X, u_ρ is a section of L_ρ satisfying a non-degeneracy condition, and the collection $\{c_m\}_m$ gives conditions to compare the sections u_ρ in different line bundles L_ρ, $\mathcal{AM}_g(X)$ is the moduli space of such data. It is an Artin stack, fibered over the moduli space of quasi-stable curves [34]. We hope to use this refined moduli to do computations for higher genus mirror principle.

On the other hand, motivated by recent progresses in open string theory, we are also trying to develop open mirror principle. Open string theory predicts formulas for the counting of holomorphic discs with boundary inside a Lagrangian submanifold, more generally of the counting of the numbers of open Riemann surfaces with boundary in Lagrangian submanifold. Linearized moduli space for such data is being constructed which gives a new compactification of such moduli spaces.

4. The Hori-Vafa Formula

In [15], Hori and Vafa generalize the world-sheet aspects of mirror symmetry to being the equivalence of $d = 2$, $N = (2, 2)$ supersymmetric field theories (i.e. without imposing the conformal invariance on the theory). This leads them to a much broader encompassing picture of mirror symmetry. Putting this in the frame work of abelian gauged linear sigma models (GLSM) of Witten enables them to link many $d = 2$ field theories together. Generalization of this setting to nonabelian GLSM leads them to the following conjecture, when the physical path integrals are interpreted appropriately mathematically:

Conjecture 4.1. The hypergeometric series for a given homogeneous space (e.g. a Grassmannian manifold) can be reproduced from the hypergeometric series of simpler homogeneous spaces (e.g. product of projective spaces). Similarly for the twisted hypergeometric series that are related to the submanifolds in homogeneous spaces.

In other words, different homogeneous spaces (or some simple quotients of them) can give rise to generalized mirror pairs. A main object to be understood in the above conjecture is the fundamental hypergeometric series

$HG[\mathbf{1}]^X(t)$ associated to the flag manifold X. Recall that in the computations of mirror principle, the existence of linearized moduli made easy the computations for toric manifolds.

An outline of how this series may be computed was given in [31] via an extended mirror principle diagram. To make clear the main ideas we will only focus on the case of Grassmannian manifolds in this article. The main problem for the computation is that there is no known good linearized moduli for Grassmannian or general flag manifolds. To overcome the difficulty we use the Grothendieck quot scheme to play the role of the linearized moduli. The method gives a complete proof of the Hori-Vafa formula in the Grassmannian case.

Let $ev: \overline{\mathcal{M}}_{0,1}(d,X) \to X$ be the evaluation map on the moduli space of stable maps with one marked point, and c the first Chern class of the tangent line at the marked point. The fundamental hypergeometric series for mirror formula is given by the push-forward:

$$ev_*[\frac{1}{\alpha(\alpha-c)}] \in H^*(X)$$

or more precisely the generating series

$$HG[1]^X(t) = e^{-tH/\alpha}\sum_{d=0}^{\infty} ev_*[\frac{1}{\alpha(\alpha-c)}]\, e^{dt}.$$

Assume the linearized moduli exists. Then functorial localization formula applied to the collapsing map: $\varphi: M_d \to N_d$, immediately gives the expression as the denominator of the hypergeometric series.

Example $X = \mathbf{P}^n$, then we have $\varphi_*(1) = 1$, functorial localization immediately gives us

$$ev_*[\frac{1}{\alpha(\alpha-c)}] = \frac{1}{\prod_{m=1}^{d}(x-m\alpha)^{n+1}}$$

where the denominators of both sides are equivariant Euler classes of normal bundles of the fixed points. Here x denotes the hyperplane class.

For $X = \mathrm{Gr}(k,n)$ or general flag manifolds, no explicit linearized moduli is known. Hori-Vafa conjectured a formula for $HG[1]^X(t)$ by which we can compute this series in terms of those of projective spaces which is the *Hori-Vafa formula for Grassmannians:*

Theorem 4.2. *We have*

$$HG[1]^{\mathrm{Gr}(k,n)}(t) = \frac{e^{(k-1)\pi\sqrt{-1}\sigma/\alpha}}{\prod_{i<j}(x_i - x_j)} \cdot \prod_{i<j}(\alpha\frac{\partial}{\partial x_i} - \alpha\frac{\partial}{\partial x_j})\Big|_{t_i=t+(k-1)\pi\sqrt{-1}} HG[1]^{\mathbf{P}}(t_1,\cdots,t_k)$$

where $\mathbf{P} = \mathbf{P}^{n-1}\times\cdots\times\mathbf{P}^{n-1}$ *is product of* k *copies of the projective spaces,* σ *is the generator of the divisor classes on* $\mathrm{Gr}(k,n)$ *and* x_i *the hyperplane class of the* i*-th copy* $\mathbf{P}^{n-1}$*:*

$$HG[1]^{\mathbf{P}}(t_1,\cdots,t_k) = \prod_{i=1}^{k} HG[1]^{\mathbf{P}^{\mathbf{n-1}}}(t_i).$$

Now we describe the ideas of the proof of the above formula. As mentioned above we use another smooth moduli space, the Grothendieck quot-scheme Q_d to play the role of the linearized moduli, and apply the functorial localization formula. Here is the general set-up:

To start, note that the Plücker embedding $\tau:\ \mathrm{Gr}(k,n)\to\mathbf{P}^N$ induces an embedding of the nonlinear moduli M_d of $\mathrm{Gr}(k,n)$ into that of $\mathbf{P}^N$. Composite of this map with the collapsing map gives us a map $\varphi:\ M_d\to W_d$ into the linearized moduli space W_d of $\mathbf{P}^N$. On the other hand the Plücker embedding also induces a map $\psi:\ Q_d\to W_d$. We have the following three crucial lemmas proved in [28].

Lemma 4.1. *The above two maps have the same image in* W_d*:* $\mathrm{Im}\,\psi = \mathrm{Im}\,\varphi$*. And all the maps are equivariant with respect to the induced circle action from* $\mathbf{P}^1$*.*

Just as in the mirror principle computations, our next step is to analyze the fixed points of the circle action induced from $\mathbf{P}^1$. In particular we need the distinguished fixed point set to get the equivariant Euler class of its normal bundle. The distinguished fixed point set in M_d is $\overline{\mathcal{M}}_{0,1}(d,\mathrm{Gr}(k,n))$ with equivariant Euler class of its normal bundle given by $\alpha(\alpha-c)$, and we know that φ is restricted to ev.

Lemma 4.2. *The distinguished fixed point set in* Q_d *is a union:* $\cup_s E_{0s}$*, where each* E_{0s} *is a fiber bundle over* $\mathrm{Gr}(k,n)$ *with fiber given by flag manifold.*

It is a complicated work to determine the fixed point sets E_{0s} and the weights of the circle action on their normal bundles. The situation for flag manifold cases are much more involved. See [28] and [35] for details.

Now let p denote the projection from E_{0s} onto $\mathrm{Gr}(k,n)$. Functorial localization formula, applied to φ and ψ, gives us the following

Lemma 4.3. *We have the equality on* $\mathrm{Gr}(k,N)$*:*

$$ev_*[\frac{1}{\alpha(\alpha-c)}] = \sum_s p_*[\frac{1}{e_T(E_{0s}/Q_d)}]$$

where $e_T(E_{0s}/Q_d)$ *is the equivariant Euler class of the normal bundle of* E_{0s} *in* Q_d.

Finally we compute $p_*[\frac{1}{e_T(E_{0s}/Q_d)}]$. There are two different approaches, the first one is by direct computations in [28], and another one is by using the well-known Euler sequences for universal sheaves [3]. The second method has the advantage of being more explicit. Note that

$$e_T(TQ|_{E_{0s}} - TE_{0s}) = e_T(TQ|_{E_{0s}})/e_T(TE_{0s}).$$

Both $e_T(TQ|_{E_{0s}})$ and $e_T(TE_{0s})$ can be written down explicitly in terms of the universal bundles on the flag bundle $E_{0s} = Fl(m_1, \cdots, m_k, S)$ over $\mathrm{Gr}(r,n)$. Here S is the universal bundle on the Grassmannian.

The push-forward by p from $Fl(m_1, \cdots, m_k, S)$ to $\mathrm{Gr}(r,n)$ is done by an analogue of family localization formula of Atiyah-Bott, which is given by a sum over the Weyl groups along the fiber which labels the fixed point sets.

In any case the final formula of degree d is given by

$$\begin{aligned} &p_*[\frac{1}{e_T(E_{0s}/Q_d)}] \\ &= (-1)^{(r-1)d} \sum_{\substack{(d_1,\ldots,d_r) \\ d_1+\ldots+d_r=d}} \frac{\prod_{1\le i<j\le r}(x_i - x_j + (d_i - d_j)\alpha)}{\prod_{1\le i<j\le r}(x_i - x_j)\prod_{i=1}^r \prod_{l=1}^{d_i}(x_i + l\alpha)^n}. \end{aligned}$$

Here $x_1, \ldots x_r$ are the Chern roots of S^*. As a corollary of our approach, we have the following:

Corollary 4.3. *The Hori-Vafa conjecture holds for Grassmannian manifolds.*

This corollary was derived in [3] by using the idea and method and also the key results in [28]. For the explicit forms of Hori-Vafa conjecture for general flag manifolds, see [35] and [4].

5. The Mariño-Vafa Conjecture

Our original motivation to study Hodge integrals was to find a general mirror formula for counting higher genus curves in Calabi-Yau manifolds. To generalize mirror principle to count the number of higher genus curves, we need to first compute Hodge integrals, i.e. the intersection numbers of the λ classes and ψ classes on the Deligne-Mumford moduli space of stable curves $\overline{\mathcal{M}}_{g,h}$. This moduli space is possibly the most famous and most interesting orbifold. It has been studied since Riemann, and by many Fields medalists for the past 50 years, from many different point of views. Still many interesting and challenging problems about the geometry and topology of these moduli spaces remain unsolved. String theory has motivated many fantastic conjectures about these moduli spaces including the famous Witten conjecture which is about the generating series of the integrals of the ψ-classes. We start with the introduction of some notations.

Recall that a point in $\overline{\mathcal{M}}_{g,h}$ consists of $(C, x_1, \ldots, x_h)$, a (nodal) curve C of genus g, and n distinguished smooth points on C. The Hodge bundle $\mathbb{E}$ is a rank g vector bundle over $\overline{\mathcal{M}}_{g,h}$ whose fiber over $[(C, x_1, \ldots, x_h)]$ is $H^0(C, \omega_C)$, the complex vector space of holomorphic one forms on C. The λ classes are the Chern Classes of $\mathbb{E}$,

$$\lambda_i = c_i(\mathbb{E}) \in H^{2i}(\overline{\mathcal{M}}_{g,h}; \mathbb{Q}).$$

On the other hand, the cotangent line $T^*_{x_i}C$ of C at the i-th marked point x_i induces a line bundle $\mathbb{L}_i$ over $\overline{\mathcal{M}}_{g,h}$. The ψ classes are the Chern classes:

$$\psi_i = c_1(\mathbb{L}_i) \in H^2(\overline{\mathcal{M}}_{g,h}; \mathbb{Q}).$$

Introduce the total Chern class

$$\Lambda_g^\vee(u) = u^g - \lambda_1 u^{g-1} + \cdots + (-1)^g \lambda_g.$$

The Mariño-Vafa formula is about the generating series of the triple Hodge integrals

$$\int_{\overline{\mathcal{M}}_{g,h}} \frac{\Lambda_g^\vee(1)\Lambda_g^\vee(\tau)\Lambda_g^\vee(-\tau-1)}{\prod_{i=1}^h (1-\mu_i\psi_i)},$$

where τ is considered as a parameter here. Later we will see that it actually comes from the weight of the group action, and also from the framing of the knot. Taking Taylor expansions in τ or in μ_i one can obtain information on the integrals of the Hodge classes and the ψ-classes. The Marinõ-Vafa conjecture asserts that the generating series of such triple Hodge integrals for all genera and any numbers of marked points can be expressed by a close formula which is a *finite* expression in terms of representations of symmetric groups, or Chern-Simons knot invariants.

We remark that the moduli spaces of stable curves have been the sources of many interests from mathematics to physics. Mumford has computed some low genus numbers. The Witten conjecture, proved by Kontsevich [20], is about the integrals of the ψ-classes.

Let us briefly recall the background of the conjecture. Mariño and Vafa [44] made this conjecture based on the large N duality between Chern-Simons and string theory. It starts from the conifold transition. We consider the resolution of singularity of the conifold X defined by

$$\left\{\begin{pmatrix} x & y \\ z & w \end{pmatrix} \in \mathbf{C}^4 : xw - yz = 0\right\}$$

in two different ways:

(1). Deformed conifold T^*S^3

$$\left\{\begin{pmatrix} x & y \\ z & w \end{pmatrix} \in \mathbf{C}^4 : xw - yz = \epsilon\right\}$$

where ϵ a real positive number. This is a symplectic resolution of the singularity.

(2). Resolved conifold by blowing up the singularity, which gives the total space

$$\tilde{X} = \mathcal{O}(-1) \oplus \mathcal{O}(-1) \to \mathbf{P}^1$$

which is explicitly given by

$$\left\{([Z_0, Z_1], \begin{pmatrix} x & y \\ z & w \end{pmatrix}) \in \mathbf{P}^1 \times \mathbf{C}^4 : \begin{matrix} (x,y) \in [Z_0, Z_1] \\ (z,w) \in [Z_0, Z_1] \end{matrix}\right\}$$

$$\begin{array}{ccc} \tilde{X} & \subset & \mathbf{P}^1 \times \mathbf{C}^4 \\ \downarrow & & \downarrow \\ X & \subset & \mathbf{C}^4 \end{array}$$

The brief history of the development of the conjecture is as follows. In 1992 Witten first conjectured that the open topological string theory on the deformed conifold T^*S^3 is equivalent to the Chern-Simons gauge theory on S^3. Such idea was pursued further by Gopakumar and Vafa in 1998, and then by Ooguri and Vafa in 2000. Based on the above conifold transition, they conjectured that the open topological string theory on the deformed conifold T^*S^3 is equivalent to the closed topological string theory on the resolved conifold $\tilde{X}$. Ooguri-Vafa only considered the zero framing case. Later Marinõ-Vafa generalized the idea to the non-zero framing case and discovered the beautiful formula for the generating series of the triple Hodge integrals. Recently Vafa and his collaborators systematically developed the theory, and for the past several years, they developed these duality ideas into the most effective tool to compute Gromov-Witten invariants on toric Calabi-Yau manifolds. The high point of their work is the theory of topological vertex. We refer to [44] and [1] for the details of the physical theory and the history of the development.

Starting with the proof of the Marinõ-Vafa conjecture [38], [39], we have developed a rather complete mathematical theory of topological vertex [26]. Many interesting consequences have been derived for the past year. Now let us see how the string theorists derived mathematical consequence from the above naive idea of string duality. First the Chern-Simons partition function has the form

$$\langle Z(U,V)\rangle = \exp(-F(\lambda,t,V))$$

where U is the holonomy of the $U(N)$ Chern-Simons gauge field around the knot $K \subset S^3$, and V is an extra $U(M)$ matrix. The partition function $\langle Z(U,V)\rangle$ gives the Chern-Simons knot invariants of K.

String duality asserts that the function $F(\lambda,t,V)$ should give the generating series of the open Gromov-Witten invariants of $(\tilde{X}, L_K)$, where L_K is a Lagrangian submanifold of the resolved conifold $\tilde{X}$ canonically associated to the knot K. More precisely by applying the t'Hooft large N expansion, and the "canonical" identifications of parameters similar to mirror formula, which at level k are given by

$$\lambda = \frac{2\pi}{k+N}, \quad t = \frac{2\pi i N}{k+N},$$

we get the partition function of the topological string theory on conifold $\tilde{X}$, and then on $\mathbf{P}^1$. which is just the generating series of the Gromov-Witten invariants. This change of variables is very striking from the point of view of mathematics.

The special case when K is the unknot is already very interesting. In non-zero framing it gives the Mariño-Vafa conjectural formula. In this case $\langle Z(U,V)\rangle$ was first computed in the zero framing by Ooguri-Vafa and in any framing $\tau \in \mathbb{Z}$ by Mariño-Vafa [44]. Comparing with Katz-Liu's computations of $F(\lambda, t, V)$ in [17], Mariño-Vafa conjectured the striking formula about the generating series of the triple Hodge integrals for all genera and any number of marked points in terms of the Chern-Simons invariants, or equivalently in terms of the representations and combinatorics of symmetric groups. It is interesting to note that the framing in the Mariño-Vafa's computations corresponds to the choice of lifting of the circle action on the pair $(\tilde{X}, L_{\text{unknot}})$ in Katz-Liu's localization computations. Both choices are parametrized by an integer τ which will be considered as a parameter in the triple Hodge integrals. Later we will take derivatives with respect to this parameter to get the cut-and-join equation.

It is natural to ask what mathematical consequence we can have for general duality, that is for general knots in general three manifolds, a first naive question is what kind of general Calabi-Yau manifolds will appear in the duality, in place of the conifold. Some special cases corresponding to the Seifert manifolds are known by gluing several copies of conifolds.

Now we give the precise statement of the Mariño-Vafa conjecture, which is an identity between the geometry of the moduli spaces of stable curves and Chern-Simons knot invariants, or the combinatorics of the representation theory of symmetric groups.

Let us first introduce the geometric side. For every partition $\mu = (\mu_1 \geq \cdots \mu_{l(\mu)} \geq 0)$, we define the triple Hodge integral to be,

$$G_{g,\mu}(\tau) = A(\tau) \cdot \int_{\overline{\mathcal{M}}_{g,l(\mu)}} \frac{\Lambda_g^\vee(1)\Lambda_g^\vee(-\tau-1)\Lambda_g^\vee(\tau)}{\prod_{i=1}^{l(\mu)}(1-\mu_i\psi_i)},$$

where the coefficient

$$A(\tau) = -\frac{\sqrt{-1}^{|\mu|+l(\mu)}}{|\operatorname{Aut}(\mu)|}[\tau(\tau+1)]^{l(\mu)-1} \prod_{i=1}^{l(\mu)} \frac{\prod_{a=1}^{\mu_i-1}(\mu_i\tau+a)}{(\mu_i-1)!}.$$

The expressions, although very complicated, arise naturally from localization computations on the moduli spaces of relative stable maps into $\mathbf{P}^1$ with ramification type μ at ∞.

We now introduce the generating series

$$G_\mu(\lambda;\tau) = \sum_{g\geq 0} \lambda^{2g-2+l(\mu)} G_{g,\mu}(\tau).$$

The special case when $g = 0$ is given by

$$\int_{\overline{\mathcal{M}}_{0,l(\mu)}} \frac{\Lambda_0^\vee(1)\Lambda_0^\vee(-\tau-1)\Lambda_0^\vee(\tau)}{\prod_{i=1}^{l(\mu)}(1-\mu_i\psi_i)} = \int_{\overline{\mathcal{M}}_{0,l(\mu)}} \frac{1}{\prod_{i=1}^{l(\mu)}(1-\mu_i\psi_i)}$$

which is known to be equal to $|\mu|^{l(\mu)-3}$ for $l(\mu) \geq 3$, and we use this expression to extend the definition to the case $l(\mu) < 3$.

Introduce formal variables $p = (p_1, p_2, \ldots, p_n, \ldots)$, and define

$$p_\mu = p_{\mu_1} \cdots p_{\mu_{l(\mu)}}$$

for any partition μ. These p_{μ_j} correspond to $\operatorname{Tr} V^{\mu_j}$ in the notations of string theorists. The generating series for all genera and all possible marked points are defined to be

$$G(\lambda;\tau;p) = \sum_{|\mu|\geq 1} G_\mu(\lambda;\tau)p_\mu,$$

which encode complete information of the triple Hodge integrals we are interested in.

Next we introduce the representation theoretical side. Let χ_μ denote the character of the irreducible representation of the symmetric group $S_{|\mu|}$, indexed by μ with $|\mu| = \sum_j \mu_j$. Let $C(\mu)$ denote the conjugacy class of $S_{|\mu|}$ indexed by μ. Introduce

$$\mathcal{W}_\mu(\lambda) = \prod_{1\leq a<b\leq l(\mu)} \frac{\sin\left[(\mu_a - \mu_b + b - a)\lambda/2\right]}{\sin\left[(b-a)\lambda/2\right]} \cdot \frac{1}{\prod_{i=1}^{l(\nu)}\prod_{v=1}^{\mu_i} 2\sin\left[(v - i + l(\mu))\lambda/2\right]}.$$

This has an interpretation in terms of quantum dimension in Chern-Simons knot theory.

We define the following generating series

$$R(\lambda;\tau;p) = \sum_{n\geq 1} \frac{(-1)^{n-1}}{n} \sum_\mu \left[\sum_{\cup_{i=1}^n \mu^i = \mu} \prod_{i=1}^n \sum_{|\nu^i|=|\mu^i|} \frac{\chi_{\nu^i}(C(\mu^i))}{z_{\mu^i}} e^{\sqrt{-1}(\tau+\frac{1}{2})\kappa_{\nu^i}\lambda/2} \mathcal{W}_{\nu^i}(\lambda)\right] p_\mu$$

where μ^i are sub-partitions of μ, $z_\mu = \prod_j \mu_j! j^{\mu_j}$ and

$$\kappa_\mu = |\mu| + \sum_i (\mu_i^2 - 2i\mu_i)$$

for a partition μ which is also standard for representation theory of symmetric groups. There is the relation $z_\mu = |\operatorname{Aut}(\mu)|\mu_1 \cdots \mu_{l(\mu)}$.

Finally we can give the precise statement of *the Mariño-Vafa conjecture:*

Conjecture 5.1. We have the identity

$$G(\lambda; \tau; p) = R(\lambda; \tau; p).$$

Before discussing the proof of this conjecture, we first give several remarks.

Remarks:

1. This conjecture is a formula: G : Geometry = R : Representations, and the representations of symmetric groups are essentially combinatorics.
2. We note that each $G_\mu(\lambda, \tau)$ is given by a finite and closed expression in terms of the representations of symmetric groups:

$$G_\mu(\lambda, \tau) = \sum_{n \geq 1} \frac{(-1)^{n-1}}{n}$$
$$\sum_{\cup_{i=1}^n \mu^i = \mu} \prod_{i=1}^n \sum_{|\nu^i| = |\mu^i|} \frac{\chi_{\nu^i}(C(\mu^i))}{z_{\mu^i}} e^{\sqrt{-1}(\tau + \frac{1}{2})\kappa_{\nu^i}\lambda/2} \mathcal{W}_{\nu^i}(\lambda).$$

   The generating series $G_\mu(\lambda, \tau)$ gives the values of the triple Hodge integrals for moduli spaces of curves of all genera with $l(\mu)$ marked points.
3. Note that an equivalent expression of this formula is the following non-connected generating series. In this situation we have a relatively simpler combinatorial expression:

$$\begin{aligned} G(\lambda; \tau; p)^\bullet &= \exp\left[G(\lambda; \tau; p)\right] \\ &= \sum_{|\mu| \geq 0} [\sum_{|\nu| = |\mu|} \frac{\chi_\nu(C(\mu))}{z_\mu} e^{\sqrt{-1}(\tau + \frac{1}{2})\kappa_\nu \lambda/2} \mathcal{W}_\nu(\lambda)] p_\mu. \end{aligned}$$

   According to Mariño and Vafa, this formula gives values for all Hodge integrals up to three Hodge classes. Lu proved that this is right if we combine with some previously known simple formulas about Hodge integrals.

4. By taking Taylor expansion in τ on both sides of the Mariño-Vafa formula, we have derived various Hodge integral identities in [40].
For examples, as easy consequences of the Mariño-Vafa formula and the cut-and-join equation as satisfied by the above generating series, we have unified simple proofs of the λ_g conjecture by comparing the coefficients in τ in the Taylor expansions of the two expressions,

$$\int_{\overline{\mathcal{M}}_{g,n}} \psi_1^{k_1} \cdots \psi_n^{k_n} \lambda_g = \binom{2g+n-3}{k_1, \ldots, k_n} \frac{2^{2g-1}-1}{2^{2g-1}} \frac{|B_{2g}|}{(2g)!},$$

for $k_1 + \cdots + k_n = 2g - 3 + n$, and the following identities for Hodge integrals:

$$\int_{\overline{\mathcal{M}}_g} \lambda_{g-1}^3 = \int_{\overline{\mathcal{M}}_g} \lambda_{g-2}\lambda_{g-1}\lambda_g = \frac{1}{2(2g-2)!} \frac{|B_{2g-2}|}{2g-2} \frac{|B_{2g}|}{2g},$$

where B_{2g} are Bernoulli numbers. And

$$\int_{\overline{\mathcal{M}}_{g,1}} \frac{\lambda_{g-1}}{1-\psi_1} = b_g \sum_{i=1}^{2g-1} \frac{1}{i} - \frac{1}{2} \sum_{\substack{g_1+g_2=g \\ g_1, g_2 > 0}} \frac{(2g_1-1)!(2g_2-1)!}{(2g-1)!} b_{g_1} b_{g_2},$$

where

$$b_g = \begin{cases} 1, & g = 0, \\ \frac{2^{2g-1}-1}{2^{2g-1}} \frac{|B_{2g}|}{(2g)!}, & g > 0. \end{cases}$$

Now let us look at how we proved this conjecture. This is joint work with Chiu-Chu Liu, Jian Zhou, see [37] and [38] for details.

The first proof of this formula is based on the *Cut-and-Join* equation which is a beautiful match of combinatorics and geometry. The details of the proof is given in [37] and [38]. First we look at the combinatorial side. Denote by $[s_1, \cdots, s_k]$ a k-cycle in the permutation group. We have the following two obvious operations:

1. *Cut*: a k-cycle is cut into an i-cycle and a j-cycle:

$$[s, t] \cdot [s, s_2, \cdots, s_i, t, t_2, \cdots t_j] = [s, s_2, \cdots, s_i][t, t_2, \cdots t_j].$$

2. *Join*: an i-cycle and a j-cycle are joined to an $(i+j)$-cycle:

$$[s,t]\cdot[s,s_2,\cdots,s_i][t,t_2,\cdots t_j]=[s,s_2,\cdots,s_i,t,t_2,\cdots t_j].$$

Such operations can be organized into differential equations which we call the cut-and-join equation.

Now we look at the geometry side. In the moduli spaces of stable maps, cut and join have the following geometric meaning:

1. *Cut*: one curve splits into two lower degree or lower genus curves.
2. *Join*: two curves are joined together to give a higher genus or higher degree curve.

The combinatorics and geometry of cut-and-join are reflected in the following two differential equations, which look like heat equation. It is easy to show that such equation is equivalent to a series of systems of linear ordinary differential equations by comparing the coefficients on p_μ. These equations are proved either by easy and direct computations in combinatorics or by localizations on moduli spaces of relative stable maps in geometry. In combinatorics, the proof is given by direct computations and was explored in combinatorics in the mid 80s and later by Zhou [37] for this case. The differential operator on the right hand side corresponds to the cut-and-join operations which we also simply denote by (CJ).

Lemma 5.1.

$$\frac{\partial R}{\partial \tau}=\frac{1}{2}\sqrt{-1}\lambda\sum_{i,j=1}^{\infty}((i+j)p_ip_j\frac{\partial R}{\partial p_{i+j}}+ijp_{i+j}(\frac{\partial R}{\partial p_i}\frac{\partial R}{\partial p_j}+\frac{\partial^2 R}{\partial p_i\partial p_j})).$$

On the geometry side the proof of such equation is given by localization on the moduli spaces of relative stable maps into the the projective line $\mathbf{P}^1$ with fixed ramifications at ∞:

Lemma 5.2.

$$\frac{\partial G}{\partial \tau}=\frac{1}{2}\sqrt{-1}\lambda\sum_{i,j=1}^{\infty}((i+j)p_ip_j\frac{\partial G}{\partial p_{i+j}}+ijp_{i+j}(\frac{\partial G}{\partial p_i}\frac{\partial G}{\partial p_j}+\frac{\partial^2 G}{\partial p_i\partial p_j})).$$

The proof of the above equation is given in [37]. Together with the following

Initial Value : $\tau = 0$,

$$G(\lambda, 0, p) = \sum_{d=1}^{\infty} \frac{p_d}{2d \sin\left(\frac{\lambda d}{2}\right)} = R(\lambda, 0, p).$$

which is precisely the Ooguri-Vafa formula and which has been proved previously for example in [51], we therefore obtain the equality which is the Mariño-Vafa conjecture by the uniqueness of the solution:

Theorem 5.2. *We have the identity*

$$G(\lambda; \tau; p) = R(\lambda; \tau; p).$$

During the proof we note that the cut-and-join equation is encoded in the geometry of the moduli spaces of stable maps. In fact we later find the convolution formula of the following form, which is a relation for the disconnected version $G^\bullet = \exp G$,

$$G^\bullet_\mu(\lambda, \tau) = \sum_{|\nu|=|\mu|} \Phi^\bullet_{\mu,\nu}(-\sqrt{-1}\tau\lambda) z_\nu K^\bullet_\nu(\lambda)$$

where $\Phi^\bullet_{\mu,\nu}$ is the generating series of double Hurwitz numbers, and z_ν is the combinatorial constant appeared in the previous formulas. Equivalently this gives the explicit solution of the cut-and-join differential equation with initial value $K^\bullet(\lambda)$, which is the generating series of the integrals of certain Euler classes on the moduli spaces of relative stable maps to $\mathbf{P}^1$. See [36] for the derivation of this formula, and see [39] for the two partition analogue.

The Witten conjecture as proved by Kontsevich states that the generating series of the ψ-class integrals satisfy infinite number of differential equations. The remarkable feature of Mariño-Vafa formula is that it gives a finite close formula. In fact by taking limits in τ and μ_i's one can obtain the Witten conjecture. A much simpler direct proof of the Witten conjecture was obtained recently by Kim and myself in [19]. We directly derived the recursion formula which implies both the Virasoro relations and the KdV equations.

The same argument as our proof of the conjecture gives a simple and geometric proof of the ELSV formula for Hurwitz numbers. It reduces to the fact that the push-forward of 1 is a constant in equivariant cohomology for a generically finite-to-one map. See [38] for more details.

We would like to briefly explain the technical details of the proof. The proof of the combinatorial cut-and-join formula is based on the Burnside formula and various simple results in symmetric functions. See [51], [27] and [38].

The proof of the geometric cut-and-join formula used the functorial localization formula in [29] and [30]. The virtual version of this formula was proved first applied to the virtual fundamental cycles in the computations of Gromov-Witten invariants in [30].

As remarked in previous sections the functorial localization formula is very effective and useful because we can use it to push computations on complicated moduli space to simpler moduli space. The moduli spaces used by mathematicians are usually the correct but complicated moduli spaces like the moduli spaces of stable maps, while the moduli spaces used by physicists are usually the simple but the wrong ones like the projective spaces. This functorial localization formula has been used successfully in the proof of the mirror formula [29], [30], the proof of the Hori-Vafa formula [28], and the easy proof of the ELSV formula [38]. Our first proof of the Mariño-Vafa formula also used this formula in a crucial way.

More precisely, let $\overline{\mathcal{M}}_g(\mathbf{P}^1, \mu)$ denote the moduli space of relative stable maps from a genus g curve to $\mathbf{P}^1$ with fixed ramification type μ at ∞, where μ is a fixed partition. We apply the functorial localization formula to the divisor morphism from the relative stable map moduli space to the projective space,

$$\mathrm{Br}: \ \overline{\mathcal{M}}_g(\mathbf{P}^1, \mu) \to \mathbf{P}^r,$$

where r denotes the dimension of $\overline{\mathcal{M}}_g(\mathbf{P}^1, \mu)$. This is similar to the set-up of mirror principle, only with a different linearized moduli space, but in both cases the target spaces are projective spaces.

We found that the fixed points of the target $\mathbf{P}^r$ precisely labels the cut-and-join operations of the triple Hodge integrals. Functorial localization reduces the problem to the study of polynomials in the equivariant cohomology group of $\mathbf{P}^r$. We were able to squeeze out a system of linear equations which implies the cut-and-join equation. Actually we derived a stronger relation than the cut-and-join equation, while the cut-and-join equation we need for the Mariño-Vafa formula is only the very first of such kind of relations. See [38] for higher order cut-and-join equations.

As was known in infinite Lie algebra theory, the cut-and-join operator is closely related to and more fundamental than the Virasoro algebras in some sense.

Recently there have appeared two different approaches to the Mariño-Vafa formula. The first one is a direct derivation of the convolution formula which was discovered during our proof of the two partition analogue of the formula [39]. See [36] for the details of the derivation in this case. The second is by Okounkov-Pandhripande [47], they gave a different approach by using the ELSV formula as initial value, and as well as the λ_g conjecture and other recursion relations from localization on the moduli spaces of stable maps to $\mathbf{P}^1$.

6. Two Partition Formula

The two partition analogue of the Mariño-Vafa formula naturally arises from the localization computations of the Gromov-Witten invariants of the open toric Calabi-Yau manifolds, as explained in [52].

To state the formula we let μ^+, μ^- be any two partitions. Introduce the Hodge integrals involving these two partitions:

$$G_{\mu^+,\mu^-}(\lambda;\tau) = B(\tau;\mu^+,\mu^-) \cdot \sum_{g\geq 0} \lambda^{2g-2} A_g(\tau;\mu^+,\mu^-)$$

where

$$A_g(\tau;\mu^+,\mu^-) = \int_{\overline{\mathcal{M}}_{g,l(\mu^+)+l(\mu^-)}} \frac{\Lambda_g^\vee(1)\Lambda_g^\vee(\tau)\Lambda_g^\vee(-\tau-1)}{\prod_{i=1}^{l(\mu^+)}(1-\mu_i^+\psi_i)\prod_{j=1}^{l(\mu^-)}\tau\left(\tau-\mu_i^-\psi_{j+l(\mu^+)}\right)}$$

and

$$B(\tau;\mu^+,\mu^-) = -\frac{(\sqrt{-1}\lambda)^{l(\mu^+)+l(\mu^-)}}{|\operatorname{Aut}(\mu^+)||\operatorname{Aut}(\mu^-)|}\left[\tau(\tau+1)\right]^{l(\mu^+)+l(\mu^-)-1}.$$

$$\prod_{i=1}^{l(\mu^+)} \frac{\prod_{a=1}^{\mu_i^+-1}(\mu_i^+\tau+a)}{(\mu_i^+-1)!} \cdot \prod_{i=1}^{l(\mu^-)} \frac{\prod_{a=1}^{\mu_i^--1}(\mu_i^-\frac{1}{\tau}+a)}{(\mu_i^--1)!}.$$

These complicated expressions naturally arise in open string theory, as well as in the localization computations of the Gromov-Witten invariants on open toric Calabi-Yau manifolds.

We introduce two generating series, first on the geometry side,

$$G^\bullet(\lambda;p^+,p^-;\tau) = \exp\left(\sum_{(\mu^+,\mu^-)\in\mathcal{P}^2} G_{\mu^+,\mu^-}(\lambda,\tau)p^+_{\mu^+}p^-_{\mu^-}\right),$$

where $\mathcal{P}^2$ denotes the set of pairs of partitions and $p^{\pm}_{\mu^{\pm}}$ are two sets of formal variables associated to the two partitions as in the last section.

On the representation side, we introduce

$$R^\bullet(\lambda; p^+, p^-; \tau) = \sum_{|\nu^\pm|=|\mu^\pm|\geq 0} \frac{\chi_{\nu^+}(C(\mu^+))}{z_{\mu^+}} \frac{\chi_{\nu^-}(C(\mu^-))}{z_{\mu^-}} \cdot e^{\sqrt{-1}(\kappa_{\nu^+}\tau + \kappa_{\nu^-}\tau^{-1})\lambda/2} \mathcal{W}_{\nu^+,\nu^-} p^+_{\mu^+} p^-_{\mu^-}.$$

Here

$$\mathcal{W}_{\mu,\nu} = q^{l(\nu)/2} \mathcal{W}_\mu \cdot s_\nu(\mathcal{E}_\mu(t))$$

$$= (-1)^{|\mu|+|\nu|} q^{\frac{\kappa_\mu + \kappa_\nu + |\mu| + |\nu|}{2}} \sum_\rho q^{-|\rho|} s_{\mu/\rho}(1, q, \dots) s_{\nu/\rho}(1, q, \dots)$$

in terms of the skew Schur functions s_μ [43]. They appear naturally in the Chern-Simons invariant of the Hopf link.

Theorem 6.1. *We have the identity:*

$$G^\bullet(\lambda; p^+, p^-; \tau) = R^\bullet(\lambda; p^+, p^-; \tau).$$

The idea of the proof is similar to that of the proof of the Mariño-Vafa formula. We prove that both sides of the above identity satisfy the same cut-and-join equation of the following type:

$$\frac{\partial}{\partial \tau} H^\bullet = \frac{1}{2}(CJ)^+ H^\bullet - \frac{1}{2\tau^2}(CJ)^- H^\bullet,$$

where $(CJ)^\pm$ denote the cut-and-join operator, the differential operator with respect to the two set of variables $p^\pm$. We then prove that they have the same initial value at $\tau = -1$:

$$G^\bullet(\lambda; p^+, p^-; -1) = R^\bullet(\lambda; p^+, p^-; -1),$$

which is again given by the Ooguri-Vafa formula [39], [52].

The cut-and-join equation can be written in a linear matrix form, and such equation follows from the convolution formula of the form

$$K^{\bullet}_{\mu^+,\mu^-}(\lambda)$$
$$= \sum_{|\nu^\pm|=\mu^\pm} G^{\bullet}_{\mu^+,\mu^-}(\lambda;\tau) z_{\nu^+} \Phi^{\bullet}_{\nu^+,\mu^+}(-\sqrt{-1}\lambda\tau) z_{\nu^-} \Phi^{\bullet}_{\nu^-,\mu^-}(\frac{-\sqrt{-1}}{\tau}\lambda)$$

where $\Phi^{\bullet}$ denotes the generating series of double Hurwitz numbers, and K_{μ^+,μ^-} is the generating series of certain integrals on the moduli spaces of relative stable maps. For more details see [39].

This convolution formula arises naturally from localization computations on the moduli spaces of relative stable maps to $\mathbf{P}^1 \times \mathbf{P}^1$ with the point (∞, ∞) blown up. So it reflects the geometric structure of the moduli spaces. Such convolution type formula was actually discovered during our search for a proof of this formula, both on the geometric and the combinatorial side, see [39] for the detailed derivations of the convolution formulas in both geometry and combinatorics.

The proof of the combinatorial side of the convolution formula is again a direct computation. The proof of the geometric side for the convolution equation is to reorganize the generating series from localization contributions on the moduli spaces of relative stable maps into $\mathbf{P}^1 \times \mathbf{P}^1$ with the point (∞, ∞) blown up, in terms of the double Hurwitz numbers. It involves careful analysis and computations.

7. The Theory of Topological Vertex

When we worked on the Mariño-Vafa formula and its generalizations, we were simply trying to generalize the method and the formula to involve more partitions, but it turned out that in the three partition case, we naturally met the theory of topological vertex. Topological vertex was first introduced in string theory by Vafa et al in [1], it can be deduced from a three partition analogue of the Mariño-Vafa formula in a highly nontrivial way. From this we were able to give a rigorous mathematical foundation for the physical theory. Topological vertex is a high point of the theory of string duality as developed by Vafa and his group for the past several years, starting from Witten's conjectural duality between Chern-Simons and open string theory. It gives the most powerful and effective way to compute the Gromov-Witten invariants for all open toric Calabi-Yau manifolds. In physics it is rare to have two theories agree up to all orders, topological vertex theory gives a very significant example. In mathematics the theory of topological vertex already has many interesting applications. Here we only briefly sketch the rough idea for the three partition analogue of the Mariño-Vafa formula. For its relation to the theory of topological vertex, we refer the reader to [26] for the details.

Given any three partitions $\overrightarrow{\mu} = \{\mu^1, \mu^2, \mu^3\}$, the cut-and-join equation in this case, for both the geometry and representation sides, has the form:

$$\begin{aligned}\frac{\partial}{\partial \tau} F^\bullet(\lambda;\tau;\mathbf{p}) = (CJ)^1 F^\bullet(\lambda;\tau;\mathbf{p}) + \frac{1}{\tau^2}(CJ)^2 F^\bullet(\lambda;\tau;\mathbf{p}) \\ + \frac{1}{(\tau+1)^2}(CJ)^3 F^\bullet(\lambda;\tau;\mathbf{p}).\end{aligned}$$

The cut-and-join operators $(CJ)^1$, $(CJ)^2$ and $(CJ)^3$ are with respect to the three partitions. More precisely they correspond to the differential operators with respect to the three groups of infinite numbers of variables $\mathbf{p} = \{p^1, p^2, p^3\}$.

The initial value for this differential equation is taken at $\tau = 1$, which is then reduced to the formulas of two partition case. The combinatorial, or the Chern-Simons invariant side is given by $\mathcal{W}_{\overrightarrow{\mu}} = \mathcal{W}_{\mu^1,\mu^2,\mu^3}$ which is a combination of the $\mathcal{W}_{\mu,\nu}$ as in the two partition case. See [26] for its explicit expression.

On the geometry side,

$$G^\bullet(\lambda;\tau;\mathbf{p}) = \exp(G(\lambda;\tau;\mathbf{p}))$$

is the non-connected version of the generating series of the triple Hodge integral. More precisely,

$$G(\lambda;\tau;\mathbf{p}) = \sum_{\overrightarrow{\mu}}[\sum_{g=0}^{\infty} \lambda^{2g-2+l(\overrightarrow{\mu})} G_{g,\overrightarrow{\mu}}(\tau)] p^1_{\mu^1} p^2_{\mu^2} p^3_{\mu^3}$$

where $l(\overrightarrow{\mu}) = l(\mu^1) + l(\mu^2) + l(\mu^3)$ and $G_{g,\overrightarrow{\mu}}(\tau)$ denotes the Hodge integrals of the following form,

$$A(\tau)\int_{\overline{\mathcal{M}}_{g,l_1+l_2+l_3}} \frac{\Lambda_g^\vee(1)\Lambda_g^\vee(\tau)\Lambda_g^\vee(-\tau-1)}{\prod_{j=1}^{l_1}(1-\mu_j^1\psi_j)\prod_{j=1}^{l_2}\tau(\tau-\mu_j^2\psi_{l_1+j})} \cdot \frac{(\tau(\tau+1))^{l_1+l_2+l_3-1}}{\prod_{j=1}^{l_3}(\tau+1)(\tau+1+\mu_j^3\psi_{l_1+l_2+j})},$$

where

$$A(\tau) = \frac{-(\sqrt{-1}\lambda)^{l_1+l_2+l_3}}{|\operatorname{Aut}(\mu^1)||\operatorname{Aut}(\mu^2)||\operatorname{Aut}(\mu^3)|} \prod_{j=1}^{l_1} \frac{\prod_{a=1}^{\mu_j^1-1}(\tau\mu_j^1+a)}{(\mu_j^1-1)!}.$$

$$\prod_{j=1}^{l_2} \frac{\prod_{a=1}^{\mu_j^1-1}((-1-1/\tau)\mu_j^2+a)}{(\mu_j^2-1)!} \prod_{j=1}^{l_3} \frac{\prod_{a=1}^{\mu_j^1-1}(-\mu_j^3/(\tau+1)+a)}{(\mu_j^3-1)!}$$

In the above expression, $l_i = l(\mu^i)$, $i = 1, 2, 3$. Despite of its complicated coefficients, these triple integrals naturally arise from localizations on the moduli spaces of relative stable maps into the blow-up of $\mathbf{P}^1 \times \mathbf{P}^1 \times \mathbf{P}^1$ along certain divisors. It also naturally appears in open string theory computations [1]. See [26] for more details.

One of our results in [26] states that $G^\bullet(\lambda;\tau;\mathbf{p})$ has a combinatorial expression $R^\bullet(\lambda;\tau;\mathbf{p})$ in terms of the Chern-Simons knot invariants $W_{\vec{\mu}}$, which is a closed combinatorial expression. More precisely it is given by

$$R^\bullet(\lambda;\tau;\mathbf{p}) = \sum_{\vec{\mu}}[\sum_{|\nu^i|=|\mu^i|} \prod_{i=1}^{3} \frac{\chi_{\nu^i}(\mu^i)}{z_{\mu^i}} q^{\frac{1}{2}(\sum_{i=1}^3 \kappa_{\nu^i}\frac{w_{i+1}}{w_i})} W_{\vec{\nu}}(q)] p^1_{\mu^1} p^2_{\mu^2} p^3_{\mu^3}.$$

Here $w_4 = w_1$ and $w_3 = -w_1 - w_2$ and $\tau = \frac{w_2}{w_1}$. Due to the complicated combinatorics in the initial values, the combinatorial expression $W_{\vec{\mu}}$ we obtained is different from the expression $\mathcal{W}_{\vec{\mu}}$ obtained by Vafa et al. Actually our expression is even simpler than theirs in some sense. The expression we obtained is more convenient for mathematical applications such as the proof of the Gopakumar-Vafa conjecture for open toric Calabi-Yau manifolds, see [48].

Theorem 7.1. *We have the equality:*

$$G^\bullet(\lambda;\tau;\boldsymbol{p}) = R^\bullet(\lambda;\tau;\boldsymbol{p}).$$

The key point to prove the above theorem is still the proof of convolution formulas for both sides which imply the cut-and-join equation. The proof of the convolution formula for $G^\bullet(\lambda;\tau;\mathbf{p})$ is much more complicated than the one and two partition cases. See [26] for details.

The most useful property of topological vertex is its gluing property induced by the orthogonal relations of the characters of the symmetric group. This is very close to the situation of two dimensional gauge theory. In fact string theorists consider topological vertex as a kind of lattice theory on Calabi-Yau manifolds. By using the gluing formula we can easily obtain closed formulas for generating series of Gromov-Witten invariants of all genera and all degrees, open or closed, for all open toric Calabi-Yau manifolds, in terms of the Chern-Simons knot invariants. Such formulas are always given by finite sum of products of those Chern-Simons type invariants $\mathcal{W}_{\mu,\nu}$'s. The magic of topological vertex is that, by simply looking at the moment map graph of the toric surfaces in the open toric Calabi-Yau, we can immediately write down the closed formula for the generating series for all genera and all degree Gromov-Witten invariants, or more precisely the Euler numbers of certain bundles on the moduli space of stable maps.

Here we only give one example to describe the topological vertex formula for the generating series of the all degree and all genera Gromov-Witten invariants for the open toric Calabi-Yau 3-folds. We write down the explicit close formula of the generating series of the Gromov-Witten invariants in this case.

Example: Consider the toric Calabi-Yau manifold which is $O(-3) \longrightarrow \mathbf{P}^2$. In this case the formula for the generating series of all degrees and all genera Gromov-Witten invariants is given by

$$\begin{aligned}&\exp\big(\sum_{g=0}^{\infty} \lambda^{2g-2} F_g(t)\big)\\ &= \sum_{\nu_1,\nu_2,\nu_3} \mathcal{W}_{\nu_1,\nu_2}\mathcal{W}_{\nu_2,\nu_3}\mathcal{W}_{\nu_3,\nu_1}(-1)^{\sum_{j=1}^{3}|\nu_j|} q^{\frac{1}{2}\sum_{i=1}^{3}\kappa_{\nu_i}}\, e^{t(\sum_{j=1}^{3}|\nu_j|)}\end{aligned}$$

where $q = e^{\sqrt{-1}\lambda}$. The precise definition of $F_g(t)$ will be given in the next section.

For general open toric Calabi-Yau manifolds, the expressions are just similar. They are all given by finite and closed formulas, which are easily read out from the moment map graphs associated to the toric surfaces, with the topological vertex associated to each vertex of the graph.

In [1] Vafa and his group first developed the theory of topological vertex by using string duality between Chern-Simons and Calabi-Yau, which is a physical theory. In [26] we established the mathematical theory of the topological vertex, and derived various mathematical corollaries, including the relation of the Gromov-Witten invariants to the equivariant index theory as motivated by the Nekrasov conjecture in string duality [37]. During the development of the mathematical theory of topological vertex we also introduced formal Calabi-Yau manifolds, see [26] for details.

8. The Gopakumar-Vafa Conjecture and the Indices of Elliptic Operators

Let $N_{g,d}$ denote the so-called Gromov-Witten invariant of genus g and degree d of an open toric Calabi-Yau 3-fold. $N_{g,d}$ is defined to be the Euler number of the obstruction bundle on the moduli space of stable maps of degree $d \in H_2(S, \mathbb{Z})$ from genus g curve into the surface base S. The open toric Calabi-Yau manifold associated to the toric surface S is the total space of the canonical line bundle K_S on S. More precisely

$$N_{g,d} = \int_{[\overline{\mathcal{M}}_g(S,d)]^v} e(V_{g,d})$$

with $V_{g,d} = R^1\pi_* u^* K_S$ a vector bundle on the moduli space induced by the canonical bundle K_S. Here $\pi : U \to \overline{\mathcal{M}}_g(S, d)$ denotes the universal curve and u can be considered as the evaluation or universal map. Let us write

$$F_g(t) = \sum_{d \geq 0} N_{g,d}\, e^{-d \cdot t}.$$

The Gopakumar-Vafa conjecture is stated as follows:

Conjecture 8.1. There exists an expression:

$$\sum_{g=0}^{\infty} \lambda^{2g-2} F_g(t) = \sum_{k=1}^{\infty} \sum_{g,d \geq 0} n_d^g \frac{1}{d} (2 \sin \frac{d\lambda}{2})^{2g-2} e^{-kd \cdot t},$$

such that n_d^g are integers, called instanton numbers.

Motivated by the Nekrasov duality conjecture between the four dimensional gauge theory and string theory, we are able to interpret the above integers n_d^g as equivariant indices of certain elliptic operators on the moduli spaces of anti-self-dual connections [37]:

Theorem 8.2. *For certain interesting cases, these n_d^g's can be written as equivariant indices on the moduli spaces of anti-self-dual connections on $\mathbb{C}^2$.*

For more precise statement, we refer the reader to [27]. The interesting cases include open toric Calabi-Yau manifolds when S is Hirzebruch surface.

The proof of this theorem is to compare fixed point formula expressions for equivariant indices of certain elliptic operators on the moduli spaces of anti-self-dual connections with the combinatorial expressions of the generating series of the Gromov-Witten invariants on the moduli spaces of stable maps. They both can be expressed in terms of Young diagrams of partitions. We find that they agree up to certain highly non-trivial "mirror transformation", a complicated variable change. This result is not only interesting for the index formula interpretation of the instanton numbers, but also for the fact that it gives the first complete examples that the Gopakumar-Vafa conjecture holds for all genera and all degrees.

Recently P. Peng [48] has given the first complete proof of the Gopakumar-Vafa conjecture for all open toric Calabi-Yau 3-folds by using our Chern-Simons expressions from the topological vertex. His method is to explore the property of the Chern-Simons expression in great detail with some clever observation about the form of the combinatorial expressions. On the other hand, Kim in [18] has derived some remarkable recursion formulas for Hodge integrals of all genera and any number of marked points, involving one λ-classes. His method is to add marked points in the moduli spaces and then follow the localization argument we used to prove the Mariño-Vafa formula.

9. Two Proofs of the ELSV Formula

In this section we describe two proofs of the ELSV formula, one is by direct localization and cut-and-join equation following our proof of the Mariño-Vafa formula, another one is to derive it from the Mariño-Vafa formula through a scaling limit. These results are contained in [40]

Given a partition μ of length $l(\mu)$, denote by $H_{g,\mu}$ the Hurwitz numbers of almost simple Hurwitz covers of $\mathbb{P}^1$ of ramification type μ by connected genus g Riemann surfaces. The ELSV formula [8], [14] states:

$$H_{g,\mu} = (2g-2+|\mu|+l(\mu))! I_{g,\mu}$$

where

$$I_{g,\mu} = \frac{1}{|\operatorname{Aut}(\mu)|} \prod_{i=1}^{l(\mu)} \frac{\mu_i^{\mu_i}}{\mu_i!} \int_{\overline{\mathcal{M}}_{g,l(\mu)}} \frac{\Lambda_g^\vee(1)}{\prod_{i=1}^{l(\mu)}(1-\mu_i\psi_i)}.$$

Define generating functions

$$\Phi_\mu(\lambda) = \sum_{g\geq 0} H_{g,\mu}\frac{\lambda^{2g-2+|\mu|+l(\mu)}}{(2g-2+|\mu|+l(\mu))!},$$
$$\Phi(\lambda;p) = \sum_{|\mu|\geq 1} \Phi_\mu(\lambda)p_\mu,$$
$$\Psi_\mu(\lambda) = \sum_{g\geq 0} I_{g,\mu}\lambda^{2g-2+|\mu|+l(\mu)},$$
$$\Psi(\lambda;p) = \sum_{|\mu|\geq 1} \Psi_\mu(\lambda)p_\mu.$$

In terms of generating functions, the ELSV formula reads

Theorem 9.1. *We have the identity*

$$\Psi(\lambda;p) = \Phi(\lambda;p).$$

We first describe a proof of this formula by using cut-and-join equations, following our proof of the Mariño-Vafa formula. It was known that $\Phi(\lambda;p)$ satisfies the following cut-and-join equation [12]:

$$\frac{\partial\Theta}{\partial\lambda} = \frac{1}{2}\sum_{i,j\geq 1}\left(ijp_{i+j}\frac{\partial^2\Theta}{\partial p_i\partial p_j} + ijp_{i+j}\frac{\partial\Theta}{\partial p_i}\frac{\partial\Theta}{\partial p_j} + (i+j)p_ip_j\frac{\partial\Theta}{\partial p_{i+j}}\right).$$

This equation was later reproved by sum formula of symplectic Gromov-Witten invariants [21].

The calculations in Section 7 and Appendix A of [37] shows that

$$\tilde{H}_{g,\mu} = (2g-2+|\mu|+l(\mu))!I_{g,\mu}$$

$$\tilde{H}_{g,\mu} = (2g-3+|\mu|+l(\mu))!\left(\sum_{\nu\in J(\mu)} I_{g,\nu} + \sum_{\nu\in C(\mu)} I_2(\nu)I_{g-1,\nu}\right.$$
$$\left. + \sum_{g_1+g_2=g}\ \sum_{\nu^1\cup\nu^2\in C(\mu)} I_3(\nu^1,\nu^2)I_{g_1,\nu_1}I_{g_2,\nu_2}\right)$$

where

$$\tilde{H}_{g,\mu} = \int_{[\overline{\mathcal{M}}_{g,0}(\mathbf{P}^1,\mu)]^{\mathrm{vir}}} \mathrm{Br}^* H^r$$

is some relative Gromov-Witten invariant of $(\mathbf{P}^1,\infty)$, and

$$C(\mu),\ J(\mu),\ I_1,\ I_2,\ I_3$$

are defined as in [21]. So we have

$$\begin{aligned}&(2g-2+|\mu|+l(\mu))I_{g,\mu}\\&=\sum_{\nu\in J(\mu)} I_{g,\nu}+\sum_{\nu\in C(\mu)} I_2(\nu)I_{g-1,\nu}+\sum_{g_1+g_2=g}\ \sum_{\nu^1\cup\nu^2\in C(\mu)} I_3(\nu^1,\nu^2)I_{g_1,\nu_1}I_{g_2,\nu_2},\end{aligned}$$

which is equivalent to the statement that the generating function $\Psi(\lambda;p)$ of $I_{g,\mu}$ also satisfies the cut-and-join equation.

Any solution $\Theta(\lambda;p)$ to the cut-and-join equation (9) is uniquely determined by its initial value $\Theta(0;p)$, so it remains to show that $\Psi(0;p)=\Phi(0;p)$. Note that $2g-2+|\mu|+l(\mu)=0$ if and only if $g=0$ and $\mu=(1)$, so

$$\Psi(0;p)=H_{0,(1)}p_1,\quad \Phi(0;p)=I_{0,(1)}p_1.$$

It is easy to see that $H_{0,(1)}=I_{0,(1)}=1$, so

$$\Psi(0;p)=\Phi(0;p).$$

One can see geometrically that the relative Gromov-Witten invariant $\tilde{H}_{g,\mu}$ is equal to the Hurwitz number $H_{g,\mu}$. This together with (9) gives a proof of the ELSV formula presented in [37] in the spirit of [14]. Note that $\tilde{H}_{g,\mu}=H_{g,\mu}$ is not used in the proof described above.

On the other hand we can deduce the ELSV formula as the limit of the Mariño-Vafa formula. By the Burnside formula, one easily gets the following expression (see e.g. [39]):

$$\begin{aligned}\Phi(\lambda;p)&=\log\left(\sum_{\mu}\left(\sum_{|\nu|=|\mu|}\frac{\chi_\nu(\mu)}{z_\mu}e^{\kappa_\nu\lambda/2}\frac{\dim R_\nu}{|\nu|!}\right)p_\mu.\right)\\&=\sum_{n\geq 1}\frac{(-1)^{n-1}}{n}\sum_{\mu}\ \sum_{\cup_{i=1}^n\mu_i=\mu}\ \prod_{i=1}^{n}\ \sum_{|\nu_i|=|\mu_i|}\frac{\chi_{\nu_i}(\mu_i)}{z_{\mu_i}}e^{\kappa_{\nu_i}\lambda/2}\frac{\dim R_{\nu_i}}{|\nu_i|!}p_\mu.\end{aligned}$$

The ELSV formula reads

$$\Psi(\lambda;p)=\Phi(\lambda;p)$$

where the left hand side is a generating function of Hodge integrals $I_{g,\mu}$, and the right hand side is a generating function of representations of symmetric groups. So the ELSV formula and the Mariño-Vafa formula are of the same type.

Actually, the ELSV formula can be obtained by taking a particular limit of the Mariño-Vafa formula $G(\lambda;\tau;p)=R(\lambda;\tau;p)$. More precisely, it is straightforward to check that

$$\begin{aligned}
&\lim_{\tau\to 0} G(\lambda\tau; \frac{1}{\tau}; (\lambda\tau)p_1, (\lambda\tau)^2 p_2, \cdots) \\
&= \sum_{|\mu|\neq 0}\sum_{g=0}^{\infty} \sqrt{-1}^{2g-2+|\mu|+l(\mu)} I_{g,\mu}\lambda^{2g-2+|\mu|+l(\mu)} p_\mu \\
&= \Psi(\sqrt{-1}\lambda; p)
\end{aligned}$$

and

$$\begin{aligned}
&\lim_{\tau\to 0} R(\lambda\tau; \frac{1}{\tau}; (\lambda\tau)p_1, (\lambda\tau)^2 p_2, \cdots) \\
&= \log\left(\sum_\mu \left(\sum_{|\nu|=|\mu|} \frac{\chi_\nu(C(\mu))}{z_\mu} e^{\sqrt{-1}\kappa_\nu\lambda/2} \lim_{t\to 0}(t^{|\nu|}V_\nu(t))\right) p_\mu\right) \\
&= \log\left(\sum_\mu \left(\sum_{|\nu|=|\mu|} \frac{\chi_\nu(C(\mu))}{z_\mu} e^{\sqrt{-1}\kappa_\nu\lambda/2} \frac{1}{\prod_{x\in\nu} h(x)}\right) p_\mu\right) \\
&= \Phi(\sqrt{-1}\lambda; p)
\end{aligned}$$

where we have used

$$\frac{1}{\prod_{x\in\nu} h(x)} = \frac{\dim R_\nu}{|\nu|!}.$$

See [40] for the notations. In this limit, the cut-and-join equation of $G(\lambda;\tau;p)$ and $R(\lambda;\tau;p)$ reduces to the cut-and-join equation of $\Psi(\lambda;p)$ and $\Phi(\lambda;p)$, respectively.

10. A Localization Proof of the Witten Conjecture

The Witten conjecture for moduli spaces states that the generating series F of the integrals of the ψ classes for all genera and any number of marked points satisfies the KdV equations and the Virasoro constraint. For example the Virasoro constraint states that F satisfies

$$L_n \cdot F = 0, \ n \geq -1$$

where L_n denote certain Virasoro operators to be given later.

Witten conjecture was first proved by Kontsevich using combinatorial model of the moduli space and matrix model, with later approaches by Okounkov-Pandhripande [47] using ELSV formula and combinatorics, by Mirzakhani [45] using Weil-Petersson volumes on moduli spaces of bordered Riemann surfaces.

I will present a much simpler proof by using functorial localization and asymptotics. This was done jointly with Y.-S. Kim in [19]. This is also motivated by methods in proving conjectures from string duality. It should have more applications.

The basic idea of our proof is to directly prove the following recursion formula which, as derived in physics by Dijkgraaf, Verlinde and Verlinde by using quantum field theory, implies the Virasoro and the KdV equation for the generating series F of the integrals of the ψ classes:

Theorem 10.1. *We have identity*

$$\langle\tilde{\sigma}_n \prod_{k\in S}\tilde{\sigma}_k\rangle_g = \sum_{k\in S}(2k+1)\langle\tilde{\sigma}_{n+k-1}\prod_{l\neq k}\tilde{\sigma}_l\rangle_g + \frac{1}{2}\sum_{a+b=n-2}\langle\tilde{\sigma}_a\tilde{\sigma}_b\prod_{l\neq a,b}\tilde{\sigma}_l\rangle_{g-1}$$

$$+\frac{1}{2}\sum_{\substack{S=X\cup Y,\\ a+b=n-2,\\ g_1+g_2=g}}\langle\tilde{\sigma}_a\prod_{k\in X}\tilde{\sigma}_k\rangle_{g_1}\langle\tilde{\sigma}_b\prod_{l\in Y}\tilde{\sigma}_l\rangle_{g_2}.$$

Here $\tilde{\sigma}_n = (2n+1)!!\psi^n$ and

$$\langle\prod_{j=1}^n\tilde{\sigma}_{k_j}\rangle_g = \int_{\overline{\mathcal{M}}_{g,n}}\prod_{j=1}^n\tilde{\sigma}_{k_j}.$$

The notation $S=\{k_1,\cdots,k_n\}=X\cup Y$.

To prove the above recursion relation, similar to the proof of the Mariño-Vafa formula, we first apply the functorial localization to the natural branch map from moduli space of relative stable maps $\overline{\mathcal{M}}_g(\mathbf{P}^1,\mu)$ to projective space $\mathbf{P}^r$ where $r=2g-2+|\mu|+l(\mu)$ is the dimension of the moduli.

As discussed in last section we easily get the cut and join equation for one Hodge integral

$$I_{g,\mu} = \frac{1}{|\text{Aut }\mu|}\prod_{i=1}^n\frac{\mu_i^{\mu_i}}{\mu_i!}\int_{\overline{\mathcal{M}}_{g,n}}\frac{\Lambda_g^\vee(1)}{\prod(1-\mu_i\psi_i)}.$$

The equation we get has the form as discussed in last section, it is trivial corollary of the fact that the push-forward of 1 in equivariant cohomology by a map between equal dimension manifolds is a constant:

$$(2g-2+|\mu|+l(\mu))I_{g,\mu}$$
$$= \sum_{\nu\in J(\mu)}I_{g,\nu} + \sum_{\nu\in C(\mu)}I_2(\nu)I_{g-1,\nu} + \sum_{g_1+g_2=g}\sum_{\nu^1\cup\nu^2\in C(\mu)}I_3(\nu^1,\nu^2)I_{g_1,\nu_1}I_{g_2,\nu_2}.$$

Note that more general formulas of such type was first found and proved by Kim in [18].

Write $\mu_i = Nx_i$. Let N go to infinity and expand in x_i, we get:

$$\sum_{i=1}^{n}\Big[\frac{(2k_i+1)!!}{2^{k_i+1}k_i!}x_i^{k_i}\prod_{j\neq i}\frac{x_j^{k_j-\frac{1}{2}}}{\sqrt{2\pi}}\int_{\overline{\mathcal{M}}_{g,n}}\prod\psi_j^{k_j}-\sum_{j\neq i}\frac{(x_i+x_j)^{k_i+k_j-\frac{1}{2}}}{\sqrt{2\pi}}$$

$$\prod_{l\neq i,j}\frac{x_l^{k_l-\frac{1}{2}}}{\sqrt{2\pi}}\int_{\overline{\mathcal{M}}_{g,n-1}}\psi^{k_i+k_j-1}\prod\psi_l^{k_l}$$

$$-\frac{1}{2}\sum_{k+l=k_i-2}\frac{(2k+1)!!(2l+1)!!}{2^{k_i}\,k_i!}x_i^{k_i}\prod_{j\neq i}\frac{x_j^{k_j-\frac{1}{2}}}{\sqrt{2\pi}}\Big[\int_{\overline{\mathcal{M}}_{g-1,n+1}}\psi_1^k\psi_2^l\prod\psi_j^{k_j}$$

$$+\sum_{\substack{g_1+g_2=g,\\ \nu_1\cup\nu_2=\nu}}\int_{\overline{\mathcal{M}}_{g_1,n_1}}\psi_1^k\prod\psi_j^{k_j}\int_{\overline{\mathcal{M}}_{g_2,n_2}}\psi_1^l\prod\psi_j^{k_j}\Big]\Big]=0.$$

Performing Laplace transforms on the x_i's, we get the recursion formula in the above theorem which implies both the KdV equations and the Virasoro constraints. For example the Virasoro constraints states that the generating series

$$\tau(\tilde{t}) = \exp\sum_{g=0}^{\infty}\langle\exp\sum_{n}\tilde{t}_n\tilde{\sigma}_n\rangle_g$$

satisfies the equations:

$$L_n\cdot\tau = 0, \qquad (n\geq -1)$$

where L_n denote the Virasoro differential operators

$$L_{-1} = -\frac{1}{2}\frac{\partial}{\partial\tilde{t}_0}+\sum_{k=1}^{\infty}(k+\frac{1}{2})\tilde{t}_k\frac{\partial}{\partial\tilde{t}_{k-1}}+\frac{1}{4}\tilde{t}_0^2$$

$$L_0 = -\frac{1}{2}\frac{\partial}{\partial\tilde{t}_1}+\sum_{k=0}^{\infty}(k+\frac{1}{2})\tilde{t}_k\frac{\partial}{\partial\tilde{t}_k}+\frac{1}{16}$$

$$L_n = -\frac{1}{2}\frac{\partial}{\partial\tilde{t}_{n-1}}+\sum_{k=0}^{\infty}(k+\frac{1}{2})\tilde{t}_k\frac{\partial}{\partial\tilde{t}_{k+n}}+\frac{1}{4}\sum_{i=1}^{n}\frac{\partial^2}{\partial\tilde{t}_{i-1}\partial\tilde{t}_{n-i}}$$

We remark the same method can be used to derive very general recursion formulas in Hodge integrals and general Gromov-Witten invariants. We hope to report these results on a later occasion.

11. Final Remarks

We strongly believe that there is a more interesting and grand duality picture between Chern-Simons invariants for three dimensional manifolds and the Gromov-Witten invariants for open toric Calabi-Yau manifolds. We hope such a duality picture will also help us solve the counting problems of higher genus curves in compact Calabi-Yau manifolds. Our proofs of the Mariño-Vafa formula, and the setup of the mathematical foundation for topological vertex theory and the results of others we have discussed above all together have just opened a small window for a more splendid picture. We can certainly expect more exciting conjectures from such duality to stimulate more developments in mathematics.

References

1. M. Aganagic, A. Klemm, M. Marino, C. Vafa, *The topological vertex*, preprint, hep-th/0305132.
2. M. Aganagic, M. Marino, C. Vafa, *All loop topological string amplitudes from Chern-Simons theory*, preprint, hep-th/0206164.
3. A. Bertram, I. Ciocan-Fontanine and B. Kim, *Two proofs of a conjecture of Hori and Vafa*, math.AG/0304403.
4. A. Bertram, I. Ciocan-Fontanine and B. Kim, *Gromov-Witten Invariants for Abelian and Nonabelian Quotients*, math.AG/0407254.
5. K. Behrend and B. Fantechi, *The intrinsic normal cone*, Invent. Math. **128** (1997), 45–88.
6. P.Candelas, X.C. de la Ossa, P.S. Green, L. Parkes *An exactly soluble superconformal theory from a mirror pair of Calabi-Yau manifolds*, Phys. Lett. B 258 (1991), no. 1-2, 118–126.
7. D.A. Cox, *The functor of a smooth toric variety*, Tôhoku Math. J. **(2)47** (1995), 251–262.
8. T. Ekedahl, S. Lando, M. Shapiro, A. Vainshtein, *Hurwitz numbers and intersections on moduli spaces of curves.* Invent. Math. **146** (2001), no. 2, 297–327.
9. C. Faber and R. Pandharipande, *Hodge integrals and Gromov-Witten theory*, Invent. Math. **139(1)** (2000), 173–199.
10. C. Faber and R. Pandharipande, *Hodge integrals, partition matrices, and the λ_g conjecture*, Ann. of Math. (2) **157** (2003), no. 1, 97–124.
11. A. Givental, *Equivariant Gromov-Witten invariants*, Internat. Math. Res. Notices 1996, no. 13, 613–663.
12. I.P. Goulden, D.M. Jackson, *Combinatorial enumeration*, John Wiley & Sons, 1983.
13. I.P. Goulden, D.M. Jackson, A. Vainshtein, *The number of ramified coverings of the sphere by the torus and surfaces of higher genera*, Ann. of Comb. **4** (2000), 27-46.
14. T. Graber, R. Vakil, *Hodge integrals and Hurwitz numbers via virtual localization*, Compositio Math. **135** (2003), no. 1, 25–36.

15. H. Hori and C. Vafa, *Mirror symmetry*, hep-th/0002222.
16. A. Iqbal, *All genus topological amplitudes and 5-brane webs as Feynman diagrams*, preprint, hep-th/0207114.
17. S. Katz, C.-C. Liu, *Enumerative geometry of stable maps with Lagrangian boundary conditions and multiple covers of the disc*, Adv. Theor. Math. Phys. **5** (2001), 1-49.
18. Y. Kim, *Computing Hodge integrals with one lambda-class*, preprint, math-ph/0501018.
19. Y. Kim, K. Liu *A simple proof of Witten conjecture through localization*, preprint, math.AG/0508384 .
20. M. Kontsevich, *Intersection theory on the moduli space of curves and the matrix Airy function.* Comm. Math. Phys. **147** (1992), no. 1, 1–23.
21. A. M. Li, G. Zhao, Q. Zheng, *The number of ramified coverings of a Riemann surface by Riemann surface*, Comm. Math. Phys. **213** (2000), no. 3, 685–696.
22. J. Li, *Stable Morphisms to singular schemes and relative stable morphisms*, J. Diff. Geom. **57** (2001), 509-578.
23. J. Li, *Relative Gromov-Witten invariants and a degeneration formula of Gromov-Witten invariants*, J. Diff. Geom. **60** (2002), 199-293.
24. J. Li, *Lecture notes on relative GW-invariants*, preprint.
25. J. Li and G. Tian, *Virtual moduli cycles and Gromov-Witten invariants of algebraic varieties*, J. Amer. Math. Soc. **11** (1998), 119–174.
26. J. Li, C.-C. Liu, K. Liu, J. Zhou, *A mathematical theory of the topological vertex* , preprint, math.AG/0411247.
27. J. Li, K. Liu, J. Zhou, *Topological string partition functions as equivariant indices*, preprint, math.AG/0412089.
28. B. Lian, C.-H. Liu, K. Liu, S.-T. Yau, *The S^1-fixed points in quot-schemes and mirror principle computations*, Contemp. Math., 322, 2003, 165-194.
29. B. Lian, K. Liu, S.-T. Yau, *Mirror Principle I*, Asian J. Math., 1, 1997, 729-763.
30. B. Lian, K. Liu, and S.-T. Yau, *Mirror principle, II*, Asian J. Math. **3** (1999), 109–146;
31. B. Lian, K. Liu, S.-T. Yau, *Mirror Principle III*, Asian J. Math., 3, 1999, 771-800.
32. B. Lian, K. Liu, and S.-T. Yau, *Mirror principle, IV*, math.AG/0007104.
33. B. Lian, K. Liu, and S.-T. Yau, *A survey of mirror principle*, math.AG/0010064.
34. C.-H. Liu, K. Liu, and S.-T. Yau, *On A-twisted moduli stack for curves from Witten's gauged linear sigma models*, math.AG/0212316.
35. C.-H. Liu, K. Liu, and S.-T. Yau, *S^1-fixed-points in hyper-Quot-schemes and an exact mirror formula for flag manifolds from the extended mirror principle diagram*, math.AG/0401367.
36. C.-C. Liu, *Formulae of one-partition and two-partition Hodge integrals*, preprint, math.AG/0502430.
37. C.-C. Liu, K. Liu, J. Zhou, *On a proof of a conjecture of Mariño-Vafa on Hodge Integrals*, Math. Res. Letters 11, 259-272(2004).
38. C.-C. Liu, K. Liu, J. Zhou, *A proof of a conjecture of Mariño-Vafa on Hodge*

Integrals, J. Differential Geometry, 65(2003) 289-340.
39. C.-C, Liu, K. Liu, J. Zhou, *A formula of two-partition Hodge integrals*, preprint, math.AG/0310272.
40. C.-C, Liu, K. Liu, J. Zhou, *Mariño-Vafa formula and Hodge integral identities*, Journal of Algebraic Geometry 2005.
41. K. Liu, *Mathematical results inspired by physics*, Proc. ICM 2002, Vol. III, 457-466.
42. K. Liu, *Modular Invariance and Characteristic Numbers*, Commun. Math. Phys. **174** (1995), 29–42.
43. I.G. MacDonald, *Symmetric functions and Hall polynomials*, 2nd edition.Claredon Press, 1995.
44. M. Mariño, C. Vafa, *Framed knots at large N*, Orbifolds in mathematics and physics (Madison, WI, 2001), 185–204, Contemp. Math., 310, Amer. Math. Soc., Providence, RI, 2002.
45. M. Mirzakhani, *Simple geodesics and Weil-Petersson volumes of moduli spaces of bordered Riemann surfaces*, preprint, 2003.
46. A. Okounkov, R. Pandharipande, *Gromov-Witten theory, Hurwitz numbers, and Matrix models, I*, preprint, math.AG/0101147
47. A. Okounkov, R. Pandharipande, *Hodge integrals and invariants of the unknots*, preprint, math.AG/0307209.
48. P. Peng, *A simple proof of Gopakumar-Vafa conjecture for local toric Calabi-Yau manifolds*, preprint, math.AG/0410540.
49. E. Witten, *Quantum field theory and the Jones polynomial*, Commun. Math. Phys. **121** (1989) 351-399.
50. E. Witten, *Two-dimensional gravity and intersection theory on moduli space.* Surveys in differential geometry (Cambridge, MA, 1990), 243–310, Lehigh Univ., Bethlehem, PA, 1991.
51. J. Zhou, *Hodge integrals, Hurwitz numbers, and symmetric groups*, preprint, math.AG/0308024.
52. J. Zhou, *A conjecture on Hodge integrals*, preprint.
53. J. Zhou, *Localizations on moduli spaces and free field realizations of Feynman rules*, preprint.

Topologization of electron liquids with Chern-Simons theory and quantum computation*

Zhenghan Wang
Microsoft Project Q c/o Kavli Institute for Theoretical Physics
University of California, Santa Barbara, CA 93106
& Department of Mathematics
Indiana University, Bloomington, IN 47405
U.S.A.
E-Mail: zhewang@indiana.edu;zhenghwa@microsoft.com

1. Introduction

In 1987 a Geometry and Topology year was organized by Prof. Chern in Nankai and I participated as an undergraduate from the University of Science and Technology of China. There I learned about M. Freedman's work on 4-dimensional manifolds. Then I went to the University of California at San Diego to study with M. Freedman in 1989, and later became his most frequent collaborator. It is a great pleasure to contribute an article to the memory of Prof. Chern based partially on some joint works with M. Freedman and others. Most of the materials are known to experts except some results about the classification of topological quantum field theories (TQFTs) in the end. This paper is written during a short time, so inaccuracies are unavoidable. Comments and questions are welcome.

There are no better places for me to start than the Chern-Simon theory. In the hands of Witten, the Chern-Simons functional is used to define TQFTs which explain the evaluations of the Jones polynomial of links at certain roots of unity. It takes great imagination to relate the Chern-Simons theory to electrons in magnetic fields, and quantum computing. Nevertheless, such a nexus does exist and I will outline this picture. No attempt has been made regarding references and completeness.

*The author is partially supported by NSF grant DMS-034772 and EIA 0130388.

2. Chern-Simons theory and TQFTs

Fix a simply connected compact Lie group G. Given a closed oriented 3-manifold M and a connection A on a principle G-bundle P over M, the Chern-Simons 3-form $\mathrm{tr}(A \wedge dA + \frac{2}{3}A^3)$ is discovered when Profs. Chern and Simons tried to derive a purely combinatorial formula for the first Pontrjagin number of a 4-manifold. Let $CS(A) = \frac{1}{8\pi^2}\int_M \mathrm{tr}(A \wedge dA + \frac{2}{3}A^3)$ be the Chern-Simons functional. To get a TQFT, we need to define a complex number for each closed oriented 3-manifold M which is a topological invariant, and a vector space $V(\Sigma)$ for each closed oriented 2-dimensional surface Σ. For a level $k \geq h^\vee + 1$, where $h^\vee$ is the dual Coxeter number of G, the 3-manifold invariant of M is the path integral $Z_k(M^3) = \int_A e^{2\pi i \cdot k \cdot CS(A)} DA$, where the integral is over all gauge-classes of connection on P and the measure DA has yet to be defined rigorously. A closely related 3-manifold invariant is discovered rigorously by N. Reshetikhin and V. Turaev based on quantum groups. To define a vector space for a closed oriented surface Σ, let X be an oriented 3-manifold whose boundary is Σ. Consider a principle G-bundle P over X, fix a connection a on the restriction of P to Σ, let $Z_{k,a} = \int_{(A,a)} e^{2\pi i \cdot k \cdot CS(A)} DA$, where the integral is over all gauge-classes of connections of A on P over X whose restriction to Σ is a. This defines a functional on all connections $\{a\}$ on the principle G-bundle P over Σ. By forming formal finite sums, we obtain an infinite dimensional vector space $S(\Sigma)$. In particular, a 3-manifold X such that $\partial X = \Sigma$ defines a vector in $S(\Sigma)$. Path integral on disks introduces relations onto the functionals, we get a finitely dimensional quotient of $S(\Sigma)$, which is the desired vector space $V(\Sigma)$. Again such finitely dimensional vector spaces are constructed mathematically by N. Reshetikhin and V. Turaev. The 3-manifold invariant of closed oriented 3-manifolds and the vectors spaces associated to the closed oriented surfaces form part of the Witten-Reshetikhin-Turaev-Chern-Simons TQFT based on G at level$=k$. Strictly speaking the 3-manifold invariant is defined only for framed 3-manifolds. This subtlety will be ignored in the following.

Given a TQFT and a closed oriented surface Σ with two connected components $\overline{\Sigma_1}, \Sigma_2$, where $\overline{\Sigma_1}$ is Σ_1 with the opposite orientation, a 3-manifold X with boundary $\partial X = \Sigma$ gives rise to a linear map from $V(\Sigma_1)$ to $V(\Sigma_2)$. Then the mapping cylinder construction for self-diffeomorphisms of surfaces leads to a projective representation of the mapping class groups of surfaces. This is the TQFT as axiomatized by M. Atiyah. Later G. Moore and N. Seiberg, K. Walker and others extended TQFTs to surfaces with boundaries. The new ingredient is the introduction of labels for the boundaries of surfaces. For the Chern-Simons TQFTs, the labels are the irreducible representations of the quantum deformation groups of G at level=k or the positive energy representations of the loop groups of G at level=k. For more details and references, see [T].

3. Electrons in a flatland

Eighteen years before the discovery of electron, a graduate student E. Hall was studying Electricity and Magnetism using a book of Maxwell. He was puzzled by a paragraph in Maxwell's book and performed an experiment to test the statement. He disproved the statement by discovering the so-called Hall effect. In 1980, K. von Klitzing discovered the integer quantum Hall effect (IQHE) which won him the 1985 Nobel Prize. Two years later, H. Stormer, D. Tsui and A. Gossard discovered the fractional quantum Hall effect (FQHE) which led to the 1998 Nobel Prize for H. Stormer, D. Tsui and R. Laughlin. They were all studying electrons in a 2-dimensional plane immersed in a perpendicular magnetic field. Laughlin's prediction of the fractional charge of quasi-particles in FQHE electron liquids is confirmed by experiments. Such quasi-particles are anyons, a term introduced by F. Wilczek. Braid statistics of anyons are deduced, and experiments to confirm braid statistics are being pursued.

The quantum mechanical problem of an electron in a magnetic field was solved by L. Landau. But the fact that there are about 10^{11} electrons per cm^2 for FQHE liquids makes the solution of the realistic Hamiltonian for such electron systems impossible, even numerically. The approach in condensed matter physics is to write down an effective theory which describes the universal properties of the electron systems. The electrons are strongly interacting with each other to form an incompressible electron liquid when the FQHE could be observed. Landau's solution for a single electron in a magnetic field shows that quantum mechanically an electron behaves like a harmonic oscillator. Therefore its energy is quantized to Landau levels. For a finite size sample of a 2-dimensional electron system in a magnetic field, the number of electrons in the sample divided by the number of flux quanta in the perpendicular magnetic field is called the Landau filling fraction ν. The state of an electron system depends strongly on the Landau filling fraction. For $\nu < 1/5$, the electron system is a Wigner crystal: the electrons are pinned at the vertices of a triangular lattice. When ν is an integer, the electron system is an IQHE liquid, where the interaction among electrons can be neglected. When ν are certain fractions such as $1/3, 1/5, \ldots$, the electrons are in a FQHE state. Both IQHE and FQHE are characterized by the quantization of the Hall resistance $R_{xy} = \nu^{-1}\frac{h}{e^2}$, where e is the electron charge and h the Planck constant, and the exponentially vanishing of the longitudinal resistance R_{xx}. There are about 50 such fractions and the quantization of R_{xy} is reproducible up to 10^{-10}. How could an electron system with so many uncontrolled factors such as the disorders, sample shapes and variations of the magnetic field strength, quantize so precisely? The IQHE has a satisfactory explanation both physically and mathematically. The mathematical explanation is based on non-commutative Chern classes. For the FQHE at filling fractions with odd denominators, the composite fermion theory based on U(1)-Chern-Simons theory is a great success: electrons combined with vortices to form composite fermions and then composite fermions, as new particles, to form their own integer quantum Hall liquids. The exceptional case is the observed FQHE $\nu = 5/2$. There are still very interesting questions about this FQH state. For more details and references see [G].

4. Topologization of electron liquids

The discovery of the fractional quantum Hall effect has cast some doubts on Landau theory for states of matter. A new concept, topological order, is proposed by Xiao-gang Wen of MIT. It is believed that the electron liquid in a FQHE state is in a topological state with a Chern-Simons TQFT as an effective theory. In general topological states of matter have TQFTs as effective theories. The $\nu = 5/2$ FQH electron liquid is still a puzzle. The leading theory is based on the Pfaffian states proposed by G. Moore and N. Read in 1991 [MR]. In this theory, the quarsi-particles are non-abelian anyons (a.k.a. plectons) and the non-abelian statistics is described by the Chern-Simons-SU(2) TQFT at level=2.

To describe the new states of matter such as the FQH electron liquids, we need new concepts and methods. Consider the following Gedanken experiment: suppose an electron liquid is confined to a closed oriented surface Σ, for example a torus. The lowest energy states of the system form a Hilbert space $V(\Sigma)$, called the ground states manifold. In an ordinary quantum system, the ground state will be unique, so $V(\Sigma)$ is 1-dimensional. But for topological states of matter, the ground states manifold is often degenerate (more than 1-dimensional), i.e. there are several orthogonal ground states with exponentially small energy differences. This ground states degeneracy is a new quantum number. Hence a topological quantum system assigns each closed oriented surface Σ a Hilbert space $V(\Sigma)$, which is exactly the rule for a TQFT. FQH electron liquid always has an energy gap in the thermodynamic limit which is equivalent to the incompressibility of the electron liquid. Therefore the ground states manifold is stable if controlled below the gap. Since the ground states manifold has the same energy, the Hamiltonian of the system restricted to the ground states manifold is 0, hence there will be no continuous evolutions. This agrees with the direct Lengendre transform form the Chern-Simons Lagrangians to Hamiltoninans. Since the Chern-Simons 3-form has only first derivatives, the corresponding Hamiltonian is identically 0. In summary, ground states degeneracy, energy gap and the vanishing of the Hamiltonian are all salient features of topological quantum systems.

Although the Hamiltonian for a topological system is identically 0, there are still discrete dynamics induced by topological changes. In this case the Schrodinger equation is analogous to the situation for a function $f(x)$ such that $f'(x) = 0$, but there are interesting solutions if the domain of $f(x)$ is not connected as then $f(x)$ can have different constants on the connected components. This is exactly why braid group representations arise as dynamics of topological quantum systems.

5. Anyons and braid group representations

Elementary excitations of FQH liquids are quasi-particles. In the following we will not distinguish quasi-particles from particles. Actually it is not inconceivable that particles are just quasi-particles from some complicated vacuum systems. Particle types serve as the labels for TQFTs. Suppose a topological quantum system confined on a surface Σ has elementary excitations localized at certain points $p_1, p_2, \cdots$ on Σ, the ground states of the system outside some small neighborhoods of p_i form a Hilbert space. This Hilbert space is associated to the surface with the small neighborhoods of p_i deleted and each resulting boundary circle is labelled by the corresponding particle type. Although there are no continuous evolutions, there are discrete evolutions of the ground states induced by topological changes such as the mapping class groups of Σ which preserve the boundaries and their labels. An interesting case is the mapping class groups of the disk with n punctures—the famous braid groups on n-strands, B_n.

Another way to describe the braid groups B_n is as follows: given a collection of n particles in the plane $\mathbb{R}^2$, and let $I = [t_0, t_1]$ be a time interval. Then the trajectories of the particles will be n disjoint curves in $\mathbb{R}^2 \times I$ if at any moment the n particles are kept apart from each other. If the n particles at time t_1 return to their initial positions at time t_0 as a set, then their trajectories form an n-braid σ. Braids can be stacked on top of each other to form the braid groups B_n. Suppose the particles can be braided adiabatically so that the quantum system would be always in the ground states, then we have a unitary transformation from the ground states at time t_0 to the ground states at time t_1. Let $V(\Sigma)$ be the Hilbert space for the ground states manifold, then a braid induces a unitary transformation on $V(\Sigma)$. Actually those unitary transformations give rise to a projective representation of the braid groups. If the n particles are of the same type, the resulting representations of the braid groups will be called the braid statistics. Note that there is a group homomorphism from the braid group B_n to the permutation groups S_n by remembering only the initial and final positions of the n particles.

The plane $\mathbb{R}^2$ above can be replaced by any space X and statistics can be defined for particles in X similarly. The braid groups are replaced by the fundamental groups $B_n(X)$ of the configuration spaces $C_n(X)$. If $X = \mathbb{R}^m$ for some $m > 2$, it is well known that $B_n(X)$ is S_n. Therefore, all particle statistics for particles in $X = \mathbb{R}^m$ will be given by representations of the permutation groups. There are two irreducible 1-dimensional representations of S_n, which correspond to bosons and fermions. If the statistics does not factorize through the permutation groups S_n, the particles are called anyons. If the images are in $U(1)$, the anyon will be called abelian, and otherwise non-abelian. The quasi-particles in the FQH liquid at $\nu = 1/3$ are abelian anyons. To be directly useful for topological quantum computing, we need non-abelian anyons. Do non-abelian anyons exist?

Mathematically are there unitary representations of the braid groups? There are many representations of the braid groups, but unitary ones are not easy to find. The most famous representations of the braid groups are probably the Burau representation discovered in 1936, which can be used to define the Alexander polynomial of links, and the Jones representation discovered in 1981, which led to the Jones polynomial of links. It is only in 1984 that the Burau representation was observed to be unitary by C. Squier, and the Jones representation is unitary as it was discovered in a unitary world [J1]. So potentially there could be non-abelian anyon statistics. An interesting question is: given a family of unitary representations of the braid groups $\rho_n : B_n \to U(k_n)$, when this family of representations can be used to simulate the standard quantum circuit model efficiently and fault tolerantly? A sufficient condition is that they come from certain TQFTs with some density on the braid group representation images, but is it necessary?

Are there non-abelian anyons in Nature? This is an important unknown question at the writing. Experiments are underway to confirm the prediction of the existence in certain FQH liquids [DFN]. Specifically the FQH liquid at $\nu = 5/2$ is believed to have non-abelian anyons whose statistics is described by the Jones representation at the 4-th root of unity. More generally N. Read and E. Rezayi conjectured that the Jones representation of the braid groups at r-th root of unity describes the non-abelian statistics for FQH liquids at filling fractions $\nu = 2 + \frac{k}{k+2}$, where $k = r - 2$ is the level [RR]. For more details and references on anyons see [Wi].

As an anecdote, a few years ago I wrote an article with others about quantum computing using non-abelian anyons and submitted it to the journal Nature. The paper was rejected within almost a week with a statement that the editors did not believe in the existence of non-abelian anyons. Fortunately the final answer has to come from Mother Nature, rather than the journal Nature.

6. Topological quantum computing

In 1980s Yu. Manin and R. Feynman articulated the possibility of computing machines based on quantum physics to compute much faster than classical computers. Shor's factoring algorithm in 1994 has dramatically changed the field and stirred great interests in building quantum computers. There are no theoretical obstacles for building quantum computers as the accuracy threshold theorem has shown. But decoherence and errors in implementing unitary gates have kept most experiments to just a few qubits. In 1997 M. Freedman proposed the possibility of TQFT computing [F]. Independently A. Kitaev proposed the idea of fault tolerant quantum computing using anyons [K]. The two ideas are essentially equivalent as we have alluded before. Leaving aside the issue of discovering non-abelian anyons, we may ask how to compute using non-abelian anyons? For more details and references see [NC].

6.1. *Jones representation of the braid groups*

Jones representation of the braid groups is the same as the Witten-Reshetikhin-Turaev-SU(2) TQFT representation of the braid groups. Closely related theories can be defined via the Kaffuman bracket. For an even level k, the two theories are essentially the same, but for odd levels the two theories are distinguished by the Frobenius-Schur indicators. However the resulting braid group representations are the samc. Therefore we will describe the braid group representations using the Kauffman bracket. The Kauffman bracket is an algebra homomorphism from the group algebras of the braid groups $\mathbb{C}[B_n]$ to the generic Temperley-Lieb algebras. For applications to quantum computing we need unitary theories. So we specialize the Kauffman variable A to certain roots of unity. The resulting algebras are reducible. Semi-simple quotients can be obtained by imposing the Jones-Wenzl idempotents. The semi-simple quotient algebras will be called the Jones algebras, which are direct sum of matrix algebras. Fix r and an A satisfying $A^4 = e^{\pm 2\pi i/r}$, the Jones representation for a braid σ is the Kauffman bracket image in the Jones algebra. To describe the Jones representation, we need to find the decomposition of the Jones algebras into their simple matrix components (irreducible sectors). The set of particle types for the Chern-Simons-SU(2) TQFT at level=k is $L = \{0, 1, \cdots, k\}$. The fusion rules are given by $a \otimes b = \oplus c$, where a, b, c satisfy

1). the sum $a + b + c$ is even,

2). $a + b \geq c, b + c \geq a, c + a \geq b$,

3). $a + b + c \leq 2k$.

A triple $(a, b, c), a, b, c \in L$ satisfying the above three conditions will be called admissible.

The Jones algebra at level=k for n-strands decomposes into irreducible sectors labeled by an integer m such that $m \in L, m = n \; mod \; 2$. Fix m, the irreducible sector has a defining representation $V_{1^n}^m$ with a basis consisting of admissible labelings of the following tree (Fig. 6.1):

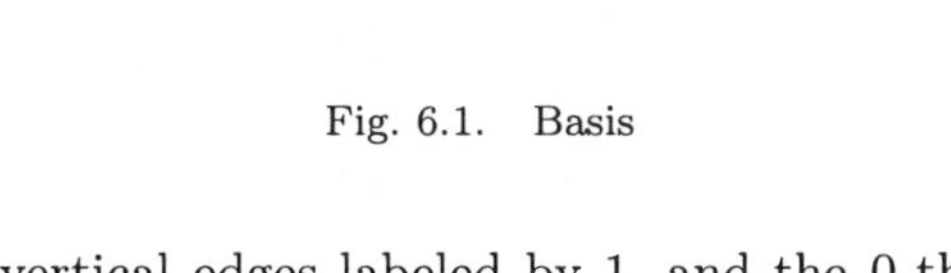

Fig. 6.1. Basis

There are n vertical edges labeled by 1, and the 0-th horizontal edge (leftmost) is always labeled by 0, and the n-th edge (rightmost) is always labeled by m. The internal $(n-1)$ edges are labeled by $a, b, c, \cdots$ such that any three labels incident to a trivalent vertex form an admissible triple. A basis with internal labelings $a, b, c, \cdots$ will be denoted by $e^m_{a,b,c,\cdots}$. The Kauffman bracket is $\sigma_i = A \cdot \mathrm{id} + A^{-1} \cdot U_i$, so it suffices to describe the matrix for U_i with basis $e^m_{a,b,c,\cdots}$ in $V^m_{1^n}$. The matrix for U_i consists of 1×1 and 2×2 blocks. Fix m and a basis element $e^m_{a,b,c,\cdots}$, suppose that the $i, i+1, i+2$ internal edges are labeled by f, g, h. If $f \neq h$, then U_i maps this basis to 0. If $f = h$, then by the fusions rules $g = f \pm 1$ (the special case is $f = 0$, then $g = 1$ only), then U_i maps $e^m_{\cdots,f,f\pm1,f,\cdots}$ back to themselves by the following 2×2 matrix:

$$\begin{pmatrix} \frac{\Delta_{f+1}}{\Delta_f} & x \\ y & \frac{\Delta_{f-1}}{\Delta_f} \end{pmatrix},$$

where Δ_k is the Chebyshev polynomial defined by $\Delta_0 = 1, \Delta_1 = d, \Delta_{k+1} = d\Delta_k + \Delta_{k-1}, d = -A^2 - A^{-2}$, and x, y satisfy $xy = \frac{\Delta_{f+1}\Delta_{f-1}}{\Delta_f^2}$.

From those formulas, there is a choice of x, y up to a scalar, and in order to get a unitary representation, we need to choose A so that the 2×2 blocks are real symmetric matrices. This forces A to satisfy $q = A^4 = e^{\pm 2\pi i/r}$. It also follows that the eigenvalues of σ_i are $-1, q$ up to scalars.

6.2. *Anyonic quantum computers*

We will use the level=2 theory to illustrate the construction of topological quantum computers. There are three particle types $\{0, 1, 2\}$. The label 0 denotes the null-particle type, which is the vacuum state. Particles of type 1 are believed to be non-abelian anyons. Consider the unitary Jones representation of B_4, the irreducible sector with $m = 0$ has a basis $\{e^0_{1,b,1}\}$, where $b = 0$ or 2. Hence this can be used to encode a qubit. For B_6, a basis consists of $e^0_{1,b_1,1,b_2,1}$, where $b_i, i = 1, 2$ is 0 or 2. Hence this can be used to encode 2-qubits. In general n-qubits can be encoded by the $m = 0$ irreducible sector of the Jones representation ρ^0_{2n+2} of B_{2n+2}. The unitary matrices of the Jones representations $\rho^0_4(B_4), \rho^0_4(B_6)$ will be quantum gates. To simulate a quantum circuit on n-qubits $U_L : (\mathbb{C}^2)^{\otimes n} \to (\mathbb{C}^2)^{\otimes n}$, we need a braid $\sigma \in B_{2n+2}$ such that the following diagram commutes:

$$\begin{array}{ccc} (\mathbb{C}^2)^{\otimes n} & \longrightarrow^{\cong} & V^0_{1^{2n+2}} \\ U_L \downarrow & & \downarrow \rho^0_{2n+2} \\ (\mathbb{C}^2)^{\otimes n} & \longrightarrow_{\cong} & V^0_{1^{2n+2}} \end{array}$$

This is not always possible because the images of the Jones representation of the braid groups at $r = 4$ are finite groups. It follows that the topological model at $r = 4$ is not universal. To get a universal computer, we consider other levels of the Chern-Simons-SU(2) TQFT. The resulting model for $r = 4$ is slightly different from the above one. To simulate n-qubits, we consider the braid group B_{4n}. The 4n edges besides the leftmost in Fig. 6.1 can be divided into n groups of 4. Consider the basis elements such every 4k-th edge is labelled by 0, and every (4k+2)-th edge can be labeled either by 0 or 2. Those 2^n basis elements will be used to encode n-qubits. The representations of the braid groups B_{4n} will be used to simulate any quantum circuits on n-qubits. This is possible for any level other than 1,2 and 4 [FLW1][FLW2].

6.3. *Measurement in topological models*

A pictorial illustration of a topological quantum computer is as follows (Fig. 6.2):

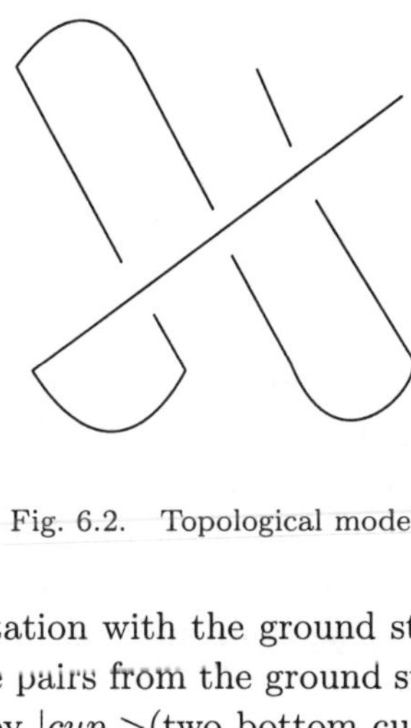

Fig. 6.2. Topological model

We start the computation with the ground states of a topological system, then create particle pairs from the ground states to encode the initial state which is denoted by $|cup>$(two bottom cups). A braid b is adiabatically performed to induce the desired unitary matrix $\rho(b)$. In the end, we annihilate the two leftmost quasi-particles (the top cap) and record the particle types of the fusion. Then we repeat the process polynomially many times to get an approximation of the probability of observing any particle type. Actually we need only to distinguish the trivial versus all other non-trivial particle types. For level=3 or $r = 5$, the probability to observe the trivial particle type 0 is $< cap|\rho^+(b) \prod_0 \rho(b)|cup >$, which is related to the Jones polynomial of the following circuit link (Fig. 6.3) by the formula:

$$p = \text{prob}(0) = \frac{1}{1+[2]^2}\left(1 + \frac{(-1)^c \cdot V_L(e^{2\pi i/5})}{[2]^{c-2}}\right),$$

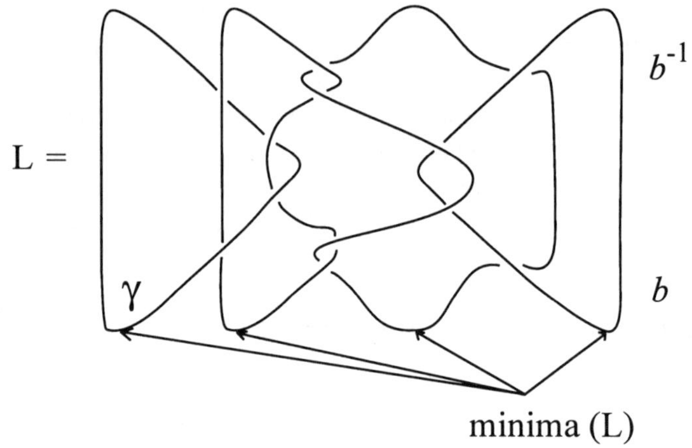

Fig. 6.3. Circuit link

where in the formula $c = c(L)$ is the number of components of the link L, $[2] = -A^2 - A^{-2}$ is quantum 2 ar $r = 5$. Our normalization for the Jones polynomial is that for the unlink with c components, the Jones polynomial is $(-[2])^c$.

To derive this formula, we assume the writhe of L is 0. Other cases are similar. In the Kauffman bracket formulation, the projector to null particle type $\prod_0$ is the same as the element $\frac{U_1}{-[2]}$ of the Jones algebras. It follows that p is just the Kauffman bracket of the tangle $b \cdot U_1 \cdot b^{-1}$ divided by $-[2]$. Now consider the Kaffman bracket $< L >$ of L, resolving the 4 crossings of L on the component γ using the Kauffman bracket results a sum of 16 terms. Simplifying, we get

$$< L >= (-[2])^c([2]^2 - 3) + (4 - [2]^2)(-[2])^c \cdot p.$$

Since the writhe is assumed to be 0, the Kauffman bracket is the same as the Jones polynomial of L. Solving for p, we obtain

$$p = \frac{3 - [2]^2}{4 - [2]^2}(1 + \frac{(-1)^c \cdot V_L(e^{2\pi i/5})}{[2]^c \cdot (3 - [2]^2)}).$$

Direct calculation using the identity $[2]^2 = 1 + [2]$ gives the desired formula. This formula shows that if non-abelian anyons exist to realize the Jones representation of the braid groups, then quantum computers will approximate the Jones polynomial of certain links. So the Jones polynomial of links are amplitudes for certain quantum processes [FKLW]. This inspired a definition of a new approximation scheme: the additive approximation which might lead to a new characterization of the computational class BQP [BFLW].

6.4. *Universality of topological models*

In order to simulate all quantum circuits, it suffices to have the closed images of the braid groups representations containing the special unitary groups for each representation space. In 1981 when Jones discovered his revolutionary unitary representation of the braid groups, he proved that the images of the irreducible sectors of his unitary representation are finite if $r = 1, 2, 3, 4, 6$ for all n and $r = 10$ for $n = 3$. For all other cases the closed images are infinite modulo center. He asked what are the closed images? In the joint work with M. Freedman, and M. Larsen [FLW2], we proved that the closed images are as large as they can be: always contain the special unitary groups. As a corollary, we have proved the universality of the anyonic quantum computers for $r \neq 1, 2, 3, 4, 6$.

The proof is interesting in its own right as we formulated a two-eigenvalue problem and found its solution [FLW2]. The question of understanding TQFT representations of the mapping class groups are widely open. Partial results are obtained in [LW].

6.5. *Simulation of TQFTs*

In another joint work with M. Freedman, and A. Kitaev [FKW], we proved that any unitary TQFT can be efficiently simulated by a quantum computer. Combined with the universality for certain TQFTs, we established the equivalence of TQFT computing with quantum computing. As corollaries of the simulation theorem, we obtained quantum algorithms for approximating quantum invariants such as the Jones polynomial. Jones polynomial is a specialization of the Tutte polynomial of graphs. It is interesting to ask if there are other partition functions in statistical mechanics such as the Potts models that can be approximated by quantum computers efficiently [Wel].

6.6. *Fault tolerance of topological models*

Anyonic quantum computers are inherently fault tolerant [K]. This is essentially a consequence of the disk axiom of TQFTs if the TQFTs can be localized to lattices on surfaces. Localization of TQFTs can also be used to establish an energy gap rigorously.

7. Classification of topological states of matter

Topological orders of FQH electron liquids are modelled by TQFTs. It is an interesting and difficult problem to classify all TQFTs, hence topological orders. In 2003 I made a conjecture that if the number of particle types is fixed, then there are only finitely many TQFTs. The best approach is based on the concept of modular tensor category (MTC) [T][BK]. A modular tensor category encodes the algebraic data inside a TQFT, and describes the consistency of an anyonic system. Modular tensor category might be a very useful concept to study topological quantum systems. In 2003 I gave a lecture at the American Institute of Mathematics to an audience of mostly condensed matter physicists. It was recognized by one of the participants, Prof. Xiao-gang Wen of MIT, that indeed tensor category is useful for physicists as his recent works have shown.

Recently S. Belinschi, R. Stong, E. Rowell and myself have achieved the classification of all MTCs up to 4 labels. The result has not been written up yet, but the list is surprisingly short. Each fusion rule is realized by either a Chern-Simons TQFT and its quantum double. For example, the fusion rules of self-dual, singly generated modular tensor categories up to rank=4 are realized by: $SU(2)$ level=1, $SO(3)$ level=3, $SU(2)$ level=2, $SO(3)$ level 5, $SU(2)$ level=3. It follows from the Ocneanu rigidity that my finiteness conjecture holds for ranks up to 4.

8. Open questions

There are many open problems in the subject and directions to pursue for mathematicians, physicists and computer scientists. We just mention a few here. The most important for the program is whether or not there are non-abelian anyons in Nature. Another question is to understand the boundary (1+1) quantum field theories of topological quantum systems. Most of the boundary QFTs are conformal field theories. What is the relation of the boundary QFT with the bulk TQFT? How do we classify them?

Quantum mechanics has been incorporated into almost every physical theory in the last century. Mathematics is experiencing the same now. Wavefunctions may well replace the digital numbers as the new notation to describe our world. The nexus among quantum topology, quantum physics and quantum computation will lead to a better understanding of our universe, and Prof. Chern would be happy to see how important a role that his Chern-Simons theory is playing in this new endeavor.

References

BK. B. Bakalov and A. Kirillov Jr., *Lectures on tensor categories and modular functions*, Amer. Math. Soc., Providence, RI, 2001.

BFLW. M. Bordewich, M. Freedman, L. Lovsz, and D. Welsh, *Approximate Counting and Quantum Computation*, Combinatorics, Probability and Computing archive Volume 14 , Issue 5-6 (November 2005).

DFN. S. Das Sarma, M. Freedman, and C. Nayak, *Topologically-Protected Qubits from a Possible Non-Abelian Fractional Quantum Hall State*, Phys. Rev. Lett. 94, 166802 (2005).

F. M. Freedman, *P/NP, and the quantum field computer.* Proc. Natl. Acad. Sci. USA 95 (1998), no. 1, 98101.

FKW. M. H. Freedman, A. Kitaev, and Z. Wang, *Simulation of topological field theories by quantum computers,* Comm. Math. Phys. 227 (2002), no.3, 587—603.

FKLW. M. H. Freedman, A. Kitaev, M. J. Larsen and Z. Wang, *Topological quantum computation, Mathematical challenges of the 21st century (Los Angeles, CA, 2000)*, Bull. Amer. Math. Soc. (N.S.) **40** (2003), no. 1, 31–38.

FLW1. M. H. Freedman, M. J. Larsen, and Z. Wang, *A modular functor which is universal for quantum computation,* Comm. Math. Phys. 227 (2002),no.3,605—622.

FLW2. M. H. Freedman, M. J. Larsen, and Z. Wang, *The two-eigenvalue problem and density of Jones representation of braid groups,* Comm. Math. Phys. **228** (2002), 177-199, arXiv: math.GT/0103200.

G. S. Girvin, *The quantum Hall effect: novel excitation and broken symmetries*, Topological aspects of low dimensional systems (Los Houches - Ecole d'Ete de Physique Theorique) (Hardcover) by A. Comtet (Editor), cond-mat/9907002.

J1. V. F. R. Jones, *Braid groups, Hecke algebras and type* II_1 *factors,* Geometric methods in operator algebras (Kyoto, 1983), 242–273, Pitman Res. Notes Math. Ser., 123, Longman Sci. Tech., Harlow, 1986.

K. A. Kitaev, *Fault-tolerant quantum computation by anyons,* Annals Phys. 303 (2003) 2-30 ,quant-ph/9707021

LW. M. J. Larsen and Z. Wang, *Density of the SO(3) TQFT representation of mapping class groups*, Comm. Math. Phys., Volume **260**(2005), Number 3, 641 - 658.

MR. G. Moore and N. Read, *Nonaelions in the fractional quantum Hall effect*, Nuc. Physc. B360(1991), 362-396.

NC. M. Nielsen and I. Chuang, *Quantum Computation and quantum information*, Cambridge University Press, 2000.

RR. N. Read and E. Rezayi, *Beyond paired quantum Hall states: parafermions and incompressible states in first excited Landau level*, Phys. Rev. B59, 8804(1999), cond-mat/9809384.

T. V. G. Turaev, *Quantum Invariants of Knots and 3-Manifolds*, de Gruyter Studies in Mathematics, 18. Walter de Gruyter & Co., Berlin, 1994.

Wel. D. Welsh, *Complexity: Knots, Colourings and Counting*, LMS Lecture Notes Series 186.

Wi. F. Wilczek, *Braid statistics and anyon superconductivity*, World Scientific Pub Co Inc (December, 1990).

Invited Contributions

Quasicrystals: Projections of 5-d Lattice into 2 and 3 Dimensions

Helen AU-YANG and Jacques H. H. PERK

Department of Physics,
Oklahoma State University,
Stillwater, OK 74078-3072, USA
E-mail: perk@okstate.edu

We show that generalized Penrose tilings can be obtained by the projection of a cut plane of a 5-dimensional lattice into two dimensions, while 3-d quasiperiodic lattices with overlapping unit cells are its projections into 3d. The frequencies of all possible vertex types in the generalized Penrose tilings, and the frequencies of all possible types of overlapping 3-d unit cells are also given here. The generalized Penrose tilings are found to be nonconvertable to kite and dart patterns, nor can they be described by the overlapping decagons of Gummelt.

1. Introduction

Quasicrystals, though originally introduced as a mathematical curiosity, have become an object of intense study by physicists and mathematicians following the startling discovery in 1984 of five- or ten-fold symmetry in diffraction patterns off certain alloys.[1] Quasicrystals have been studied most often by filling the space aperiodically with nonoverlapping tiles, such as in Penrose tilings.[2–4] However, in the mid 1990s, Gummelt[5] proposed a new description of the regular Penrose tiling in terms of the overlapping of decorated decagons. Further research[6–9] has shown that this may be a more sensible way to understand quasicrystalline materials—made of overlapping unit cells sharing atoms of nearby neighbors.[7]

We shall use de Bruijn's multigrid method to produce a new example of 3-dimensional overlapping unit cells.[10] Moreover, we shall use the pentagrid method to obtain generalized Penrose tilings, which cannot be converted to kite and dart patterns, nor do they satisfy the inflation and deflation rules. Therefore, since Conway's cartwheels, which are in fact the overlapping decagons of Gummelt, are constructed from kite and dart patterns,[3] they cannot be used to describe the generalized Penrose tilings.

2. Grids and the 'Cut and Projection Method'

It is well-known that a Penrose tiling can be obtained by the projection of a particularly 'cut' slice of the 5-d euclidian lattice onto a 2-d plane $\mathcal{D}$,[4,11,12] and that its diffraction pattern,[13–15] therefore, has five- or ten-fold symmetry. It is also known that not all lattice points $\boldsymbol{k}$ in $\mathbb{Z}^5$ can be mapped onto vertices of a Penrose tiling; only those points in a particular 'cut' slice whose projections into the 3-dimensional orthogonal space $\mathcal{W}$ are inside the window of acceptance,[11,16] contribute. The window has been shown[11] to be the projection of the 5-d unit cell Cu(5) with 2^5 vertices into this 3-d space $\mathcal{W}$.

If $\boldsymbol{d}_j$ are the generators of the plane $\mathcal{D}$ and $\boldsymbol{w}_j$ are the generators of its orthogonal space $\mathcal{W}$, then the projection operators are the matrices

$$\boldsymbol{D}^{\mathrm{T}} = (\boldsymbol{d}_0, \cdots, \boldsymbol{d}_4), \quad \boldsymbol{W}^{\mathrm{T}} = (\boldsymbol{w}_0, \cdots, \boldsymbol{w}_4) \tag{2.1}$$

satisfying $\boldsymbol{D}^{\mathrm{T}}\boldsymbol{W} = \boldsymbol{W}^{\mathrm{T}}\boldsymbol{D} = 0$, where the superscript T denotes matrix transposition. More specifically, we choose

$$\boldsymbol{d}_j^{\mathrm{T}} = (\cos j\theta, \sin j\theta), \quad \boldsymbol{w}_j^{\mathrm{T}} = (\cos 2j\theta, \sin 2j\theta, 1) = (\boldsymbol{d}_{2j}^{\mathrm{T}}, 1), \tag{2.2}$$

where $j = 0, \cdots, 4$ and $\theta = 2\pi/5$. Using notations and ideas introduced by de Bruijn,[4] we consider the 2-d or 3-d pentagrid consisting of five grids of either equidistant lines given by

$$x \cos j\theta + y \sin j\theta + \gamma_j = \boldsymbol{d}_j^T \boldsymbol{r} + \gamma_j = k_j, \quad \boldsymbol{r}^{\mathrm{T}} = (x, y), \tag{2.3}$$

or equidistant planes defined by

$$x \cos 2j\theta + y \sin 2j\theta + z + \gamma_j = \boldsymbol{w}_j^{\mathrm{T}} \boldsymbol{R} + \gamma_j = k_j, \quad \boldsymbol{R}^{\mathrm{T}} = (x, y, z), \tag{2.4}$$

for $j = 0, \cdots, 4$, and with the five $k_j \in \mathbb{Z}$. In (2.3) and (2.4), the γ_j are real numbers which shift the grids from the origin. We denote their sum by

$$\gamma_0 + \gamma_1 + \gamma_2 + \gamma_3 + \gamma_4 = c, \quad 0 \le c < 1. \tag{2.5}$$

Without loss of generality, we may restrict c to $0 \le c < 1$, as we can see from (2.3) that $c \to c - n$ if we let $k_0 \to k_0 + n$. Obviously, such a relabeling cannot change the 2-d or 3-d quasiperiodic patterns.

It has been shown by de Bruijn[4] that the Penrose tiling associated with a 2-d pentagrid has simple matching rules only for $c = 0$. In other words, for $0 < c < 1$ the corresponding generalized Penrose tilings do not satisfy simple matching rules, and have different sets of vertices for different intervals of c.[17] Nevertheless, the diffraction patterns are believed to be the same for all values of c.[18,19]

Let the integer k_j be assigned to all points sandwiched between the grid lines or planes defined by $k_j - 1$ and k_j. This k_j can be found by

$$K_j(\boldsymbol{r}) = \lceil \boldsymbol{d}_j^{\mathrm{T}} \boldsymbol{r} + \gamma_j \rceil, \quad \forall \boldsymbol{r} \in \mathbb{R}^2 \tag{2.6}$$

$$\tilde{K}_j(\boldsymbol{R}) = \lceil \boldsymbol{w}_j^{\mathrm{T}} \boldsymbol{R} + \gamma_j \rceil, \quad \forall \boldsymbol{R} \in \mathbb{R}^3 \tag{2.7}$$

for $j = 0, \cdots, 4$, (and $\lceil x \rceil$ is the smallest integer greater than or equal to x). A mesh in $\mathbb{R}^2$ is an interior area, enclosed by grid lines, containing points with the same five integers $K_j(\boldsymbol{r})$, while a mesh in $\mathbb{R}^3$ is now an interior volume, enclosed by grid planes, containing points with the same five integers $\tilde{K}_j(\boldsymbol{R})$. One then maps each mesh in $\mathbb{R}^2$ to a vertex in $\mathcal{D}$ by

$$\boldsymbol{f}(\boldsymbol{r}) = \sum_{j=0}^{4} K_j(\boldsymbol{r}) \boldsymbol{d}_j = \boldsymbol{D}^{\mathrm{T}} \boldsymbol{K}(\boldsymbol{r}), \quad \boldsymbol{K}^{\mathrm{T}}(\boldsymbol{r}) = (K_0(\boldsymbol{r}), \cdots, K_4(\boldsymbol{r})), \tag{2.8}$$

and each mesh in $\mathbb{R}^3$ to a vertex in $\mathcal{W}$ by

$$\boldsymbol{g}(\boldsymbol{R}) = \sum_{j=0}^{4} \tilde{K}_j(\boldsymbol{R}) \boldsymbol{w}_j = \boldsymbol{W}^{\mathrm{T}} \tilde{\boldsymbol{K}}(\boldsymbol{R}), \quad \tilde{\boldsymbol{K}}^{\mathrm{T}}(\boldsymbol{R}) = (\tilde{K}_0(\boldsymbol{R}), \cdots, \tilde{K}_4(\boldsymbol{R})). \tag{2.9}$$

The resulting sets of vertices $\mathcal{I} = \{\boldsymbol{f}(\boldsymbol{r}) | \boldsymbol{r} \in \mathbb{R}^2\}$ and $\mathcal{L} = \{\boldsymbol{g}(\boldsymbol{R}) | \boldsymbol{R} \in \mathbb{R}^3\}$ are, respectively, the two- and three-dimensional quasiperiodic lattices.

3. Window of Acceptance

Given a point $\mathbf{k}^T = (k_0, \ldots, k_4)$ in the five-dimensional lattice,* one may ask whether there is a mesh in the pentagrid (or the 3-d multigrid) such that $K_j(\boldsymbol{r}) = k_j$ (or $\tilde{K}_j(\boldsymbol{R}) = k_j$) for $j = 0, \ldots, 4$. As seen from (2.6) (or (2.7)), this is equivalent to asking whether it is possible to find points $\boldsymbol{r}$ in $\mathbb{R}^2$ (or $\boldsymbol{R}$ in $\mathbb{R}^3$), and points $\boldsymbol{\lambda}^{\mathrm{T}} = (\lambda_0, \ldots, \lambda_4)$ with $0 \leq \lambda_j < 1$, such that

$$\boldsymbol{D}\boldsymbol{r} + \boldsymbol{\gamma} + \boldsymbol{\lambda} = \mathbf{k},, \qquad (\boldsymbol{W}\boldsymbol{R} + \boldsymbol{\gamma} + \boldsymbol{\lambda} = \mathbf{k}), \tag{3.1}$$

where $\boldsymbol{\gamma}^{\mathrm{T}} = (\gamma_0, \cdots, \gamma_4)$ and where $\boldsymbol{\lambda}$ lies inside the 5-d unit cube Cu(5). Whenever (3.1) holds, the point $\mathbf{k}$ in $\mathbb{Z}^5$ is said to satisfy the mesh condition. Since $\boldsymbol{W}^T \boldsymbol{D} = \boldsymbol{D}^T \boldsymbol{W} = 0$, the above equations become

$$\boldsymbol{W}^T [\mathbf{k} - \gamma] = \boldsymbol{W}^T \boldsymbol{\lambda}, \tag{3.2}$$

$$\boldsymbol{D}^T [\mathbf{k} - \gamma] = \boldsymbol{D}^T \boldsymbol{\lambda}, \tag{3.3}$$

*For a formulation for more general cases, see Ref. 11.

such that $\boldsymbol{D}^T\boldsymbol{k} \in \mathcal{I}$ if (3.2) holds, or $\boldsymbol{W}^T\boldsymbol{k} \in \mathcal{L}$ if (3.3) holds. Thus, $\boldsymbol{W}^T\boldsymbol{\lambda}$ is the window of acceptance for projections into 2d and $\boldsymbol{D}^T\boldsymbol{\lambda}$ for 3d. They are respectively the interiors of the convex hulls of the points $\boldsymbol{W}^{\mathrm{T}}\boldsymbol{n}_i$ and $\boldsymbol{D}^{\mathrm{T}}\boldsymbol{n}_i$, where the $\boldsymbol{n}_i$ are the 2^5 vertices of the 5-d unit cube Cu(5).

We choose the 32 $\boldsymbol{n}_i$'s as follows

$$\begin{aligned}
&\mathbf{n}_0^T=(0,0,0,0,0),\ \mathbf{n}_1^T=(1,0,0,0,0),\ \mathbf{n}_2^T=(0,0,0,1,0),\ \mathbf{n}_3^T=(0,1,0,0,0),\\
&\mathbf{n}_4^T=(0,0,0,0,1),\ \mathbf{n}_5^T=(0,0,1,0,0),\ \mathbf{n}_6^T=(1,0,0,1,0),\ \mathbf{n}_7^T=(0,1,0,1,0),\\
&\mathbf{n}_8^T=(0,1,0,0,1),\ \mathbf{n}_9^T=(0,0,1,0,1),\ \mathbf{n}_{10}^T=(1,0,1,0,0),\ \mathbf{n}_{11}^T=(1,1,0,0,0),\\
&\mathbf{n}_{12}^T=(0,0,0,1,1),\ \mathbf{n}_{13}^T=(0,1,1,0,0),\ \mathbf{n}_{14}^T=(1,0,0,0,1),\ \mathbf{n}_{15}^T=(0,0,1,1,0),\\
&\mathbf{n}_{16}^T=(1,1,0,0,1),\ \mathbf{n}_{17}^T=(0,0,1,1,1),\ \mathbf{n}_{18}^T=(1,1,1,0,0),\ \mathbf{n}_{19}^T=(1,0,0,1,1),\\
&\mathbf{n}_{20}^T=(0,1,1,1,0),\ \mathbf{n}_{21}^T=(1,1,0,1,0),\ \mathbf{n}_{22}^T=(0,1,0,1,1),\ \mathbf{n}_{23}^T=(0,1,1,0,1),\\
&\mathbf{n}_{24}^T=(1,0,1,0,1),\ \mathbf{n}_{25}^T=(1,0,1,1,0),\ \mathbf{n}_{26}^T=(1,1,0,1,1),\ \mathbf{n}_{27}^T=(0,1,1,1,1),\\
&\mathbf{n}_{28}^T=(1,1,1,0,1),\ \mathbf{n}_{29}^T=(1,0,1,1,1),\ \mathbf{n}_{30}^T=(1,1,1,1,0),\ \mathbf{n}_{31}^T=(1,1,1,1,1).
\end{aligned}\tag{3.4}$$

The projection of these 32 points into $\mathcal{W}$ is a polytope $\mathcal{P}$ having 20 faces and 40 edges connecting the 22 vertices, as is shown in Fig. 3.1. We let $\boldsymbol{P}_i = \boldsymbol{W}^{\mathrm{T}}\boldsymbol{n}_i$ for $i = 0, \cdots, 31$. The bottom is $\boldsymbol{P}_0 = (0,0,0)$ and top is $\boldsymbol{P}_{31} = (0,0,5)$; they are called the tips of the polytope. The remaining twenty vertices of $\mathcal{P}$ are

$$\begin{aligned}
\boldsymbol{P}_{j+1} &= (\boldsymbol{d}_j, 1), \quad \boldsymbol{P}_{j+6} = (\boldsymbol{d}_j + \boldsymbol{d}_{j+1}, 2),\\
\boldsymbol{P}_{j+21} &= (-\boldsymbol{d}_{j-2} - \boldsymbol{d}_{j-1}, 3), \quad \boldsymbol{P}_{j+26} = (-\boldsymbol{d}_{j-1}, 4),
\end{aligned}\tag{3.5}$$

for $j = 0, \cdots, 4$. The other 10 points $\boldsymbol{P}_{11}, \cdots, \boldsymbol{P}_{20}$ are in the interior of the polytope and are given by

$$\boldsymbol{P}_{11+j} = (\boldsymbol{d}_j + \boldsymbol{d}_{j+2}, 2), \quad \boldsymbol{P}_{16+j} = (-\boldsymbol{d}_{j+1} - \boldsymbol{d}_{j-1}, 3), \tag{3.6}$$

again for $j = 0, \cdots, 4$.

The orthogonal projection of the 32 points $\boldsymbol{n}_i$ into $\mathcal{D}$ is a decagon $\mathcal{Q}$ with 10 edges connecting the 10 vertices. Let $\boldsymbol{Q}_i = \boldsymbol{D}^{\mathrm{T}}\boldsymbol{n}_i$, for $i = 0, \cdots, 31$. Then the vertices of the decagon are

$$\boldsymbol{Q}_{11+j} = -p\boldsymbol{d}_{3-2j}, \quad \boldsymbol{Q}_{16+j} = p\boldsymbol{d}_{5-2j}, \tag{3.7}$$

with $j = 0, \cdots, 4$, and $p = (\sqrt{5}+1)/2$ is the golden ratio. The remaining 22 points $\boldsymbol{Q}_0, \cdots, \boldsymbol{Q}_{10}, \boldsymbol{Q}_{21}, \cdots, \boldsymbol{Q}_{31}$ are in the interior; they are given by

$$\begin{aligned}
\boldsymbol{Q}_0 = \boldsymbol{Q}_{31} = 0, \quad \boldsymbol{Q}_{j+1} &= \boldsymbol{d}_{5-2j}, \quad \boldsymbol{Q}_{26+j} = -\boldsymbol{d}_{2-2j},\\
\boldsymbol{Q}_{j+6} = p^{-1}\boldsymbol{d}_{4-2j}&, \quad \boldsymbol{Q}_{21+j} = -p^{-1}\boldsymbol{d}_{3-2j}.
\end{aligned}\tag{3.8}$$

The decagons are shown in Fig. 3.2. Thus if orthogonal projection $D^{\mathrm{T}}(\boldsymbol{k}-\gamma)$ is in $\mathcal{Q}$, then its projection $\boldsymbol{W}^{\mathrm{T}}\boldsymbol{k}$ is in $\mathcal{L}$.

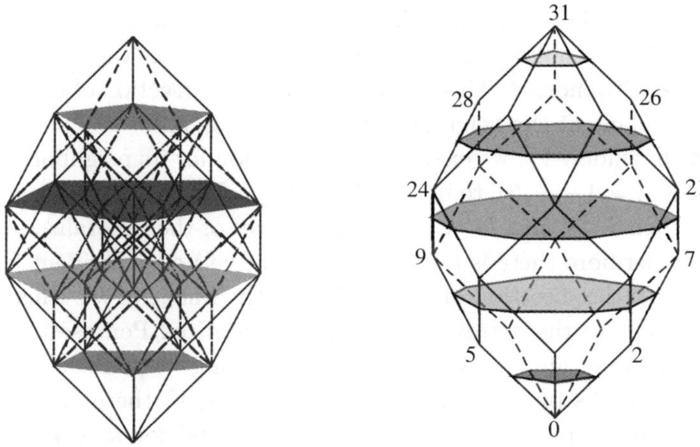

(a) $c = 0$ (b) $c \neq 0$

Fig. 3.1. The projection of the 5-dimensional unit cube into the orthogonal 3-space $\mathcal{W}$. The polytopes with 22 vertices are tilted 10 degree with respect to the vertical, so that the intersections V_I with the planes $z = I - c$ can be seen. In (a), for $c = 0$, we show the projection of the 32 points, 10 of which are in the interior, and the V_I are all pentagons. In (b), for $c \neq 0$, the V_I are pentagons for $I = 1, 5$, and decagons for $I = 2, 3, 4$.

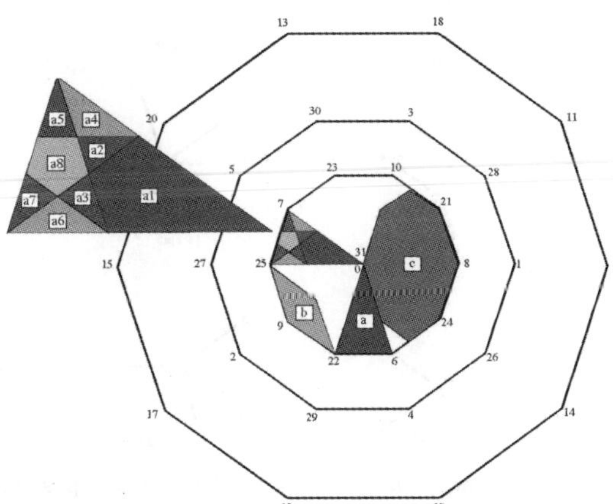

Fig. 3.2. The projection of the 5-d unit cube Cu(5) into the orthogonal 2-d space $\boldsymbol{D}$. The window is a decagon $\mathcal{Q}$ whose vertices are given by (3.7). Those n_i which are mapped to interior points (vertices) of $\mathcal{P}$ in Fig. 3.1, are mapped into the boundary vertices of $\mathcal{Q}$.

4. Generalized Penrose Tilings

Using (2.1) and (2.2), we may rewrite the three components of (3.2) as

$$\sum_{j=0}^{4}(k_j - \gamma_j) = I - c = \sum_{j=0}^{4}\lambda_j, \quad \sum_{j=0}^{4}(k_j - \gamma_j)\boldsymbol{d}_{2j} = \sum_{j=0}^{4}\lambda_j \boldsymbol{d}_{2j}, \tag{4.1}$$

where $I \equiv \sum k_j$ is the index of $\mathbf{k}$, an integer in the interval $[1, 5]$ for $0 < c < 1$. ($I = 5$ does not occur for $c = 0$.) Eq. (4.1) defines the window V_I for accepting $\mathbf{k}$ with index I. This window V_I is the intersection of the polytope $\mathcal{P}$ with the plane at the height $I - c$ shown in Fig. 3.1.

For $\boldsymbol{k}$ in window V_I, we examine the condition for its neighbor $\boldsymbol{k}'$, (with $\boldsymbol{k}' = \boldsymbol{k} \pm \boldsymbol{n}_j$, $j = 1, \cdots, 5$), to be in window $V_{I\pm1}$. Whenever this condition is satisfied, then $\boldsymbol{D}^T\boldsymbol{k}$ and $\boldsymbol{D}^T\boldsymbol{k}'$ are both vertices of the generalized Penrose tiling. Furthermore there is a 'positive' ('negative') edge incident from the image of $\boldsymbol{k}$ in the direction of $\boldsymbol{d}_{3j}$ ($-\boldsymbol{d}_{3j}$) to the image of $\boldsymbol{k}'$. This way we can determine all the vertex types of the generalized Penrose tiling for a given c. Denoting all vertices with index I having n 'positive' edges and n' 'negative' edges by $[n, n']_I$, we find that for $\boldsymbol{k} \in V_1$ there are only three kinds of vertex types $[5, 0]_1$, $[4, 0]_1$, and $[3, 0]_1$, and for $\boldsymbol{k} \in V_5$ there are also only three kinds of vertex types $[0, 5]_5$, $[0, 4]_5$, and $[0, 3]_5$, shown in Fig. 4.1(a).

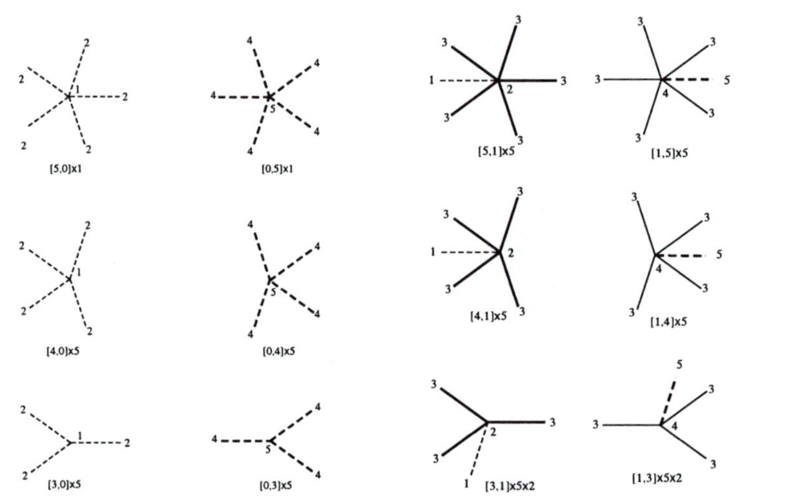

(a) vertex types $[n, 0]_1$ and $[0, n']_5$ (b) vertex types $[n, 1]_2$ and $[1, n']_4$

Fig. 4.1. (a) Edges connecting two sites with indices 1 and 2 are represented by thin dashed lines, while edges connecting sites with indices 5 and 4 are represented by thick dashed lines. (b) A few examples of vertex types $[n, 1]_2$ and $[n, 1]_4$ are given here. Edges connecting sites with indices 2 and 3 are denoted by thick lines, and edges connecting sites with indices 3 and 4 by thin lines. We use $[n, n'] \times 5$ to indicate the 5-fold multiplicity under 72° rotations allowed for the vertex, and $[n, n'] \times 5 \times 2$ to indicate the additional reflection symmetry when it is present.

If the probability of finding a vertex of type $[n, n']_I$ is denoted by

$A_I(n,n')/5p$, then

$$A_1[5,0] = \tfrac{1}{2}(p^{-1}+p)p^{-3}(1-c)^2,$$
$$A_1[4,0] = \tfrac{5}{2}p^{-4}(1-c)^2,\ A_1[3,0] = \tfrac{5}{2}p^{-3}(1-c)^2, \tag{4.2}$$

while $A_5[0,n]$ is given by replacing $1-c$ in $A_1[n,0]$ by c.

There are nine different vertex types for $I = 2,4$, see Fig. 4.1(b) for some examples of each type. Their frequencies are

$$A_2[5,0] = \tfrac{1}{2}(p^{-1}+p)[\theta(p^{-2}-c)(p^{-3}+c)^2 + \theta(c-p^{-2})p^{-4}(2-c)^2],$$
$$A_2[5,1] = \theta(p^{-2}-c)\tfrac{5}{2}(p^{-5}+c)p(p^{-2}-c),$$
$$A_2[5,2] = \theta(p^{-2}-c)\tfrac{5}{2}p^{-1}(p^{-2}-c)^2,$$
$$A_2[4,0] = \theta(c-p^{-2})\tfrac{5}{2}(c-p^{-2})[p^{-1}(1-c)+p^{-3}(2-c)],$$
$$A_2[4,1] = \theta(p^{-2}-c)\tfrac{5}{2}p^{-1}c^2 + \theta(c-p^{-2})\tfrac{5}{2}p^{-3}(1-c)^2,$$
$$A_2[3,2] = \theta(p^{-2}-c)\tfrac{5}{2}p^2(p^{-2}-c)^2,$$
$$A_2[3,1] = 5p^{-2}(1-c)^2 - \theta(p^{-2}-c)5p^2(p^{-2}-c)^2,$$
$$A_2[3,0] = \tfrac{5}{2}c^2 - \theta(c-p^{-2})5(c-p^{-2})^2,$$
$$A_2[2,1] = \tfrac{5}{2}p^{-1}(1-c)^2, \tag{4.3}$$

where $\theta(x)$ is the Heaviside function, *i.e.*, $\theta(x) = 1$ for $x \geq 0$, and zero otherwise. We find that the open interval $0 < c < 1$ is split into two intervals $0 < c < p^{-2}$ and $p^{-2} < c < 1$. Inside the former interval, $A_2(4,0) = 0$, and only eight kinds of vertices are allowed; inside the latter, $A_2(5,2) = A_2(5,1) = A_2(3,1) = 0$, allowing only six vertex types. At the boundary $c = 0$ or $c = p^{-2}$, there are only five allowed vertex types. We find that $A_4[n,n']$ can be obtained from $A_2[n',n]$ by $c \to 1-c$. Now for c in the interval $0 < c < p^{-1}$ there are six nonvanishing vertex types, while inside the interval $p^{-1} < c < 1$, there are eight nonvanishing vertex types.

There are many vertex types $[n,n']_3$. Twelve out of twenty of their frequency functions $A_3[n,n']$ are given as

$$A_3[0,5] = \theta(p^{-3}-c)\tfrac{1}{2}(p^{-1}+p)(p^{-3}-c)^2,$$
$$A_3[1,5] = \theta(p^{-3}-c)\tfrac{5}{2}(p^{-3}-c)^2,$$
$$A_3[2,5] = \theta(p^{-2}-c)\tfrac{5}{2}p^2(p^{-2}-c)^2 - \theta(p^{-3}-c)5p^2(p^{-3}-c)^2,$$
$$A_3[3,5] = \theta(2p^{-3}-c)[\tfrac{5}{2}c^2 - \theta(c-p^{-3})5p^2(c-p^{-3})^2$$
$$+\theta(c-p^{-2})5p^3(c-p^{-2})^2],$$
$$A_3[4,5] = \theta(c-p^{-3})[\theta(p^{-2}+p^{-4}-c)\tfrac{5}{2}p^3(p^{-2}+p^{-4}-c)^2$$
$$-\theta(2p^{-3}-c)5p^3(2p^{-3}-c)^2 + \theta(p^{-2}-c)5p^2(p^{-2}-c)^2],$$
$$A_3[5,5] = \theta(c-p^{-3})[\theta(2p^{-2}-c)\tfrac{1}{2}(p+p^{-1})(2p^{-2}-c)^2$$
$$-\theta(p^{-2}+p^{-4}-c)\tfrac{5}{2}p^3(p^{-2}+p^{-4}-c)^2 + \theta(2p^{-3}-c)\tfrac{5}{2}p^3(2p^{-3}-c)^2],$$
$$A_3[3,4] = \theta(p^{-1}-c)[\tfrac{5}{2}p^{-3}c^2$$
$$-\theta(c-p^{-2})5p(c-p^{-2})^2 + \theta(c-2p^{-3})\tfrac{5}{2}p^3(c-2p^{-3})^2],$$

$$A_3[4,4] = \theta(p^{-1}-c)[\theta(c-p^{-2})5(c-p^{-2})^2 - \theta(c-2p^{-3})5p^3(c-2p^{-3})^2 + \theta(c-p^{-2}-p^{-4})5p^3(c-p^{-2}-p^{-4})^2],$$
$$A_3[3,3] = 5p^{-4}(1-c)^2 - \theta(p^{-1}-c)5p^{-1}(p^{-1}-c)^2 + \theta(p^{-2}-c)5p^{-1}(p^{-2}-c)^2,$$
$$A_3[2,3] = 5p^{-3}(1-c)^2 - \theta(p^{-2}-c)\tfrac{5}{2}p^{-1}(p^{-2}-c)^2,$$
$$A_3[2,2] = 10p^{-1}c(1-c), \quad A_3[1,2] = \tfrac{5}{2}p^{-1}(1-c)^2. \tag{4.4}$$

The remaining eight $A_3[n',n]$ can be obtained from $A_3[n,n']$ by letting $c \to 1-c$. They are continuous functions of c.

We plot in Fig. 5.1 generalized Penrose tilings for $c = p^{-2} = 0.3819660098$ and $c = 0.5$. We find that the number of vertices of index 1 increases, and of index 5 decreases, as c increases.

5. Overlapping polytope

Consider now the projection of $\mathbb{Z}^5$ into the 3-d space $\mathcal{W}$. It is easy to find the conditions for both $\boldsymbol{k}$ and its neighbors $\boldsymbol{k}+\boldsymbol{n}_j$, for $j = 1 \cdots 5$, to satisfy their mesh conditions, so that they are vertices of quasiperiodic lattice $\mathcal{L}$.

We find that every point inside the innermost decagon $\hat{\mathcal{Q}}$ in Fig. 3.2 corresponds to a point in $\mathcal{L}$ that is connected with its 10 neighbors, and is in fact a tip of a polytope. This innermost decagon $\hat{\mathcal{Q}}$ is further divided into 10 triangles. Each point inside a triangle corresponds to a polytope in $\mathcal{L}$ having exactly four interior points which are also in $\mathcal{L}$. Points in the same triangle correspond to polytopes having the same four interiors points, but for different triangles the polytopes have different sets of interior points. Thus each unit cell contains 26 'atoms,' 22 exterior and 4 interior sites.

Each of the triangles in $\hat{\mathcal{Q}}$ is further divided into eight regions shown in Fig. 3.2. The points inside the quadrilateral denoted by (a1) in Fig. 3.2, correspond to a polytope intersecting with four other polytopes and sharing with each a polyhedron $\mathcal{J}$ with six faces; inside the two triangles denoted by (a2) and (a3), each point corresponds to a polytope intersecting with five other polytopes and sharing with one of them a polyhedron $\mathcal{K}$ with twelve faces and with the other four polyhedra $\mathcal{J}$; inside the two other triangles (a4) and (a6), each point corresponds to a polytope intersecting with four neighboring polytopes sharing with one of them a polyhedron $\mathcal{K}$ and with the other three polyhedra $\mathcal{J}$; inside the two remaining triangles (a5) and (a7), a polytope intersects with five other polytopes, sharing with two of them a polyhedron $\mathcal{K}$ and with the other three a polyhedron $\mathcal{J}$; inside the pentagon (a8), a polytope intersects with six other polytopes sharing with two of them a polyhedron $\mathcal{K}$ and with the other four a polyhedron $\mathcal{J}$. Their relative frequencies are related to the ratio of their areas and are

$1 : p^{-3} : p^{-2} : p^{-3} : \frac{1}{2}(p^{-2} + p^{-4})$. These frequencies are independent of c.

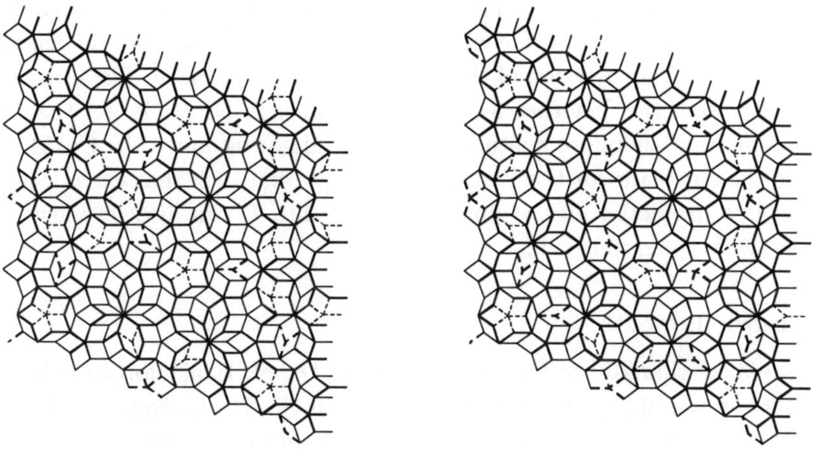

(a) $c = p^{-2}$ (b) $c = 0.5$

Fig. 5.1. Generalized Penrose tilings: There are four kinds of edges. Edges connecting two sites with index 1 and index 2 are represented by a thin dashed line; edges connecting sites with index 4 and index 5 by a thick dashed line; edges connecting sites with index 2 and index 3 by a thick line; and edges connecting sites with index 3 and index 4 by a thin line. Even though no arrows are drawn on the edges, the 'positive' (connecting I to $I+1$ sites) or 'negative' (connecting I to $I-1$ sites) direction of an edge, is completely determined by the indices of the sites at the two ends of an edge.

The 3-d quasiperiodic lattice $\mathcal{L}$ can be further shown to be periodic in the z-direction, which is the direction of the line joining the bottom and the top of the polytopes $\mathcal{P}$, and aperiodic in the xy-directions.[10]

6. Conclusion

The generalized Penrose tilings of thin and fat rhombs cannot be converted to tilings of kites and darts. This can be seen as follows: Four thin rhombs and one fat rhomb is the only way to fit the vertex of type $[3,1]_2$ in Fig. 4.1(b), which can be easily seen to be nonconvertable to a tiling of darts and kites. On the other hand, for $c = 0$, the kite-and-dart patterns of the Penrose tiling[5] can be viewed as single repeating cartwheels,[3] which overlap with their neighbors. These cartwheels are the overlapping quasi-unit-cells of Gummelt,[5–9] and are larger than the decagons which are the projections of the 5-d unit cells onto 2 dimensions.[17] The generalized Penrose tilings are shown to be inequivalent to kite-and-dart patterns, nor do they satisfy the inflation and deflation rules. Therefore, the method of Gummelt cannot be used here. It can be seen from Fig. 5.1 that in the neighborhood of the star vertices $[5,0]_3$ or $[0,5]_4$, only parts of decagons which are the projections of the 5-d unit cells onto 2 dimensions[17] are in $\mathcal{L}$. This is not like the case for $c=0$ or for the preojection of the 5-d lattice onto 3-d space. The difference may be due to the fact that 3-d cut hyperplanes in 5d are larger than 2-d cut planes and therefore contain most of neighboring unit cells Cu(5). For $c=0$, the cut plane for the Penrose tiling is special such that each decagon which is a projection of the unit cell Cu(5) into $\mathcal{D}$ can also be viewed as quasi-overlapping unit cell.

Acknowledgments

We are most thankful to Dr. M. Widom and Dr. M. Baake for providing us with many useful references and to Dr. Molin Ge, Dr. Chengming Bai and the Nankai Institute of Mathematics for their hospitality and support.

References

1. D. Shechtman, I. Blech, D. R. Gratias and J. W. Cahn, *Phys. Rev. Lett.* **53**, 1951–1953 (1984).
2. R. Penrose, in *Introduction to the Mathematics of Quasicrystals*, Aperiodicity and Order, Vol. 2, M. V. Jarić, ed., (Academic, Boston, 1989), pp. 53–79.
3. B. Grünbaum and G. C. Shephard, *Tilings and Patterns*, (W. H. Freeman and Co., New York, 1987), Ch. 10.
4. N. G. de Bruijn, *Indagationes Mathematicae* **84**, 38–52, 53–66 (1981).
5. P. Gummelt, *Geometriae Dedicata* **62**, 1–17 (1996).
6. P. J. Steinhardt and H. C. Jeong, *Nature* **382**, 433–435 (1996).
7. P. J. Steinhardt, H. C. Jeong, K. Saitoheong, M. Tanaka, E. Abe and A. P. Tsai, *Nature* **396**, 55–57 (1996).
8. P. J. Lord and S. Ranganathan, *Acta Crystallogr.* **A57**, 531–539 (2001).
9. P. J. Lord, S. Ranganathan and U. D. Kulkarni, *Philosophical Magazine* **A81**, 2645–2651 (2001).
10. H. Au-Yang and J. H. H. Perk, preprint cond-mat/0507117.
11. N. G. de Bruijn, *Indagationes Mathematicae* **A89**, 123–152 (1986).
12. F. Gähler and J. Rhyner, *J. Phys.* **A19**, 267–277 (1986).
13. V. Elser, *Acta Crystallogr.* **A42**, 36–43 (1986).
14. M. Duneau and A. Katz, *Phys. Rev. Lett.* **54**, 2688–2691 (1985).
15. A. L. Mackay, *Physica* **A114**, 609–613 (1982).
16. M. Baake, D. Joseph, P. Kramer and M. Schlottmann, *J. Phys.* **A23**, L1037–L1041 (1990).
17. H. Au-Yang and J. H. H. Perk, *Generalized Penrose Tilings*, to be published.
18. D. Levine and P. J. Steinhardt, *Phys. Rev.* **B34**, 596–616 (1986).
19. J. E. S. Socolar and P. J. Steinhardt, *Phys. Rev.* **B34**, 617–647 (1986).

Theoretical Analysis of the Double Spin Chain Compound $KCuCl_3$

M. T. Batchelor, X.-W. Guan and N. Oelkers

Department of Theoretical Physics,
Research School of Physical Sciences & Engineering and
Department of Mathematics, Mathematical Sciences Institute,
Australian National University, Canberra ACT 0200, Australia

We investigate thermal and magnetic properties of the double spin chain compound $KCuCl_3$ via an exactly solved ladder model with strong rung interaction. Results from the analysis of the thermodynamic Bethe Ansatz equations suggests the critical field values $H_{c1} = 22.74\,\mathrm{T}$ and $H_{c2} = 51.34\,\mathrm{T}$, in good agreement with the experimental observations. The temperature dependent magnetic properties are directly evaluated from the exact free energy. Good overall agreement is seen between the theoretical and experimental susceptibility curves. Our results suggest that this compound lies in the strong dimerized phase with an energy gap $\Delta \approx 35\,\mathrm{K}$ at zero temperature.

1. Introduction

It is believed that the compounds $KCuCl_3$, $TlCuCl_3$ and NH_4CuCl_3 exhibit a double spin chain structure,[1–13] along the lines of Fig. 1.1 In the double chain structure, coupling constants $J_\perp$ ($J_\parallel$) denote the interchain (intrachain) spin exchange interactions, with J_d a diagonal interaction. However, there appears to be no uniform agreement on the values of these coupling constants for the double chain compounds. In particular, the coupling constants for the compound $KCuCl_3$ are uncertain. Several theoretical models have been proposed to describe this material, including a double chain model with strong antiferromagnetic dimerization,[3,4] a ladder model with additional diagonal interactions[1,2] and a three-dimensional coupled spin-dimer system.[6–10] None of these models provide an overall fit for all thermal and magnetic properties, see, e.g., the review by Dagotto.[14] Measurements of the high field magnetization[5,6] and the susceptibility[1] indicate that $KCuCl_3$ exhibits a singlet ground state with an energy gap $\Delta \approx 31\,\mathrm{K}$ at $T = 1.7\,\mathrm{K}$. Nevertheless, it has been difficult to fix all of the coupling

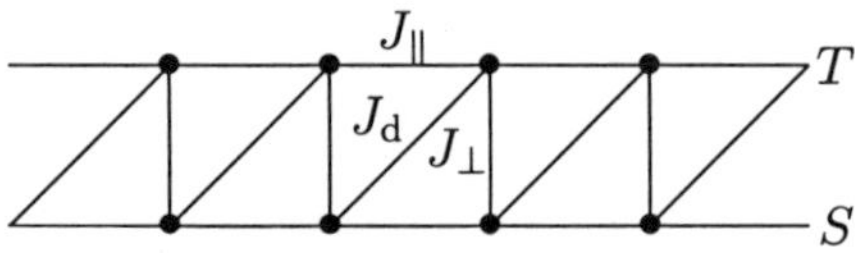

Fig. 1.1. Schematic picture of the structure of double chain compounds such as $KCuCl_3$. Here $J_\perp$ ($J_{\|}$) is the interchain (intrachain) interaction. J_d is the spin exchange interaction in the diagonal direction.

parameters of the model by fitting to only one physical property at a time. At very low temperatures, $T < 5\,K$, the compound $KCuCl_3$ exhibits three-dimensional magnetic ordering due to complex structural magnetic interaction paths.[7–10]

In this communication we investigate the critical fields, magnetization and susceptibility of the compound $KCuCl_3$ via an integrable ladder model. The results are used to examine the values of the coupling constants for the double chain structure. The results for the ladder model with strong rung coupling are seen to be in good agreement with the experimental results for the energy gap, critical fields, susceptibility and magnetization.

2. The integrable ladder model

It has been shown that integrable (exactly solved) ladder models can be used to describe real ladder compounds with strong rung interaction.[15–17] These integrable ladder models enjoy the nice property that thermal and magnetic quantities can be obtained exactly via well developed methods from integrable systems, such as the Thermodynamic Bethe Ansatz (TBA),[18] the Quantum Transfer Matrix (QTM),[19] T-systems[20] and the High Temperature Expansion (HTE) of Non Linear Integral Equations (NLIE).[21–24]

The simplest integrable two-leg spin-$\frac{1}{2}$ ladder model is constructed from the integrable $su(4)$ spin chain with singlet rung interaction. The Hamiltonian is given by[25]

$$\mathrm{H} = J_{\|}\mathrm{H}_{\mathrm{leg}} + J_\perp \sum_{j=1}^{L} \vec{S}_j \cdot \vec{T}_j - \mu_{\mathrm{B}} g H \sum_{j=1}^{L} (S_j^z + T_j^z), \tag{2.1}$$

where

$$\mathrm{H}_{\mathrm{leg}} = \sum_{j=1}^{L} \left(\vec{S}_j \cdot \vec{S}_{j+1} + \vec{T}_j \cdot \vec{T}_{j+1} + 4(\vec{S}_j \cdot \vec{S}_{j+1})(\vec{T}_j \cdot \vec{T}_{j+1}) \right). \tag{2.2}$$

Here L is the number of rungs with $\vec{S}_j = (S_j^x, S_j^y, S_j^z)$ and $\vec{T}_j = (T_j^x, T_j^y, T_j^z)$ spin-$\frac{1}{2}$ operators acting on site j. The Bohr magneton is μ_B and g is the Landé factor. Periodic boundary conditions, $\vec{S}_{L+1} = \vec{S}_1$, $\vec{T}_{L+1} = \vec{T}_1$, are imposed.

In contrast to the standard Heisenberg ladder model the integrable ladder model features an additional biquadratic spin interaction term in the definition (2.2) of H_{leg}. This term causes a shift in the critical value of the rung coupling $J_\perp$ at which the energy gap closes, and it also causes a rescaling of the parameter $J_\parallel$ for strong rung coupling. In the strong coupling limit $J_\perp \gg J_\parallel$ the rung interaction dominates the ground state and low-lying excitations. The integrable model then lies in the same phase as the standard Heisenberg ladder, motivating its analysis.

The ground state properties at zero temperature may be obtained from the TBA equations.[15,26–28] Details of the derivation can be found in Ref. 17. In the strong coupling limit the integrable spin-$\frac{1}{2}$ ladder model exhibits three quantum phases: a gapped phase in the regime $H < H_{c1}$, a fully polarized phase for $H > H_{c2}$ and a Luttinger liquid magnetic phase in the regime $H_{c1} < H < H_{c2}$. The exact values for the critical fields are[15] $H_{c1} = J_\perp - 4J_\parallel$ and $H_{c2} = J_\perp + 4J_\parallel$.

On the other hand, the temperature dependent free energy has been calculated via the exact HTE of the NLIE.[16,17] The free energy of the integrable spin ladder (2.1) is given in the form[16,17]

$$-\frac{1}{T}f(T,H) = \ln Q_1^{(1)} + \sum_{n=1}^{\infty} c_{n,0}^{(1)} \left(\frac{J_\parallel}{T}\right)^n \tag{2.3}$$

where $Q^{(1)}$ and the first few coefficients $c_{n,0}^{(1)}$ are given explicitly in Refs. 16, 17. These terms are functions of the rung coupling $J_\perp$, $\mu_B g H$ and the temperature. Most importantly, the exact expression (2.3) for the free energy can be used to examine physical properties such as the magnetization, susceptibility and magnetic specific heat via the standard thermodynamic relations

$$M = -\left.\frac{\partial f(T,H)}{\partial H}\right|_T, \quad \chi = -\left.\frac{\partial^2 f(T,H)}{\partial H^2}\right|_T, \quad C = -T\left.\frac{\partial^2 f(T,H)}{\partial T^2}\right|_H.$$

3. The compound $KCuCl_3$

In this section we examine the low temperature properties of the compound $KCuCl_3$. Experimental measurements of the high field magnetization[5,6] show that magnetic anisotropies are negligible, because the critical fields are almost the same for the external field in different directions. However, the susceptibility curves for the external magnetic field along the different directions are influenced by different g-factors.[1] In this way magnetic anisotropies may lead to different critical fields for external magnetic fields along different directions. This can be easily seen from the TBA analysis. For instance, if the rung interaction along the z-axis is increased, i.e., by adding an extra term $\Delta_z = \sum_{j=1}^{L} S_j^z T_j^z$ to the rung interaction, the critical fields for the magnetic field along the z-direction are given by

$$\begin{aligned} H_{c1} &= J_\perp + \frac{1}{2}\Delta_z - 4J_\parallel, \\ H_{c2} &= J_\perp + \frac{1}{2}\Delta_z + 4J_\parallel. \end{aligned} \tag{3.1}$$

For the magnetic field along the x-direction they are given by

$$\begin{aligned} H_{c1} &= \sqrt{(J_\perp + \frac{1}{2}\Delta_z - 4J_\parallel)(J_\perp - 4J_\parallel)}, \\ H_{c2} &= \sqrt{(J_\perp + \frac{1}{2}\Delta_z + 4J_\parallel)(J_\perp + 4J_\parallel)}. \end{aligned} \tag{3.2}$$

The experimental results[1,5,6] indicate that Δ_z is negligible. Analysis of such anisotropic behaviour can be found in Ref. 28. We therefore take the high field magnetization curves for the external field along the perpendicular and parallel directions to the cleavage plane as evidence that the double chain ladder model is magnetically isotropic along the chain direction. In the strong coupling case two components of the triplet never contribute to the ground state at zero temperature, due to the strong single component contribution along the rungs. It has been suggested[29] that the triplet excitation can be considered as an analogue of Bose-Einstein condensation for magnons[30–33] for this class of compounds. The strongly coupled spin ladder with magnon excitations for strong magnetic fields can be mapped to a one-dimensional XXZ-Heisenberg chain with an effective magnetic field. In this case the TBA equations reduce to only one level. The experimental magnetization curves[5,6] suggest an energy gap $\Delta \approx 31.1\,\mathrm{K}$ and the critical field values $H_{c1} \approx 20\,\mathrm{T}$ and $H_{c1} \approx 50\,\mathrm{T}$ at $T = 1.3\,\mathrm{K}$. Fitting the zero temperature TBA critical fields and susceptibility to the experimental curves[1] gives the coupling constants $J_\parallel = 5.5\,\mathrm{K}$ and $J_\perp = 57\,\mathrm{K}$ for the integrable spin ladder model (2.1).

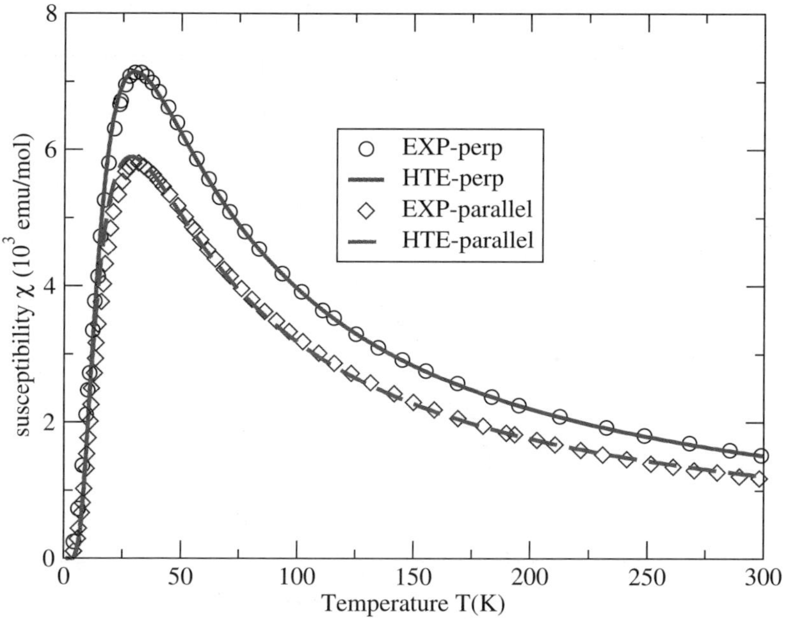

Fig. 3.1. Comparison between theoretical and experimental susceptibility curves versus temperature for the compound $KCuCl_3$. Circles and diamonds denote the experimental data extracted from Ref. 1 for an external field perpendicular or parallel to the chain direction. The solid and dashed curves are the corresponding susceptibility curves evaluated directly from the HTE at $H = 0\,\mathrm{T}$. Fitting results in the coupling constants $J_\perp = 57\,\mathrm{K}$ and $J_\| = 5.5\,\mathrm{K}$ with $g = 2.29$ (perpendicular), $g = 2.05$ (parallel) and $\mu_B = 0.672$ K/T. The conversion constant is $\chi_{\mathrm{HTE}} \approx 0.40615\chi_{\mathrm{EXP}}$ fixed in Ref. 15.

3.1. *Susceptibility*

The application of the HTE (2.3) for the free energy of Hamiltonian (2.1) indicates that the coupling constants $J_\| = 5.5\,\mathrm{K}$ and $J_\perp = 57\,\mathrm{K}$ also give excellent fits to the susceptibility. The temperature dependence of the susceptibility curves is shown in Fig. 3.1. The solid and dashed lines denote the susceptibility for the external field perpendicular and parallel to the double chain direction, as derived from the free energy expression (2.3) with up to fifth order HTE. Here the Landé factors $g = 2.29$ (perpendicular) and $g = 2.05$ (parallel) for the external field direction were used. A rounded peak at $T = 28.5\,\mathrm{K}$ in the zero magnetic field susceptibility curve indicates typical antiferromagnetic behaviour. The overall agreement with the experimental susceptibility curves is excellent. The susceptibility

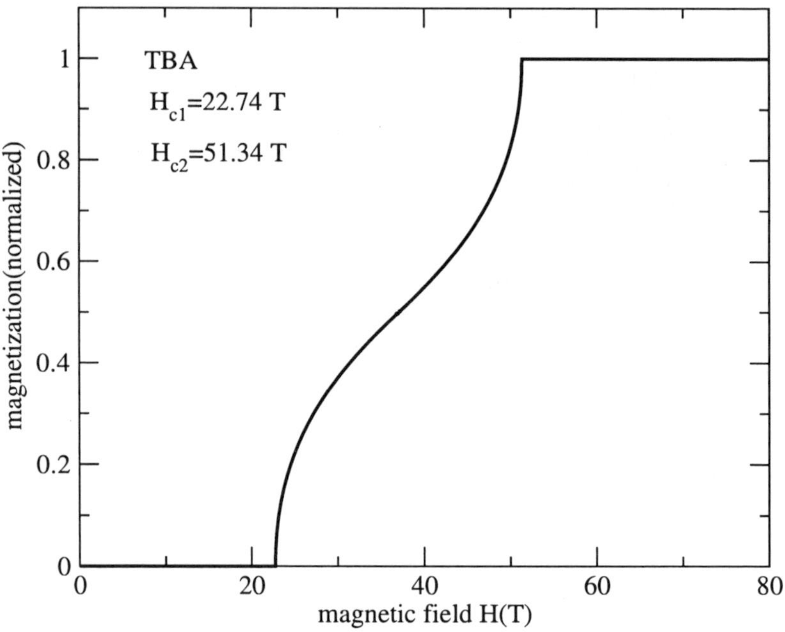

Fig. 3.2. Magnetization versus magnetic field for the compound $KCuCl_3$ with the same constants as in Fig. 3.1. This curve indicates the nature of the high field quantum phase diagram. The stiffness in the vicinities of the critical fields H_{c1} and H_{c2} is softened by increasing temperature. The critical fields predicted by the TBA are in good agreement with the experimental values.

for the external field parallel to the chain direction has been examined via different theoretical models.[3] Their conclusion favours a dimerized Heisenberg ladder structure with additional diagonal spin interactions, with the suggested coupling constants $J_{\|} = J_{\rm d} = 8.35\,{\rm K}$ and $J_{\perp} = 50.1\,{\rm K}$ for the double chain structure compound. However, their fitting constants result in an energy gap $\Delta \approx 38\,{\rm K}$, which is much larger than the experimental value. We conclude that it is not necessary to introduce diagonal spin exchange interaction due to the strong dimerization along the rungs. The diagonal spin exchange interaction has only a weak effect on the low temperature behaviour. Moreover, the leg interaction is also suppressed by the relatively strong rung dimerization.

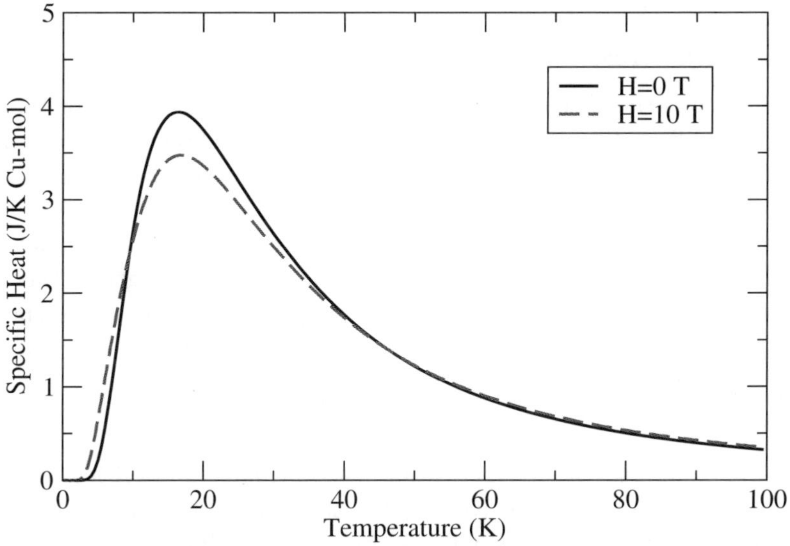

Fig. 3.3. Specific heat versus temperature for different magnetic field strengths for the compound $KCuCl_3$ with the same set of coupling constants $J_\perp$, $J_\parallel$ and $g = 2.29$. The solid and dashed curves are evaluated from the HTE for $H = 0\,\mathrm{T}$ and $H = 10\,\mathrm{T}$. The conversion constant is $C_{\mathrm{HTE}} \approx 4.515 C_{\mathrm{EXP}}$.

3.2. *Magnetization*

The magnetization is a particular interesting quantity to study as the field dependent magnetization curve leads to the prediction of the low temperature phase diagram as well as magnetization plateaux. The high field magnetization curve evaluated from the TBA at zero temperature is shown in Fig. 3.2. By the nature of the high temperature expansion, we are unable to produce these very low temperature, $T < 5.5\,\mathrm{K}$, magnetization curves from the free energy (2.3) for this particular compound. This highlights the complementary role of the TBA and HTE approaches. The magnetization curve indicates that the rung singlets form a nonmagnetic ground state if the magnetic field is less than the critical field value $H_{c1} = 22.74\,\mathrm{T}$. The gap closes at this critical point. If the magnetic field is above the critical point, the lower component of the triplet becomes involved in the ground state. The magnetization increases almost linearly with the field towards the critical point $H_{c2} = 51.34\,\mathrm{T}$, at which the ground state becomes fully polarized. This is in good agreement with the experimental values $H_{c1} \approx 20\,\mathrm{T}$ and

$H_{c2} \approx 50\,\mathrm{T}$.[5,6]

3.3. *Specific Heat*

Fig. 3.3 shows the specific heat curves obtained from the HTE for the free energy at different magnetic field strengths. In the absence of a magnetic field the rounded peak indicates short range ordering with a large gap. At temperatures less than $T = 17\,\mathrm{K}$ the exponential decay signals an ordered phase. The magnetic field is seen to only weakly affect the magnetic specific heat at low temperatures, mainly because of the strength of the rung singlets. As yet there appears to be no experimental data for the specific heat.

4. Conclusions

We have examined the magnetization, susceptibility and critical fields of the double chain compound $KCuCl_3$ via the integrable spin ladder model (2.1). The theoretical results obtained from Thermodynamic Bethe Ansatz and High Temperature Expansion calculations are seen to lead to good agreement with the experimental measurements for these quantities. We conclude that this compound exhibits strong rung coupling which leads to dimerized rung spins. This is consistent with the experimental analysis.[1,2] We have also presented the specific heat curves for different magnetic fields.

Acknowledgments

M.T.B. and X.W.G. thank Mo-Lin Ge and the Nankai Institute of Mathematics for their kind hospitality. This work has been supported by the Australian Research Council. N.O. has been partially supported by DAAD.

References

1. H. Tanaka, K. Takatsu, W. Shiramura and T. Ono, *J. Phys. Soc. Japan* **65** 1945 (1996)
2. H. Tanaka, K.-I. Takatsu, W. Shiramura and T. Ono, *Physica* **B246-247** 545 (1998)
3. T. Nakamura and K. Okamoto, *Phys. Rev.* **B58** 2441 (1998)
4. M. Muller and H. J. Mikeska, *J. Phys: Condens. Matter* **12** 7633 (2000)
5. W. Shiramura *et al.*, *J. Phys. Soc. Japan* **66** 1900 (1997)
6. A. Oosawa *et al.*, *Phys. Rev.* **B66** 104405 (2002)
7. N. Cavadini *et al.*, *Physica* **B335** 37 (2003)
8. N. Cavadini *et al.*, *Eur. Phys. J.* **B7** 519 (1999)

9. N. Cavadini *et al.*, *J. Phys: Condens. Matter* **12** 5463 (2000)
10. M. Muller, H. J. Mikeska and N. Cavadini, *J. Phys: Condens. Matter* **15** 8513 (2003)
11. M. Fujisawa, T. Osakabe, K. Kakurai and H. Tanaka, *J. Phys. Soc. Japan* **72** 1026 (2003); A. Oosawa, H. A. Katori and H. Tanaka, arXiv:cond-mat/0010383
12. M. Hagiwara, H. A. Katori, U. Schollwöck and H.-J. Mikeska, *Phys. Rev.* **B62** 1051 (2000)
13. M. Matsumoto, B. Normand, T. M. Rice and M. Sigrist, *Phys. Rev. Lett.* **89** 077203 (2002)
14. E. Dagotto, *Rep. Prog. Phys.* **62** 1525 (1999)
15. M. T. Batchelor, X.-W. Guan, A. Foerster and H.-Q. Zhou, *New J. Phys.* **5** 107 (2003)
16. M. T. Batchelor, X.-W. Guan, N. Oelkers, K. Sakai, Z. Tsuboi and A. Foerster, *Phys. Rev. Lett.* **91** 217202 (2003)
17. M. T. Batchelor, X.-W. Guan, N. Oelkers and Z. Tsuboi, *Integrable models and quantum spin ladders: comparison between theory and experiment for the strong coupling compounds*, arXiv:cond-mat/0512489
18. M. Takahashi, *Thermodynamics of One-Dimensional Solvable Models* (Cambridge University Press, Cambridge, 1999)
19. M. Suzuki, *Phys. Rev.* **B31** 2957 (1985)
A. Klümper, *Ann. Physik* **1** 540 (1992)
G. Jüttner, A. Klümper and J. Suzuki, *Nucl. Phys.* **B487** 650 (1997)
20. A. Kuniba, T. Nakanishi and J. Suzuki, *Int. J. Mod. Phys.* **A9** 5215 (1994); *Int. J. Mod. Phys.* **A9** 5267 (1994)
21. Z. Tsuboi, *J. Phys.* **A36** 1493 (2003)
22. Z. Tsuboi, *J. Phys.* **A37** 1747 (2004)
23. Z. Tsuboi and M. Takahashi, *J. Phys. Soc. Japan* **74** 898 (2005)
24. Z. Tsuboi, *Nucl. Phys.* **B737** 261 (2006)
25. Y. Wang, *Phys. Rev.* **B60** 9236 (1999)
26. M. T. Batchelor, X.-W. Guan, N. Oelkers and Z.-J. Ying, *J. Stat. Phys.* **116** 571 (2004)
27. Z. J. Ying, I. Roditi, A. Foerster and B. Chen, *Eur. Phys. J.* **B41** 67 (2004)
28. Z.-J. Ying, A. Foerster, X.-W. Guan, B. Chen and I. Roditi, *Eur. Phys. J.* **B38** 535 (2004); Z.-J. Ying, X.-W. Guan, A. Foerster, I. Roditi and B. Chen, arXiv:cond-mat/0309718
29. G. Chaboussant *et al.*, *Eur. Phys. J.* **B6** 167 (1998)
30. A. Oosawa, M. Ishii and H. Tanaka, *J. Phys: Condens. Matter* **11** 265 (1999); T. Nikuni, M. Oshikawa, A. Oosawa and H. Tanaka, *Phys. Rev. Lett.* **84** 5868 (2000)
31. J. Sirker, A. Weiße and O. P. Sushkov, *Physica* **B359-361** 1318 (2005)
32. T. Nikuni, M. Oshikawa, A. Oosawa and H. Tanaka, *Phys. Rev. Lett.* **84** 5868 (2000)
33. K. Kudo *et al.*, arXiv:cond-mat/0312063

Applications of Geometric Cluster Algorithms

H. W. J. BLÖTE

Faculty of Applied Sciences, Delft University of Technology,
P.O. Box 5046, 2600 GA Delft, The Netherlands
and
Lorentz Institute, Leiden University,
P.O. Box 9506, 2300 RA Leiden, The Netherlands
E-mail: bloete@tnw.tudelft.nl

Y. DENG

Department of Physics, New York University,
4 Washington Place, New York, NY 10003, USA

J. R. HERINGA

Faculty of Applied Sciences, Delft University of Technology,
P.O. Box 5046, 2600 GA Delft, The Netherlands

We formulate the geometric cluster algorithm in terms of geometric symmetry operations, and specify the conditions that make it possible to formulate a proof of detailed balance. The nonlocal nature of this algorithm allows the construction of algorithms with a reduced critical slowing down. We discuss the possibilities that arise for the construction of efficient algorithms and review some phenomena that can be investigated in more detail than is possible by simulation algorithms of a local nature. In particular we focus on systems with a conserved quantity which are subject to the Fisher renormalization phenomenon.

1. Introduction

Monte Carlo simulations of lattice models using local updates[1] tend to become time consuming when large-scale correlations exist, such as in critical systems. A key parameter is the dynamic exponent z which describes the autocorrelation time τ_L of a critical system of linear size L as

$$\tau_L \propto L^z \,. \tag{1.1}$$

Typically, models in d dimensions require of order L^d operations to update every particle in the system. For each new statistically independent state one thus needs of order L^{d+z} operations. While still dependent on the static universality class, one often finds that the dynamic exponent has a value $z \approx 2$ for simulation algorithms that use local updates. As a consequence of the positive value of z, the critical-slowing-down phenomenon occurs and makes it difficult to simulate large critical systems.

The so-called cluster algorithms alleviate this problem, but they are not generally applicable. The possibility to construct such a nonlocal algorithm for a specific model system depends sensitively on the symmetries of the system[2]. A nonlocal cluster flip is actually a symmetry operation applied to the cluster. While the first generation of cluster algorithms[3–6,8,7] employed spin up-down or spin permutation symmetries, later algorithms were devised [9,2] that employed lattice symmetries. The relative efficiency of cluster algorithms is due to the fact that cluster algorithms apply configuration changes in regions of appreciable size. More generally, it is determined by the cluster-size distribution. For instance, if some cluster algorithm generates a distribution such that most clusters cover the whole system except some small regions, this would result only in trivial changes of the spin configuration. For reasons of efficiency, it may be argued that the percolation threshold of the cluster-formation process should coincide with the critical point. This leads to a wide distribution of cluster sizes, so that updates occur on all length scales. In a number of cases, such as the Swendsen-Wang algorithm applied to the Potts model, this coincidence can be proved, while in other cases the critical point lies well within the percolating region of the cluster-formation process. In that case the clusters tend to be too large for maximum efficiency, but even in such cases a cluster algorithm may still be much faster than a Metropolis-like algorithm[10].

As one of the useful applications of the geometric cluster algorithm we mention the puzzling results presented a few years ago by Yamagata[11,12]. These results concerned three-dimensional lattice gases with nearest-neighbor exclusion on two bipartite lattices, namely the simple cubic and the body-centered cubic lattice. These results, obtained by means of a local update algorithm on a supercomputer, suggested that the phase transitions of these two models did not belong to the Ising universality class. This result was difficult to understand in view of the Ising-like symmetry (*i.e.*, the symmetry of the two sublattices) and the fact that only short-range interactions are present. If Yamagata's result were correct, this would imply that our present understanding of universality is seriously flawed. To obtain a more clear numerical picture, it was necessary to include corrections to scaling in the analysis. This proved feasible on the basis of results obtained by means of the geometric cluster algorithm[9,13]. It was found that the correction-to-scaling amplitudes are rather large in these two models, and they can therefore not be neglected. Furthermore it was found that the critical exponents agree accurately with the known values for the three-dimensional Ising universality class.

Since geometric cluster algorithms do not change the lattice variables, but only move them over the lattice, they are suitable to simulate models under a constraint, such as lattice gases with a conserved number of lattice-gas particles. While local algorithms are available for this purpose, cluster algorithms may be more efficient by several orders of magnitude. For the above-mentioned geometric cluster simulations of three-dimensional lattice gases, it was however deemed necessary to remove the constraint and leave the particle density as a freely fluctuating variable. In order to solve the problem, the results of the new simulations should be compatible with the existing results that used local updates. For this reason, the cluster simulations were supplemented by Metropolis-like sweeps. Although it is to be expected that this re-introduces some critical slowing down, the combined algorithms were still found to be very efficient in comparison with simulations restricted to local updates.

To perform geometric cluster simulations with maximum efficiency, one may choose to work without local updates and thus to keep the particle density fixed. However, it is important to realize that the conservation of particle density has considerable impact on the critical behavior of the system. The theory for such constrained systems[14] is known as 'Fisher renormalization'. This theory, which was formulated for the thermodynamic limit, shows that the critical singularities of the temperature-like variables, such as the specific heat, are strongly suppressed. The key factor is here that the particle density is a temperature-like variable (and, in the cases of the aforementioned lattice gases, it is simply the variable that is conjugate to the reduced chemical potential, which parametrizes the temperature). According to Fisher renormalization, magnetic observables such as the susceptibility are not affected by the constraint.

The outline of this paper is as follows. In Sec. 2 we describe the geometric cluster algorithm and demonstrate that it satisfies the condition of detailed balance. We present some numerical data for the constrained specific heat in Sec. 3, and discuss the observed behavior in Sec. 4.

2. The geometric cluster algorithm

The need for efficient algorithms to simulate various model systems is obviously stimulating efforts to devise new cluster algorithms. However, these nonlocal algorithms are not as easy to generalize as local (Metropolis-type) algorithms and thus restricted to a limited range of applicability.

The success of such algorithms obviously depends on two conditions: first one needs a proof of detailed balance, so that one can be assured to obtain an unbiased sample of the pertinent ensemble; and second, the algorithm has to be efficient in comparison with local algorithms.

In the proof of detailed balance of the single-cluster variant[5] of the Swendsen-Wang algorithm[3] algorithm, two essential conditions are that the cluster flip corresponds with a global symmetry of the Hamiltonian, and that the symmetry operation is self-inverse. For the Swendsen-Wang and related algorithms, this is the Potts permutation symmetry, including the Ising spin up-down symmetry. It is thus interesting to consider other symmetries, such as geometric symmetries of the lattice, to serve as the basis of a cluster algorithm.

In order to investigate hard-core gases in continuous space, Dress and Krauth[15] developed a cluster method using such geometric operations on the particle positions. Unfortunately, for hard disks, the percolation threshold of the cluster formation process occurs at some distance from the phase transition of the model[15], so that the resulting algorithm is not very efficient in suppressing critical slowing down.

For bipartite lattice gases with nearest-neighbor exclusion, the situation is more favorable because the critical density is much lower. Indeed the geometric cluster algorithm, when applied to these systems, leads to a wide distribution of cluster sizes, and the critical distributions for different system sizes collapse on a single curve[9]. This indicates that the percolation process that forms the geometric clusters itself is also critical.

We define the process forming a geometric cluster such as to expose the analogy with the Wolff algorithm. Let sites i, j and k map on i', j' and k' under the geometric symmetry. Let us now interchange a neighbor k of i with k', and consider the consequences for the energy of the pair of bonds (ik) and $(i'k')$. The change of the reduced energy (*i.e.*, divided by kT) due to this interchange is denoted Δ_{ik}. For instance, in the case of the Ising model we have, using obvious notation, $\Delta_{ik} = K(s_i s_k + s_{i'} s_{k'} - s_i s_{k'} - s_{i'} s_k)$. Then we proceed as follows:

(1) choose a random lattice site i; i and i' belong to the cluster.

(2) interchange s_i and $s_{i'}$
(3) for all neighbor sites k of i which do not (yet) belong to the cluster, do the following:
   (a) if $\Delta_{ik} > 0$ do the following with probability $1 - e^{-\Delta_{ik}}$:
      i. interchange s_k and $s_{k'}$ (k and k' included in cluster);
      ii. write k in a list of addresses (the stack).
   (b) if $\Delta_{ik} \leq 0$, do nothing.
(4) read an address j from the stack;
(5) execute the steps listed under 3, substituting j for i;
(6) erase the address j from the stack;
(7) repeat steps 4-6 until the stack is empty.

When the stack is empty, the cluster is completed and moved.

2.1. *The proof of detailed balance*

The validity of the Swendsen-Wang algorithm[3] may be shown on the basis of the Kasteleyn-Fortuin random-cluster decomposition[16] of the Potts model. This can be viewed as a probabilistic process that splits the lattice in groups of sites called random clusters. While all spins in a random cluster have the same value, spins in different clusters are uncorrelated. This property can thus be used to randomly assign new spin values to the clusters. More generally the proof of validity of a Monte Carlo algorithm relies on two conditions, which are ergodicity and detailed balance. Ergodicity guarantees that, after a sufficient number of Monte Carlo moves, all configurations are generated with a nonzero probability. Detailed balance says that the ratio of the transition probabilities between two states must be equal to the ratio of the Boltzmann weights. The proof of ergodicity is simple in most cases, and here we focus on the proof of detailed balance. We formulate this proof for the case that there exists a geometric lattice symmetry, *i.e.*, the Hamiltonian is invariant under this symmetry operation. The symmetry must be self-inverse, such as lattice inversions and translations over half the system size in the case of periodic boundary conditions.

Let us now consider the probability $T(S', S)$ of a cluster move which transforms a spin configuration S into S' by moving the spins contained in the geometric cluster $\mathcal{C}$ according to the pertinent lattice symmetry.

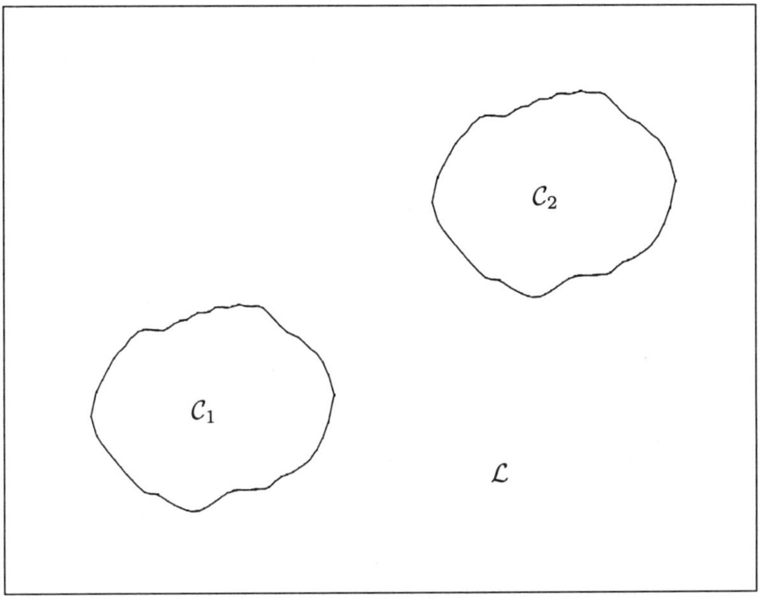

Fig. 2.1. Illustration of a cluster move involving a geometric cluster $\mathcal{C} \equiv \mathcal{C}_1 \cup \mathcal{C}_2$. The move shown here projects $\mathcal{C}_1$ and $\mathcal{C}_2$ on one another, as a result of a translation over half the diagonal system size. In general, the cluster $\mathcal{C}$ may or may not consist of two disjoint regions. The proof of detailed balance does not depend on this possibility.

As indicated in Fig. 2.1, it is well possible that the cluster $\mathcal{C}$ actually consists of two disjoint parts $\mathcal{C}_1$ and $\mathcal{C}_2$. The cluster flip then simply replaces $\mathcal{C}_1$ by $\mathcal{C}_2$ and it vice versa. This move transforms the original spin configuration S into a new state S'. We consider the case that the Hamiltonian contains only pair interactions. The probability $T(S', S)$ of this cluster flip can be written as $T_{\rm i}(\mathcal{C}, S)T_{\rm b}(\mathcal{C}, S)$ where $T_{\rm i}$ denotes the internal probability that the cluster formation process connects all the sites inside $\mathcal{C}$, and $T_{\rm b}$ denotes the probability that no site outside the boundary of $\mathcal{C}$ is included in the cluster. Since the cluster flip corresponds with a global symmetry of the Hamiltonian, the change of the reduced energy due to the cluster move comes only from the bond pairs (i, k) and (i', k') crossing the boundary of $\mathcal{C}$. This energy change can be written as $\sum \Delta_{ik} = \sum^+ \Delta_{ik} + \sum^- \Delta_{ik}$ where $\sum^+$ counts only the bond pairs that increase in energy and $\sum^-$ those that decrease in energy. The cluster formation rules given above imply that

$$T_{\rm b}(\mathcal{C}, S) = \exp\left[-\sum\nolimits^+ \Delta_{ik}\right] . \tag{2.1}$$

Next, we consider the probability $T(S', S) = T_{\rm i}(\mathcal{C}, S')T_{\rm b}(\mathcal{C}, S')$ of the reverse cluster flip $S' \to S$. In view of the symmetry we have $T_{\rm i}(\mathcal{C}, S) = T_{\rm i}(\mathcal{C}, S')$. The boundary probability $T_{\rm b}(\mathcal{C}, S')$ is now determined by the bond pairs whose energy increases due to the reverse move. These contributions add up to $-\sum^{-}\Delta_{ik}$ where the sum is defined for the original move $S \to S'$. Thus

$$T_{\rm b}(\mathcal{C}, S') = \exp\left[+\sum^{-}\Delta_{ik}\right] . \tag{2.2}$$

Combining these results one finds

$$\frac{T(S', S)}{T(S, S')} = \exp\left[\sum -\Delta_{ik}\right] \tag{2.3}$$

which is the condition of detailed balance for Boltzmann statistics.

3. Some applications

The Fisher renormalization approach, while originally formulated for the thermodynamic limit, can also be applied to derive the finite-size scaling behavior of the constrained specific heat[17]:

$$C(L) = C_\infty + aL^{-|2y_t - d|} + \cdots . \tag{3.1}$$

Thus, if the unconstrained specific heat diverges ($2y_t - d > 0$), the singularity is inverted. For two-dimensional Ising-like models, the unconstrained specific heat diverges as $-\ln|T - T_c|$, and the constrained specific heat is predicted as

$$C(L) = C_\infty + a/\ln L + \cdots . \tag{3.2}$$

This prediction was tested for the case of the two-dimensional Blume-Capel model, *i.e.* the spin-1 Ising model with variable chemical potential D of the vacancies. For this purpose, an arbitrary critical point was determined[18] as $K = 1$, $D = 1.70271780\,(3)$ by means of a transfer-matrix analysis. The corresponding vacancy density is $\rho_c = 0.3495830\,(2)$. The critical constrained specific heat was determined by means of geometric cluster simulations of finite Blume-Capel systems. We used systems with periodic boundary conditions, and spatial inversions about a randomly chosen center as a symmetry operation. Since the particle density is quantized in finite systems, data were averaged between two particle densities. Fig. 3.1 shows the numerical results. Remarkably, the singular part of the heat capacity appears to be proportional to $1/(\ln L)^2$ instead of $1/\ln L$ as in Eq. 3.2.

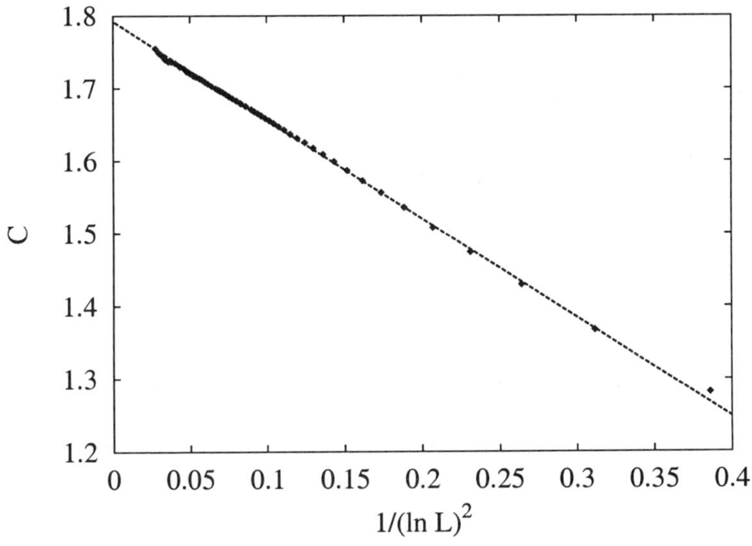

Fig. 3.1. Constrained specific heat of the critical Blume-Capel model vs. L.

As another test, we have simulated the hard-hexagon model[19]. This two-dimensional model has a temperature exponent $y_t = 6/5$, and thus a diverging specific heat. Thus on the basis of the Fisher renormalization mechanism one expects the following finite-size dependence at the critical point:

$$C(L) = C_\infty + aL^{d-2y_t} = C_\infty + aL^{-2/5} . \tag{3.3}$$

The results for the specific heat. obtained by geometric cluster simulations are shown in Fig. 3.2. Also in this case we find that the specific heat does not follow the prediction obtained from the Fisher renormalization mechanism. The singular part of the specific heat appears to be proportional to $L^{-4/5}$, not to $L^{-2/5}$ as predicted by Eq. 3.1.

4. Discussion

We investigated two two-dimensional models subject to a constraint, and determined the finite-size dependence of the critical specific heat. These results were compared with predictions obtained from applications of the Fisher renormalization procedure to finite systems. In both cases, the observed critical singularity is the the square of the predicted one. To explain

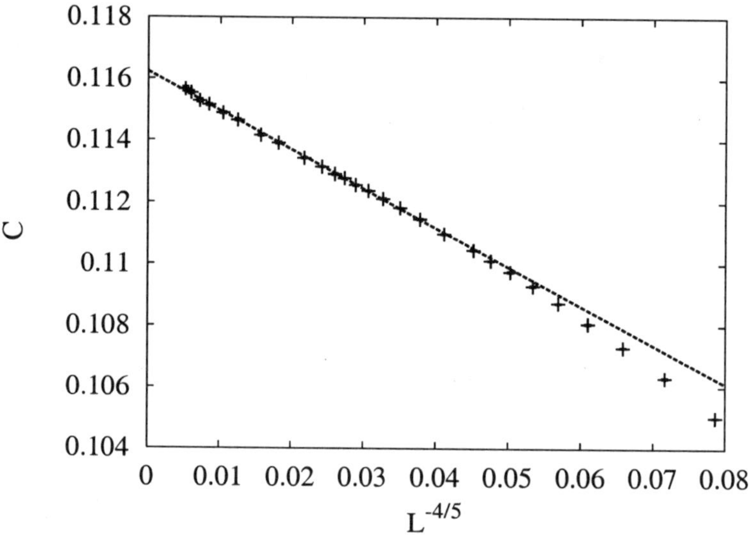

Fig. 3.2. Constrained specific heat of the critical hard-hexagon model vs. L.

this discrepancy, one may consider the possibility that the leading singularity accidentally vanishes. The constrained specific heat contains contributions from the analytic part as well as from the singular part of the unconstrained free energy. However, we see no obvious reason why different terms should cancel.

Thus one may consider other reasons that may explain the discrepancy. Here we mention the fact that the Fisher renormalization procedure substitutes the constrained system by an unconstrained system, with the conjugate parameter of the particle density adjusted such that the density of both systems coincides. In the case of finite systems, this is not completely correct, because the density of the unconstrained system is still allowed to fluctuate about its average. It is thus plausible that the critical singularities of constrained systems are suppressed even further than the predictions obtained from Fisher renormalization.

Acknowledgments

We are much indebted to Profs. M. E. Fisher and J. M. J. van Leeuwen for valuable comments. This work is part of the research program of the "Stichting voor Fundamenteel Onderzoek der Materie (FOM)", which is financially supported by the "Nederlandse Organisatie voor Wetenschappelijk Onderzoek (NWO)".

References

1. N. Metropolis, A. Rosenbluth, M. Rosenbluth, A. Teller, E. Teller, *J. Chem. Phys.* **21**, 1087 (1953).
2. J. R. Heringa and H. W. J. Blöte, *Phys. Rev.* **E57**, 4976 (1998).
3. R. H. Swendsen and J.-S. Wang, *Phys. Rev. Lett.* **58**, 86 (1987).
4. U. Wolff, *Phys. Rev. Lett.* **60**, 1461 (1988).
5. U. Wolff, *Phys. Rev. Lett.* **62**, 361 (1989).
6. C. F. Baillie and P. D. Coddington, *Phys. Rev.* **B43**, 10617 (1991).
7. M. A. Novotny and H. G. Evertz in *Computer Simulation Studies in Condensed-matter Physics* **VI**, eds. D. P. Landau, K. K. Mon and H.-B. Schüttler (Springer, Berlin 1993), 188.
8. E. Luijten and H. W. J. Blöte, *Int. J. Mod. Phys.* **C6**, 359 (1995).
9. J. R. Heringa and H. W. J. Blöte, *Physica* **A232**, 369 (1996).
10. H. W. J. Blöte, J. R. Heringa and E. Luijten, *Computer Physics Communications* **147**, 58 (2002).
11. A. Yamagata, *Physica* **A222**, 119 (1995).
12. A. Yamagata, *Physica* **A231**, 495 (1996).
13. J. R. Heringa and H. W. J. Blöte, *Physica* **A251**, 224 (1998).
14. M. E. Fisher, *Phys. Rev.* **176**, 257 (1968).
15. C. Dress and W. Krauth, *J. Phys.* **A28**, L597 (1995).
16. P. W. Kasteleyn and C. M. Fortuin, *J. Phys. Soc. Jpn. (Suppl.)* **26S**, 11 (1969).
17. Y. Deng, J. R. Heringa and H. W. J. Blöte, *Phys. Rev.* **E70**, 046111 (2004).
18. X. Qian, Y. Deng and H. W. J. Blöte, *Phys. Rev.* **E72**, 056132 (2005).
19. R. J. Baxter, *J. Phys. A* **13**, L61, (1980); *J. Stat. Phys.* **26**, 427, (1981).

Equivariant Cohomology and Localization for Lie Algebroids and Applications

U. Bruzzo

Scuola Internazionale Superiore di Studi Avanzati,
Via Beirut 4, 34100 Trieste, Italy

Istituto Nazionale di Fisica Nucleare, Sezione di Trieste

Let A be a Lie algebroid on a differentiable manifold M, and assume that A is equipped with an infinitesimal G-action compatible with a G-action on M, where G is a compact Lie group. We define an equivariant cohomology associated with these data and prove a localization formula together with a Bott-type formula.

1. Introduction

The tangent bundle TM to a differentiable manifold M has a remarkable property, namely, the space of its global sections has a Lie algebra structure. The notion of Lie algebroid generalizes this fact: a Lie algebroid is basically a vector bundle A whose space of global sections $\Gamma(A)$ has a Lie algebra structure. To keep contact with the geometry of the base manifold, one requires the existence of a vector bundle morphism $A \to TM$, called the *anchor*, that when evaluated on global sections is a Lie algebra homomorphism. (A further assumption, a Leibniz rule for the bracket on $\Gamma(A)$, is also imposed.)

Standard examples of Lie algebroids are provided by integrable distributions in TM (that is, regular foliations of M), which is the case when the anchor is an injective morphism; by Poisson manifolds, where A is the cotangent bundle, and the bracket is the one induced on differential 1-forms by the Poisson tensor; and a very interesting example is given by the bundle of first order differential operators on a vector bundle E with scalar symbol, where the anchor is the natural projection onto the vector fields, and the bracket is given by the commutator of differential operators (this is the so-called *Atiyah algebroid*).

One should also mention the fact that the datum of a Lie algebroid A on a differentiable manifold M is equivalent to the specification of a supermanifold $(M, \mathcal{F})$ (where the structure sheaf $\mathcal{F}$ is the sheaf of germs of sections of the exterior algebra bundle $\Lambda^\bullet A^*$) together with an odd supervector field squaring to zero.[16]

Recently there has been a surge of interest for Lie algebroids, for instance in connection with integrable systems, as a basic structure for defining new field-theoretic models,[15] as a tool for generalizing several constructions (e.g., connections) to singular settings,[10] and in relation with integrability properties of formal deformation spaces.[2]

Every Lie algebroid intrinsically defines a cohomology theory. A natural question arises when the Lie algebroid carries the action of some group G: can we define an equivariant Lie algebroid cohomology? The answer to this question comes from the general theory of G-differential complexes developed by Ginzburg.[11] It is interesting to note that from a physical viewpoint this equivariant Lie algebroid cohomology may be identified with the BRST cohomology: this happens for instance when we consider some Lie algebroids which are naturally defined on the moduli space of instantons. The relevant physical theory in this case is a version of topological supersymmetric Yang-Mills theory.[13,5]

Once the equivariant Lie algebroid cohomology is defined, a natural further step is to study localization formulas that generalize the usual formula for equivariant (de Rham) cohomology. The purpose of this review is indeed to descrive such a formula, together with a related Bott-type formula.

As far as the structure of this paper is concerned, in Section 2 I review the basic definitions and some constructions concerning Lie algebroid cohomology. Section 3 introduces the equivariant Lie algebroid cohomology. Moreover I describe there the localization formula for Lie algebroids. Section 4 is devoted to the description of a Bott-type localization formula. Results are just stated and commented; proofs may be found in Reference 4.

This paper reports on joint research made with L. Cirio, P. Rossi and V. Roubsov,[4] whom I thank for allowing me to reproduce here our joint results.

2. Lie algebroid cohomology

Let M be a smooth manifold, and denote by $\mathfrak{X}(M)$ the space of vector fields on M. The basic idea underlying the notion of Lie algebroid is to lift the Lie algebra structure on $\mathfrak{X}(M)$ given by the usual Lie bracket $[\,,]$ to a Lie algebra structure on the space of global sections of some vector bundle on M.

Definition 2.1. A Lie algebroid over M is a vector bundle on A equipped with the following structures:

(1) a vector bundle morphism $a\colon A \to TM$, called the anchor;
(2) a Lie algebra structure on the space of global sections $\Gamma(A)$, such that $a\colon \Gamma(A) \to \mathfrak{X}(M)$ is a Lie algebra homomorphism, and the following Leibniz rule holds true for every α, $\beta \in \Gamma(A)$ and every function f:

$$\{\alpha, f\beta\} = f\{\alpha, \beta\} + a(\alpha)(f)\,\beta$$

(let us denote by $\{\,,\}$ the bracket in $\Gamma(A)$).

□

The cohomology complex $(C_A^\bullet, \delta)$ associated with a Lie algebroid A is defined as $C_A^\bullet = \Gamma(\Lambda^\bullet A^*)$ with differential δ

$$\begin{aligned}(\delta\xi)(\alpha_1, \ldots, \alpha_{p+1}) = &\sum_{i=1}^{p+1}(-1)^{i-1}a(\alpha_i)(\xi(\alpha_1, \ldots, \hat{\alpha}_i, \ldots, \alpha_{p+1})) \\ &+ \sum_{i<j}(-1)^{i+j}\xi(\{\alpha_i, \alpha_j\}, \ldots, \hat{\alpha}_i, \ldots, \hat{\alpha}_j, \ldots, \alpha_{p+1})\end{aligned}$$

if $\xi \in C_A^p$. The cohomology of this complex will be denoted by $H^\bullet(A)$.

Cohomology classes in $H^\bullet(A)$ are not apt to be integrated on the base manifold. For this purpose we need a version of cohomology which is twisted by an "orientation bundle".[9] This is defined as the line bundle $Q_A = \det(A) \otimes \Omega_M^m$, where $m = \dim M$, and Ω_M^m is the bundle of differential m-forms on M. For every $s \in A$ one defines a map $L_s = \{s, \cdot\} =\colon \Gamma(\wedge^\bullet A) \to \Gamma(\wedge^\bullet A)$ by letting

$$L_s(s_1 \wedge \cdots \wedge s_k) = \sum_{i=1}^{k} s_1 \wedge \cdots \wedge \{s, s_i\} \wedge \cdots \wedge s_k .$$

Furthermore one defines a map $D\colon \Gamma(Q_A) \to \Gamma(A^* \otimes Q_A)$ by letting

$$D\tau(s) = L_s(X) \otimes \mu + X \otimes \mathcal{L}_{a(s)}\mu$$

if $\tau = X \otimes \mu$ and $s \in \Gamma(A)$. The twisted cohomology complex is defined as $\tilde{C}_A^\bullet = \Gamma(\wedge^\bullet A^* \otimes Q_A)$ where the differential $\tilde{\delta}$ is defined by

$$\tilde{\delta}(\xi \otimes \tau) = \delta\xi \otimes \tau + (-1)^{\deg(\xi)}\xi \otimes D\tau .$$

The resulting cohomology is denoted by $H^\bullet(Q_A)$.

If M is compact and oriented there is a nondegenerate pairing $C_A^k \otimes \tilde{C}_A^{r-k} \to \mathbb{R}$ defined as

$$\xi \otimes (\psi \otimes X \otimes \mu) \mapsto \int_M (\xi \wedge \psi, X)\, \mu$$

which descends to cohomology, yielding a bilinear map

$$H^\bullet(A) \otimes H^{r-\bullet}(Q_A) \to \mathbb{R}\,. \tag{2.1}$$

However this pairing may be degenerate in general.

One also has a natural morphism $C_A^\bullet \otimes \tilde{C}_A^\bullet \to \tilde{C}_A^\bullet$ which is compatible with the degrees. Again this descends to cohomology and provides a cup product

$$H^i(A) \otimes H^j(Q_A) \to H^{i+j}(Q_A)\,. \tag{2.2}$$

3. Equivariant Lie algebroid cohomology and localization

In this section we introduce an equivariant cohomology for Lie algebroids, basically following the pattern exploited in Reference 11 to define equivariant cohomology for Poisson manifolds.

Assume that a Lie group G (whose Lie algebra we denote by $\mathfrak{g}$) has an action ρ on M, and that there is a Lie algebra map $b\colon \mathfrak{g} \to \Gamma(A)$ such that the diagram

$$\begin{array}{ccc} \mathfrak{g} & \xrightarrow{\;b\;} & \Gamma(A) \\ & \searrow^{\tilde{\rho}} & \downarrow{\scriptstyle a} \\ & & \mathfrak{X}(M) \end{array} \tag{3.1}$$

commutes, where $\tilde{\rho}\colon \mathfrak{g} \to \mathfrak{X}(M)$ is the Lie algebra homomorphism

$$\xi \mapsto \xi^* = \frac{d}{dt} \rho_{\exp(-t\xi)}|_{t=0}.$$

One should note that the cohomology complex and cohomology groups resulting from this construction may depend on b unless the anchor is injective.

If for a $\xi \in \mathfrak{g}$ the point $x \in M$ is a zero of ξ^* one has endomorphisms

$$L_\xi\colon T_xM \to T_xM, \qquad \mathbb{L}_\xi\colon A_x \to A_x$$

given by

$$L_\xi(v) = [\xi^*, v], \qquad \mathbb{L}_\xi(\omega) = \{b(\xi), \omega\}.$$

We consider the complex

$$\mathfrak{A}^\bullet = \mathrm{Sym}^\bullet(\mathfrak{g}^*) \otimes \Gamma(\wedge^\bullet A^*)$$

with the grading

$$\deg(\mathcal{P} \otimes \beta) = 2\deg(\mathcal{P}) + \deg(\beta)$$

and define the equivariant differential $\delta_{\mathfrak{g}} : \mathfrak{A}^\bullet \to \mathfrak{A}^{\bullet+1}$

$$(\delta_{\mathfrak{g}}(\mathcal{P} \otimes \beta))(\xi) = \mathcal{P}(\xi)\left(\delta(\beta) - i_{b(\xi)}\beta\right).$$

If we denote $\mathfrak{A}^\bullet_G = \ker \delta^2_{\mathfrak{g}}$, then $(\mathfrak{A}^\bullet_G, \delta_{\mathfrak{g}})$ is a cohomology complex, whose cohomology we denote $H^\bullet_G(A)$ and call the *equivariant cohomology* of the Lie algebroid A.

By considering the complex

$$\mathfrak{Q}^\bullet = \mathfrak{A}^\bullet \otimes \Gamma(Q_A) = \mathrm{Sym}^\bullet(\mathfrak{g}^*) \otimes \Gamma(\wedge^\bullet A^* \otimes Q_A)$$

with a differential $\tilde{\delta}_{\mathfrak{g}}$ obtained by coupling $\delta_{\mathfrak{g}}$ with the differential D, and letting $\mathfrak{Q}^\bullet_G = \ker \tilde{\delta}^2_{\mathfrak{g}}$, one also has a twisted equivariant cohomology $H^\bullet_G(Q_A)$, and there is a cup product

$$H^i_G(A) \otimes H^k_G(Q_A) \to H^{i+k}_G(Q_A).$$

We shall now write a localization formula for the equivariant Lie algebroid cohomology. Let h be a G-invariant metric on M, and denote by μ the Riemannian measure associated with h. We introduce the skew-symmetric linear morphism $\bar{\mathbb{L}}_\xi(p) \colon A_p \to A^*_p$ as the composition

$$A_p \xrightarrow{\mathbb{L}_\xi} A_p \xrightarrow{a} T_pM \xrightarrow{h} T^*_pM \xrightarrow{a^*} A^*_p$$

We call the exterior power $\Lambda^{m/2}\bar{\mathbb{L}}_\xi(p)$ the *Pfaffian* of $\bar{\mathbb{L}}_\xi(p)$, and denote it $\mathrm{Pf}_a(\bar{\mathbb{L}}_\xi(p)))$; it is an element in $\Lambda^m(A^*_p)$. If $\gamma = \sum_i \omega_i \otimes X_i \otimes \mu$, to each point $p \in M_\xi$ we may thus attach the real number (residue)

$$R_{\gamma,\xi}(p) = (-1)^r \sum_i \left\{\left[\mathrm{Pf}_a(\bar{\mathbb{L}}_\xi(p)) \wedge \omega_i(\xi)\right] \rfloor X_i\right\}_{[0]}(p)$$

which turns out to be independent of the choice of the metric h.

Theorem 3.1. *Let M be a compact oriented m-dimensional manifold over which a compact Lie group G acts. Assume that $\xi \in \mathfrak{g} = Lie\,(G)$ is such that the associated fundamental vector field ξ^* has only isolated zeroes. Let A be a rank r Lie algebroid on M, and assume that a Lie algebra homomorphism $b\colon \mathfrak{g} \to \Gamma(A)$ exists making the diagram (3.1) commutative. Finally, let $\gamma \in \mathfrak{Q}^\bullet$ be equivariantly closed, $\tilde{\delta}_{\mathfrak{g}}\gamma = 0$.*

Then if $r < m$ one has $\int_M \gamma(\xi) = 0$, while if $r \geq m$ the following localization formula holds:

$$\int_M \gamma(\xi) = (-2\pi)^{m/2} \sum_{p \in M_\xi} R_{\gamma,\xi}(p).$$

□

In the case of the "trivial" algebroid given by the tangent bundle with the identity map as anchor, this reduces to the ordinary localization formula for the equivariant de Rham cohomology (see e.g. Reference 1).

4. A Bott-type formula

One may show that Theorem 3.1 is a quite general statement that incorporates a number of by now classical results. For instance, one can deduce from it Carell's localization formula[7] for the actions of lifts of holomorphic vector fields on holomorphic vector bundles and the equivariant Riemann-Roch theorem.[6] It also implies a result which generalizes the classical Bott formula[3] together with similar results by Cenkl and Kubarski.[8,12]

Let A be a rank r Lie algebroid over a compact oriented manifold M, with anchor a. If $p \in M$ is a zero of the vector field $a(\alpha)$ for some $\alpha \in \Gamma(A)$, one can define the Chern classes $c_i(\mathbb{L}_{\alpha,p})$ of the endomorphism $\mathbb{L}_{\alpha,p} \colon A_p \to A_p$ by letting

$$\sum_{i=0}^{r} c_i(\mathbb{L}_{\alpha,p})\,\lambda^i = \det(1 + \lambda\,\mathbb{L}_{\alpha,p})$$

(cfr. Reference 3). By means of these classes one can define the real numbers

$$\Phi(\alpha, p) = \Phi(c_1(\mathbb{L}_{\alpha,p}), \dots, c_r(\mathbb{L}_{\alpha,p}))\,.$$

Note that since A is a real vector bundle, Chern classes of odd order vanish identically.

The polynomial Φ also allow us to attach a real number to the Lie algebroid A. By using a G-invariant metric h on M, and a G-invariant fibre metric H on A, which is compatible with h via the anchor map, one can construct an element $\omega \in H^0(Q_A)$. We define the real number

$$\Phi(A) = \int_M \Phi(\lambda_1(A), \dots, \lambda_r(A)) \cup \omega$$

where the λ_i are the Chern-type characteristic classes of the Lie algebroid A.[10,4] We will show that this number only depends on the Lie algebroid A.

We also define the element in $\mathrm{Sym}^\bullet(\mathfrak{g}^*)$

$$\Phi^{\mathfrak{g}}(A) = \int_M \Phi(\lambda_1^{\mathfrak{g}}(A), \ldots, \lambda_r^{\mathfrak{g}}(A)) \cup \omega \tag{4.1}$$

$$= \int_M \Phi(\varsigma_1(R_\eta + \mu), \ldots, \varsigma_r(R_\eta + \mu)) \cup \omega\,. \tag{4.2}$$

Theorem 4.1. *Let A be a Lie algebroid on a compact oriented manifold M, and let $\alpha \in \Gamma(A)$ be any section such that the vector field $a(\alpha)$ has compact integral curves and has isolated zeroes. Let Φ be a homogeneous polynomial in $r = rk(A)$ variables. Then if $r = \dim M$ one has*

$$\Phi(A) = (-2\pi)^{\frac{m}{2}} \sum_p \frac{\Phi(a(\alpha), p)}{\det^{1/2} L_{a(\alpha),p}} \tag{4.3}$$

(where the sum runs over the zeroes of α at which the anchor is an isomorphism), while $\Phi(A) = 0$ if the condition $r = \dim M$ does not hold.

□

Here we have set

$$\Phi(a(\alpha), p) = \Phi(c_1(L_{a(\alpha),p}), \ldots, c_r(L_{a(\alpha),p}))\,.$$

As claimed before, this result shows the independence of the *characteristic number* $\Phi(A)$ on the fibred metric H on A. The contributions in the right-hand side of Eq. (4.3) are the same as in the usual Bott formula for the vector field $a(\alpha)$, but the sum is done on a smaller set of fixed points.

Remark 4.1. If $r = \dim M$ but $2\deg(\Phi) \neq \dim M$, the left-hand side of Eq. (4.3) vanishes by dimensionality reasons, and this provides identities among the terms in the right-hand side. △

Acknowledgments

The contents of this paper reflect the talk I gave at the 23rd International Conference on Differential Geometric Methods in Theoretical Physics, Nankai University, August 20 to 26, 2005. I am exceedingly grateful to the organizers of the conference, notably Professors Bai Chengming, Ge Mo-Lin, and Zhang Weiping for the unsurpassed organization and hospitality. Participation in the conference was made possible by support from the local organizers and by a grant of the Italian Institute for Higher Mathematics (INDAM). This work has been partially supported by the European Union through the FP6 Marie Curie Research and Training Network ENIGMA (Contract number MRTN-CT-2004-5652).

References

1. N. Berline, E. Getzler, and M. Vergne, *Heat kernels and Dirac operators*, Grundlehren der Mathematischen Wissenschaften, vol. 298, Springer-Verlag, Berlin, 1992.
2. J. Block, *Duality and equivalence of module categories in noncommutative geometry I*, `math.QA/0509284`
3. R. Bott, *Vector fields and characteristic numbers*, Michigan Math. J. **14** (1967), 231–244.
4. U. Bruzzo, L. Cirio, P. Rossi , and V. Rubtsov, *Equivariant cohomology and localization for Lie algebroids,* `math.DG/0506392`
5. U. Bruzzo, F. Fucito, J. F. Morales, and A. Tanzini, *Multi-instanton calculus and equivariant cohomology*, J. High Energy Physics **05** (2003), 054–076.
6. U. Bruzzo and V. Rubtsov, *Holomorphic Atiyah algebroids and localization,* in preparation.
7. J.B. Carell and D.I. Lieberman, *Vector fields and Chern numbers*, Math. Ann. **225** (1977), 263–273.
8. B. Cenkl, *Zeros of vector fields and characteristic numbers*, J. Diff. Geom. **8** (1973), 25–46.
9. S. Evens, J.-H. Lu, and A. Weinstein, *Transverse measures, the modular class and a cohomology pairing for Lie algebroids*, Quart. J. Math. Oxford **50** (1999), 417–436.
10. R. Loja Fernandes, *Lie algebroids, holonomy and characteristic classes*, Adv. Math. **170** (2002), 119–179.
11. V. L. Ginzburg, *Equivariant Poisson cohomology and a spectral sequence associated with a moment map*, Internat. J. Math. **10** (1999), 977–1010.
12. J. Kubarski, *Bott's vanishing theorem for regular Lie algebroids*, Trans. Amer. Math. Soc. **348** (1996), 2151–2167.
13. N. A. Nekrasov, *Seiberg-Witten potential from instanton counting*, Avd. Theor. Math. Phys. **7** (2003), 831–864.
14. V. Rubtsov, *Cohomology of Der-complex*, Russian Math. Surv. **35(4)** (1980), 190–191.
15. T. Strobl, *Algebroid Yang-Mills theory*, `arXiv:hep-th/0406215`.
16. A. Vaintrob, *Lie algebroids and homological vector fields*, Russ. Math. Surv. **52** (1997), 428–429.

Directed Percolation in Two Dimensions: An Exact Solution

L. C. Chen

Department of Mathematics
Fu-Jen Catholic University, Hsinchuang , Taipei 24205, Taiwan

F. Y. Wu

Department of Physics
Northeastern University, Boston, Massachusetts 02115, U.S.A.

We consider a directed percolation process on an $\mathcal{M} \times \mathcal{N}$ rectangular lattice whose vertical edges are directed upward with an occupation probability y and horizontal edges directed toward the right with occupation probabilities x and 1 in alternate rows. We deduce a closed-form expression for the percolation probability $P(x, y)$, the probability that one or more directed paths connect the lower-left and upper-right corner sites of the lattice. It is shown that $P(x, y)$ is critical in the aspect ratio $\alpha = \mathcal{M}/\mathcal{N}$ at a value $\alpha_c(x, y) = [1 - y^2 - x(1-y)^2]/2y^2$ where $P(x, y)$ is discontinuous, and the critical exponent of the correlation length for $\alpha < \alpha_c(x, y)$ is $\nu = 2$.

Key words: Directed Percolation, Critical behavior.

An outstanding unsolved problem in stochastic processes is the consideration of directed percolation [1,2]. Directed percolation is a Markovian bond percolation process in which bonds are directed such that only clusters with a "flow" are relevant. Very few exact results of directed percolation are known. In 1981 Domany and Kinzel [3] solved one version of a directed percolation where the occupation probability is fixed at unity in one spatial direction of a rectangular lattice. The problem was subsequently reformulated and solved as a random walk by one of us and Stanley [4]. However, the Domany-Kinzel model is essentially of a one-dimensional nature due to the restricted freedom in one spatial direction. To uncover the genuine nature of a two-dimensional directed percolation it is necessary to relax this uni-directional restriction.

As a first step toward this goal we consider in this paper a directed percolation in which the unity percolation probability occurs in every *other* row of a rectangular lattice. We deduce a closed-form expression for the percolation probability and analyze its critical properties for large lattices.

We first describe our model. Consider a 2-dimensional rectangular net of $(M+1) \times (2N+1)$ sites with an aspect ratio

$$\alpha = M/2N. \tag{1}$$

Number the sites by (m, n) with $m = 0, 1, \cdots M$, $n = 0, 1, \cdots 2N$ as shown in Fig. 1. Consider a bond percolation process on the lattice with vertical edges occupied with a probability $p_y = y$ and horizontal edges in the n-th row occupied with a probability

$$\begin{aligned} p_x &= 1, \quad n = \text{odd} \\ &= x, \quad n = \text{even}. \end{aligned} \tag{2}$$

Direct edges in the upward direction and toward the right. Occupied edges form directed paths if traced along the arrows. In ensuing discussions we shall refer to percolation configurations as *bond configurations.* A bond configuration is *percolating* if it contains one or more directed paths connecting the two opposite corner sites $(0,0)$ and $(M, 2N)$. A typical percolating configuration is shown in Fig. 1.

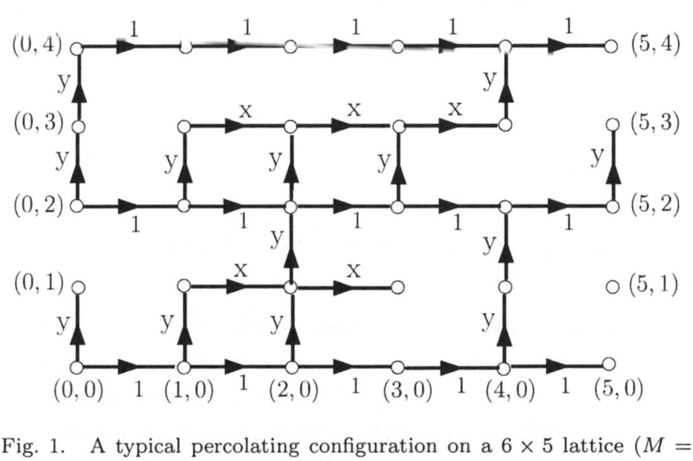

Fig. 1. A typical percolating configuration on a 6×5 lattice ($M = 5, N = 2$). Open circles denote lattice sites. Oriented edges are occupied with weights shown. Empty edges carry weights $1 - x$ and $1 - y$ in horizontal and vertical directions respectively.

In a bond configuration there are n_x (*resp.* $MN - n_x$) occupied (*resp.* empty) horizontal edges, and n_y (*resp.* $2(M+1)N - n_y$) occupied (*resp.* empty) vertical edges. Then the *percolation probability*, the probability that a bond configuration is percolating, is

$$P_{M,2N}(x,y) = \sum_{perc\ conf} x^{n_x}(1-x)^{MN-n_x} y^{n_y}(1-y)^{2(M+1)N-n_y} \qquad (3)$$

where the summation is restricted to percolating bond configurations. It is clear that $0 \le P_{M,2N}(x,y) \le 1$ since the summation (3) is identically 1 if unrestricted. It is also clear that $P_{\infty,2N}(x,y) = 1$ and $P_{0,2N}(x,y) = 0$. Our interest is to investigate how does P change from 1 to 0 as α varies, and whether the change is a sharp transition.

We state the main result as a Proposition:

Proposition:

For any $x \in [0,1]$ and $y \in (0,1)$, there exists a critical aspect ratio

$$\alpha_c(x,y) = [1 - y^2 - x(1-y)^2]/2\,y^2 \qquad (4)$$

such that

$$\lim_{N\to\infty} P_{2\alpha N,2N}(x,y) = \begin{cases} 1 & \textit{if}\ \ \alpha > \alpha_c(x,y) \\ 0 & \textit{if}\ \ \alpha < \alpha_c(x,y) \\ \frac{1}{2} & \textit{if}\ \ \alpha = \alpha_c(x,y). \end{cases} \qquad (5)$$

Moreover, for $\alpha < \alpha_c(x,y)$, we have the asymptotic behavior

$$P(2\alpha N, 2N) \sim e^{-2N/\xi} \qquad (6)$$

where

$$\xi \sim (\alpha_c - \alpha)^{-2}. \qquad (7)$$

Remarks:

1. Equation (6) defines ξ as the correlation length and Eq. (7) gives the correlation length critical exponent $\nu = 2$.

2. For $x = 1$ our model reduces to the Domany-Kinzel model [3,4] on an $(M+1)\times(2N+1)$ lattice and (4) leads to $\alpha_c = (1-y)/y$ in agreement with previous result. For $x = 0$ our model is again a Domany-Kinzel model but on an $(M+1)\times(N+1)$ lattice with a vertical edge occupation probability y^2. Our result gives the critical aspect ratio $2\alpha_c = (1-y^2)/y^2$ again in agreement with [3,4].

Proof of the Proposition:

The main body of this paper is the proof of the Proposition.

There are $2N$ rows of vertical edges in the lattice. Number these rows from 1 to $2N$ starting from the bottom. An occupied vertical edge in a bond configuration is *wet* if it lies on a percolating path connecting $(0,0)$ and $(M,2N)$, and is *primary* wet if it is the first wet edge (in a row of vertical edges) counting from the left. In the bottom row of vertical edges in Fig. 1, for example, there are two wet edges and the primary wet edge is the one connecting sites $(1,0)$ and $(1,1)$. In a percolating configuration there is one primary wet edge in every row and these edges carry an overall occupation probability y^{2N}. Since a bond configuration is percolating whenever a vertical edge in the $2N$-th row is primary wet, which can occur at any of the m-th horizontal positions $m = 0, 1, \cdots, M$, we have

$$P_{M,2N}(x,y) = y^{2N} \sum_{m=0}^{M} w_{m,2N}. \tag{8}$$

Here $y^{2n} w_{m,2n}$ is the probability that the primary wet edge in the $(2n)$-th row occurs at the horizontal position m.

We first establish a Lemma:

Lemma:

$$w_{m,2n} = \frac{1}{2\pi i} \oint \frac{dt}{t^{m+1}(1 - at + bt^2)^n} \tag{9}$$

where the contour of integration is around the unit circle and

$$a = 1 - y^2 + x(1-y)^2, \qquad b = x(1-y)^2. \tag{10}$$

Proof of the Lemma:

It is not difficult to see that the function $w_{m,2n}(x,y)$ satisfies the recursion relation

$$w_{m,2n} = \sum_{k=0}^{m} w_{k,2} w_{m-k,2n-2} \tag{11}$$

and the initial condition

$$w_{m,0} = \delta_{\mathrm{Kr}}(m, 0). \tag{12}$$

Define generating functions

$$W_1(t) = \sum_{m=0}^{\infty} w_{m,2} t^m \tag{13}$$

$$W_2(t,s) = \sum_{m=0}^{\infty}\sum_{n=0}^{\infty} w_{m,2n} t^m s^n. \tag{14}$$

Substituting (11) into (14) and changing the order of summation by using $\sum_{m=0}^{\infty}\sum_{n=0}^{m} = \sum_{n=0}^{\infty}\sum_{m=n}^{\infty}$, we obtain after some rearrangement and the use of (12),

$$W_2(t,s) = 1 + s\, W_1(t) W_2(t,s)$$

which yields

$$W_2(t,s) = \frac{1}{1 - sW_1(t)}. \tag{15}$$

We can now invert (14) to obtain

$$\begin{aligned} w_{m,2n} &= \frac{1}{(2\pi i)^2} \oint \frac{dt}{t^{n+1}} \oint \frac{ds}{s^{n+1}} \left(\frac{1}{1 - sW_1(t)} \right) \\ &= \frac{1}{2\pi i} \oint \frac{dt}{t^{m+1}} \Big[W_1(t) \Big]^n, \end{aligned} \tag{16}$$

where the contour of integration is around the unit circle.

To compute $W_1(t)$ we need to evaluate $w_{m,2}(x,y)$ for an $(m+1)\times 3$ lattice. There are now 2 rows of vertical edges. As aforementioned $y^2 w_{m,2}$ is the probability that $(0,0)$ is connected to $(m,2)$ with the primary wet vertical edge in the top row occurring at m. However the primary wet vertical edge in the bottom row can be at any j in $0 \le j \le m$. Denote the probability for this to occur by $y^2 \lambda_j (1-y)^{m-j} x^{m-j}$. Then we have

$$w_{m,2} = \sum_{j=0}^{m} \lambda_j (1-y)^{m-j} x^{m-j}, \tag{17}$$

where the factor $(1-y)^{m-j}x^{m-j}$ ensures that the primary wet edge in the top row is at m as shown in Fig 2(a). Particularly, we have $w_{0,2} = \lambda_0 = 1$.

The factor λ_j in (17) satisfies a recursive relation which can be written as

$$\lambda_j = (1-y)\lambda_{j-1} + y(1-x)(1-y)\, w_{j-1,2}, \quad j = 1, 2, \cdots, m. \tag{18}$$

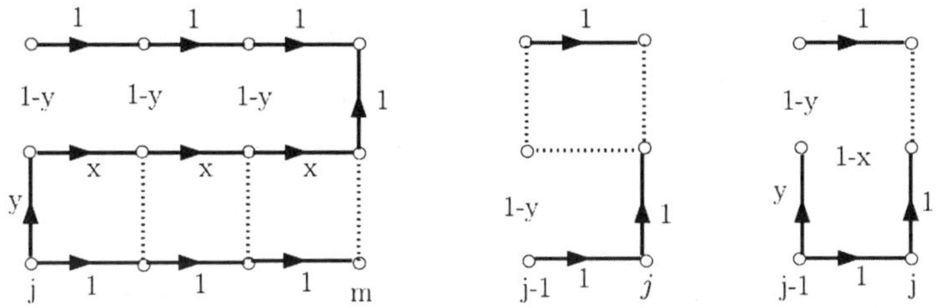

Fig. 2. Construction of recursion relations. (a) Construction of (17). (b) Construction of (18). Occupied edges are shown as oriented edges; dotted edges can be either occupied or empty. To each row of vertical edges there is an additional factor y not shown.

The two terms on the right-hand side of (18) arise from the two possibilities that the vertical edge connecting $(j-1,0)$ and $(j-1,1)$ is either empty (with probability $1-y$) or occupied (with probability y) as shown in the two panels in Fig. 2(b). In the latter case the factor $(1-x)(1-y)$ ensures that the site $(j-1,1)$ is not on a percolating path.

To solve the coupled recursion relations (17) and (18), define the generating function

$$\Lambda(t) = \sum_{j=0}^{\infty} \lambda_j \, t^j .$$

Multiplying (17) and (18) by t^m and t^{j-1}, respectively, and summing over m and $j-1$ from 0 to ∞, we obtain after some manipulation

$$\begin{aligned} W_1(t) &= \frac{y^2 \Lambda(t)}{1-x(1-y)t}, \\ \frac{1}{t}\left(\Lambda(t)-1\right) &= (1-y)\Lambda(t) + y(1-x)(1-y)W_1(t). \end{aligned} \tag{19}$$

This gives

$$W_1(t) = \frac{1}{1-at+bt^2} \tag{20}$$

after eliminating $\Lambda(t)$ where a, b are given in (10). The substitution of (20) into (16) establishes the Lemma.

We now continue the proof of the Proposition.

Substitute (9) into (8) and carry out the summation in m. This leads to

$$P_{M,2N}(x,y) = \frac{y^{2N}}{2\pi i} \oint_{C+} \frac{dt}{(t-1)(1-a\,t+b\,t^2)^N} \left(1 - \frac{1}{t^{M+1}}\right) \tag{21}$$

where the contour $C+$ is a circle enclosing the unit circle. Let t_1 and t_2 be the two roots of $1 - at + bt^2 = 0$, both of which are real. We have $t_1 t_2 = 1/b$, $t_1 + t_2 = a/b$, and hence

$$\begin{aligned}(t_1 - 1)(t_2 - 1) &= t_1 t_2 - (t_1 + t_2) + 1 \\ &= \frac{1}{b} - \frac{a}{b} + 1 = \frac{y}{(1-x)^2} > 0,\end{aligned}$$

so both t_1 and t_2 lie outside the unit circle. We can therefore choose the radius of $C+$ to be greater than 1 but smaller than both t_1 and t_2 so that $C+$ encloses only the simple pole $t = 1$ in (21). It follows that the first term on the right-hand side of (21) picks up only the residue at $t = 1$ which is

$$\frac{y^{2N}}{(1-a+b)^N} = 1,$$

and we obtain

$$P_{M,2N}(x, y) = 1 - I_{M,N}$$

where

$$I_{M,N} = \frac{y^{2N}}{2\pi i} \oint_{C+} \frac{dt}{(t-1)t^{M+1}(1-at+bt^2)^N}. \tag{22}$$

Note that since $|t| > 1$ along $C+$ (22) leads to the expected result $P_{\infty,2N} = 1$.

To further evaluate $I_{M,N}$ we introduce $z = 1/t$ to write

$$I_{M,N} = \frac{y^{2N}}{2\pi i} \oint_{C-} \frac{z^{M+2N} dz}{(z-1)(z^2 - az + b)^N} \tag{23}$$

where the contour $C-$ is now within the unit circle.

For M, N large and fixed aspect ratio $\alpha = M/2N$, we can rewrite (23) as

$$I_{M,N} = \frac{1}{2\pi i} \oint_{C-} \frac{dz}{z-1} \Big[f_\alpha(z)\Big]^N \tag{24}$$

where

$$f_\alpha(z) = \frac{y^2 z^{2+\alpha}}{z^2 - az + b}.$$

The integral $I_{M,N}$ can be evaluated using the method of steepest descent [5,6] by deforming the contour to pass a point $z = z_0$ where $f_\alpha(z)$ is stationary. To the leading order this gives $I_{M,N} \sim [f_\alpha(z_0)]^N$. Moreover, since $I_{M,N} \leq 1$, we must have $f_\alpha(z_0) \leq 1$ with the equal sign holding at $f_\alpha(z_0) = 1$. Thus a transition occurs at $z_0 = 1$.

Now

$$f'_\alpha(z) = \frac{y^2 z^{1+\alpha}}{(z^2 - az + b)^2} \Big[\alpha z^2 - (1+\alpha)a\, z + (2+\alpha)b \Big]$$

and the stationary point z_0 is determined by

$$\alpha z_0^2 - (1+\alpha)a\, z_0 + (2+\alpha)b = 0.$$

The critical condition $z_0 = 1$ now gives

$$\alpha = \frac{a - 2b}{1 - a + b} = \alpha_c(x, y) \qquad (25)$$

where $\alpha_c(x, y)$ is given in (4). It is readily verified that we have $(d\alpha/dz)_{z=1} < 0$ along (25). Thus, for $\alpha > \alpha_c(x, y)$, the stationary point z_0 lies within the unit circle so we can deform $C-$ continuously to pass z_0, and obtain $I_{2\alpha N,N} = [f_\alpha(z_0)]^N \sim 0$. This gives $P_{2\alpha N,N}(x, y) \sim 1$ which establishes the first line of (5).

On the other hand, for $\alpha < \alpha_c(x, y)$, z_0 occurs outside the unit circle and when the contour $C-$ is deformed to pass z_0 it must cross the simple pole at $z = 1$ and picks up the residue at the pole, which is equal to 1. This gives $I_{2\alpha N,N} \sim 1 - [f_\alpha(z_0)]^N$ and $P_{2\alpha N,N}(x, y) \sim [f_\alpha(z_0)]^N \sim 0$ for large N. This establishes the second line of (5).

For $\alpha = \alpha_c(x, y)$, z_0 is on the unit circle so the crossing of the contour at $z = 1$ picks up only half of the residue, namely, 1/2. This establishes the third line of (5).

Finally, for $\alpha < \alpha_c(x, y)$, the method of steepest decent [5,6] dictates that we have

$$\big[f_\alpha(z_0)\big]^N = e^{N \ln[f_\alpha(z_0)]} \sim e^{-NC_1(x,y)(z_0-1)^2} \sim e^{-NC_2(x,y)(\alpha-\alpha_c)^2}$$

where expressions of $C_1(x, y)$ and $C_2(x, y)$, which do not affect our conclusions, can be explicitly evaluated. This establishes the asymptotic behavior (6) with $\xi = 2/[C_2(x, y)(\alpha - \alpha_c)^2]$.

We have completed the proof of the Proposition.

In summary, we have obtained a closed-form expression for the percolation probability $P_{M,2N}(x, y)$ for the directed percolation process in which the occupation probability is y in the vertical direction and alternately x

and 1 in the horizontal direction. For M, N large, the percolation probability exhibits a critical behavior at $\alpha = \alpha_c$. The correlation length ξ for $\alpha < \alpha_c$ is found to diverge with the critical exponent $\nu = 2$. While these properties are similar to those found in the Domany-Kinzel model [3,4], our analysis permits the relaxation of the restriction of unit occupation probability in one spatial direction. It is hoped that the analysis serves as the first step of further relaxation in percolation probabilities, eventually leading to an understanding of genuine 2-dimensional directed percolation processes.

This work was initiated while FYW was at the National Center of Theoretical Sciences (NCTS) in Taipei. The support of the NCTS is gratefully acknowledged. Work of LCC has been supported in part by the National Science Council, Taiwan.

References

1. R. Durrent, Oriented percolation in two dimensions, Annals of Probability **12**, 999-1040 (1984).
2. B. D. Hughes, Random walks and random environments, Vol. 2 (Oxford University Press, New York 1996) p. 59.
3. E. Domany and W. Kinzel, Directed percolation in 2 dimensions: Numerical analysis and an exact solution, Phys. Rev. Lett. **47**, 5-8 (1981).
4. F. Y. Wu and H. E. Stanley, Domany-Kinzel model of directed percolation: Formulation as a random-walk problem and some exact results, Phys. Rev. Lett. **48**, 775-778 (1982).
5. P. M. Morse and H. Feshbach, Methods of thoretical physics, Vol. 1 (McGraw-Hill, New York 1953) p. 434.
6. P. Dennery and A. Krzywicki, Mathematics for physicists (Harper and Row, New York 1967), reprinted (Dover, New York 1996) p. 87.

Generalized Drinfeld Polynomials for Highest Weight Vectors of the Borel Subalgebra of the sl_2 Loop Algebra

Tetsuo Deguchi

Department of Physics, Ochanomizu University,
2-1-1 Ohtsuka, Bunkyo-ku
Tokyo, 112-8610, Japan
E-mail: deguchi@phys.ocha.ac.jp

In a Borel subalgebra $U(B)$ of the sl_2 loop algebra, we introduce a highest weight vector Ψ. We call such a representation of $U(B)$ that is generated by Ψ *highest weight.* We define a generalization of the Drinfeld polynomial for a finite-dimensional highest weight representation of $U(B)$. We show that every finite-dimensional highest weight representation of the Borel subalgebra is irreducible if the evaluation parameters are distinct. We also discuss the necessary and sufficient conditions for a finite-dimensional highest weight representation of $U(B)$ to be irreducible.

1. Introduction

In the classical analogue of the Drinfeld realization of the quantum sl_2 loop algebra, $U_q(L(sl_2))$, the Drinfeld generators, $\bar{x}_k^{\pm}$ and $\bar{h}_k$ for $k \in \mathbf{Z}$, satisfy the following defining relations[1,2,9]:

$$[\bar{h}_j, \bar{x}_k^{\pm}] = \pm 2\bar{x}_{j+k}^{\pm} , \quad [\bar{x}_j^{+}, \bar{x}_k^{-}] = \bar{h}_{j+k} ,$$
$$[\bar{h}_j, \bar{h}_k] = 0 , \quad [\bar{x}_j^{\pm}, \bar{x}_k^{\pm}] = 0 , \quad \text{for } j, k \in \mathbf{Z}. \tag{1.1}$$

In a representation of $U(L(sl_2))$, a vector Ω is called *a highest weight vector* if Ω is annihilated by generators $\bar{x}_k^{+}$ for all integers k and such that Ω is a simultaneous eigenvector of every generator of the Cartan subalgebra, $\bar{h}_k$ ($k \in \mathbf{Z}$) [1,2]. We call a representation of $U(L(sl_2))$ *highest weight* if it is generated by a highest weight vector. For a finite-dimensional irreducible representation we associate a unique polynomial through the highest weight $\bar{d}_k^{\pm}$. It is shown that any given irreducible highest weight representation is finite-dimensional if and only if it has the Drinfeld polynomial [1].

Recently it was shown that the XXZ spin chain at roots of unity has the sl_2 loop algebra symmetry [5,7,10–12]. Fabricius and McCoy has conjectured

[12] that every Bethe ansatz eigenstate should be highest weight of the sl_2 loop algebra, and also that the Drinfeld polynomial can be derived from the Bethe state. It is explicitly shown that regular XXZ Bethe states in some sectors are indeed highest weight [5]. However, it is still nontrivial how to connect the highest weight vector with the Drinfeld polynomial. In fact, the Drinfeld polynomial is defined for an irreducible representation not for a highest weight vector [1]. Furthermore, there exist finite-dimensional highest weight representations that are reducible and indecomposable. It has been shown that a given highest weight representation is irreducible if the evaluation parameters are distinct [3,6]. Here, we shall define evaluation parameters in §3. Thanks to the theorem, we solve the connection problem at least for the case of distinct evaluation parameters.

In this paper, we discuss a generalization of the theorem to the case of a highest weight representation of a Borel subalgebra of $U(L(sl_2))$. The generalization should play a key role in the study of the spectral degeneracy of the XXZ spin chain under twisted boundary conditions [4,14,8]. Let us consider the subalgebra generated by generators h_0, x_0^+ and x_1^- satisfying the relations (1.1). We call it a Borel subalgebra of $U(L(sl_2))$, and denote it by $U(B)$. It has the following generators:

$$h_k \, , \; x_k^+ \text{ for } k \in \mathbf{Z}_{\geq 0} \, , \quad x_k^- \text{ for } k \in \mathbf{Z}_{>0} . \tag{1.2}$$

We define a highest weight vector of the Borel subalgebra $U(B)$ by such a vector Ψ that satisfies the following relations:

$$x_k^+ \Psi = 0 \, , \quad h_k \Psi = d_k \Psi \, , \quad \text{for } k \in \mathbf{Z}_{\geq 0} \, . \tag{1.3}$$

We call the representation of $U(B)$ generated by Ψ *highest weight* and the set $\{d_k\}$ the highest weight. Here we note that d_0 is not necessarily an integer, since x_{-1}^+ does not exist in $U(B)$. In §2 of the present paper, we derive a useful recursive relation of $x_k^- \Psi$ for $k \in \mathbf{Z}_{>0}$. In §3 we introduce a generalization of the Drinfeld polynomial for a finite-dimensional highest weight representation of the Borel subalgebra $U(B)$. In §4 we show that every highest weight representation of the Borel subalgebra with distinct and nonzero evaluation parameters is irreducible.

Throughout the paper, we denote by Ψ a highest weight vector of the Borel subalgebra $U(B)$ with highest weight d_k and by V_B the representation generated by it, i.e. $V_B = U(B)\Psi$. We also assume that V_B is finite-dimensional.

2. Sectors of V_B and nilpotency

Lemma 2.1. *Let us define the sector of $h_0 = d_0 - 2n$ in V_B for an integer $n \geq 0$ by the subspace consisting of vectors $v_n \in V_B$ such that $h_0 v_n = (d_0 - 2n) v_n$. Here we recall $h_0 \Psi = d_0 \Psi$. Then, V_B is given by the direct sum of such sectors. Any vector v_n in the sector of $h_0 = d_0 - 2n$ is expressed as a linear combination of monomial vectors $x_{j_1}^- \cdots x_{j_n}^- \Psi$.*

Proof. It is clear from the PBW theorem [13]. □

We note that generator x_1^- is nilpotent in any V_B.

Definition 2.1. We say that generator x_1^- is nilpotent of degree r in V_B, if $(x_1^-)^{r+1}\Psi = 0$, while $(x_1^-)^j \Psi \neq 0$ for $0 < j \leq r$.

The degree r of nilpotency for generator x_1^- gives the largest n for non-vanishing sectors of $h_0 = d_0 - 2n$, as shown in the next proposition.

Proposition 2.1. *If generator x_1^- is nilpotent of degree r, then the sector of $h = d_0 - 2r$ is one-dimensional: every monomial vector in the sector is proportional to $(x_1^-)^r \Psi$ with some constant $C_{k_1,\ldots,k_r}$:*

$$x_{k_1}^- \cdots x_{k_r}^- \Psi = C_{k_1,\ldots,k_r} (x_1^-)^r \Psi \,, \quad \text{for } k_1, \ldots, k_r \in \boldsymbol{Z}_{>0} \,. \tag{2.1}$$

Furthermore, sectors of $h = d_0 - 2n$ for $n > r$ are of zero-dimensional. For instance, we have $x_{k_1}^- \cdots x_{k_{r+1}}^- \Psi = 0$ for $k_1, \cdots, k_{r+1} \in \boldsymbol{Z}_{>0}$.

Proof. Setting $m = r$ in lemma 2.3, we have eq. (2.1). For the case of $n > r$ we show it from lemma 2.3 where we set $m = n$. □

Let B_+ be such a subalgebra of $U(B)$ that is generated by x_k^+ for $k \in \mathbf{Z}_{>0}$. We define $(X)^{(n)}$ by $X^n = X^n/n!$.

Lemma 2.2. *Let m and t be integers satisfying $0 \leq t \leq m+1$. In the Borel subalgebra $U(B)$, for $k_1, \ldots, k_t, n \in \boldsymbol{Z}_{>0}$, and $\ell \in \boldsymbol{Z}_{\geq 0}$, we have*

$$\begin{aligned}
& x_\ell^+ (x_n^-)^{(m+1-t)} x_{k_1}^- \cdots x_{k_t}^- \\
&= -x_{\ell+2n}^- (x_n^-)^{(m-t-1)} x_{k_1}^- \cdots x_{k_t}^- + (x_n^-)^{(m-t)} x_{k_1}^- \cdots x_{k_t}^- h_{\ell+n} \\
&\quad + \sum_{j=1}^{t} (x_n^-)^{(m+1-t)} \prod_{i=1, i\neq j}^{t} x_{k_i}^- \cdot h_{\ell+k_j} + (-2) \sum_{j=1}^{t} (x_n^-)^{(m-t)} x_{\ell+n+k_j}^- \prod_{i=1; i\neq j}^{t} x_{k_i}^- \\
&\quad + (-2) \sum_{1\leq j_1 < j_2 \leq t} (x_n^-)^{(m+1-t)} x_{\ell+k_{j_1}+k_{j_2}}^- \prod_{i=1; i\neq j_1, j_2}^{t} x_{k_i}^- \mod U(B) B_+
\end{aligned} \tag{2.2}$$

Lemma 2.3. *Suppose that x_1^- is nilpotent of degree r in V_B, and m be an integer with $m \geq r$. Let us take a positive integer p satisfying $p \leq m$. We have*

$$(x_1^-)^{m-p} x_{k_1}^- \cdots x_{k_p}^- \Psi = A_{k_1,\cdots,k_p}^{(r)} (x_1^-)^m \Psi \,, \tag{2.3}$$

for any set of positive integers $k_1, \ldots, k_p$.

Proof. We prove (2.3) by induction on p by making use of eq. (2.2). □

Lemma 2.4. *The following recursive formulas on n hold for $n > 0$:*

(A_n): $(x_0^+)^{(n-1)}(x_1^-)^{(n)} = \sum_{j=1}^{n}(-1)^{j-1} x_j^- (x_0^+)^{(n-j)}(x_1^-)^{(n-j)}$ *mod* $U(B)B_+$.

(B_n):

$$n\,(x_0^+)^{(n)}(x_1^-)^{(n)} = \sum_{j=1}^{n}(-1)^{j-1} h_j (x_0^+)^{(n-j)}(x_1^-)^{(n-j)} \ \ mod\ U(B)B_+$$

.

(C_n): $[h_1, (x_0^+)^{(m)}(x_1^-)^{(m)}] = 0$ *mod* $U(B)B_+$ *for* $m \leq n$.

Making use of (B_n) of lemma 2.4 inductively, we show that Ψ is a simultaneous eigenvector of operators $(x_0^+)^{(n)}(x_1^-)^{(n)}$ for $n > 0$. For a given positive integer k, we denote by λ_k the eigenvalue: $(x_0^+)^{(k)}(x_1^-)^{(k)}\Psi = \lambda_k \Psi$.

Lemma 2.5. *If x_1^- is nilpotent of degree r in V_B, we have*

$$x_{r+1}^- \Psi = \sum_{j=1}^{r} (-1)^{r-j} \lambda_{r+1-j} x_j^- \Psi \,. \tag{2.4}$$

Moreover, it leads to the following:

$$x_{r+1+p}^- \Psi = \sum_{j=1}^{r} (-1)^{r-j} \lambda_{r+1+p-j} x_{j+p}^- \Psi \,, \quad \text{for } p \in \mathbf{Z}_{\geq 0}. \tag{2.5}$$

Proof. Relation (2.4) is derived from (A_{r+1}) of lemma 2.4. Making use of $x_{r+1+n}^- = (-2)^{-1}[h_n, x_{r+1}^-]$ and (2.4), we derive (2.5). □

Proposition 2.2. *Suppose that x_1^- is nilpotent of degree r in V_B. In the sector of $h_0 = d_0 - 2n$ with $0 \leq n \leq r$, every vector is expressed as a sum of monomial vectors $x_{k_1}^- \cdots x_{k_n}^- \Psi$ for integers $k_1, k_2, \ldots, k_n$ satisfying $1 \leq k_1 \leq k_2 \leq \cdots \leq k_n \leq r$.*

Proof. It is clear from (2.5). □

3. Generalized Drinfeld Polynomials $P_\Psi(u)$ for V_B

Definition 3.1. Suppose that x_1^- is nilpotent of degree r in V_B. We define a polynomial $P_\Psi(u)$ by

$$P_\Psi(u) = \sum_{k=0}^{r} \lambda_k (-u)^k . \tag{3.1}$$

Definition 3.2. If polynomial $P_\Psi(u)$ of V_B is factorized as

$$P_\Psi(u) = \prod_{k=1}^{s} (1 - a_k u)^{m_k} , \tag{3.2}$$

where $a_1, a_2, \ldots, a_s$ are distinct, and their multiplicities are given by $m_1, m_2, \ldots, m_s$, respectively, then we call a_j the *evaluation parameters* of highest weight vector Ψ. We denote by $\boldsymbol{a}$ the set of s parameters, $a_1, a_2, \ldots, a_s$.

We note that r is given by the sum: $r = m_1 + \cdots + m_s$. Let us define parameters $\hat{a}_i$ for $i = 1, 2, \ldots, r$, as follows:

$$\hat{a}_i = a_k \quad \text{if } m_1 + m_2 + \cdots + m_{k-1} < i \leq m_1 + \cdots + m_{k-1} + m_k . \tag{3.3}$$

Then, the set $\hat{\boldsymbol{a}} = \{\hat{a}_j \,|\, j = 1, 2, \ldots, r\}$ corresponds to the set of evaluation parameters a_j with multiplicities m_j for $j = 1, 2, \ldots, r$.

4. Generators with parameters

4.1. *Loop algebra generators with parameters*

Let A be a set of parameters such as $\{\alpha_1, \alpha_2, \ldots, \alpha_m\}$. We define generators with m parameters $x_m^{\pm}(A)$ and $h_m(A)$ as follows [6]:

$$x_m^{\pm}(A) = \sum_{k=0}^{m} (-1)^k x_{m-k}^{\pm} \sum_{\{i_1, \ldots, i_k\} \subset \{1, \ldots, m\}} \alpha_{i_1} \alpha_{i_2} \cdots \alpha_{i_k} ,$$

$$h_m(A) = \sum_{k=0}^{m} (-1)^k h_{m-k}^{\pm} \sum_{\{i_1, \ldots, i_k\} \subset \{1, \ldots, m\}} \alpha_{i_1} \alpha_{i_2} \cdots \alpha_{i_k} . \tag{3.1}$$

In terms of generators with parameters we generalize the defining relations of the sl_2 loop algebra. Let A and B are arbitrary sets of m and n parameters, respectively. The operators with parameters satisfy the following:

$$[x_m^+(A), x_n^-(B)] = h_{m+n}(A \cup B) , \quad [h_m(A), x_n^{\pm}(B)] = \pm 2 x_{m+n}^{\pm}(A \cup B) . \tag{3.2}$$

By using the relations (3.2), it is straightforward to show the following:

$$
\begin{aligned}
[x_\ell^+(A), (x_m^-(B))^{(n)}] &= (x_m^-(B))^{(n-1)} h_{\ell+m}(A \cup B) \\
&\quad - x_{\ell+2m}^-(A \cup B \cup B)(x_m^-(B))^{(n-2)} , \\
[h_\ell(A), (x_m^\pm(B))^{(n)}] &= \pm 2 (x_m^\pm(B))^{(n-1)} x_{\ell+m}^\pm(A \cup B) . \qquad (3.3)
\end{aligned}
$$

Here the symbol $(X)^{(n)}$ denotes the nth power of operator X divided by the n factorial, i.e. $(X)^{(n)} = X/n!$.

Let the symbol $\boldsymbol{\alpha}$ denote a set of m parameters, α_j for $j = 1, 2, \ldots, m$. We denote by A_j the set of all the parameters except for α_j, i.e. $A_j = \boldsymbol{\alpha} \setminus \{\alpha_j\} = \{\alpha_1, \ldots, \alpha_{j-1}, \alpha_{j+1}, \ldots, \alpha_m\}$. We introduce the following symbol:

$$
\rho_j^\pm(\boldsymbol{\alpha}) = x_{m-1}^\pm(A_j) \quad \text{for} \quad j = 1, 2, \ldots, m. \qquad (3.4)
$$

Here we note the following:

Lemma 4.1. *If $x_n^-(A)\Omega = 0$ for some set of n parameters, A, then we have $x_{n+m}^-(A \cup B)\Omega = 0$ for any set of m parameters, B.*

Hereafter, we denote by $a_j^{\otimes m}$ the set of parameter a_j with multiplicity m, i.e. $a_j^{\otimes m} = \{a_j, a_j, \ldots, a_j\}$. Moreover, in the case of $m = 1$, we write $x_1^\pm(a_j^{\otimes 1})$ simply as $x_1^\pm(a_j)$.

4.2. *Borel subalgebra generators with parameters*

In the case of the Borel subalgebra $U(B)$, we do not have generator x_0^- in $U(B)$. In order to introduce generators with parameters for $U(B)$, we thus need some trick.

For a given set of m parameters, α_j for $j = 1, 2, \ldots, m$, we introduce the extended set of parameters as follows:

$$
\boldsymbol{\alpha}^{(n)} = \boldsymbol{\alpha} \cup \{0^{\otimes n}\} . \qquad (3.5)
$$

Here we recall that $a^{\otimes n}$ denotes the set of a with multiplicity n. We also introduce the following symbols:

$$
\rho_j^\pm(\boldsymbol{\alpha}^{(1)}) = x_m^\pm(A_j^{(1)}) \quad \text{for} \quad j = 1, 2, \ldots, m. \qquad (3.6)
$$

It is easy to show

$$
\sum_{j=1}^{n} \frac{\rho_j^\pm(\boldsymbol{\alpha}^{(1)})}{\prod_{k=1;k\neq j}^{m} \alpha_{kj}} = x_{m+1-n}^\pm(\{\alpha_{n+1}, \ldots, \alpha_m\} \cup \{0\}) \quad (1 \le n \le m) . \qquad (3.7)
$$

It follows inductively on n that x_k^- for $1 \le k \le m$ are expressed in terms of linear combinations of $\rho_j^-(\boldsymbol{\alpha}^{(1)})$ with $1 \le j \le m$.

The reduction relation (2.4) is expressed as $x^-_{r+1}(\hat{\boldsymbol{a}}^{(1)})\Psi = 0$. However, if we have

$$x^-_{s+1}(\boldsymbol{a}^{(1)})\Psi = 0\,, \tag{3.8}$$

making use of (3.8), we can express monomial vector $x^-_{j_1}x^-_{j_2}\cdots x^-_{j_n}\Psi$ of any set of positive integers, $j_1,\ldots,j_n$, as a linear combination of $\rho^-_{k_1}(\boldsymbol{a}^{(1)})\rho^-_{k_2}(\boldsymbol{a}^{(1)})\cdots\rho^-_{k_n}(\boldsymbol{a}^{(1)})\Psi$ over some sets of integers with $1 \le k_1,\ldots,k_n \le r$.

5. Highest weight representations

5.1. *The case of distinct evaluation parameters*

Let us discuss the case where all the evaluation parameters a_j have multiplicity 1, i.e. $m_j = 1$ for $j = 1,\ldots,s$. We call it the case of distinct evaluation parameters. Here we note that $s = r$. We therefore have

$$x^-_{s+1}(\boldsymbol{a}^{(1)})\Psi = 0\,. \tag{3.1}$$

Lemma 5.1. *If all evaluation parameters $\hat{a}_j$ are distinct ($m_j = 1$ for all j), we have*

$$\left(\rho^-_j(\boldsymbol{a}^{(1)})\right)^2 \Psi = 0\,. \tag{3.2}$$

Proof. First, we show

$$x^+_0\,(\rho^-_j(\boldsymbol{a}^{(1)}))^2\,\Psi = 0\,. \tag{3.3}$$

From eq. (3.3) we have

$$x^+_0\,(\rho^-_j(\boldsymbol{a}^{(1)}))^{(2)}\Psi = x^-_s(A^{(1)}_j)h_s(A^{(1)}_j)\Psi - x^-_{2s}(A^{(1)}_j \cup A^{(1)}_j)\Psi\,.$$

We set $a_0 = 0$. In terms of $a_{kj} = a_k - a_j$, we have

$$h_s(A^{(1)}_j)\Psi = \prod_{k=0;k\neq j}^{s} a_{jk}\,\Psi\,,$$

and using eq. (3.1) and lemma 4.1 we have

$$x^-_{2s}(A^{(1)}_j \cup A^{(1)}_j)\Psi = a_{j0}\prod_{k=1;\neq j}^{s} a_{jk}\,x^-_s(A^{(1)}_j)\Psi\,.$$

We thus obtain eq. (3.3). Secondly, we apply $(x_0^+)^{(r-1)}(x_1^-(a_j))^{(r-1)}$ to $(\rho_j^-(\boldsymbol{a}^{(1)}))^2\Psi$. The product is given by zero since it is out of the sectors of V_Ψ due to the fact that $(r-1)+2>r$ and proposition 2.1:

$$(x_0^+)^{(r-1)}(x_1^-(a_j))^{(r-1)}\,(\rho_j^-(\boldsymbol{a}^{(1)}))^2\,\Psi = 0\,.$$

We then show that the left-hand-side is given by

$$\rho_j^-(\boldsymbol{a}^{(1)})^2\,(x_0^+)^{(r-1)}(x_1^-(a_j))^{(r-1)}\,\Psi = \prod_{k=1;k\neq j}^{r} a_{kj}\,\times\,(\rho_j^-(\boldsymbol{a}^{(1)}))^2\,\Psi\,.$$

Here, through induction on n and using B_n of lemma (2.4), we show

$$[(x_0^+)^{(n)}(x_1^-(a_j))^{(n)}, \rho_j^-(\boldsymbol{a}^{(1)})^2]\Psi = 0 \quad (n\leq r-1).$$

Since $a_{kj}\neq 0$ for $k\neq j$, we obtain eq. (3.2). □

Lemma 5.2. *Let x_1^- be nilpotent of degree r in V_B. In the sector of $h_0 = d_0-2n$ for an integer n with $0\leq n\leq r$, every vector v_n is written as*

$$v_n = \sum_{1\leq j_1<\cdots<j_n\leq s} C_{j_1,\cdots,j_n}\prod_{t=1}^{n}\rho_{j_t}^-(\boldsymbol{a}^{(1)})\,\Psi\,. \tag{3.4}$$

Suppose that $\lambda_r\neq 0$. Then, if v_n is zero, all the coefficients $C_{j_1,\cdots,j_n}$ in (3.4) are given by zero.

Proof. In terms of $\rho_j^-(\boldsymbol{a}^{(1)})$, any vector in the sector is expressed as a linear combination of $\rho_{j_1}^-(\boldsymbol{a}^{(1)})\cdots\rho_{j_n}^-(\boldsymbol{a}^{(1)})\,\Psi$. From lemma 5.1 we may assume $1\leq j_1<\cdots<j_n\leq s$. For a set of integers with $1\leq i_1,\ldots,i_n\leq s$, multiplying both sides of eq. (3.4) with $\rho_{i_1}^+(\boldsymbol{a}^{(1)})\cdots\rho_{i_n}^+(\boldsymbol{a}^{(1)})$, we have

$$\rho_{i_1}^+(\boldsymbol{a}^{(1)})\cdots\rho_{i_n}^+(\boldsymbol{a}^{(1)})v_n = C_{i_1,\cdots,i_n}\prod_{t=1}^{n}\prod_{k=0;k\neq i_t}^{s} a_{i_t k}^2\,\times\,\Psi$$

Therefore, if $v_n=0$, all the coefficients $C_{j_1,\cdots,j_n}$ are given by zero. □

From lemmas 5.1, 5.2 and proposition 2.1 we have the following:

Proposition 5.1. If evaluation parameters $\hat{a}_j$ of Ψ are distinct, the set of vectors $\prod_{t=1}^{n}\rho_{j_t}^-(\boldsymbol{a}^{(1)})\,\Psi$ for $1\leq j_1<\cdots<j_n\leq s$ gives a basis of the sector of $h_0=d_0-2n$ in V_B.

Theorem 5.1. *Let V_B denotes the finite-dimensional representation of $U(B)$ generated by a highest weight vector Ψ. If x_1^- is nilpotent of degree r in V_B and Ψ has distinct and nonzero evaluation parameters $a_1,\ldots,a_r$, then V_B is irreducible.*

Proof. We show that every nonzero vector of V_B has such an element of the loop algebra that maps it to Ψ. Suppose that there is a nonzero vector v_n in the sector of $h_0 = d_0 - 2n$ that has no such element. Then, we have

$$x_{k_1}^+ \cdots x_{k_n}^+ v_n = 0 \tag{3.5}$$

for all monomial elements $x_{k_1}^+ \cdots x_{k_n}^+$. Here v_n is expressed in terms of the basis vectors $\rho_{j_1}^-(\boldsymbol{a}^{(1)}) \cdots \rho_{j_n}^-(\boldsymbol{a}^{(1)})\Psi$ with coefficients $C_{j_1,\ldots,j_n}$ and $1 \le j_1 < \cdots < j_n \le s$, as in (3.4). Then, by the same argument as in lemma 5.2 we show that all the coefficients $C_{j_1,\ldots,j_n}$ vanish. However, this contradicts with the assumption that v_n is nonzero. It therefore follows that v_n has such an element that maps it to Ψ. We thus obtain the theorem. □

5.2. *The case of degenerate evaluation parameters*

Let us discuss a general criteria for a finite-dimensional highest weight representation to be irreducible.

Theorem 5.2. *Recall that V_B is a finite-dimensional representation of the Borel subalgebra $U(B)$ generated by a highest weight vector Ψ that has evaluation parameters a_j with multiplicities m_j for $j = 1, 2, \ldots, s$. Suppose that x_1^- is nilpotent of degree r and the evaluation parameters are nonzero, i.e. $a_1 a_2 \cdots a_s \neq 0$. We also recall that $\boldsymbol{a}$ denotes the set of evaluation parameters: $\boldsymbol{a} = \{a_1, a_2, \ldots, a_s\}$. Then, V_B is irreducible if and only if $x_{s+1}^-(\boldsymbol{a}^{(1)})\Psi = 0$.*

We prove it by generalizing the proof of theorem 5.1 (cf. Ref. [6]).

Theorem 5.2 plays an important role when we discuss the spectral degeneracy of the twisted XXZ spin chain at roots of unity associated with the Borel subalgebra $U(B)$ of the sl_2 loop algebra. Here the spin chain satisfies the twisted boundary conditions. We show in some sectors that a regular Bethe ansatz eigenvector $|R; \Phi\rangle$ is a highest weight vector of the Borel subalgebra $U(B)$ for some twist angle Φ [5,8]. It is nontrivial whether the highest weight representation V_B generated by $|R; \Phi\rangle$ is irreducible or not. Suppose that x_1^- is nilpotent of degree r in V_B, $|R; \Phi\rangle$ has nonzero evaluation parameters a_j with multiplicities m_j for $j = 1, 2, \ldots, s$, where $m_1 + \cdots + m_s = r$, and we have the following relation:

$$x_{s+1}^-(\boldsymbol{a}^{(1)})\, |R; \Phi\rangle = 0\,, \tag{3.6}$$

where $\boldsymbol{a}$ denotes the set of evaluation parameters $a_1, a_2, \ldots, a_s$. Then, it follows from theorem 5.2 that V_B is irreducible, and the degenerate multiplicity of $|R; \Phi\rangle$ is given by $(m_1 + 1)(m_2 + 1) \cdots (m_s + 1)$.

Acknowledgments

The author would like to thank the organizers for their kind hospitality and invitation to the 23rd International Conference of Differential Geometric Methods in Theoretical Physics, August 20-26, 2005, Nankai Institute of Mathematics, Tianjin, China. He is also like to thank Dr. A. Nishino for his interest in the research. This work is partially supported by Grant-in-Aid for Scientific Research (C) No. 17540351.

References

1. V. Chari and A. Pressley, Quantum Affine Algebras, Commun. Math. Phys. **142** (1991) 261–283.
2. V. Chari and A. Pressley, Quantum Affine Algebras at Roots of Unity, Representation Theory **1** (1997) 280–328.
3. V. Chari and A. Pressley, Weyl modules for classical and quantum affine algebras, Representation Theory **5** (2001) 191–223.
4. T. Deguchi, The sl_2 loop algebra symmetry of the twisted transfer matrix of the six-vertex model at roots of unity, J. Phys. A **37** (2004) 347–358.
5. T. Deguchi, Regular XXZ Bethe states as highest weight vectors of the sl_2 loop algebra, cond-mat/0503564.
6. T. Deguchi, The Six-Vertex Model at Roots of Unity and some Highest Weight Representations of the sl_2 Loop Algebra, to appear in Ann. Henri Poincaré (2006 Birkhäuser Verlag Basel/Switzerland) (cond-mat/0603112).
7. T. Deguchi, K. Fabricius and B. M. McCoy, The sl_2 Loop Algebra Symmetry of the Six-Vertex Model at Roots of Unity, J. Stat. Phys. **102** (2001) 701–736.
8. T. Deguchi and K. Kudo, Spectral degeneracy of the twisted XXZ spin chain, in preparation.
9. V.G. Drinfeld, A new realization of Yangians and quantized affine algebras, Soviet Math. Doklady **36** (1988) 212–216.
10. K. Fabricius and B. M. McCoy, Bethe's Equation Is Incomplete for the XXZ Model at Roots of Unity, J. Stat. Phys. **103**(2001) 647–678.
11. K. Fabricius and B. M. McCoy, Completing Bethe's Equations at Roots of Unity, J. Stat. Phys. **104**(2001) 573–587.
12. K. Fabricius and B. M. McCoy, Evaluation Parameters and Bethe Roots for the Six-Vertex Model at Roots of Unity, in Progress in Mathematical Physics Vol. 23 (*MathPhys Odyssey 2001*), edited by M. Kashiwara and T. Miwa, (Birkhäuser, Boston, 2002) 119–144.
13. N. Jacobson, *Lie algebras* (Wiley, New York, 1962)
14. C. Korff, The twisted XXZ chain at roots of unity revisited, J. Phys. A **37** (2004) 1681–1689.

On the Physical Significance of q-deformation in Many-body Physics

J. P. Draayer[1], K. D. Sviratcheva[1], C. Bahri[1], and
A. I. Georgieva[2]

[1] *Louisiana State University, Department of Physics and Astronomy, Baton Rouge, Louisiana, 70808-4001 USA*
[2] *Institute of Nuclear Research and Nuclear Energy, Bulgarian Academy of Sciences, Sofia 1784, Bulgaria*

A quantum extension of an algebraic $\mathfrak{sp}(4)$ model is applied to a study of pairing correlations in nuclei with mass $40 \leq A \leq 100$. While a reasonable overall description of certain nuclear properties is achieved in the nondeformed limit of the theory, the q-deformation brings forward superior results and plays a significant role in understanding nonlinear effects in many-body physics.

1. Introduction

The concept of quantum (or q-) deformation, formulated by Drinfeld and Jimbo[1–3], arose in physics. Originally, the q-analog of SU(2) appeared in the application of the quantum inverse scattering method to 2-dimensional models in quantum field theory and statistical mechanics[4]. Thereafter, especially following the introduction of the q-deformed harmonic oscillator[5,6], considerable attention has been focused on studies based on the novel and promising approach of quantum deformation in various fields of physics[3,7]. In recent years, in addition to purely mathematical examinations of quantum algebraic concepts (see e.g.[8]), and particularly of quantum symplectic algebras[9–12], studies of interest include applications in string/brane theory, conformal field theory, statistical/quantum mechanics, and metal clusters[13], as well as in nuclear physics[14,15].

The earliest applications of the quantum algebraic concept to nuclear structure were related to an $\mathrm{SU}_q(2)$ description of rotational bands in axially deformed nuclei[16]. In the realm of the pairing correlations models the quantum deformation concept was introduced first for like-particle pairing[17] based on an $\mathfrak{su}_q(2)$ approach, which was later extended to $\mathfrak{so}_q(5)$ to include

pn pairing correlations[18]. Even though optimum values of the q-parameter have achieved an overall improved fit to the experimental energies, the question on the physical nature of q-deformation when applied to the nuclear many-body problem remains open.

Pairing, introduced in physics for describing superconductivity, is fundamental to condensed matter, nuclear, and astrophysical phenomena of recent interest. In nuclear physics, the "quasi-spin" symplectic Sp(4) group [with a Lie algebra isomorphic to $\mathfrak{so}(5)$[19–21]] together with its dual Sp(2Ω), for 2Ω shell degeneracy, use the seniority quantum number[22,23] to classify the nuclear energy spectra. A two-body microscopic model with Sp(4) dynamical symmetry allows one to focus on like-particle (pp and nn) and proton-neutron pn isovector (isospin $T = 1$) pairing correlations and, in addition, to include a pn isoscalar ($T = 0$) interaction. While nuclear properties are generally well-described within this framework[24,25], nonlinear local deviations due to many-body interactions can be modelled[26] by a q-deformed extension of $\mathfrak{sp}(4)$. In general, many-body interactions are rather complicated to handle, nevertheless, they introduce an overall improvement of the theory[27]. An important property of the q-deformed model is that it does not violate physical laws fundamental to a quantum mechanical nuclear system and conserves the angular momentum, the total number of particles, and the isospin projection.

2. Nonlinear pairing model

Mathematically, a deformation parameter (q) is used to realize a mapping of c-numbers (or operators) X into their q-equivalents: $[X]_p \doteq \frac{q^{pX}-q^{-pX}}{q^p-q^{-p}} \stackrel{q\to 1}{\to} X$ (denoted $[X]$ when $p = 1$) and hence $[X]_p$ is nonlinear in X. A feature of any quantum algebra is that in the $q \to 1$ ($\varkappa \to 0$, $q = e^{\varkappa}$) limit, one recovers the nondeformed results.

The $\mathfrak{sp}_q(4)$ deformed algebra[11,28,29] is realized in terms of q-deformed fermion operators, $\alpha^\dagger_{\nu=\{jm\sigma\}}$ and α_ν, each of which creates and annihilates a nucleon with isospin σ ($\pm\frac{1}{2}$ for proton/neutron) in a single-particle state of total angular momentum j (half-integer) with third projection m. The q-operators are defined through their anticommutation relations[29],

$$\begin{array}{ll} \{\alpha_{jm\sigma}, \alpha^\dagger_{j'm'\sigma}\}_{q^{\pm 1}} = q^{\pm\frac{N_{2\sigma}}{2\Omega}}\delta_{j,j'}\delta_{m,m'}, & \{\alpha_{jm\sigma}, \alpha^\dagger_{j'm'\sigma'}\} = 0, \sigma \neq \sigma', \\ \{\alpha^\dagger_{jm\sigma}, \alpha^\dagger_{j'm'\sigma'}\} = 0, & \{\alpha_{jm\sigma}, \alpha_{j'm'\sigma'}\} = 0, \end{array} \tag{2.1}$$

where the q-anticommutator is $\{A, B\}_{q^p} = AB + q^p BA$ and $2\Omega = \sum_j (2j+1)$ is the space dimension for given σ. The $N_{2\sigma=\pm 1}$ proton/neutron number operators in (2.1) belong to the $\mathfrak{sp}_q(4)$ basis operators set, yet they remain undeformed ($N_{2\sigma} = \sum_{j,m} c^\dagger_{jm\sigma} c_{jm\sigma}$, where $c^{(\dagger)}_\nu$ are nondeformed fermion operators). Hence, the quantum approach retains the physical meaning of observables such as the total nucleon number operator, $N = N_{+1} + N_{-1}$, and the isospin projection, $T_0 = \frac{1}{2}(N_{+1} - N_{-1})$. In addition, number conservation requires counting of q-deformed particles to proceed in the same fashion as in the $q \to 1$ limit,

$$[N_{2\sigma'}, \alpha^\dagger_{jm\sigma}] = \delta_{\sigma,\sigma'} \alpha^\dagger_{jm\sigma}, \quad [N_{2\sigma'}, \alpha_{jm\sigma}] = -\delta_{\sigma,\sigma'} \alpha_{jm\sigma}. \tag{2.2}$$

The set of anticommutation relations for the q-deformed fermion operators can be chosen from among various possibilities (for example, see [11,30]), each of them suitable for a certain mathematical application. However, if we start with the usual q-deformed anticommutation relations for fermions, which is analogous to the q-deformed commutation relations for n creation (annihilation) boson system that realizes the standard Drinfeld-Jimbo quantum $\mathfrak{u}_q(n)$ algebra[1,2], namely, $\alpha_{jm\sigma}\alpha^\dagger_{jm\sigma} + q^{\pm 1}\alpha^\dagger_{jm\sigma}\alpha_{jm\sigma} = q^{\pm N_{jm,2\sigma}}$, with $N_{jm,2\sigma} = c^\dagger_{jm\sigma} c_{jm\sigma}$ counting the particles of type σ in a (j,m)-state, the relation follows, $\alpha^\dagger_{jm\sigma}\alpha_{jm\sigma} = [N_{jm,2\sigma}]$. Clearly, this relation turns out to be undeformed due to the fermion nature of the nucleons, that is $N_{jm,2\sigma} = 0$ or 1, and essentially leads back to a non-deformed algebra in contrast to our definitions (2.1). The anticommutation relations (2.1) for two conjugate fermion operators, $\alpha_{jm\sigma}$ and $\alpha^\dagger_{jm\sigma}$, yield $\alpha^\dagger_{jm\sigma}\alpha_{jm\sigma} = [\frac{N_{2\sigma}}{2\Omega}]$ and hence $\sum_{j,m} \alpha^\dagger_{jm\sigma}\alpha_{jm\sigma} = 2\Omega[\frac{N^{(j)}_{2\sigma}}{2\Omega}]$. Such a relation coincides with the nondeformed definition of the total particle number operator in the $q \to 1$ limit as $\alpha^{(\dagger)}_{jm\sigma} \to c^{(\dagger)}_{jm\sigma}$ and hence justifies the introduction of the $1/(2\Omega)$ factor into the novel set of anticommutations (2.1).

The basis operators[29], $T_\pm$ and $A^{(\dagger)}_{1,0,-1}$, of the $\mathfrak{sp}_q(4)$ algebra are constructed as eight bilinear products of the fermion q-operators coupled to total angular momentum and parity $J^\pi = 0^+$,

$$A^\dagger_{k=\sigma+\sigma'} = \frac{1}{\sqrt{2\Omega(1+\delta_{\sigma\sigma'})}} \sum_{jm} (-1)^{j-m} \alpha^\dagger_{jm\sigma}\alpha^\dagger_{j,-m,\sigma'} = (A_{-k})^*, \tag{2.3}$$

$$A_{-k} = \frac{1}{\sqrt{2\Omega(1+\delta_{\sigma\sigma'})}} \sum_{jm} (-1)^{j-m} \alpha_{j,-m,\sigma}\alpha_{jm\sigma'}, \tag{2.4}$$

$$T_\pm = \frac{1}{\sqrt{2\Omega}} \sum_{jm} \alpha^\dagger_{jm,\pm 1/2}\alpha_{jm,\mp 1/2}, \tag{2.5}$$

in addition to the two Cartan operators $N_{\pm 1}$ of $\mathfrak{sp}(4)$. In the $q \to 1$ limit, $T_{0,\pm}$ are associated with isospin and $A^{(\dagger)}_{\mp 1,0,\pm 1}$ annihilate (create) a proton-proton, proton-neutron, or neutron-neutron $J = 0$ pair. The latter construct a q-deformed basis, $|n_1, n_0, n_{-1})_q = (A^\dagger_1)^{n_1}(A^\dagger_0)^{n_0}(A^\dagger_{-1})^{n_{-1}}|0\rangle$, specified by the n_1 proton-proton pair number, n_0 proton-neutron pair number, and n_{-1} neutron-neutron pair number.

While an explicit form for the second-order Casimir operator of $\mathfrak{sp}_q(4)$ for other q-deformed schemes can be given[10], this is not a simple task for the case of the deformation (2.1) suitable for nuclear physics applications because it includes, by construction, a dependence on the shell structure. Nevertheless, we are able to find a q-deformed second-order operator[31], $O_2(\mathfrak{sp}_q(4))$, that is diagonal in the q-deformed basis and that in the $q \to 1$ limit reverts to the second-order $\mathfrak{sp}(4)$ Casimir invariant. It can be expressed through the C_2 Casimir invariants of the $\mathfrak{su}_q^k(2)$ subalgebras of $\mathfrak{sp}_q(4)$ with $k = \pm 1, 0, T$ denoting like-particle, proton-neutron and isospin symmetries[29],

$$O_2(\mathfrak{sp}_q(4)) = \textstyle\sum_k \beta_k \frac{C_2(\mathfrak{su}_q^k(2))}{\Omega} - \frac{\beta_1}{2}\left[\frac{2}{\Omega}\right] \rho \{ \left[\frac{N_{+1}-\Omega}{2}\right]^2_{\frac{1}{\Omega}} + \left[\frac{N_{-1}-\Omega}{2}\right]^2_{\frac{1}{\Omega}} \}, \tag{2.6}$$

where $\rho = \frac{[2]+[2]_{\frac{1}{2\Omega}}}{4}$. The β's in (2.6) are q-functions of the pair numbers,

$$\beta_0 = \frac{\left[2_{n_0+n_1-\Omega-\frac{1}{2}}\right]_{\frac{1}{2\Omega}}\left[2_{n_0+n_{-1}-\Omega+\frac{1}{2}}\right]_{\frac{1}{2\Omega}} + \left[2_{n_0+n_1-\Omega+\frac{1}{2}}\right]_{\frac{1}{2\Omega}}\left[2_{n_0+n_{-1}-\Omega-\frac{1}{2}}\right]_{\frac{1}{2\Omega}}}{2\left[2_{n_1-\Omega-\frac{1}{2}}\right]_{\frac{1}{2\Omega}}\left[2_{n_{-1}-\Omega-\frac{1}{2}}\right]_{\frac{1}{2\Omega}}},$$

$$\beta_{\pm 1} \equiv \beta_1 = \frac{\Phi(n_0-1)\Phi(n_0-2)}{\frac{2\sqrt{\rho_+\rho_-}}{[2]}\sum_{k=1}^{n_0-1} S_q(k)} \overset{q\to 1}{\to} 2, \quad \beta_T = 1, \tag{2.7}$$

where we define

$$S_q(k) \doteq [2_{k-\Omega-\frac{1}{2}}]_{\frac{1}{2\Omega}} \textstyle\sum_{i=0}^{k-1} \frac{[2]^i}{2^i}[2_{k-1-i}]_{\frac{1}{2\Omega}} \overset{q\to 1}{\to} 4k, \; [2_X]_k \doteq \frac{[2X]_k}{[X]_k} \overset{q\to 1}{\to} 2,$$

$$\Phi(n_0) \doteq \sum_{k=0}^{n_0} \frac{[2]^k}{2^k}[2_{n_0-k}]_{\frac{1}{2\Omega}} \overset{q\to 1}{\to} 2(n_0+1), \; \rho_\pm \doteq \frac{q^{\pm 1}+q^{\pm\frac{1}{2\Omega}}}{2} \overset{q\to 1}{\to} 1. \tag{2.8}$$

The eigenvalue of $O_2(\mathfrak{sp}_q(4))$ (2.6) in the q-deformed basis is

$$\begin{aligned}\langle O_2(\mathfrak{sp}_q(4))\rangle &= \beta_1 2\rho\left[\tfrac{1}{\Omega}\right]\left[\tfrac{\Omega-n_0}{2}\right]_{\frac{1}{\Omega}}\left[\tfrac{\Omega-n_0}{2}+1\right]_{\frac{1}{\Omega}} \\ &-\frac{\beta_1}{2}\left[\tfrac{2}{\Omega}\right]\rho\{\left[\tfrac{2n_++n_0-\Omega}{2}\right]^2_{\frac{1}{\Omega}} + \left[\tfrac{2n_-+n_0-\Omega}{2}\right]^2_{\frac{1}{\Omega}}\} \\ &+2\beta_0\left[\tfrac{1}{2\Omega}\right]\left[\tfrac{2\Omega-2(n_1+n_{-1})}{2}\right]_{\frac{1}{2\Omega}}\left[\tfrac{2\Omega-2(n_1+n_{-1})}{2}+1\right]_{\frac{1}{2\Omega}} + \left[\tfrac{1}{\Omega}\right][n_1-n_{-1}]^2_{\frac{1}{2\Omega}} \\ &+\frac{\Phi(n_0-1)(\Psi(n_0-1,n_1)+\Psi(n_0-1,n_{-1}))+\Phi(n_0)(\Psi(n_0,n_1-1)+\Psi(n_0,n_{-1}-1))}{4[2]\Omega},\end{aligned} \tag{2.9}$$

with $\Psi(n_0, n_{\pm 1}) \doteq 2\sqrt{\rho_+\rho_-}\,[n_\pm + 1]_{\frac{1}{2\Omega}}\left[2_{n_0+n_\pm+1/2-\Omega}\right]_{\frac{1}{2\Omega}} \overset{q\to 1}{\to} 4(n_{\pm 1}+1)$.

The second-order operator (2.6) is a Casimir invariant only in the non-deformed limit of the theory. Nevertheless, its importance in the q-deformed case is obvious. It is an operator that consists of number preserving products of all ten q-deformed generators, and the q-deformed pair basis states are its eigenvectors. Its zeroth-order approximation commutes with the generators of the q-deformed symplectic symmetry. It also gives a direct relation between the expectation values of the second-order products of the operators that build $O_2(\mathfrak{sp}_q(4))$. Hence the result can be used to provide for an exact solution of a q-deformed model Hamiltonian.

As for the microscopic nondeformed approach, the most general Hamiltonian[24] with q-deformed symplectic dynamical symmetry ($\mathfrak{sp}_q(4) \supset \mathfrak{su}_q(2)$) that conserves proton and neutron particle numbers is

$$\begin{aligned} H_q = &-\varepsilon_q N - G_q \textstyle\sum_{k=-1}^{1} A_k^\dagger A_{-k} - F_q A_0^\dagger A_0 - \frac{E_q}{2\Omega}\left(\mathbf{T}^2 - \Omega\left[\frac{N}{2\Omega}\right]\right) \\ &- D_q\Omega\left[\frac{1}{\Omega}\right][T_0]^2_{\frac{1}{2\Omega}} - C_q 2\Omega\left[\frac{1}{\Omega}\right]\left[\frac{N}{2}\right]_{\frac{1}{2\Omega}}\left[\frac{N}{2} - 2\Omega\right]_{\frac{1}{2\Omega}}, \end{aligned} \tag{2.10}$$

where $\mathbf{T}^2 = \Omega(\{T_+, T_-\} + \left[\frac{1}{\Omega}\right][T_0]^2_{\frac{1}{2\Omega}})$. In principle, the deformation parameters $\gamma_q = \{\varepsilon_q, G_q, F_q, E_q, D_q, C_q\}$ can differ from their nondeformed counterparts $\gamma = \{\varepsilon, G, F, E, D, C\}$, which we assume to be constant within a major shell. The model describes the behavior of N_{+1} valence protons and N_{-1} valence neutrons in the mean-field of a doubly-magic nuclear core.

The nondeformed Hamiltonian H, $H_q \overset{q\to 1}{\to} H$, is an effective two-body interaction that includes isovector pairing (parameter G) and a so-called symmetry term (E), which together with the N^2-term arise naturally from a general two-body rotational and isospin invariant microscopic interaction. Both the C and E-terms account for an isoscalar pn interaction that is diagonal in an isospin basis. These interactions govern the lowest 0^+ isobaric analog states of light and medium mass even-A nuclei ($40 \leq A \leq 100$) with protons and neutrons occupying the same major shell, where the seniority zero limit is approximately valid[24,25,32]. For these states, the nondeformed model has already proven to provide a reasonable overall description for a total of 136 nuclei[24]. This includes a remarkable reproduction of the energy of the states and their detailed structure reflecting observed $N_{+1} = N_{-1}$ irregularities and staggering patterns[25]. As a consequence, any deviation within a nucleus from the reference global behavior can be attributed to local effects which although typically small can be important for determining the detailed structure of individual nuclei[27]. As a group theoretical approach, the quantum extension of H includes many-body interactions in a very prescribed way, retaining the simplicity of the exact solution.

Moreover, the quantum model not only has the $\mathfrak{sp}_q(4) \supset \mathfrak{su}_q(2)$ dynamical symmetry, it contains the original dynamical Sp(4) symmetry.

3. *q*-Deformation and many-body interactions

From an undeformed perspective, the deformation introduces higher-order, many-body terms into a theory that starts with only one-body and two-body interactions. The way in which the higher-order effects enter into the theory is governed by the $[X]$ form. In terms of $\varkappa$, everything is tied to the deformation with $[X] = \frac{\sinh(\varkappa X)}{\sinh(\varkappa)} = X(1 + \varkappa^2 \frac{X^2-1}{6} + \varkappa^4 \frac{3X^4-10X^2+7}{360} + \ldots) \overset{\varkappa \to 0}{\to} X$. An illustrative example is the expansion in $\varkappa$ of the last term in H_q (2.10), $-C_q 2\Omega \left[\frac{1}{\Omega}\right] \left[\frac{N}{2}\right]_{\frac{1}{2\Omega}} \left[\frac{N}{2} - 2\Omega\right]_{\frac{1}{2\Omega}} = -2C_q \frac{N}{2}(\frac{N}{2} - 2\Omega) - \frac{C_q \varkappa^2 \{(16\Omega^2-24\Omega+5)(V^{(1)}+V^{(2)})+6V^{(2)}+(6-8\Omega)V^{(3)}+V^{(4)}\}}{96\Omega^2} - \ldots$, with $V^{(1)} = \sum_{\nu_1} c^\dagger_{\nu_1} c_{\nu_1}$, $V^{(2)} = \sum_{\nu_1\nu_2} c^\dagger_{\nu_1} c^\dagger_{\nu_2} c_{\nu_2} c_{\nu_1}$, $V^{(3)} = \sum_{\nu_1\nu_2\nu_3} c^\dagger_{\nu_1} c^\dagger_{\nu_2} c^\dagger_{\nu_3} c_{\nu_3} c_{\nu_2} c_{\nu_1}$, and $V^{(4)} = \sum_{\nu_1\nu_2\nu_3\nu_4} c^\dagger_{\nu_1} c^\dagger_{\nu_2} c^\dagger_{\nu_3} c^\dagger_{\nu_4} c_{\nu_4} c_{\nu_3} c_{\nu_2} c_{\nu_1}$. The zeroth-order approximation corresponds to the nondeformed two-body force and coincides with it for a strength C_q equal to C, and the higher-order terms introduce many-body interactions. The latter may not be negligible, for example, we find that the contribution of the four-body interaction in the expansion above can reach a magnitude of several MeV in nuclei in the $1f_{5/2}2p_{1/2}2p_{3/2}1g_{9/2}$.

Similarly, the zeroth-order term of H_q (2.10) coincides with the H nondeformed interaction only if the strength parameters are equal, $\gamma_q = \gamma$. This term must remain unchanged when deformation is introduced, since H has been shown to reproduce reasonably well the overall behavior common for all the nuclei in a shell. This is why we fix the values of the parameters $\gamma_q = \gamma$ and allow only $\varkappa$ to vary. The decoupling of the deformation from the γ parameters that are used to characterize the two-body interaction itself, means that the latter can be assigned best-fit global values for the model space under consideration without compromising overall quality of the theory. This in turn underscores the fact that the deformation represents something fundamentally different, a feature that cannot be "mocked up" by allowing the strengths of the nondeformed interaction to absorb its effect. In short, the q-deformation adds to the theory, which describes quite well the overall nuclear behavior, a mean-field correction along with two-, three-, and many-body interactions of a local character that can be responsible for residual single-particle and many-body effects.

The possible presence of local effects built over the global properties of the 0^+ states under consideration can be recognized within an individual nucleus by the deviation of the predicted nondeformed energy $\langle H \rangle$ from

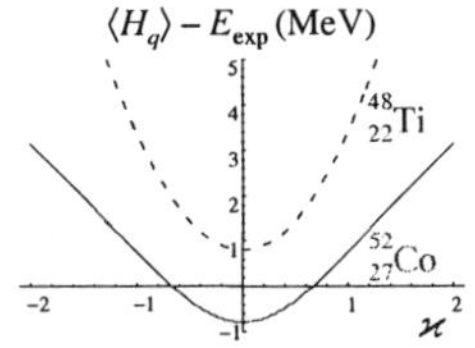

Fig. 3.1. Theoretical and experimental energy difference vs. the $\varkappa$ parameter for a typical near-closed shell nucleus (solid line) and for a mid-shell nucleus (dashed line).

the experimental value E_{exp}, namely, the solution of the equation $\langle H_q\rangle = E_{\text{exp}}$ provides a rough estimate for $\varkappa$ (see Fig. 3.1). However, in nuclei where $\langle H\rangle \geq E_{\text{exp}}$ there is no solution (see Fig. 3.1) and the theoretical prediction closest to experiment occurs at the nondeformed point, $\varkappa = 0$. The observed smooth behavior of discrete solutions for $\varkappa$ (Fig. 3.2(a)) reveals its functional dependence on the model quantum numbers. This result, even though qualitative, underscores the fact that the q-deformation as prescribed by the $\mathfrak{sp}_q(4)$ model is not random in character but rather fundamentally related to the very nature of the nuclear interaction.

This, in turn, allows us to assign a parametrized functional dependence of the deformation parameter on the total particle number N and the isospin projection T_0,

$$\varkappa(N,T_0) = \xi_1(\tfrac{N}{2\Omega}-1)(\tfrac{N}{2\Omega}+\xi_2-2\theta(N-2\Omega))e^{-0.5(\frac{2T_0}{\xi_3})^2} + \xi_4\theta(N-2\Omega)|T_0|\sqrt{\tfrac{N}{2\Omega}-1},\quad \theta(x)=\begin{cases}1, & x\geq 0\\ 0, & x<0\end{cases}, \tag{3.1}$$

which reflects the complicated development of nonlinear effects observed in Figure 3.2(a). As a next step, we use the $\varkappa(N,T_0)$ deformation function (3.1) to fit the minimum eigenvalues of H_q (2.10) to the relevant experimental energies of the even-even nuclei in the $1f_{7/2}$ and $1f_{5/2}2p_{1/2}2p_{3/2}1g_{9/2}$ shells. In doing this, we minimize any renormalization of the q-deformed parameter due to a possible influence of other local effects that are not present in the model. In the fitting procedure, only the four parameters $(\xi_{1,2,3,4})$ of $\varkappa(N,T_0)$ in (3.1) are varied. Determined statistically, they provide an estimate for the overall significance of q-deformation within a shell.

The $q \neq 1$ results are uniformly superior to those of the nondeformed limit. In the $1f_{5/2}2p_{1/2}2p_{3/2}1g_{9/2}$ shell, for example, the q-deformed model, $SOS_q = 130.21$ MeV^2 ($\chi_q = 1.28$ MeV)*, clearly improves the nondeformed

* SOS is defined as the sum of the squared differences in the theoretical and experimental

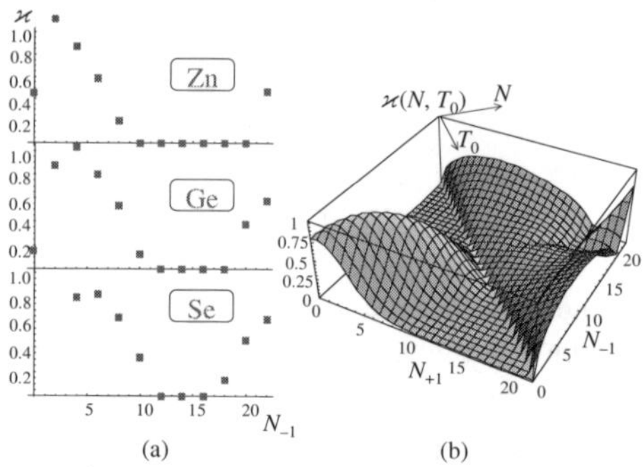

Fig. 3.2. $\varkappa$-Parameter estimation: (a) within each nucleus, and (b) $\varkappa(N, T_0)$ within the $1f_{5/2}2p_{1/2}2p_{3/2}1g_{9/2}$ shell (global parameters: $\varepsilon = 13.851$, $G/\Omega = 0.296$, $F/\Omega = 0.056$, $E/(2\Omega) = -0.489$, $D = -0.307$, and $C = 0.190$ in MeV).

theory, $SOS = 271.63$ MeV^2 ($\chi = 1.79$ MeV). The optimum results are achieved for $\xi_1 = -2.13$, $\xi_2 = 0.37$, $\xi_3 = 3.07$, $\xi_4 = 0.15$. The behavior of the q deformation (as prescribed by (3.1)) is consistent in both of the regions considered (shells $1f_{7/2}$ and $1f_{5/2}2p_{1/2}2p_{3/2}1g_{9/2}$). As a whole, the model with the local q improves the energy prediction compared to the nondeformed global model and reproduces more closely the experiment numbers (see Fig. 3.3). One reason may be that the q-deformed fermions, unlike usual quasiparticles, indeed obey the fundamental laws.

The many-body nature of the interaction is most important away from mid-shell and for many even-even nuclei tends to peak [with significant values of q] when $N_{+1} = N_{-1}$ where strong pairing correlations are expected (see Fig. 3.2). Values of the deformation parameter $q \approx 1$ may be found in nuclei with only one or two particle/hole pairs from a closed shell. For these nuclei the number of particles is insufficient to sample the effect of higher-order terms in a deformed interaction and the nondeformed limit gives a good description. Around mid-shell ($N \approx 2\Omega$) the deformation adds little improvement to the $\varkappa = 0$ theory. This suggests that for these nuclei the many-body interactions as prescribed by $\varkappa(N, T_0)$ in (3.1) are negligible and the model is not sufficient to describe other types of local effects that may be present. The results imply that even though the q-parameter gives additional freedom for all the nuclei, it only improves the model around regions of dominant pairing correlations. In short, the pair formation favors the nonnegligible higher-order interactions between the pair constituents

---

energies, and χ^2 is the averaged SOS per a degree of freedom in the statistics.

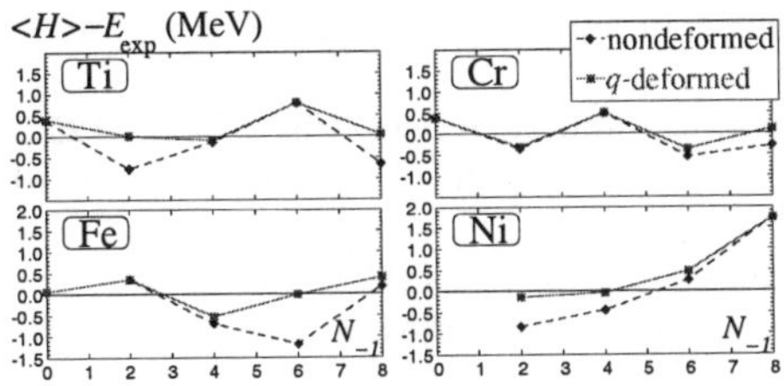

Fig. 3.3. The q-deformed and nondeformed energies compared to experimental values for even-even isotopes in the $1f_{7/2}$ shell (global parameters: $\varepsilon = 13.149$, $G/\Omega = 0.453$, $F/\Omega = 0.072$, $E/(2\Omega) = -1.120$, $D = 0.149$, and $C = 0.473$ in MeV).

that are detected via the $\mathfrak{sp}_q(4)$ model.

In summary, a q-deformed nonlinear extension of the Sp(4) model, which is the underlying symmetry for describing isovector pairing correlations and pn isoscalar interactions in atomic nuclei, was constructed. When compared to experimental data, the theory shows a smooth functional dependence of the deformation parameter q on the proton and neutron numbers. In addition, q-deformation yields results uniformly superior to those of the nondeformed limit and detects the local presence and importance of many-body interactions accompanying dominant pairing correlations in nuclei. The outcome suggests that q-deformation has physical significance extending to the very nature of the nuclear interaction itself and beyond what can be achieved by simply tweaking the parameters of a two-body interaction. The role of q-deformation is not model limited, it can extend to include a description of various many-body effects.

This work was supported by the US National Science Foundation, Grant Number 0140300.

References

1. M. Jimbo, Lett. Math. Phys. **10**, 63 (1985); **11**, 247 (1986).
2. V. Drinfeld, Proc. Int. Congress on Math, Vol. 1 (UCPress, Berkeley: 1986).
3. L. D. Faddeev, N. Y. Reshetikhin and L. A. Takhtajan, Algebr. Anal. **1**, 129 (1988).
4. P.P. Kulish and N.Yu. Reshetikhin, J. Soviet Math. **23**, 2435 (1983); E. Sklyanin, L. Takhatajan and L. Faddeev, Theor. Math. Phys. **40**, 688 (1979).
5. L. C. Biedenharn, J. Phys. A: Math. Gen. **22**, L873 (1989); A. J. Macfarlane, J. Phys. A: Math. Gen. **22**, 4581 (1989).
6. P. P. Kulish and E. V. Damaskinski, J. Phys. A: Math. Gen. **23**, L415 (1990).
7. L. C. Biedenharn and M. A. Lohe, *Quantum Group Symmetry and q-Tensor Algebras* (World Scientific Publishing Co. Pte. Ltd., 1995).

8. Lianrong Dai, Feng Pan and J. P. Draayer, J. Phys. A: Math. Gen. **34**, 6585 (2001); 6595 (2001).
9. J. Lukierski, A. Nowicki and H. Ruegg, Phys. Lett. **B271**, 321 (1991).
10. B. Abdesselam, D. Arnaudon and A. Chakrabarti, J. Phys. A: Math. Gen. **28**, 3701 (1995).
11. T. Hayashi, Commun. Math. Phys. **127**, 129 (1990).
12. J. P. Draayer, A. I. Georgieva and M. I. Ivanov, J. Phys. A: Math. Gen. **34**, 2999 (2001).
13. A. Ballesteros, N.R. Bruno and F.J. Herranz, Phys. Lett. **B574**, 276 (2003); F.A. Bais, B.J. Schroers, and J.K. Slingerland, Phys. Rev. Lett. **89**, 181601 (2002); A. Algin, M. Arik, and A.S. Arikan, Phys. Rev. E **65**, 026140 (2002); Jian-zu Zhang, Phys. Lett. **B477**, 361 (2000); D. Bonatsos et al., Phys. Rev. A **62**, 013203 (2000).
14. D. Bonatsos, B.A. Kotsos, P.P. Raychev, and P.A. Terziev, Phys. Rev. C **66**, 054306 (2002).
15. A. Ballesteros, O. Civitarese, F.J. Herranz, and M. Reboiro, Phys. Rev. C **66**, 064317 (2002); A. Ballesteros, O. Civitarese, and M. Reboiro, Phys. Rev. C **68**, 044307 (2003)
16. P.P. Raychev, R.P. Roussev, and Yu.F. Smirnov, J. Phys. G: Nucl. Part. Phys. **16**, L137 (1990); D. Bonatsos, E. N. Argyres, S. B. Drenska, P. P. Raychev, R. P. Roussev and Yu. F. Smirnov, Phys. Lett. **251B**, 477 (1990).
17. S. Shelly Sharma, Phys. Rev. C **46**, 904 (1992); D. Bonatsos, J. Phys. A: Math. Gen. **25**, L101 (1992).
18. S. Szpikowski, W. Berej and L. Prochniak, *Symmetries in Science X*, (Plenum Press, NY, 1997).
19. A.K. Kerman, Ann. Phys. (NY) **12**, 300 (1961).
20. K. Helmers, Nucl. Phys. **23**, 594 (1961).
21. K.T. Hecht, Nucl. Phys. **63**, 177 (1965); Phys. Rev. **139**, B794 (1965); Nucl. Phys. A **102**, 11 (1967); J.N. Ginocchio, Nucl. Phys. **74**, 321 (1965).
22. G. Racah, Phys. Rev **63**, 367 (1943).
23. B.H. Flowers, Proc. Roy. Soc.(London)**A212**, 248 (1952).
24. K.D. Sviratcheva, A.I. Georgieva, and J.P. Draayer, J. Phys. G: Nucl. Part. Phys. **29**, 1281 (2003).
25. K.D. Sviratcheva, A.I. Georgieva, and J.P. Draayer, Phys. Rev. C **69**, 024313 (2004).
26. K. D. Sviratcheva, C. Bahri, A. I. Georgieva, and J.P. Draayer, Phys. Rev. Lett. **93**, 152501 (2004).
27. P. Navrátil and B.R. Barrett, Phys. Rev. C **59**, 1906 (1999); P. Navrátil and W.E. Ormand, Phys. Rev. C **68**, 034305 (2003); Feng Pan, V.G. Gueorguiev, and J.P. Draayer, Phys. Rev. Lett. **92**, 112503 (2004).
28. B. Abdesselam, D. Arnaudon, and A. Chakrabarti, J. Phys. A: Math. Gen. **28**, 3701 (1995).
29. K. D. Sviratcheva, A. I. Georgieva, V. G. Gueorguiev, J. P. Draayer and M. I. Ivanov, J. Phys. A: Math. Gen. **34**, 8365 (2001).
30. V. Rittenberg and M. Scheunert, J. Math. Phys. **33**, 436 (1992); W. Pusz, *Rep. Math. Phys.* **27**, 349 (1989); G. Fiore, *Clifford Algebras and Their Ap-*

plications in Mathematical Physics: Algebra and Physics, ed. R. Ablamowicz and B. Fauser, (Birkhaeuser, Basle), Vol.1, 269 (2000).
31. K. D. Sviratcheva, A. I. Georgieva and J. P. Draayer, J. Phys. A: Math. Gen. **36**, 7579 (2003).
32. J. Engel, K. Langanke, and P. Vogel, Phys. Lett. **B389**, 211 (1996).

A Matrix Product Ansatz Solution of an Exactly Solvable Interacting Vertex Model*

A. A. Ferreira and F. C. Alcaraz

Universidade de São Paulo,
Instituto de Física de São Carlos, CP 369, 13560-970, São Carlos, SP, Brazil

A special family of solvable five-vertex model is introduced on a square lattice. Beyond the usual nearest neighbor interactions, the vertices defining the model also interact along one of the diagonals of the lattice. This family of models includes on a special limit the standard six-vertex model [1,2]. The exact solution of these models gives the first application of the matrix product ansatz introduced recently and applied successfully in the solution of quantum chains.

1. Introduction

We are going to introduce and solve a special family of five-vertex models where besides the usual nearest-neighbour interactions, imposed by their connectivity, there exist additional interactions among more distant vertices. We are going to show that this family of models is exactly integrable and contains as a special case the standard six-vertex model. The solution of these models will be obtained by the exact diagonalization of the diagonal-to-diagonal transfer matrix. The exact solution of transfer matrices associated to vertex models or quantum hamiltonians are usually obtained through the Bethe ansatz [3] on its several formulations. This ansatz asserts that the amplitudes of the eigenfunctions of these operators are given by a sum of appropriate plane waves. Instead of using the Bethe ansatz, the solution we are going to derived will be obtained through a matrix product ansatz introduced recently [4]. In this ansatz, the amplitudes of the eigenfunctions are given in terms of a matrix product of matrices obeying special algebraic relations. The present paper presents the first application of the matrix product ansatz for the exact solution of a transfer matrix.

* This work was supported in part by FAPESP and CNPq (Brazilian Agencies)

2. The interacting five-vertex model

The family of vertex models we are going to introduce and solve are defined on a square lattice with M horizontal lines and L vertical rows (see Fig. 1a).

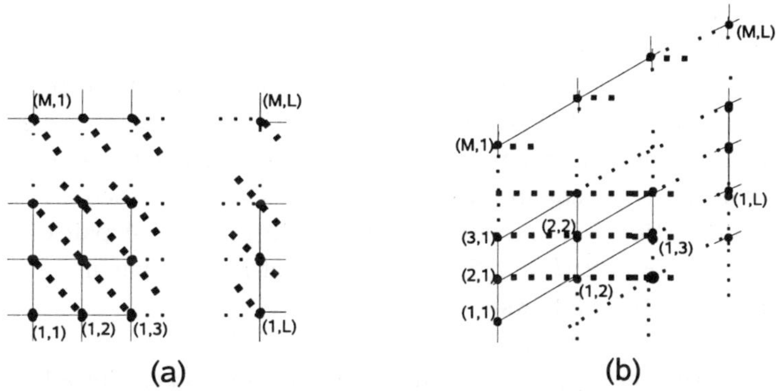

Fig. 1. (a) Square lattice with M horizontal lines and L vertical rows. The extra interactions are along the dashed diagonals. (b)The distorted diagonal lattice where the extra interactions are along the horizontal.

As in the six-vertex model we impose that the allowed arrow configurations only contain vertices satisfying the ice rules, namely, the fugacity of a given vertex is infinite unless among its four arrows two of of them point inward and two of them point outward of its center. According to this ice rule we have the allowed 6 vertex configurations shown in Fig. 2, with their respective fugacities $c_1, c_2, b_1, b_2, a_0, a_1$.

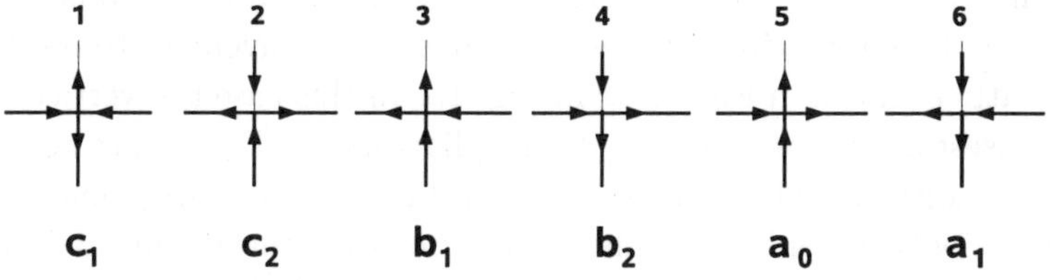

Fig. 2. Vertex configurations with respective fugacities for the six-vertex model.

The partition function is given by the sum of all possible vertex configurations with the Boltzmann weights given by the product of the fugacities of the vertices.

The family of models we are going to consider are interacting five-vertex models where besides having interactions (infinite or zero) imposed by the lattice connectivity also contains interactions among pairs of vertices at larger distances. The allowed vertex configurations, with their respective configurations are the first five configurations shown in Fig. 2. Distinctly from the six-vertex model the vertex configurations with fugacity a_1 is forbidden (zero fugacity).

These interacting five-vertex models are labeled by a fixed positive integer t that may take the values $t = 1, 2, 3, \ldots$. This parameter specify the additional interactions among the vertices. These interactions occur along the diagonals of the square lattice that go from the top left to the bottom right direction (see the dashes diagonals in Fig. 1a). A pair of vertices at distance $D = l\sqrt{2}$ ($l = 1, 2, \ldots$), in units of lattice spacing, along this diagonal interacts as follows

a) the interaction energy is zero if $l > t$

b) if one of the vertices is a_0 the interaction energy is zero for all values of l

c) if neither of the vertices is a_0 the interaction energy is infinite if $l \leq t$, except on the special case where $l = t$ and c_2 is on the left of c_1. In this case the interaction energy e_I is finite and produces a Boltzmann weight c_I given by[†]

$$c_I = e^{-\beta e_I} = \frac{a_1}{c_1 c_2} \tag{2.1}$$

3. The diagonal-to-diagonal transfer matrix

Following Bariev [5] in order to construct the diagonal-to-diagonal transfer matrix for the interacting five-vertex models it is convenient to distort the square lattice shown in Fig. 1a as in Fig. 1b. In this case the vertices which are at closest distances along the dashed diagonals of Fig. 1a are now at the the closest distance along the horizontal direction. We are going to solve the model with toroidal boundary conditions on the distorted lattice of Fig. 1b.

The vertices configurations on the distorted lattice are show in Fig. 3. We also present in this figure a convenient representation of the vertices where we only draw the arrows pointing to the botton.

[†]This notation is chosen in order to compare these interacting five-vertex models with the standard six-vertex model

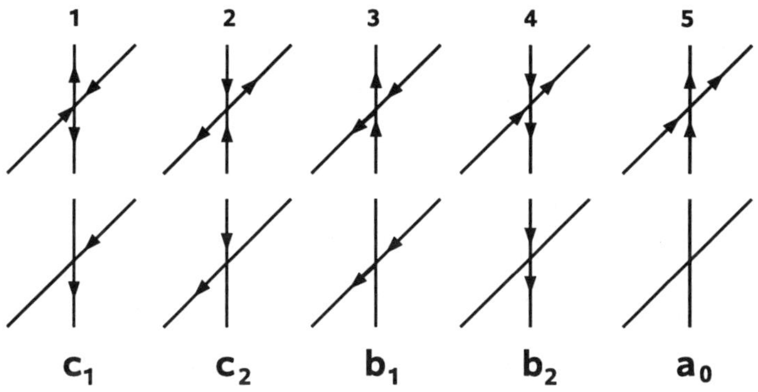

Fig. 3. Vertices configurations on the distorted lattice. A representation where only the arrows pointing to the bottom are drawn on the second line.

Using the vertex representation of the second line of Fig. 3, it is interesting to observe that in the distorted lattice the allowed arrow configurations on the horizontal lines beside having a fixed number of arrows can be interpreted as if the arrows would have an effective size $s = 2t + 1$ $s = 3, 5, \ldots$, in lattice spacing units of the distorted lattice. An arrow on a given link has a hard-core interaction that exclude the occupation of other arrows at the link itself as well as the $2t$-nearest links on its right.

The interpretation where the arrow have an effective size allows us a simple extension of our model to case where $t = 0$. In this case the arrows have a unit size and the extra hard-core interaction among the vertices with fugacities c_2 and c_1 occurs when the arrows are at the same site, giving an extra vertex configuration with total contribution $c_1c_2\frac{a_1}{c_1c_2} = a_1$ and the model reduces to the well known six-vertex model, with fugacites given in Fig. 2.

4. The matrix product ansatz and the diagonalization of the transfer matrix T_{D-D}

As a consequence of the arrow conservation and the translation invariance of the arrow configurations on the horizontal lines of Fig. 1b the matrix T_{D-D} split into block disjoint sectors labeled by the number n ($n = 0, 1, \ldots, 2L$) of arrows and momentum $p = \frac{2\pi}{L}j$ ($j = 0, 1 \ldots, L-1$). We want to solve, in each of these sectors, the eigenvalue equation

$$\Lambda_{n,p}|\Psi_{n,p}> = T_{D-D}|\Psi_{n,p}>, \tag{4.1}$$

where $\Lambda_{n,p}$ and $|\Psi_{n,p}>$ are the eigenvalues and eigenvectors of T_{D-D}, respectively. These eigenvectors in general can be written as

$$|\Psi_{n,p}>=\sum_{\{x\}}\sum_{\{\alpha\}}^{(*)}\phi^{p}_{\alpha_1,\ldots,\alpha_n}(x_1,\ldots,x_n)|x_1,\alpha_1;\ldots;x_n,\alpha_n>, \quad (4.2)$$

where $\phi^{p}_{\alpha_1,\ldots,\alpha_n}(x_1,\ldots,x_n)$ is the amplitude corresponding to the arrow configuration where n arrows of type $(\alpha_1,\ldots,\alpha_n)$ are located at $(x_1,\ldots,x_n)$, respectively. The symbol (*) in (4.2) means that the sums of $\{x\}$ and $\{\alpha\}$ are restricted to the sets obeying the hard-core exclusions, for agiven interacting parameter t:

$$\begin{aligned}&x_{i+1}\geq x_i+2t+1-\delta_{\alpha_i,1}\delta_{\alpha_{i+1},2},\\&2t+1-\delta_{\alpha_1,1}\delta_{\alpha_n,2}\leq x_n-x_1\leq L-2t-1+\delta_{\alpha_n,1}\delta_{\alpha_1,2}.\end{aligned} \quad (4.3)$$

Since $|\Psi_{n,p}>$ is also an eigenvalue with momentum p the amplitudes also satisfy

$$\frac{\phi^{p}_{\alpha_1,\ldots,\alpha_n}(x_1,\ldots,x_n)}{\phi^{p}_{\alpha_1,\ldots,\alpha_n}(x_1+m,\ldots,x_n+m)}=e^{-imp},\quad m=0,1,\ldots,L-1. \quad (4.4)$$

The exact solution of (4.1) is obtained by an appropriate ansatz for the unknown amplitude $\phi^{p}_{\alpha_1,\ldots,\alpha_n}(x_1,\ldots,x_n)$. As shown in last section in the case where $t=0$ $(s=1)$ our model reduces to the standard six-vertex model and on this case an appropriate coordinate Bethe ansatz is known [6] that solve the eingenvalue equation (4.1). In this case, as usual, the amplitudes $\{\phi^{p}_{\alpha_1,\ldots,\alpha_n}(x_1,\ldots,x_n)\}$ are given by a combination of plane waves whose wavenumbers are fixed by the eigenvalue equation (4.1).

In this paper we are going to solve (4.1) for general values of t ($t=0,1,2,\ldots$) or $s=2t+1$ ($s=1,3,5,\ldots$) by using a distinct ansatz. The matrix product ansatz we are going to use was introduced in [4] for quantum integrable chains. We present in this paper the first application of this matrix product ansatz for transfer matrices. According to this ansatz the amplitudes $\phi^{p}_{\alpha_1,\ldots,\alpha_n}(x_1,\ldots,x_n)$ are obtained in terms of a matrix product of matrices satisfying an unknown associative algebra. The model is exact integrable if the eigenvalue equations fix consistently the algebraic relations among the matrices.

In order to formulate the matrix product ansatz we make a one-to-one correspondence between configurations of arrows and products of matrices.

The matrix product associated to a given arrow configuration is obtained by associating to the sites with no arrow a matrix E, to the sites with a single arrow of type α ($\alpha = 1,2$) a matrix A^{α}, and finally to the sites with two arrows‡ we associate the matrix $A^{(1)}E^{-1}A^{(2)}$. The matrix product ansatz imposes that the unknown amplitudes in (4.2) are given by the traces

$$\phi^{p}_{\alpha_1,\ldots,\alpha_n}(x_1,\ldots,x_n) = \mathrm{Tr}[E^{x_1-1}A^{(\alpha_1)}E^{x_2-x_1-1}A^{(\alpha_2)}\ldots \\ E^{x_n-x_{n-1}-1}A^{(\alpha_n)}E^{L-x_n}\Omega_p]. \tag{4.5}$$

The matrix Ω_p is introduced in order to fix the momentum p of the eigenstate $|\Psi_{n,p}>$.

The constraints imposed by the eigenvalue equation (4.1) on the sector with a fixed number n of arrows and momentum p are solved by identifying the matrices $A^{(\alpha)}$ of the ansatz (4.5) as composed by n spectral dependent matrices A_{k_i} ($i = 1,\ldots,n$),

$$A^{\alpha} = \sum_{i}^{n} \phi^{i}_{\alpha} A_{k_i} E^{1-2t}, \quad \alpha = 1,2, \tag{4.6}$$

satisfying the algebraic relation

$$EA_{k_i} = e^{ik_i}A_{k_i}E, \quad A_k\Omega_p = e^{-ipt}\Omega_p A_k, \\ A_{k_i}A_{k_j} = s(k_j,k_i)A_{k_j}A_{k_i}, \quad A^2_{k_i} = 0, \quad i = 1,\ldots,n, \tag{4.7}$$

where $s(k_i,k_j)$ is given by (4.8).

$$s(k_j,k_i) = \\ -\frac{\Lambda_1(k_i)\Lambda_1(k_j)b_1 - \Lambda_1(k_j)(b_2b_1 - c_2c_1 + a_1) + a_1b_2}{\Lambda_1(k_i)\Lambda_1(k_j)b_1 - \Lambda_1(k_i)(b_2b_1 - c_2c_1 + a_1) + a_1b_2}. \tag{4.8}$$

and

$$\Lambda_1(k) = \Lambda_1^{(l)}(k) = \\ \frac{1}{2}(b_2 + b_1e^{ik} + l[(b_2 + b_1e^{ik})^2 - 4e^{ik}(b_2b_1 - c_2c_1)]^{\frac{1}{2}}), \tag{4.9}$$

with $l = \pm 1$. Since we can always factorize one of the fugacities on the partition function we have chosen in the previous expressions $a_0 = 1$.

‡This last case only is allowed in the case of the six-vertex model where $t = 0$ and $s = 1$.

The eigenvalues of the transfer matrices and momenta are given by

$$\Lambda_n(k_1,\ldots,k_n)=\prod_{i=n}^{n}\Lambda_1(k_i),\quad p=\sum_{i=1}^{n}k_i. \tag{4.10}$$

The spectral parameters $\{k_i\}$ are fixed by the cyclic property of the trace of the matrix products appearing on the ansatz, and are given by the the solution of the nonlinear set of equations

$$e^{ik_jL}=-\prod_{l=1}^{n}s(k_j,k_l)\left(\frac{e^{ik_j}}{e^{ik_j}}\right)^{2t},\quad j=1,\ldots,n, \tag{4.11}$$

with $s(k_j,k_l)$ given by (4.8). This last equation reproduces for $t=0$ the spectral parameter equations for the six-vertex model obtained in [6] by using the coordinate Bethe ansatz.

Since $p=\sum_{j=1}^{n}k_j$ we can rewrite the spectral parameter equation (4.11) on the sector with a number n ($n=1,2,\ldots$) of arrows and momentum $p=\frac{2\pi j}{L}$ ($j=0,1,\ldots,L-1$) as

$$e^{ik_j(L-2tn)}e^{-2ipt}=-\prod_{l=1}^{n}s(k_j,k_l), \tag{4.12}$$

which implies that the eigenvalues belonging to the sector labeled by (n,p) of T_{D-D} of the interacting five-vertex model with a parameter t ($t=0,1,2,\ldots$) is related to those of the standard six-vertex model ($t=0$) on a lattice size $L^{'}=L-2nt$ and with a seam[§] along the vertical direction of Fig. 1b, that depends on the momentum p. The same phenomena also happens on quantum hamiltonians with hard-exclusion effects [8].

5. Roots of the spectral parameter equations

In order to complete the solution of any integrable model we need to find the roots of the associated spectral parameter equations (Eq.(4.11) in our case). The solution of those equations is in general a quite difficult problem for finite L. However numerical analysis on small lattices allow us to conjecture for each problem the particular distribution of roots that corresponds to the most important eigenvalues in the bulk limit ($L\to\infty$). Those are

[§]The phase e^{-i2pt} in (4.12) could be obtained by considering a six-vertex model on the geometry of Fig. 1b, but with a seam with distinct vertex fugacities along the vertical direction.

the eigenvalues with the higher absolute values in the case of transfer matrix calculations. The equations we obtained in the last section were never analyzed previously either for finite or infinite values of L. Even on the simplest case $t = 0$, where the model reduces to the six-vertex model, the spectral parameter equations obtained in [6] throught the Bethe ansatz were not analyzed.

In our general solution of last section we have, for arbitrary values of the interacting range t ($t = 0, 1, 2, \ldots$), five free parameters: $(a_1, b_1, b_2, c_1, c_2)$. The particular case where we have no interactions along the diagonals ($a_1 = 0$) is special and is not going to be considered here (see [7] for a discussion of the parametrization on this case).

In order to simplify our analysis we are going hereafter to restrict ourselves to a symmetric version of our model with only three free parameters (δ, b, c), namely,

$$a_0 = 1, \quad a_1 = \delta^2, \quad b_1 = b_2 = b\delta, \quad c_1 = c_2 = c\delta. \tag{5.1}$$

The parameter δ give us, in the case where $t = 0$, the contribution to the fugacity due to an electric field on the symmetric six-vertex model, and for general values of t, $\mu = -\ln \delta$ plays the role of a chemical potential controlling the number of arrows in the thermodynamic limit.

Instead of writing the spectral parameters equations in terms of the spectral parameters $(k_1, \ldots, k_n)$ as in (4.12) it is more convenient to write these equations in terms of the variables $\lambda_j \equiv \frac{\Lambda_1(k_j)}{\delta}$, with $\Lambda_1(k_j)$ given by (4.9). In this case the eigenvalues of T_{D-D} are given by

$$\Lambda_n = \delta^n \lambda_1 \cdots \lambda_n, \tag{5.2}$$

where $\{\lambda_j\}$ satisfy

$$\left(\frac{\lambda_j(b-\lambda_j)}{b(b-\lambda_j)-c^2}\right)^{L-2tn} e^{-2ipt} = (-)^{n+1} \prod_{l=1}^{n} \frac{\lambda_l\lambda_j - 2\Delta\lambda_j + 1}{\lambda_l\lambda_j - 2\Delta\lambda_l + 1}, \quad j = 1, \ldots, n, \tag{5.3}$$

where we have introduced the anisotropy parameter

$$\Delta = \frac{b^2 - c^2 + 1}{2b}. \tag{5.4}$$

We see from (5.2)-(5.4) that we have now only two free parameters: b and c. The interacting parameter δ that gives the contribution due to the interaction among the vertices along the diagonal, does not appear on the equations (5.3) and (5.4), it only gives an overall scale for the eigenvalues as shown in (5.2). Inspired on the usual parametrization of the six-vertex model [2] we express the parameters b and c in terms of the parameters σ and γ:

$$b = b(\gamma,\sigma) = \frac{\sin\sigma}{\sin(\gamma-\sigma)}, \quad c = c(\gamma,\sigma) = \frac{\sin\gamma}{\sin(\gamma-\sigma)},$$
$$\Delta = -\cos\gamma. \tag{5.5}$$

The fact that Δ is a real number imply that γ is real for $-1 \leq \Delta \leq 1$ and pure imaginary for $|\Delta| > 1$. Since the right-hand side of (5.3) is the same for all values of the parameter t it is interesting, as in the six-vertex model [2], to make the change of variables $\lambda_j \to \sigma_j$, where,

$$\lambda_j = \frac{\sinh(i\gamma - \sigma_j)}{\sinh\sigma_j}, \quad j = 1, \ldots, n. \tag{5.6}$$

In terms of these new variables $\{\sigma_j\}$ the spectral parameter equations (5.3) becomes

$$\left(\frac{\sinh(i\gamma-\sigma_j)\sinh(i\gamma-i\sigma-\sigma_j)}{\sinh(\sigma_j)\sinh(i\gamma+\sigma_j)}\right)^{L-2tn} e^{-2ipt} =$$
$$-\prod_{l=1}^{n}\frac{\sinh(\sigma_j-\sigma_l+i\gamma)}{\sinh(\sigma_j-\sigma_l-i\gamma)}, \quad j = 1, \ldots, n. \tag{5.7}$$

These equations are quite distinct from the corresponding spectral parameter equations derived for the row-to-row transfer matrix of the six-vertex model. Since no numerical analysis of the roots for this type of equations is reported on the literature we made an extensive numerical study of these equations for finite values of L and several values of the anisotropy Δ. In the particular case where $\Delta = 0$ ($\gamma = \frac{\pi}{2}$) these equations can be solved analytically. Solutions of these equations are obtained by the Newton method by using the distribution of roots $\{\sigma_i\}$ at $\Delta = 0$ as the starting point to obtain the corresponding roots at other values of $\Delta \neq 0$. Our numerical analysis shows that the eigenspectrum of T_{D-D} are formed by real or complex-conjugated pairs of roots ensuring that the partition function is a real number. We verified that the eigenvalue with highest modulus belonging to the sector with n arrows is real and corresponds to a zero momentum eigenstate ($p = 0$). The distribution of roots $\{\sigma\}$ corresponding to these eigenvalues have a fixed imaginary part, that depends on σ and γ, and a symmetrically distributed real part, i. e.,

$$\mathrm{Im}(\sigma_j) = \frac{\gamma - \sigma}{2}, \quad \mathrm{Real}(\sigma_j) = \mathrm{Real}(\sigma_{n-j}),$$
$$j = 1, \ldots, n. \qquad (5.8)$$

We have also verified for all sectors the occurrence of several other real eigenvalues. In these cases the corresponding roots $\{\sigma_i\}$ have imaginary parts given either by $\frac{\gamma-\sigma}{2}$ or $\frac{\gamma-\sigma}{2} - \frac{\pi}{2}$.

Due to space limitations we do present here the free energy calculations in the thermodynamic limit and the phase diagram of the models. These calculations are going to be presented on an extended version of these notes [9].

References

1. E. H. Lieb, *Phys. Rev.* **162**, 162 (1967).
2. R. J. Baxter, *Exactly Solved Models in Statistical Mechanics* (New Yoirk: Academic) (1982).
3. H. .A. Bethe, *Z. Phys.* **71**, 205 (1931).
4. F. C. Alcaraz and M. J. Lazo, *J. Phys. A: Math. Gen* **37**, L1 (2004); *J. Phys. A: Math. Gen.* **37**, 4149 (2004); *Braz. J. Phys.* **33**, 533 (2003).
5. R. Z. Bariev, *Mat. Fiz.* **49**, 261 (1981).
6. T. T. Truong and K. D. Schotte, *Nucl. Phys.* **B20[FS8]**, 77 (1983).
7. H. Y. Huang, F. Y. Wu, H. Kunz and D. Kim, *Physica A* **228**, 1 (1996).
8. F. C. Alcaraz and R. Z. Bariev, *Phys. Rev. E* **60**, 79 (1999).
9. A. A. Ferreira and F. C. Alcaraz, to be published.

A 2h-dimensional Model with Virasoro Symmetry

P. Furlan

Dipartimento di Fisica Teorica dell'Università di Trieste, Italy,
and
Istituto Nazionale di Fisica Nucleare (INFN), Sezione di Trieste, Italy

V.B. Petkova

Institute for Nuclear Research and Nuclear Energy, Sofia, Bulgaria

The set of partial differential equations for the Appell hypergeometric function in two variables $F_4(\alpha, \beta, \gamma, \alpha+\beta-\gamma+2-h; x, y)$ is shown to arise as a null vector decoupling relation in a higher dimensional generalisation of the Coulomb gas model. It corresponds to a level two singular vector of an intrinsic Virasoro algebra.

Dedicated to the memory of Mitko Stoyanov

1. Introduction

The hypergeometric function is an ubiquitous object of the two-dimensional conformal field theories, providing examples of 4-point correlation functions of various models. The reason behind this is that it is the simplest example of a solution of null vector decoupling equations associated with singular vectors in Virasoro algebra Verma modules. Thus the second order hypergeometric equation appears as a differential operator realisation of a singular vector at level two [1].

Our aim in this paper is to demonstrate that a hidden Virasoro algebra plays a similar role in a higher dimensional conformal model. In particular the singular vector at level two gives rise in even 2h-dimensional space-time to a pair of second order linear partial differential equations. These are the

Appell - Kampé de Fériet (AK) equations, [2,3]

$$\Big(x(1-x)\,\frac{\partial^2}{\partial x^2} - \; y^2\,\frac{\partial^2}{\partial y^2} - 2xy\,\frac{\partial^2}{\partial x \partial y} + \gamma\frac{\partial}{\partial x} \\ -(\alpha+\beta+1)(x\frac{\partial}{\partial x} + y\frac{\partial}{\partial y}) - \alpha\beta\Big)F = 0 \quad \text{(1.1a)}$$

$$\Big(y(1-y)\frac{\partial^2}{\partial y^2} - \; x^2\,\frac{\partial^2}{\partial x^2} - 2xy\,\frac{\partial^2}{\partial x \partial y} + \gamma'\frac{\partial}{\partial y} \\ -(\alpha+\beta+1)(x\frac{\partial}{\partial x} + y\frac{\partial}{\partial y}) - \alpha\beta\Big)F = 0 \quad \text{(1.1b)}$$

satisfied by the Appell hypergeometric functions of type (here $(\alpha)_n = \Gamma(\alpha + n)/\Gamma(\alpha))$

$$F_4(\alpha, \beta, \gamma, \gamma'; x, y) = \sum_{m,n=0}^{+\infty} \frac{(\alpha)_{m+n}(\beta)_{m+n}}{(\gamma)_m(\gamma')_n m! n!} x^m y^n\,, \tag{1.2}$$

$$\text{with} \quad \alpha + \beta - \gamma - \gamma' = h - 2\,. \tag{1.3}$$

The two variables $x = \frac{r_{13}r_{24}}{r_{14}r_{23}}\,,\, y = \frac{r_{12}r_{34}}{r_{14}r_{23}}\,;\quad r_{ij} := x_{ij}^2$ are the two anharmonic ratios, made of the coordinates $x_i \in \mathbb{R}^{(2h)}$ of a 4-point conformal invariant*. The model is a 2h-dimensional generalisation of the two-dimensional Coulomb gas model with a charge at infinity [5], described by a (sub)-canonical field with logarithmic propagator,

$$\langle \phi(x_1)\phi(x_2)\rangle \sim ((-\Box)^h)^{-1} = -\frac{1}{(4\pi)^h\,\Gamma(h)}\,\log x_{12}^2\,, \tag{1.4}$$

and scalar fields realised by vertex operators $V_\alpha(x) = e^{i\alpha\phi(x)}$; it was studied in [6]. In the two-dimensional case the system of equations (1.1) reduces, after proper change of variables, to a linear combination of two (chiral) hypergeometric equations.

The appearance of a Virasoro algebra in a four-dimensional context was pointed out many years ago by Dimitar (Mitko) Stoyanov [7], who was studying the infinite dimensional Lie algebras preserving the solutions of the Laplace equation; one of the two algebras he had constructed, contains a subalgebra isomorphic to the Virasoro algebra; see [8] for a more recent development. Stoyanov was also among the first, who advocated the relevance

*With a different interpretation of the parameters the Appell function F_4 appears in a conformal context [4] as describing the conformal partial waves.

of the logarithmic field ϕ (1.4) in various models in two and four dimensions, see e.g. [9], where this field is used in the study of the four-dimensional massless QED, leading in particular to anomalous dimensions of the spinor fields.

1.1. *The 4-point function*

We recall here briefly the construction in [6]. Consider the 4-point function in dimension $2h$ described by only one integration. It is written in terms of vertex operators (VO) as

$$\int \mathrm{d}^{2h}\, x_5\, \langle V_{\alpha_+}(x_5)\, V_{\alpha_4}(x_4)\, V_{\alpha_3}(x_3)\, V_{\alpha_2}(x_2)\, V_{\alpha_1}(x_1)\,\rangle = \tag{1.5}$$

$$\pi^h \prod_i \frac{1}{\Gamma(\delta_i)} \prod_{1\le i<j\le 4} r_{ij}^{2\alpha_i\alpha_j}\, r_{12}^{h-\delta_1-\delta_2}\, r_{13}^{h-\delta_1-\delta_3}\, r_{23}^{\delta_1-h}\, r_{14}^{-\delta_4} F(x,y)\,,$$

$\delta_i = -2\alpha_+\,\alpha_i$, and the conformal invariance imposes the condition

$$\sum_{i=1}^{4} \delta_i = 2h \quad \leftrightarrow \quad \sum_{i=1}^{4} \alpha_i + \alpha_+ = 2\alpha_0\, (= \alpha_+ - \frac{h}{\alpha_+})\,. \tag{1.6}$$

The charges are parametrised by two arbitrary parameters J and t as in the two-dimensional case

$$\alpha^J = J\sqrt{\frac{h}{t}} = -J\alpha_+\,, \quad 2\alpha_0 = \sqrt{h}(\sqrt{t} - \frac{1}{\sqrt{t}})\,, \tag{1.7}$$

and the scaling dimension is $d = 2\triangle(\alpha) = 2\alpha(\alpha - 2\alpha_0)$, or,

$$\triangle(\alpha^J) = h\,\triangle_J\,, \ \ \triangle_J = J(J+1-t)/t = \triangle_{t-1-J}\,. \tag{1.8}$$

Following [10], $F(x,y)$ is given by the two fold Mellin integral

$$F(x,y) = \frac{1}{(2\pi i)^2} \int_\uparrow \mathrm{d}s \int_\uparrow \mathrm{d}t\, x^s\, y^t\, \Gamma(-s)\,\Gamma(-t) \tag{1.9}$$

$$\Gamma(\delta_4+s+t)\,\Gamma(h-\delta_1+s+t)\,\Gamma(\delta_1+\delta_2-h-t)\,\Gamma(\delta_1+\delta_3-h-s)$$

with the paths of integration running parallel to the imaginary axis. Closing the contours to the right and taking into account the poles of the gamma factors produces a linear combination of four infinite sums, that can be identified with the four linearly independent solutions of the AK equations with parameters

$$\alpha = \delta_4\,, \quad \beta = h - \delta_1\,, \quad \gamma = 1 + h - \delta_1 - \delta_3\,, \quad \gamma' = 1 + h - \delta_1 - \delta_2\,, \tag{1.10}$$

satisfying (1.3) as a result of the constraint (1.6).

2. Fock space quantisation of the 2h-dimensional sub-canonical field

We choose complex Euclidean coordinates $z_a = e^{i\tau} n_a , n \in S^{2h-1} \subset \mathbb{R}^{2h} , \tau \in \mathbb{C}$, $z^2 = \sum_{a=1}^{2h} z_a^2$. For real τ one recovers the compactified Minkowski space $S^1 \times S^{2h-1}$. We shall mostly use the real Euclidean coordinates $x_a = e^t n_a$ with $t = i\tau$ - real; both notations z_a and x_a will appear throughout the paper. The field $\phi(z)$ satisfying $\Box^h \phi(z) = (\sum_a \partial^2_{z_a})^h \phi(z) = 0$ admits the mode expansion

$$\phi(z) = 2q - i\, b_0 \log z^2 + 2i \sum_{n\neq 0} \frac{b_n(z)}{n} = 2q - i\, b_0 \log z^2 + 2i \sum_{n\neq 0} (z^2)^{\frac{-n}{2}} \frac{b_n(\hat{z}))}{n} \tag{2.1}$$

with commutation relations

$$[b_n(z_1), b_{-m}(z_2)] = n \cos n\theta_{12}\, \delta_{nm} \left(\frac{z_2^2}{z_1^2}\right)^{\frac{n}{2}} , \quad [b_0, q] = -i\, . \tag{2.2}$$

Here $\hat{z} = z/\sqrt{z^2}$, $\cos\theta_{12} = \hat{z_1} \cdot \hat{z_2}$ and $b_n(\rho\, z) = \rho^{-n} b_n(z)$. The one-dimensional projection of (2.2) with $z_i = \sqrt{z_i^2}\, e \, , e^2 = 1 \, , i = 1, 2$ (so that $\cos\theta_{12} = 1$) reads

$$[b_n(e), b_{-m}(e)] = n\, \delta_{nm}\, . \tag{2.3}$$

It is assumed that $b_n(z)\,|0\rangle = 0\, , \ \langle 0|\, b_{-n}(z) = 0\, , \ n \geq 0\, .$

2.1. *Relation to the free field quantisation*

For simplicity of presentation we restrict here to the four-dimensional case, $2h = 4$. The modes $b_n(z)\, , n \neq 0$ can be constructed as linear combinations of the free field modes $a_n(z)$ described in [11]

$$[a_n(z_1), a_{-m}(z_2)] = \frac{1}{z_1^2} [a^*_{-n}(\frac{z_1}{z_1^2}), a_{-m}(z_2)] = \delta_{nm} \frac{1}{z_1^2} \left(\frac{z_2^2}{z_1^2}\right)^{\frac{n-1}{2}} C^{(1)}_{n-1}(\hat{z}_1 \cdot \hat{z}_2)\, , \tag{2.4}$$

where $n > 0$ and $C^{(1)}_n(\cos\theta) = \frac{\sin(n+1)\theta}{\sin\theta}$. The modes $a_n(z)$ are homogeneous $a_n(\rho\, z) = \rho^{-n-1} a_n(z)$, harmonic variables $\Box\, a_n(z) = 0$. For $n > 0$, $a_{-n}(z)\, , a^*_{-n}(z)$ are polynomials, realising an irrep of $SO(4)$ of dim n^2 (i.e., $a_{-n-1}(z) = z_{\mu_1} \ldots . z_{\mu_n}\, a_{\mu_1 \ldots \mu_n}$, where $a_{\mu_1 \ldots \mu_n}$ are symmetric, traceless tensors), while $a_n(z) := \frac{1}{z^2}\, a^*_{-n}(\frac{z}{z^2})$. We take two independent free fields, i.e., two commuting copies $\{a_n\}\, , \{a'_n\}$, $[a_n, a'_m] = 0$, each set satisfying (2.4) and define

$$b_{-n}(z) = \sqrt{\frac{n}{2}} (a_{-n-1}(z) + z^2\, a'_{-n+1}(z))\, , \ n > 0$$

$$b_n(z) = \sqrt{\frac{n}{2}}(z^2\, a_{n+1}(z) - a'_{n-1}(z))\,,\ n > 0 \tag{2.5}$$

so that

$$\Box^2 b_n(z) = 0\,, \quad b_n(\rho\, z) = \rho^{-n}\, b_n(z)\,.$$

Indeed (2.5) is the unique decomposition of the homogeneous polynomial of degree n, subject to this equation, into a sum of homogeneous harmonic polynomials. The commutation relations (2.4) then imply (2.2). The generalisation to $h > 2$ is straightforward with the Gegenbauer polynomials $C^{(h-1)}_{n-h+1}(\cos\theta)$ appearing in the r.h.s. of (2.4). In the two-dimensional case $2h = 2$ the free field modes $b_n(z)$ split into a sum of chiral pieces.

2.2. *Vertex operators*

Let

$$V_\alpha(z) =: e^{i\alpha\phi(z)} := (e^{i2\alpha q}\, e^{i\alpha\phi_<(z)})\,((z^2)^{\alpha\ b_0}\, e^{i\alpha\phi_>(z)}) = V^-_\alpha(z)\; V^+_\alpha(z) \tag{2.6}$$

where $\phi_{\substack{<\\>}}(z) = \mp 2i\sum_{k>0}\frac{b_{\mp k}(z)}{k}$. The commutation relations (2.2) imply

$$[b_n(z_1), V_\alpha(z_2)] = 2\alpha\,(\frac{z_2^2}{z_1^2})^{\frac{n}{2}}\ \cos n\theta_{12}\, V_\alpha(z_2)\,, \tag{2.7}$$

$$V^+_{\alpha_1}(z_1)\, V^-_{\alpha_2}(z_2) = \big(z^2_{12}\big)^{2\alpha_1\,\alpha_2}\, V^-_{\alpha_2}(z_2)\, V^+_{\alpha_1}(z_1)\,, \tag{2.8}$$

and then the operator product expansion

$$V_{\alpha_1}(z_1)\, V_{\alpha_2}(z_2) = \big(z^2_{12}\big)^{2\alpha_1\,\alpha_2}\, V_{\alpha_1+\alpha_2}(z_2) + \ldots\ldots \tag{2.9}$$

It is consistent with the 2-point function

$$\langle 2\alpha_0|V_{2\alpha_0-\alpha}(z_1)\, V_\alpha(z_2)|0\rangle = (z^2_{12})^{-2\Delta(\alpha)}\,, \quad \Delta(\alpha) = \alpha(\alpha - 2\alpha_0)\,, \tag{2.10}$$

where $2\alpha_0$ parametrises the charge at infinity, i.e., we reproduce (1.8). The (normalised) bra and ket states are determined from the vertex operators as

$$|\alpha\rangle = V_\alpha(0)|0\rangle = e^{2i\,\alpha\, q}|0\rangle\,, \tag{2.11a}$$

$$\langle\alpha| = \langle 0|e^{-2i\,\alpha q} = \lim_{x\to\infty}(x^2)^{2\Delta(\alpha)}\langle 2\alpha_0|\, V_{2\alpha_0-\alpha}(x)\,. \tag{2.11b}$$

Having (2.6) one computes the matrix elements (with $\alpha_{p+1} = 2\alpha_0 - \sum_{i=1}^p \alpha_i$)

$$\langle 2\alpha_0 - \alpha_{p+1}|V_{\alpha_p}(z_p)\ldots V_{\alpha_2}(z_2)|\alpha_1\rangle = \prod_{1\le i<j\le p}(z^2_{ij})^{2\alpha_i\alpha_j}\,. \tag{2.12}$$

The charge conservation condition in (2.12) implies the identities

$$\sum_{i=1}^{p+1} \Delta(\alpha_i) = - \sum_{1\le i<j\le p+1} 2\alpha_i\alpha_j \Leftrightarrow \sum_{i=1}^{p} \Delta(\alpha_i) - \Delta(\alpha_{p+1}) = - \sum_{1\le i<j\le p} 2\alpha_i\alpha_j . \tag{2.13}$$

The integral of the VO with charge $\alpha = \alpha_+$, or $\alpha = 2\alpha_0 - \alpha_+$, i.e., scaling dimension $2\Delta(\alpha_+) = 2h$, provides the 2h-dimensional analog of the screening charge operator. In a 4-point matrix element with one screening charge we shall use the notation $x_5, \alpha_5 = \alpha_+$ keeping the index 4 for the last field in the 4-point function. Then the matrix element is related to the $x_4 \to \infty$, $x_1 \to 0$ limit of the 4-point function in (1.5) according to

$$\int d^{2h} x_5 \langle \alpha_4 | V_{\alpha_+}(x_5) V_{\alpha_3}(x_3) V_{\alpha_2}(x_2) | \alpha_1 \rangle =: \int \mathrm{d}^{2h} x_5 \, \mathcal{A} \tag{2.14}$$
$$= (x_3^2)^{-\Delta} \, x^{-2\alpha_2(\alpha_3-\alpha_1)} \, y^{2\alpha_1(2\alpha_0-\alpha_2)} \; F(x,y) \, , \quad x = \frac{x_3^2}{x_{23}^2} \, , \quad y = \frac{x_2^2}{x_{23}^2} \, ,$$

where in agreement with (2.13) and using that $2\alpha_1 = -\alpha_+$

$$\Delta := \sum_{i=1}^{3} \Delta(\alpha_i) - \Delta(2\alpha_0 - \sum_{i=1}^{3} \alpha_i - \alpha_+) = 2\alpha_1(\alpha_2 + \alpha_3 - 2\alpha_0) - 2\alpha_2\alpha_3 \, . \tag{2.15}$$

2.3. *A Virasoro algebra*

Analogously to the one-dimensional case one can construct generators which close, using the commutation relations (2.3) for collinear vectors, a Virasoro algebra

$$L_n(e) = \frac{1}{2} \sum_{k\neq 0,n} b_{n-k}(e)\, b_k(e) + \frac{1}{\sqrt{2h}}\, b_n b_0 - \sqrt{\frac{2}{h}}\alpha_0 (n+1) b_n(e) \, , \quad n \neq 0 \, ,$$
$$L_0(e) = \sum_{n>0} b_{-n}(e)\, b_n(e) + \frac{b_0}{2h}(\frac{b_0}{2} - 2\alpha_0) \, . \tag{2.16}$$

For $n \neq 0$, using

$$[b_k(\hat{z}_1), L_{-n}(\hat{z}_2)] = \Big(b_{k-n}(\hat{z}_2) + ((\frac{1}{\sqrt{2h}} - 1) b_0 + \frac{2\alpha_0}{\sqrt{2h}}(n-1)) \, \delta_{k,n} \Big) k \, \cos k\theta_{12} \, , \tag{2.17}$$

we obtain, denoting $w := \sqrt{z^2}$,

$$[L_{-n}(\hat{z}_1), V_\alpha(z_2)] = 2\alpha \textstyle\sum_k w_2^{k-n} \, \cos(n-k)\theta_{12} \; : b_{-k}(\hat{z}_1)\, V_\alpha(z_2) :$$
$$-2\Delta(\alpha)(n-1) w_2^{-n} \, \cos n\theta_{12}\, V_\alpha(z_2) + 2\alpha(\tfrac{1}{\sqrt{2h}} - 1)\big(b_{-n}(\hat{z}_1)\, V_\alpha(z_2)$$
$$+ w_2^{-n} \, \cos n\theta_{12} V_\alpha(z_2)(b_0 + 2\alpha(n-1))\big)$$

$$+\alpha^2\, w_2^{-n}\big((n-1)\cos n\theta_{12} - \tfrac{\sin(n-1)\theta_{12}}{\sin\theta_{12}}\big) V_\alpha(z_2)\,, \qquad (2.18)$$

where the meaning of the normal product is

$$: b_k\, V := b_k\, V\,, \text{ for } k<0\,, \quad : b_k\, V := V\, b_k \text{ for } k\geq 0\,, \qquad (2.19)$$

(we notice that for $n=\pm 1$ the very last term in (2.18) vanishes), while for $n=0$, from

$$[b_k(\hat{z}_1), L_0(\hat{z}_2)] = b_k(\hat{z}_2) k \cos k\theta_{12}\,, \qquad (2.20)$$

we obtain

$$[L_0(\hat{z}_1), V_\alpha(z_2)] = \qquad (2.21)$$
$$2\alpha \sum_k w_2^k \cos k\theta_{12}\ : b_{-k}(\hat{z}_1) V_\alpha(z_2) : + \tfrac{\triangle(\alpha)}{h} V_\alpha(z_2) + 2\alpha(\tfrac{1}{2h}-1) V_\alpha(z_2)\, b_0\,.$$

The eigenvalues of L_0 and the central charge operator do not depend on h and coincide with the eigenvalues in the one-dimensional case, (cf. (1.7))

$$L_0|\alpha^J\rangle = \tfrac{\triangle(\alpha^J)}{h}|\alpha^J\rangle = \triangle_J|\alpha^J\rangle\,, \quad c = 1 - \tfrac{24}{h}\alpha_0^2 = 13 - 6(t+\tfrac{1}{t}) \qquad (2.22)$$

and we can use all the standard expressions for the singular vectors, as e.g., the singular vector at level two

$$\Big(t\, L_{-1}^2 - L_{-2}\Big)\, |\alpha^{J=\frac{1}{2}}\rangle\,. \qquad (2.23)$$

3. The null vector decoupling condition

Let $2\alpha_1 = \sqrt{\frac{h}{t}}$ so that $\triangle(\alpha_1) = h(\frac{3}{4t} - \frac{1}{2})$. By a straightforward application of the commutator formulae derived above one proves the null vector decoupling identity

$$\langle \alpha_{p+1}|V_{\alpha_p}(z_p)\ldots V_{\alpha_2}(z_2)\big(tL_{-1}^2(\hat{z}) - L_{-2}(\hat{z})\big)|\alpha_1\rangle = 0\,, \qquad (3.1)$$

which holds true for any z.

We can partially express the matrix element (3.1) in terms of differential operators (here $\langle\alpha_{p+1},\ldots,\alpha_2| := \langle\alpha_{p+1}|V_{\alpha_p}(z_p)\cdots V_{\alpha_2}(z_2)\rangle$)

$$\frac{2h}{t}\langle\alpha_{p+1},\ldots,\alpha_2|\big(tL_{-1}^2(z_0) - L_{-2}(z_0)\big)|\alpha_1\rangle = \Big(\big(z_0\cdot\sum_{k=2}^{p}\partial_{z_k}\big)^2 - \qquad (3.2)$$
$$\sum_{i=2}^{p}\delta_{0i}\, z_0\cdot\sum_{k=2}^{p}\partial_{z_k} - \big(\sum_{i=2}^{p} 4\alpha_i\alpha_1\,\frac{w_0}{w_i}\cos\theta_{0i}\big)^2 + \sum_{i=2}^{p} 4\alpha_i\alpha_1\,\frac{w_0^2}{w_i^2}\cos 2\theta_{0i}\Big)\,\mathcal{A}\,.$$

We consider first the 3-point null vector decoupling condition, which determines the possible "fusions" with the fundamental field with $\triangle(\alpha^{J=\frac{1}{2}})$.

The 3-point matrix element (with one screening charge) is determined by the L_0 Ward identity as

$$\int d^{2h}x_5 \langle\alpha_3|V_{\alpha_+}(x_5)V_{\alpha_2}(x_2)|\alpha_1\rangle = w_2^{2(\triangle(\alpha_3)-\triangle(\alpha_1)-\triangle(\alpha_2))}\,. \tag{3.3}$$

Choosing the argument of the Virasoro generators as $z_0 = x_2$ we can represent (3.2) fully in terms of derivatives and integrating out the derivative terms with respect to x_5 gives for the null vector equation

$$2a\left(\tfrac{2h}{t}(1-t)+2a\right) - \tfrac{4h}{t}\triangle(\alpha_2) = 0\,. \tag{3.4}$$

This equation for a does not depend on the charge α_2 itself but rather on the scaling dimension $2\triangle(\alpha_2)$ and is the same as the one for the 3-point matrix element without a screening charge. We obtain as in the one-dimensional case $2h = 1$ two solutions for $\triangle(\alpha_3)$

$$\triangle(\alpha_3) = \triangle(\alpha_2+\alpha_1)\,, \qquad \triangle(\alpha_3) = \triangle(\alpha_2-\alpha_1) = \triangle(\alpha_2+\alpha_1+\alpha_5)\,; \tag{3.5}$$

the first corresponds to the screeningless case, the second to the matrix element (3.3).

We shall now apply the relations (3.2) for the 4-point matrix element with one screening charge. In this case we can specialise the argument of the generators $L_{-n}(x_j)$ in (3.2) to the coordinate of each of the two middle vertex operators $x_j = x_2$, or $x_j = x_3$ and thus obtain two identities

$$\begin{aligned} 0 = \tfrac{2h}{t}\int \mathrm{d}^{2h}x_5 \langle\alpha_4|V_{\alpha_+}(x_5)\,V_{\alpha_3}(x_3)\,V_{\alpha_2}(x_2)\left(tL_{-1}^2(x_i) - L_{-2}(x_i)\right)|\alpha_1\rangle \\ \equiv \mathcal{D}_i \int \mathrm{d}^{2h}x_5\ \mathcal{A} + \int \mathrm{d}^{2h}x_5\ I_i\,, \quad i = 2,3\,. \end{aligned} \tag{3.6}$$

Here $\mathcal{A}$ is the matrix element in (2.14) and $\mathcal{D}_i$ are differential operators

$$\begin{aligned} &\mathcal{D}_2 = \mathcal{D}(x_2, x_3; \alpha_3) = (x_2\cdot D)^2 - x_2\cdot D - (x_2\cdot\partial_{x_2})^2 - \rho^2(x_3\cdot\partial_{x_3})^2 - \\ &2\rho\cos\theta\, x_2\cdot\partial_{x_2}x_3\cdot\partial_{x_3} + (2h-4)(x_2\cdot D - x_2\cdot\partial_{x_2} - \rho\cos\theta x_3\cdot\partial_{x_3}) + \\ &x_2\cdot\partial_{x_2} + \rho^2 x_3\cdot\partial_{x_3} + \left(2+\frac{2h}{t}(1-t)\right)\left((1+\rho\cos\theta)(x_2\cdot D - x_2\cdot\partial_{x_2}) - \right. \\ &\left.(\rho^2+\rho\cos\theta)x_3\cdot\partial_{x_3}\right) + 4\tilde\alpha\tilde\beta\, x\,\rho^2\sin^2\theta + \frac{4h}{t}\triangle(\alpha_3)\rho^2\sin^2\theta\,. \end{aligned} \tag{3.7a}$$

and $\mathcal{D}_3 = \mathcal{D}(x_3, x_2; \alpha_2)$ with

$$D = \partial_{x_2} + \partial_{x_3}\,,\quad \rho^2 = \frac{x_2^2}{x_3^2} = \frac{y}{x}\,,\quad 2\rho\cos\theta = 2\frac{x_2\cdot x_3}{x_3^2} = \frac{x+y-1}{x}\,.$$

Furthermore these operators are expressed as

$$\frac{x_j^2}{x_{23}^2}\mathcal{D}_i = (x_3^2)^{-\triangle}\left((x+y-1)^2 - 4xy\right)D_i(x,y)\,(x_3^2)^{\triangle}\,,\quad i,j = 2,3\,, i\neq j\,. \tag{3.8}$$

The operators $D_i(x,y)$ here are given by formulae analogous to the two differential operators in (1.1), with the parameters $\alpha, \beta, \gamma, \gamma'$ in (1.10) replaced by $\tilde\alpha, \tilde\beta, \tilde\gamma, \tilde\gamma'$

$$\begin{aligned} \tilde\alpha &= -2\alpha_2\alpha_3 + h + \tfrac{2h}{t}(1-t) - 2\Delta\,, \quad \tilde\beta = 2\alpha_2\alpha_3\,, \\ \tilde\gamma &= \ \ 1 + \tfrac{h}{t}(1-t) - 2\Delta\,, \quad \ \ \tilde\gamma' = 1 + \tfrac{h}{t}(1-t)\,, \end{aligned} \tag{3.9}$$

plus the additional terms

$$-\frac{h}{t}\,\frac{d(\alpha_3)}{x} := \frac{\Delta(\Delta - \frac{h}{t}(1-t)) - \frac{h}{t}\Delta(\alpha_3)}{x}\,, \qquad -\frac{h}{t}\frac{\Delta(\alpha_2)}{y} \tag{3.10}$$

respectively. These operators are precisely the AK operators (1.1) when the latter are rewritten on the matrix element $(x_3^2)^\Delta \int \mathcal{A}$, which according to (2.14) differs by a prefactor from F(x,y).

The integrands I_i of the remaining integrals in (3.6) are expected to be expressible as full derivatives in the integration variable so that these integrals vanish identically. Indeed we have checked this for the linear combination

$$\begin{aligned} &\int \mathrm{d}^{2h}x_5 \big(I_1 - \tfrac{r_2}{r_3} I_2\big) = \tfrac{2h}{t} \int \mathrm{d}^{2h}x_5 \Big(\big(\partial_{x_5} \cdot x_2 \tfrac{x_3\cdot x_5}{r_5} - \partial_{x_5} \cdot x_3 \tfrac{x_2\cdot x_5}{r_5}\big) \tfrac{x_{52}\cdot x_{53}}{r_3} \\ &+ \partial_{x_5} \cdot x_{52} \big(\tfrac{x_3\cdot x_5}{r_5} \tfrac{x_2\cdot x_3}{r_3} - \tfrac{x_2\cdot x_5}{r_5}\big) - \partial_{x_5} \cdot x_{53} \big(\tfrac{x_2\cdot x_5}{r_5} \tfrac{x_2\cdot x_3}{r_3} - \tfrac{x_3\cdot x_5}{r_5} \tfrac{r_2}{r_3}\big)\Big) \mathcal{A} = 0\,. \end{aligned}$$

In the screeningless case one recovers the same operators (3.7) but with different value $\Delta \to \Delta' = -2\alpha_1\alpha_2 - 2\alpha_1\alpha_3 - 2\alpha_2\alpha_3$ to be inserted in (3.8). Changing back variables, this correlation function corresponds to a constant factor $F(x,y)$ with parameters $\alpha'\,\beta' = 0$, trivially satisfying (1.1).

4. Discussion

We have revealed a hidden Virasoro symmetry in a 2h-dimensional model and have demonstrated that it leads to differential equations for the 4-point correlation functions. This generalises a basic property believed so far to be intrinsically restricted to the 2-dimensional theories. This symmetry also allows to determine the leading short distance behaviour of the higher dimensional models purely algebraically, without having to perform the complicated multiple Mellin integral computation of the Symanzik method.

The main features of the 2d theories - chiral factorisability, explicit simple realisation of the Virasoro generators in terms of differential operators, simple behaviour under projective transformations – all turned out not to be crucial for the derivation. A weak point however of our investigation so far, which needs a further effort, is the treatment of the screening charge operators.

The four-dimensional model considered here is still an unrealistic, toy model. In particular the four-dimensional analogs of the two-dimensional $c < 1$ minimal models are non-unitary [6]. On the other hand despite the impressive recent developments in the perturbative supersymmetric N=4 theories, there are still few exact results on conformal points beyond the perturbation theory. It remains to be seen whether the generalised Coulomb gas model, or some related extension, could be used as a building block in more realistic applications.

Acknowledgments

We thank N. Nikolov, G. Sotkov and I. Todorov for the interest in this work and for a useful discussion. P.F. acknowledges the support of the Italian Ministry of Education, University and Research (MIUR). V.B.P. acknowledges the hospitality of INFN, Trieste and the University of Northumbria, Newcastle, UK. This research is supported by the TMR Network EUCLID, contract HPRN-CT-2002-00325, and by the Bulgarian National Council for Scientific Research, grant F-1205/02.

References

1. A.A. Belavin, A.M. Polyakov and A.M. Zamolodchikov, Nucl. Phys. B241 (1984) 333.
2. P. Appell, J. Kampé de Fériet: *Fonctions Hypergéométriques et Hypersphériques - Polynomes d' Hermite*, Gauthier - Villars, Paris, 1926.
3. H. Bateman: *Higher Transcendental Functions*, vol. I, McGraw-Hill, New York, 1953.
4. S. Ferrara, A.F. Grillo, R. Gatto and G. Parisi, Il Nuovo Cimento, 19 A, No. 4, (1974) 667.
5. Vl.S. Dotsenko and V.A. Fateev, Nucl. Phys. B240 [FS 12] (1984) 312.
6. P. Furlan and V.B. Petkova, Mod. Phys. Lett. A4 (1989) 227.
7. D.T. Stoyanov, Infinite dimensional Lie algebras connected with the four-dimensional Laplace operator, ISAS preprint 10/85/E.P., and in the Proceedings of the *Symposium on Conformal Groups and Structures*, Clausthal, 1985, Eds. A.O. Barut and H.D. Doebner, Lecture Notes in Physics, 261 (1986) 379, Springer-Verlag, Berlin.
8. N. M. Nikolov, Ya.S. Stanev and I.T. Todorov, J. Phys. A35 (2002) 2985.
9. G.M. Sotkov and D.T. Stoyanov, J. Phys. A13 (1980) 2807; J. Phys. A16 (1983) 2817.
10. K. Symanzik, Lettere al Nuovo Cimento, 3 (1972) 734.
11. I.T. Todorov, Infinite dimensional Lie algebras in conformal QFT models, in the Proceedings of the *Symposium on Conformal Groups and Structures*, Clausthal, 1985, Eds. A.O. Barut and H.D. Doebner, Lecture Notes in Physics, vol. 261, (1986) 387, Springer-Verlag, Berlin.

3-dimensional Integrable Lattice Models and the Bazhanov-Stroganov Model

G. VON GEHLEN

Physikalisches Institut, Universität Bonn, Nussallee 12, D-53115 Bonn
E-mail: gehlen@th.physik.uni-bonn.de

S. PAKULIAK

Bogoliubov Laboratory of Theoretical Physics,
Joint Institute for Nuclear Research, Dubna 141980, Russia
E-mail: pakuliak@thsun1.jinr.ru

S. SERGEEV

Department of Theoretical Physics, Building 59,
The Australian National University, Canberra ACT 0200, Australia
E-mail: sergey.sergeev@anu.edu.au

After reviewing the construction of 3D integrable generalized Zamolodchikov-Bazhanov-Baxter models starting from the Sergeev mapping operator, we show how the L-operator of the 2D-integrable Bazhanov-Stroganov model follows from a Linear Problem by imposing quasi-periodicity. The 3D classical mapping and the associated 3D parametrization is used to derive isospectral transformations for the inhomogenous classical and quantum 2D-Bazhanov-Stroganov model transfer matrices.

1. Introduction

Whereas there is a good understanding how to construct systematically 2D integrable lattice systems, the construction of 3D integrable lattice models still is relying on special solutions of tetrahedron equations (TE) which guarantee the integrability. Most of the models studied during the past two decades are generalizations of Zamolodchikov's 1981 construction[1] and its 1992 generalization by Bazhanov and Baxter[2] using cyclic root-of-unity structures. Only very recently a new solution of the TE based on a q-oscillator algebra involving $U_q(\widehat{sl}(n))$ structures has been found[3]. Considering a 3D integrable model, quite immediately one may obtain a related

2D integrable model by imposing quasi-periodical boundary conditions in the third spacial direction.

In this talk we consider the recent generalized version of the 3D integrable Zamolodchikov-Bazhanov-Baxter (ZBB) model[5] and its 2D reduction, the Bazhanov-Stroganov (BS) model[7]. We show explicitly how the 3D model can be used to derive intertwining relations and isospectral transformations of the 2D BS model. It seems to be hard to find these features without our insight from the 3D structure.

2. Vertex formulation of the generalized ZBB-model

In the vertex formulation of the ZBB-model[4,5] the quantum variables are attached to the links i of a 3D oriented lattice. They are taken to be elements $(\mathbf{u}_i, \mathbf{w}_i)$ of an ultra-local affine Weyl algebra at root of unity:

$$\mathbf{u}_j \cdot \mathbf{w}_j = \omega^{\delta_{i,j}}\, \mathbf{w}_j \cdot \mathbf{u}_j; \qquad \omega^N = 1; \qquad N \in \mathbb{Z}; \quad N \geq 2. \qquad (2.1)$$

At each link i there shall also be a scalar κ_i and we define $\mathfrak{w}_i = (\mathbf{u}_i, \mathbf{w}_i, \kappa_i)$. In the formulation of Ref.[5] the basic object of the model is the operator $\mathcal{R}_{123}$

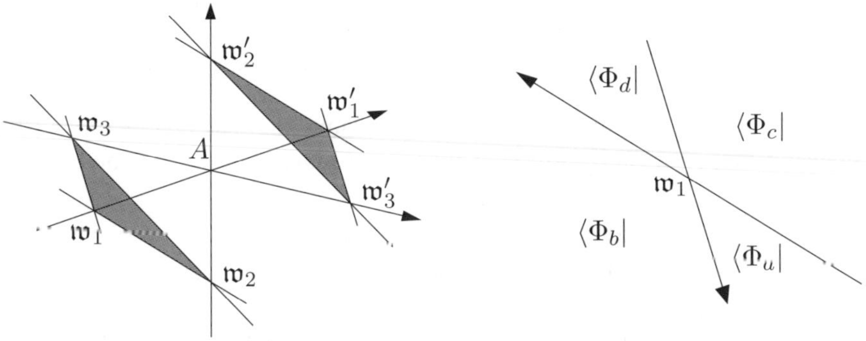

Fig. 2.1. Left: The six links of the basic oriented lattice forming a vertex A, and (shaded) the auxiliary planes through the initial variables $\mathfrak{w}_1$, $\mathfrak{w}_2$, $\mathfrak{w}_3$ and through the final variables $\mathfrak{w}'_1$, $\mathfrak{w}'_2$, $\mathfrak{w}'_3$. On the right hand side of the Figure we show the four co-currents in the four sections of the initial auxiliary plane around $\mathfrak{w}_1$.

(defined to be invertible, rational and canonical) which maps the triple of the dynamical variables $\mathfrak{w}_1$, $\mathfrak{w}_2$, $\mathfrak{w}_3$ on the incoming links onto the triple $\mathfrak{w}'_1$, $\mathfrak{w}'_2$, $\mathfrak{w}'_3$ on the outgoing links: For any rational function Ψ of the $\mathbf{u}_1, \ldots, \mathbf{w}_3$ we define

$$(\mathcal{R}_{123} \circ \Psi)(\mathbf{u}_1, \mathbf{w}_1, \mathbf{u}_2, \ldots, \mathbf{w}_3) \stackrel{\text{def}}{=} \Psi(\mathbf{u}'_1, \mathbf{w}'_1, \mathbf{u}'_2, \ldots, \mathbf{w}'_3). \qquad (2.2)$$

We further postulate at each variable $\mathfrak{w}_i$ the Linear Problem

$$0 = \langle\Phi_a| + \omega^{1/2}\langle\Phi_b|\,\mathbf{u}_i + \langle\Phi_c|\,\mathbf{w}_i + \kappa_i\langle\Phi_d|\,\mathbf{u}_i\,\mathbf{w}_i\,, \tag{2.3}$$

which relates the four co-currents $\langle\Phi_a|, \ldots, \langle\Phi_d|$ defined around the variable $\mathfrak{w}_i$ as shown in Fig.2.1. Demanding a Baxter Z-invariance, the Linear Problem (2.3) leads to the following unique expression for the mapping $\mathcal{R}_{123}$:

$$\begin{aligned}
\kappa_2 \mathbf{u}_1'^{\,-1} &= \kappa_1 \mathbf{u}_2^{-1} + \kappa_3 \mathbf{u}_1^{-1}\mathbf{w}_2^{-1}\mathbf{w}_3 - \omega^{1/2}\kappa_1\kappa_3\mathbf{u}_2^{-1}\mathbf{w}_2^{-1}\mathbf{w}_3;\\
\mathbf{u}_2'^{\,-1} &= \mathbf{u}_1^{-1} - \omega^{1/2}\mathbf{u}_1^{-1}\mathbf{w}_1\mathbf{u}_3^{-1} + \kappa_1\mathbf{w}_1\mathbf{u}_2^{-1}\mathbf{u}_3^{-1};\\
\mathbf{w}_1' &= \mathbf{w}_1\mathbf{w}_2\mathbf{w}_3^{-1} - \omega^{1/2}\mathbf{w}_2\mathbf{u}_3\mathbf{w}_3^{-1} + \kappa_3\mathbf{u}_3;
\end{aligned}$$

$$\mathbf{w}_1'\mathbf{w}_2' = \mathbf{w}_2\mathbf{w}_1; \qquad \mathbf{u}_3'\mathbf{u}_2' = \mathbf{u}_2\mathbf{u}_3; \qquad \mathbf{w'}_3^{-1}\mathbf{u}_1' = \mathbf{u}_1\mathbf{w}_3^{-1}. \tag{2.4}$$

Let us represent the root of unity affine Weyl elements by $N \times N$-matrices:

$$\mathbf{u} \equiv u\,\mathbf{X}; \quad \mathbf{w} \equiv w\,\mathbf{Z}; \quad u,\, w \in \mathbb{C}; \quad \mathbf{X}\,\mathbf{Z} = \omega\,\mathbf{Z}\,\mathbf{X}; \quad \mathbf{X}^N = \mathbf{Z}^N = 1\,.$$

$$\mathbf{X}\,|\beta\rangle = \omega^\beta\,|\beta\rangle\,; \qquad \mathbf{Z}\,|\beta\rangle = |\beta+1\rangle\,; \qquad \langle\alpha|\beta\rangle = \delta_{\alpha,\beta}\,, \tag{2.5}$$

so that (2.4) become relations of $N^3 \times N^3$ matrices. The N-th powers of the Weyl elements are centers and we write:

$$\mathbf{u}_j^N = u_j^N \equiv U_j; \quad \mathbf{w}_j^N = w_j^N \equiv W_j; \quad (\mathbf{u} + \mathbf{w})^N = U + W. \tag{2.6}$$

Now, taking the Nth powers of (2.4) and using the last of Eqs.(2.6), the quantum mapping $\mathcal{R}_{123}$ induces a *functional mapping* $\mathcal{R}_{123}^{(f)}$ of the centers ($K_j \equiv \kappa_j^N$):

$$\begin{aligned}
\frac{W_1}{W_1'} &= \frac{W_2'}{W_2} = \frac{W_1\,W_3}{W_1\,W_2 + U_3\,W_2 + K_3\,U_3\,W_3}\,;\\
\frac{U_2'}{U_2} &= \frac{U_3}{U_3'} = \frac{U_1\,U_3}{U_2\,U_3 + U_2\,W_1 + K_1\,U_1\,W_1}\,;\\
\frac{U_1'}{U_1} &= \frac{W_3'}{W_3} = \frac{K_2\,U_2\,W_2}{K_1\,U_1\,W_2 + K_3\,U_2\,W_3 + K_1\,K_3\,U_1\,W_3}\,.
\end{aligned} \tag{2.7}$$

Taking Nth roots and fixing some phases in order to obtain a relation for $u_1, \ldots, w_3'$ instead of for the $U_1, \ldots, W_3'$, we write this as

$$\left(\mathcal{R}_{123}^{(f)} \circ \psi\right)(u_1,\, w_1,\, u_2,\, \ldots,\, w_3) \stackrel{def}{=} \psi(u_1',\, w_1',\, u_2',\, \ldots\,, w_3')\,. \tag{2.8}$$

A remarkable feature arises : $\mathcal{R}_{123}$ decomposes into a matrix conjugation $\mathbf{R}_{123}$ (this is a $N^3 \times N^3$-matix) and the purely functional mapping $\mathcal{R}_{123}^{(f)}$:

$$\mathcal{R}_{123} \circ \Psi = \mathbf{R}_{123}\left(\mathcal{R}_{123}^{(f)} \circ \Psi\right)\mathbf{R}_{123}^{-1}\,. \tag{2.9}$$

With little effort it can be shown[5,6] that $\mathcal{R}_{ijk}$ satisfies the TE

$$\mathcal{R}_{123}\,\mathcal{R}_{145}\,\mathcal{R}_{246}\,\mathcal{R}_{356} \;=\; \mathcal{R}_{356}\,\mathcal{R}_{246}\,\mathcal{R}_{145}\,\mathcal{R}_{123},$$

and $\mathcal{R}_{ijk}^{(f)}$ obeys the analogous functional TE. So (2.8) defines a classical *integrable* mapping. $\mathbf{R}_{123}$ can be written as the weighted cross ratio of the Bazhanov-Baxter cyclic functions $w_p(n)$, in components:

$$\mathbf{R}_{i_1 i_2 i_3}^{j_1 j_2 j_3} \;=\; \delta_{i_2+i_3,j_2+j_3}\,\omega^{(j_1-i_1)j_3}\,\frac{w_{p_1}(i_2-i_1)\,w_{p_2}(j_2-j_1)}{w_{p_3}(j_2-i_1)\,w_{p_4}(i_2-j_1)} \tag{2.10}$$

where

$$\frac{w_p(n)}{w_p(n-1)} \;=\; \frac{y}{1\;-\;\omega^n\,x}\,; \qquad x^N+y^N \;=\; 1\,; \qquad n\in\mathbb{Z}_N \tag{2.11}$$

with $p=(x,y)$. In (2.10) $p_1,\ldots,p_4$ are four points on the Fermat curve determined by the affine parameters of the initial and final Weyl variables:

$$x_1=\frac{u_2}{\omega^{1/2}\kappa_1\,u_1}; \quad x_2=\frac{\kappa_2\,u_2'}{\omega^{1/2}u_1'}; \quad x_3=\frac{u_2'}{\omega\,u_1}; \quad x_1x_2=\omega x_3x_4. \tag{2.12}$$

Up to now we have just mentioned the mapping $\mathcal{R}_{123}$ at a single vertex. In order to construct a 3D integrable system we have to consider such a mapping at each vertex of an extended 3D lattice. Let $\mathbf{e}_1$, $\mathbf{e}_2$, $\mathbf{e}_3$ be vectors spanning the unit cell of the lattice. We label the variables on the links by vectors $\mathbf{n} \;=\; n_1\mathbf{e}_1+n_2\mathbf{e}_2+n_3\mathbf{e}_3$. So, instead of (2.7), for the functional part of the mapping we get the system of equations[9]

$$\frac{W_{1,\mathbf{n}}}{W_{1,\mathbf{n}+\mathbf{e}_1}} \;=\; \frac{W_{2,\mathbf{n}+\mathbf{e}_2}}{W_{2,\mathbf{n}}} \;=\; \frac{W_{1,\mathbf{n}}W_{3,\mathbf{n}}}{W_{1,\mathbf{n}}W_{2,\mathbf{n}}+U_{3,\mathbf{n}}W_{2,\mathbf{n}}+K_{3,n_1,n_2}U_{3,\mathbf{n}}W_{3,\mathbf{n}}};$$

$$\frac{U_{2,\mathbf{n}+\mathbf{e}_2}}{U_{2,\mathbf{n}}} \;=\; \frac{U_{3,\mathbf{n}}}{U_{3,\mathbf{n}+\mathbf{e}_3}} \;=\; \ldots; \qquad \frac{U_{1,\mathbf{n}+\mathbf{e}_1}}{U_{1,\mathbf{n}}} \;=\; \frac{W_{3,\mathbf{n}+\mathbf{e}_3}}{W_{3,\mathbf{n}}} \;=\; \ldots \tag{2.13}$$

which relates the classical variables along the 3D lattice. We change variables[9] introducing for each link $\mathbf{n}$ three functions $\tau_{1,\mathbf{n}}$, $\tau_{2,\mathbf{n}}$, $\tau_{3,\mathbf{n}}$ and three complex pairs $\mathcal{X}_{n_1}=(X_{n_1}',X_{n_1})$, $\mathcal{Y}_{n_2}=(Y_{n_2}',Y_{n_2})$, $\mathcal{Z}_{n_3}=(Z_{n_3}',Z_{n_3})$:

$$U_{1,\mathbf{n}}=(-1)^N\frac{Z_{n_3}'-Y_{n_2}}{Y_{n_2}-Z_{n_3}}\frac{\tau_{2,\mathbf{n}}}{\tau_{2,\mathbf{n}+\mathbf{e}_3}}; \quad W_{1,\mathbf{n}}=(-1)^N\frac{Z_{n_3}-Y_{n_2}'}{Z_{n_3}-Y_{n_2}}\frac{\tau_{3,\mathbf{n}+\mathbf{e}_2}}{\tau_{3,\mathbf{n}}};$$

$$U_{2,\mathbf{n}}=(-1)^N\frac{Z_{n_3}'-X_{n_1}}{X_{n_1}-Z_{n_3}}\frac{\tau_{1,\mathbf{n}}}{\tau_{1,\mathbf{n}+\mathbf{e}_3}}; \quad W_{2,\mathbf{n}}=(-1)^N\frac{Z_{n_3}-X_{n_1}'}{Z_{n_3}-X_{n_1}}\frac{\tau_{3,\mathbf{n}}}{\tau_{3,\mathbf{n}+\mathbf{e}_1}};$$

$$K_{1,n_2,n_3}=-\left\{\begin{matrix} Y_{n_2}' & Y_{n_2} \\ Z_{n_3}' & Z_{n_3}\end{matrix}\right\}; \quad \ldots\;; \quad \left\{\begin{matrix} A & B \\ C & D\end{matrix}\right\}\equiv\frac{(A-C)(B-D)}{(A-D)(B-C)}. \tag{2.14}$$

Then (2.13) become trilinear Hirota equations for the τ-functions:

$$\begin{aligned}(X_\alpha - X_\beta)(X'_\beta - X'_\gamma)(X_\gamma - X_\alpha)\tau_{\alpha,\mathbf{n}+\mathbf{e}_\beta+\mathbf{e}_\gamma}\tau_{\beta,\mathbf{n}}\tau_{\gamma,\mathbf{n}} \\ + (X_\alpha - X'_\beta)(X_\beta - X_\gamma)(X'_\gamma - X_\alpha)\tau_{\alpha,\mathbf{n}}\tau_{\beta,\mathbf{n}+\mathbf{e}_\gamma}\tau_{\gamma,\mathbf{n}+\mathbf{e}_\beta} \\ = (X_\alpha - X_\beta)(X'_\beta - X_\gamma)(X'_\gamma - X_\alpha)\tau_{\alpha,\mathbf{n}+\mathbf{e}_\beta}\tau_{\beta,\mathbf{n}+\mathbf{e}_\gamma}\tau_{\gamma,\mathbf{n}} \\ + (X_\alpha - X'_\beta)(X_\beta - X'_\gamma)(X_\gamma - X_\alpha)\tau_{\alpha,\mathbf{n}+\mathbf{e}_\gamma}\tau_{\beta,\mathbf{n}}\tau_{\gamma,\mathbf{n}+\mathbf{e}_\beta}\,,\end{aligned} \qquad (2.15)$$

with $\{\alpha,\beta,\gamma\}$ any even permutation of $\{1,2,3\}$; $X_1 = X_{n_1}$, $X_2 = Y'_{n_2}$, etc. Eqs. (2.15) can be solved in terms of the rational g-soliton functions[9] H:

$$\tau_{\alpha,\mathbf{n}} = H\left(\left\{f_j^{(\alpha,\mathbf{n})}\right\}\right) \quad \text{with} \quad H(\{f_j\}) = \frac{\det|\,P_j^i - f_j P_j'^{\,i}\,|_{i,j=0}^{g-1}}{\prod_{i>j}(P_i - P_j)}. \qquad (2.16)$$

The functions H have g arguments f_j and $2g$ parameters $\mathcal{P}_j = (P'_j,\, P_j)$. They solve (2.15) because they satisfy to a double Fay-type identity. The arguments $f_j^{(\alpha,\mathbf{n})}$ are factorized rational expressions

$$f_j^{(1,\mathbf{n})} = \mathbf{f}_j(X_{n_1})\,\sigma_j(\mathcal{Y}_{n_2})\,I_{j,\mathbf{n}}; \qquad f_j^{(3,\mathbf{n})} = \mathbf{f}_j(Z_{n_3})\sigma_j(\mathcal{Y}_{n_2})\,I_{j,\mathbf{n}};$$

$$f_j^{(2,\mathbf{n})} = \mathbf{f}_j(Y_{n_2})\,I_{j,\mathbf{n}} \quad \text{with} \quad \mathbf{f}_j(A) = \frac{P_j - A}{P'_j - A}\,f_j; \quad \sigma_j(\mathcal{A}) = \begin{Bmatrix} P'_j & P_j \\ A' & A \end{Bmatrix};$$

$$I_{j,\mathbf{n}} = \prod_{\ell=0}^{n_1-1}\sigma_j(\mathcal{X}_\ell)\prod_{m=0}^{n_2-1}\sigma_j^{-1}(\mathcal{Y}_m)\prod_{n=0}^{n_3-1}\sigma_j(\mathcal{Z}_n)\,; \qquad I_{j,\vec{0}} = 1. \qquad (2.17)$$

Inserting (2.16) and (2.17) into (2.14) all classical variables $U_{1,\mathbf{n}}, U_{2,\mathbf{n}},\ldots,$ $W_{3,\mathbf{n}}$ can be written in terms of the function $\mathcal{V}$ defined as

$$\mathcal{V}(\{f_j\},\mathcal{A},\mathcal{B}) = -(-1)^N\frac{A-B'}{A-B}\,H(\{\mathbf{f}_j(A)\})\Big/H(\{\mathbf{f}_j(A)\,\sigma_j(\mathcal{B})\});$$

$$U_{1,\mathbf{n}} = \mathcal{V}(\{f_j I_{j,\mathbf{n}}\},\mathcal{Y}_{n_2},\mathcal{Z}_{n_3}); \quad U_{2,\mathbf{n}} = \mathcal{V}(\{f_j\sigma_j(\mathcal{Y}_{n_2})I_{j,\mathbf{n}}\},\mathcal{X}_{n_1},\mathcal{Z}_{n_3});$$

$$U_{3,\mathbf{n}} = \mathcal{V}(\{f_j I_{j,\mathbf{n}}\},\mathcal{X}_{n_1},\mathcal{Y}_{n_2}); \quad W_{1,\mathbf{n}} = -\mathcal{V}(\{f_j I_{j,\mathbf{n}}\},\mathcal{Z}_{n_3},\mathcal{Y}_{n_2}); \quad \textit{etc.} \qquad (2.18)$$

We get $W_{i,\mathbf{n}}$ from $U_{i,\mathbf{n}}$ permuting the last arguments and changing sign. Moving away from the origin $\mathbf{n} = 0$, the first argument of the $U_{1,\mathbf{n}},\ldots,W_{3,\mathbf{n}}$ at each step picks up one more factor $\sigma_j(\mathcal{X}_\ell)$, $\sigma_j^{-1}(\mathcal{Y}_m)$ or $\sigma_j(\mathcal{Z}_n)$ via $I_{j,\mathbf{n}}$. The simplest choice is $g = 0$, i.e. to take $H(\{f_j\}) \equiv 1$, so that $\mathcal{R}^{(f)}$ becomes trivial. In (2.7) this amounts to take $U'_i = U_i$ and $W'_i = W_i$. In this case the solution to (2.7) can be expressed in terms of 3 parameters and the $\mathbf{R}_{123}$ reduce to the ZBB Boltzmann weights.

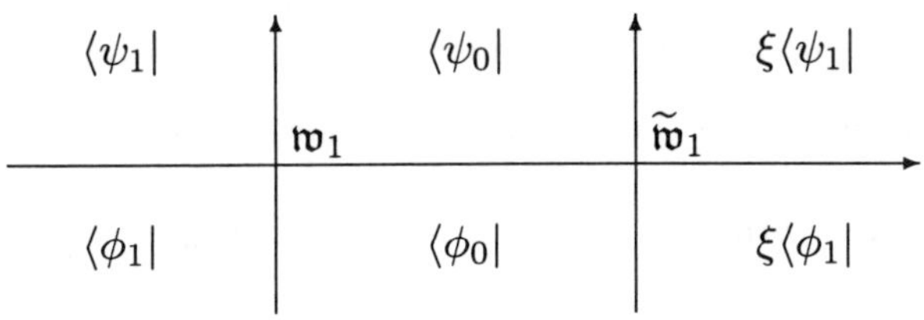

Fig. 3.1. Auxiliary plane in the neighborhood of the Weyl variables $\mathfrak{w}_1$ and $\tilde{\mathfrak{w}}_1$. Quasi-periodic boundary conditions are assumed between left and right co-currents.

3. Bazhanov-Stroganov L from the Linear Problem

Let us write the Linear Problem (2.3) for the variables $\mathfrak{w}_1$ and $\tilde{\mathfrak{w}}_1$, see Fig.3.1, imposing quasi-periodic b.c. between the left and right hand columns:

$$0 = \langle\psi_0| + \xi\omega^{1/2}\langle\psi_1|\tilde{\mathbf{u}}_1 + \langle\phi_0|\tilde{\mathbf{w}}_1 + \xi\tilde{\kappa}_1\langle\phi_1|\tilde{\mathbf{u}}_1\tilde{\mathbf{w}}_1\ ;$$

$$0 = \langle\psi_1| + \ \omega^{1/2}\langle\psi_0|\mathbf{u}_1 + \langle\phi_1|\mathbf{w}_1 \ + \ \kappa_1\langle\phi_0|\mathbf{u}_1\mathbf{w}_1\ .$$

In matrix form, writing $\langle\phi| = (\langle\phi_0|, \langle\phi_1|)$, $\langle\psi| = (\langle\psi_0|, \langle\psi_1|)$, this becomes

$$\langle\psi|\,(\omega\,\xi\,\mathbf{u}_1\,\tilde{\mathbf{u}}_1 \ - \ 1)\tilde{\mathbf{w}}_1^{-1} = \langle\phi|\cdot L_1(\xi)$$

with

$$L_1(\xi) \ = \ \begin{pmatrix} 1-\omega^{1/2}\xi\mathbf{u}_1\tilde{\mathbf{u}}_1\kappa_1\mathbf{w}_1\tilde{\mathbf{w}}_1^{-1} & -\mathbf{u}_1(\omega^{1/2}-\kappa_1\mathbf{w}_1\tilde{\mathbf{w}}_1^{-1}) \\ \xi\tilde{\mathbf{u}}_1\,(\tilde{\kappa}_1-\omega^{1/2}\mathbf{w}_1\tilde{\mathbf{w}}_1^{-1}) & -\omega^{1/2}\xi\mathbf{u}_1\tilde{\mathbf{u}}_1\tilde{\kappa}_1+\mathbf{w}_1\tilde{\mathbf{w}}_1^{-1} \end{pmatrix}. \qquad (3.1)$$

Only the three Weyl elements $\mathbf{w}_1\tilde{\mathbf{w}}_1^{-1}$, $\mathbf{u}_1$, $\tilde{\mathbf{u}}_1$ appear. So we can use the N-dim. representation with $\mathbf{X}$ and $\mathbf{Z}$ as in (2.5)

$$\mathbf{w}_1\tilde{\mathbf{w}}_1^{-1} = w_1{\tilde{w}_1}^{-1}\,\mathbf{Z}\,; \qquad \mathbf{u}_1 = u_1\,\mathbf{X}\,; \qquad \tilde{\mathbf{u}}_1 = \tilde{u}_1\,\mathbf{X}^{-1}. \qquad (3.2)$$

Apart from some rescaling and a gauge transformation, (3.1) is the L operator proposed by Bazhanov and Stroganov[7]:

$$L(\lambda;\,q,q') = \begin{pmatrix} 1+\lambda\dfrac{y_q y_{q'}}{\mu_q\mu_{q'}}\,\mathbf{Z} & \lambda\,\mathbf{X}^{-1}\left(x_q - \dfrac{y_{q'}}{\mu_q\mu_{q'}}\,\mathbf{Z}\right) \\ \mathbf{X}\left(\omega x_{q'} - \dfrac{y_q}{\mu_q\mu_{q'}}\,\mathbf{Z}\right) & \lambda\,\omega\,x_q x_{q'} + \dfrac{1}{\mu_q\mu_{q'}}\,\mathbf{Z} \end{pmatrix} \qquad (3.3)$$

if we put $\lambda \ = \ (\omega\,u_1\tilde{u}_1\,x_q\,y_q\,\xi)^{-1}$ and

$$\kappa_1 = \omega^{1/2}\frac{x_q}{y_{q'}};\quad \tilde{\kappa}_1 = \omega^{-1/2}\frac{y_q}{x_{q'}};\quad \frac{w_1}{\tilde{w}_1} = \omega^{-1}\frac{y_q y_{q'}}{x_q x_{q'}\,\mu_q\mu_{q'}}\,, \qquad (3.4)$$

where the variables x_q, y_q, μ_q satisfy the Chiral Potts (CP) Baxter relations

$$x_q^N + y_q^N \ = \ k\,(x_q^N\,y_q^N + 1); \quad k\,x_q^N = 1 - k'\mu_q^{-N}; \quad k\,y_q^N = 1 - k'\mu_q^N \quad (3.5)$$

(same for $x_{q'}$, $y_{q'}$, $\mu_{q'}$). k' is the temperature parameter and $k^2 = 1 - k'^2$.

The transfer matrix of the periodic BS-quantum chain of length Q is

$$\mathbf{T} = \mathrm{Tr}_{\mathbb{C}^2}\mathbf{M}; \quad \text{with} \quad \mathbf{M} = L(\lambda; q_0, q_0') \ \ldots \ L(\lambda; q_{Q-1}, q_{Q-1}'), \tag{3.6}$$

where each L has its pair of rapidities q_i, q_i', but k shall be the same for all L. The BS quantum chain is integrable[7,8] because of the intertwining

$$\sum_{j_1,j_2,\beta} R_{i_1j_1,i_2j_2} L_{j_1k_1}^{\alpha_1\beta}(\lambda) L_{j_2k_2}^{\beta\alpha_2}(\nu) = \sum_{j_1,j_2,\beta} L_{i_2j_2}^{\alpha_1\beta}(\nu) L_{i_1j_1}^{\beta\alpha_2}(\lambda) R_{j_1k_1,j_2k_2}. \tag{3.7}$$

Here we have written L with matrix indices: L is a 2×2 matrix (latin indices taking the values $0,1$), whose entries are $N \times N$ matrices (greek indices running over $0,1,\ldots,N-1$). R is a twisted six-vertex R-matrix. The great interest in the model defined by (3.3) is due to the second intertwining relation in the N-dim. (quantum) space:

$$\begin{aligned}\sum_{\beta_1\beta_2,k} \mathsf{S}_{\alpha_1\alpha_2}^{\beta_1\beta_2}(p,p',q,q')\ L_{i_1k}^{\beta_1\gamma_1}(\lambda;p,p')\ L_{k\,i_2}^{\beta_2,\gamma_2}(\lambda;q,q') \\ = \sum_{\beta_1\beta_2,k} L_{i_1k}^{\alpha_2\beta_2}(\lambda;q,q')\ L_{k\,i_2}^{\alpha_1\beta_1}(\lambda;p,p')\ \mathsf{S}_{\beta_1,\beta_2}^{\gamma_1,\gamma_2}(p,p',q,q'),\end{aligned} \tag{3.8}$$

since S turns out to be the product of four CP Boltzmann weights[7]:

$$\mathsf{S}_{\alpha_1\alpha_2}^{\beta_1\beta_2}(p,p',q,q') = W_{pq'}(\alpha_1-\alpha_2)W_{p'q}(\beta_2-\beta_1)\overline{W}_{pq}(\beta_2-\alpha_1)\overline{W}_{p'q'}(\beta_1-\alpha_2).$$

Consequences of this relation have been essential in solving the CP-model[10,11]. Using the 3D interpretation of the operators L, one can also interpret (3.8) in 3D, obtaining S as the product of two operators (2.10)[12]:

$$\mathsf{S}_{\alpha_1\alpha_2;\beta_1\beta_2}(p,p',q,q') = \sum_{\sigma,\tau\in\mathbb{Z}_N} \mathbf{R}_{\alpha_1,\alpha_2,\sigma}^{\beta_1,\beta_2,\tau}\ \widetilde{\mathbf{R}}_{-\alpha_1,-\alpha_2,\tau}^{-\beta_1,-\beta_2,\sigma}. \tag{3.9}$$

$\mathbf{R}$ and $\widetilde{\mathbf{R}}$ depend on four Fermat points each $(p_1,\ldots,p_4;\ \tilde{p_1},\ldots,\tilde{p_4})$, which are related to the CP variables of S by

$$x_1 = \frac{y_{q'}}{\omega x_p}; \quad x_2 = \frac{x_q}{y_{p'}}; \quad x_3 = \frac{x_q}{\omega x_p}; \quad \tilde{x}_1 = \frac{x_{q'}}{y_p}; \quad \tilde{x}_2 = \frac{y_q}{\omega x_{p'}}; \quad etc.$$

With $p_i = (x_i, y_i)$; $Op_i = (\omega^{-1}x_i^{-1},\ \omega^{-1/2}x_i^{-1}y_i)$ we find

$$W_{pq'}(\alpha_1-\alpha_2) \equiv \frac{w_{\tilde{p}_1}(\alpha_1-\alpha_2)}{w_{O p_1}(\alpha_1-\alpha_2)}; \quad \overline{W}_{pq}(\beta_2-\alpha_1) \equiv \frac{w_{O\tilde{p}_3}(\beta_2-\alpha_1)}{w_{p_3}(\beta_2-\alpha_1)}.$$

4. The classical BS-model, isospectral transformations

An important application of the 3D approach to the BS model is the derivation of an isospectral transform of the BS-transfer matrix. We first show an isospectrality of the *classical* inhomogenous version of the BS-model. We define the classical counterpart A^{class} of an operator A as its root-of-unity sum (each matrix element summed separately):

$$A^{class}(\xi^N) = \prod_{i\in\mathbb{Z}_N} A(\xi\omega^i).$$

So, from (3.1) we define the classical BS-L-operator $\mathcal{L}$, writing $\Lambda = \xi^N$:

$$\mathcal{L}(\Lambda) \equiv L_1^{class}(\xi) = \begin{pmatrix} 1+\Lambda U_1\tilde{U}_1 K_1 W_1 \tilde{W}_1^{-1} & U_1(1+K_1 W_1\tilde{W}_1^{-1}) \\ \Lambda\tilde{U}_1(\tilde{K}_1+W_1\tilde{W}_1^{-1}) & W_1\tilde{W}_1^{-1}+\Lambda\tilde{K}_1 U_1\tilde{U}_1 \end{pmatrix}. \tag{4.1}$$

To define the classical BS-transfer matrix in the 3D framework, we build the monodromy along the $\mathbf{e}_2$-direction. The 3D lattice is taken quasiperiodic in the $\mathbf{e}_3$-direction after two steps as in Fig. 3.1. We consider only one layer in the $\mathbf{e}_1$ direction. So we get the monodromy $\mathcal{M}$ and transfer matrix $\mathcal{T}$:

$$\mathcal{T}(\Lambda) = \mathrm{Tr}_{\mathbb{C}_2}\mathcal{M}(\Lambda); \qquad \mathcal{M}(\Lambda) = \mathcal{L}_0(\Lambda)\,\mathcal{L}_1(\Lambda)\cdots\mathcal{L}_{Q-1}(\Lambda) \tag{4.2}$$

with
$$\mathcal{L}_n = \begin{pmatrix} 1+\Lambda\, U_{1,n}\tilde{U}_{1,n}K_{1,n}V_{1,n} & U_{1,n}(1+K_{1,n}V_{1,n}) \\ \Lambda\,\tilde{U}_{1,n}\,(\tilde{K}_{1,n}+V_{1,n}) & V_{1,n}+\Lambda\,\tilde{K}_{1,n}U_{1,n}\tilde{U}_{1,n} \end{pmatrix} \tag{4.3}$$

where we have abbreviated $U_{1,n\mathbf{e}_2} = U_{1,n}$, $U_{1,n\mathbf{e}_2+\mathbf{e}_3} = \tilde{U}_{1,n}$, $V_{1,n} = W_{1,n\mathbf{e}_2}/W_{1,n\mathbf{e}_2+\mathbf{e}_3}$ $K_{1:n\mathbf{e}_2} = K_{1,n}$, $K_{1:n\mathbf{e}_2+\mathbf{e}_3} = \tilde{K}_{1,n}$.

In order to derive an isospectrality of $\mathcal{T}$, we commute an auxiliary operator $\mathcal{L}_n^{aux}$ through $\mathcal{M}$ such that

$$\mathcal{L}_0^{aux}(\Lambda)\,\mathcal{M}(\Lambda) = \mathcal{M}^\star(\Lambda)\,\mathcal{L}_Q^{aux}(\Lambda) \quad \text{with} \quad \mathcal{L}_0^{aux}(\Lambda) = \mathcal{L}_Q^{aux}(\Lambda), \tag{4.4}$$

since then $\mathrm{Tr}_{\mathbb{C}_2}\mathcal{M} = \mathrm{Tr}_{\mathbb{C}_2}\mathcal{M}^\star$. We shall see that the 3D-functional mapping $\mathcal{R}_{123}^{(f)}$ Eqs.(2.7) can be used to solve the problem (4.4). We take the initial auxiliary operator $\mathcal{L}_0^{aux}$ to be of the same form as $\mathcal{L}_0$ in (4.3) but the index 2 replacing the index 1.

Starting to commute the $\mathcal{L}_0^{aux}$ through $\mathcal{M}$, in the first step $\mathcal{L}_0^{aux}\mathcal{L}_0 = \mathcal{L}_0^\star\,\mathcal{L}_1^{aux}$ we assume that also both matrices on the right hand side have the form (4.3): for $\mathcal{L}_0^\star$ with the variables $U_{1,0}^\star$, $W_{1,0}^\star$, $\tilde{U}_{1,0}^\star$, $\tilde{W}_{1,0}^\star$ and for $\mathcal{L}_1^{aux}$ with the variables $U_{2,0}^\star$, $W_{2,0}^\star$, $\tilde{U}_{2,0}^\star$, $\tilde{W}_{2,0}^\star$. We claim that the mapping

$$\begin{aligned} S_{12}^{(f)}:\quad & U_{1,0}, W_{1,0}, U_{2,0}, W_{2,0}, \tilde{U}_{1,0}, \tilde{W}_{1,0}, \tilde{U}_{2,0}, \tilde{W}_{2,0} \\ & \mapsto U_{1,0}^\star, W_{1,0}^\star, U_{2,0}^\star, W_{2,0}^\star, \tilde{U}_{1,0}^\star, \tilde{W}_{1,0}^\star, \tilde{U}_{2,0}^\star, \tilde{W}_{2,0}^\star \end{aligned} \tag{4.5}$$

solving $\mathcal{L}_0^{aux}\mathcal{L}_0 = \mathcal{L}_0^\star\,\mathcal{L}_1^{aux}$ is obtained by composing the two mappings

$$\mathcal{R}_{123}^{(f)}:\quad U_{1,0},W_{1,0},U_{2,0},W_{2,0},U_3,W_3 \;\mapsto\; U_{1,0}^\star,W_{1,0}^\star,U_{2,0}^\star,W_{2,0}^\star,U_3',W_3'$$
$$\mathcal{R}_{\tilde{1}\tilde{2}3}^{(f)}:\quad \tilde{U}_{1,0},\tilde{W}_{1,0},\tilde{U}_{2,0},\tilde{W}_{2,0},U_3',W_3' \;\mapsto\; \tilde{U}_{1,0}^\star,\tilde{W}_{1,0}^\star,\tilde{U}_{2,0}^\star,\tilde{W}_{2,0}^\star,U_3^\star,W_3^\star,$$

(using in $\mathcal{R}_{123}^{(f)}$ the constants K_1, K_2, K_3 and in $\mathcal{R}_{\tilde{1}\tilde{2}3}^{(f)}$ then $\tilde{K}_1, \tilde{K}_2, K_3$) eliminating the auxiliary variables U_3 and W_3 by imposing the periodic conditions $U_3^\star = U_3$ and $W_3^\star = W_3$. The proof and the detailed somewhat lengthy formulas for (4.5) are given in Ref.[12]. These become quite simple in the parametrization (2.18) which we shall use in the following.

Using (2.18), the entries of (4.3) depend on the variables $\mathcal{Y}_0,\ldots,\mathcal{Y}_n$ and $\mathcal{Z}_0$, $\mathcal{Z}_1$, but not on $\mathcal{X}$ (apart from the f_j and $\mathcal{P}_j$ which we shall not indicate explicitly). All $\mathcal{L}_n$ and $\mathcal{L}_n^{aux}$ depend on $\mathcal{Z}_0$ from $U_{j,n}$ and $W_{j,n}$ and on $\mathcal{Z}_1$ from $\tilde{U}_{j,n}$ and $\tilde{W}_{j,n}$. So we shall also not indicate the dependence on $\mathcal{Z}_0$, $\mathcal{Z}_1$ explicitly and we write just $\mathcal{L}_n\left(\{\mathbf{f}_j(Y_n)I_{j,n\mathbf{e}_2}\},\mathcal{Y}_n\right)$ or even shorter $\mathcal{L}_n\left(\{\mathbf{f}_j\,I_{j,n\mathbf{e}_2}\},\mathcal{Y}_n\right)$ in place of (4.3) and analogously $\mathcal{L}_0^{aux}\left(\{\mathbf{f}_j\sigma_j(\mathcal{Y}_0)\},\mathcal{X}\right)$ in place of $\mathcal{L}_0^{aux}\left(\{\mathbf{f}_j(X)\sigma_j(\mathcal{Y}_0)\},\mathcal{X}\right)$. In this notation the intertwining and the mapping (4.5) are found[12] to take the form

$$\mathcal{L}_0^{aux}(\{\mathbf{f}_j\,\sigma_j(\mathcal{Y}_0)\},\mathcal{X})\;\mathcal{L}_0(\{\mathbf{f}_j\},\mathcal{Y}_0) \;=\; \mathcal{L}_0(\{\mathbf{f}_j\sigma_j(\mathcal{X})\},\mathcal{Y}_0)\;\mathcal{L}_1^{aux}(\{\mathbf{f}_j\},\mathcal{X}),$$

where the periodicity requirement $U_3^\star = U_3$ imposes g constraint equations

$$\sigma_j(\mathcal{Z}_0)\,\sigma_j(\mathcal{Z}_1) \;\equiv\; \left\{\begin{matrix} P_j' & P_j \\ Z_0' & Z_0 \end{matrix}\right\}\left\{\begin{matrix} P_j' & P_j \\ Z_1' & Z_1 \end{matrix}\right\} \;=\; 1 \tag{4.6}$$

on the suppressed variables $\mathcal{P}_j$. We see that in the intertwining $\mathcal{L}_0$ picks up a factor $\sigma_j(\mathcal{X})$ and the argument of $\mathcal{L}_1^{aux}$ gets divided by $\sigma_j(\mathcal{Y}_0)$. Moving the auxiliary operator all through the monodromy then leads to

$$\mathcal{L}_0^{aux}(\{\mathbf{f}_j\sigma_j(\mathcal{Y}_0)\},\mathcal{X})\;\mathcal{M} \;=\; \mathcal{M}^\star\,\mathcal{L}_Q^{aux}\left(\left\{\mathbf{f}_j\prod_{i=1}^{Q-1}\sigma_j^{-1}(\mathcal{Y}_i)\right\},\mathcal{X}\right)$$

with $\quad\mathcal{M} = \mathcal{L}_0(\{\mathbf{f}_j\},\mathcal{Y}_0)\;\ldots\;\mathcal{L}_{Q-1}\left(\left\{\mathbf{f}_j\prod_{i=0}^{Q-2}\sigma_j^{-1}(\mathcal{Y}_i)\right\},\mathcal{Y}_{Q-1}\right);$

$$\mathcal{M}^\star = \mathcal{L}_0(\{\mathbf{f}_j\,\sigma_j(\mathcal{X})\},\mathcal{Y}_0)\ldots\mathcal{L}_{Q-1}\left(\left\{\mathbf{f}_j\prod_{i=0}^{Q-2}\sigma_j^{-1}(\mathcal{Y}_i)\sigma_j(\mathcal{X})\right\},\mathcal{Y}_{Q-1}\right).$$

Imposing periodicity (4.4) leads to the g equations $I_{j,\,Q\mathbf{e}_2} = 1$ or explicitly

$$\prod_{i=0}^{Q-1}\sigma_j^{-1}(\mathcal{Y}_i) \;\equiv\; \prod_{i=0}^{Q-1}\left\{\begin{matrix} P_j' & P_j \\ Y_j & Y_j' \end{matrix}\right\} \;=\; 1\,. \tag{4.7}$$

Eqs. (4.6) and (4.7) together fix the soliton parameters P_j', P_j of the functions H. If these Eqs. are fulfilled, $\mathrm{Tr}\,\mathcal{M}$ and $\mathrm{Tr}\,\mathcal{M}^\star$ are *isospectral*. Since $\mathcal{M}^\star$ is obtained from $\mathcal{M}$ by the substitution $f_j \to f_j^\star = f_j\sigma_j(\mathcal{X})$, we get a non-trivial isospectrality of the inhomogenous (all $\mathcal{Y}_i$ may be chosen to be different) classical BS-chain if some $f_j \neq 0$.

In order to derive an isospectrality for the inhomogenous *quantum* BS transfer matrix (3.6), again we pull an auxiliary L-operator through the monodromy. So we need to know the quantum intertwining operator $\mathbf{S}_{12}$

$$\begin{aligned}\mathbf{S}_{12}\, L(\xi;\, u_2,\, \tilde{u}_2,\, w_2,\, \tilde{w}_2,\, \kappa_2,\, \tilde{\kappa}_2) \cdot L(\xi;\, u_1,\, \tilde{u}_1,\, w_1,\, \tilde{w}_1,\, \kappa_1,\, \tilde{\kappa}_1) \\ = L(\xi;\, u_1^\star,\, \tilde{u}_1^\star,\, w_1^\star,\, \tilde{w}_1^\star,\, \kappa_1,\, \tilde{\kappa}_1) \cdot L(\xi;\, u_2^\star,\, \tilde{u}_2^\star,\, w_2^\star,\, \tilde{w}_2^\star,\, \kappa_2,\, \tilde{\kappa}_2)\, \mathbf{S}_{12}. \quad (4.8)\end{aligned}$$

Here the $u_1^\star, \ldots, \tilde{w}_2^\star$ shall be related to the $u_1, \ldots, \tilde{w}_2$ by the classical mapping (4.5). In order to use our previous classical intertwining results we use the quantum operators L in the form (3.1) rather than (3.3). If we take the mapping (4.5) to be trivial, we get back to (3.8) and (3.9). However, for non-trivial $\mathcal{R}^{(f)}$ it is easy to see[12] using (2.9) that the solution to the fully dynamical equation (4.8) is $\mathbf{S}_{12} = \mathrm{Tr}_3(\mathcal{R}_{123}^{(f)} \circ \mathbf{R}_{\tilde{1}\tilde{2}3}) \cdot \mathbf{R}_{123}$, where the Fermat parameters $x_1, x_2, x_3, \tilde{x}_1, \tilde{x}_2, \tilde{x}_3$ determining the matrices $\mathbf{R}$ follow from (2.12) and (2.18). In the parametrization (2.18) we can write (4.8) as

$$\begin{aligned}\mathbf{S}\left(\{f_j\}, \mathcal{X}, \mathcal{Y}\right)\, L\left(\{f_j \sigma_j(\mathcal{Y})\}, \mathcal{X}\right) L\left(\{f_j\}, \mathcal{Y}\right) \\ = L\left(\{f_j \sigma_j(\mathcal{X})\}, \mathcal{Y}\right)\, L\left(\{f_j\}, \mathcal{X}\right)\, \mathbf{S}\left(\{f_j\}, \mathcal{X}, \mathcal{Y}\right). \quad (4.9)\end{aligned}$$

This is analogous to the classical case, except that also a matrix conjugation by $\mathbf{S}$ appears. Choosing $L\left(\{f_j \sigma_j(\mathcal{Y})\}, \mathcal{X}\right)$ as the initial auxiliary operator, we get isospectrality of the transfer matrices $\mathbf{T}$ and $\mathbf{T}^\star$ where

$$\mathbf{K}\,\mathbf{T} = \mathbf{T}^\star \mathbf{K}; \quad \mathbf{K} = \mathrm{Tr}_{\mathbb{C}^N} \mathbf{S}\left(\{f_j\}, \mathcal{X}, \mathcal{Y}_0\right) \ldots \mathbf{S}\left(\{f_j\, I_{j,(Q-1)\mathbf{e}_2}\}, \mathcal{X}, \mathcal{Y}_{Q-1}\right).$$

We can see directly that each quantum operator L_n of the form (3.1) with (3.2), when parameterized according to (2.18), depends on the rational functions $H\left(\{f_j\}\right)$ only via u_n and the ratio $w_n \tilde{w}_n^{-1}$. In the product $u_n \tilde{u}_n$ the H drop out because of (4.6). Now u_n and $\tilde{u}_n$ occur only multiplying $\mathbf{X}_n$ resp. $\mathbf{X}_n^{-1}$ and $w_n \tilde{w}_n^{-1}$ always multiplies $\mathbf{Z}_n$. So we can absorb[12] all the dependence on the $H\left(\{f_j\}\right)$ in a change of normalization of the $\mathbf{X}_n$ and $\mathbf{Z}_n$ which preserves $\mathbf{X}_n \mathbf{Z}_m = \omega^{\delta_{n,m}} \mathbf{Z}_m \mathbf{X}_n$. The spectrum of resulting redefined transfer matrix becomes independent of the choice of the f_j.

Acknowledgements

This work is supported by INTAS-OPEN-03-51-3350 and by the Australian Research Council. GvG takes the pleasure to thank Profs. Mo-Lin Ge and Chengming Bai for creating a unique warm ambience for discussions at Nankai University. He is grateful for their invitation to present this work.

References

1. A.B.Zamolodchikov, *Comm. Math. Phys.* **79**, 489 (1981)
2. V.V.Bazhanov and R.J.Baxter, *J. Stat. Phys.* **69**, 453 (1992); **71**, 839 (1993)
3. V.V.Bazhanov and S.Sergeev, hep-th/0509181
4. S.M.Sergeev, V.V.Mangazeev, Yu.G.Stroganov, *J. Stat. Phys.* **82**, 31 (1996)
5. S.M.Sergeev, *J. Phys.* **A32**, 5693 (1999); *Phys. Part. Nuclei* **35**, 1 (2004)
6. G.von Gehlen, S.Pakuliak and S.Sergeev, *J. Phys.* **A36**, 975 (2003); **A37**, 1159 (2004)
7. V.V.Bazhanov and Yu.G.Stroganov, *J. Stat. Phys.* **59**, 799 (1990)
8. I.G.Korepanov, *Zapiski Nauchn. Semin. POMI* **215**, 163 (1994)
9. S.Pakuliak and S.Sergeev, *Int. J. Math. and Math. Sc.* **31**, 513 (2002); *Theor. Math. Phys.* **136**, 917 (2003)
10. R.J.Baxter, V.V.Bazhanov, J.H.H.Perk, *Int. J. Mod. Phys.* **B**4, 803 (1990)
11. R.J.Baxter, *Phys. Lett.* **A146**, 110 (1990); *J. Stat. Phys.* **120**, 1 (2005)
12. G.von Gehlen, S.Pakuliak and S.Sergeev, *J. Phys.* **A38**, 7269 (2005)

Exact Solution of Two Planar Polygon Models

Anthony J. Guttmann and Iwan Jensen
ARC Centre of Excellence for Mathematics and Statistics of Complex Systems
Department of Mathematics and Statistics
The University of Melbourne, Victoria 3010, Australia

Using a simple transfer matrix approach we have derived long series expansions for the perimeter generating functions of both *three-choice polygons* and *punctured staircase polygons*. In both cases we find that all the known terms in the generating function can be reproduced from a linear Fuchsian differential equation of order 8. We report on an analysis of the properties of the differential equations.

1. Introduction

A well-known long standing problem in combinatorics and statistical mechanics is the enumeration by perimeter of self-avoiding polygons (or walks) on a two- or three-dimensional lattice. Recently, we have gained a greater understanding of the difficulty of this problem, as Rechnitzer[14] has *proved* that the (anisotropic) generating function for square lattice self-avoiding polygons is not differentiably finite[15]. This property had been *conjectured*, on numerical grounds[5], but not proved. So the generating function cannot be expressed as a solution of an ordinary differential equation with polynomial coefficients. There are many simplifications of this problem that are solvable[1], but these simpler models impose an effective directedness or other constraint that reduce them, in essence, to one-dimensional problems.

One model, that of *three-choice polygons*, has remained unsolved despite the knowledge that its solution must be D-finite. Recent numerical work[7] resulted in an exact differential equation apparently satisfied by the perimeter generating function of three-choice polygons. Similarly for another model, that of *punctured staircase polygons*, that is a staircase polygon with an arbitrary staircase puncture. Again we found[8] that the perimeter generating function is apparently satisfied by an exact differential equation. While our results do not constitute rigorous mathematical proofs the numerical evidence is overwhelmingly compelling.

The next two sections consider these two models, in turn.

2. Three-choice polygons

Three-choice self-avoiding walks on the square lattice, $\mathbb{Z}^2$, were introduced by Manna[13] and can be defined as follows: Starting from the origin one can step in any direction; after a step upward or downward one can head in any direction (except backward); after a step to the left one can only step forward or head downward, and after a step to the right one can continue forward or turn upward. Alternatively put, one cannot make a right-hand turn after a horizontal step. Whittington[17] showed that the growth constant for three-choice walks is exactly 2, so that if w_n denotes the number of such walks of n steps on an infinite lattice, equivalent up to a translation, then $w_n \sim 2^{n+o(n)}$. It is perhaps surprising that the best known result for the sub-dominant term is $2^{o(n)}$, but attempts to improve on this have been unsuccessful. Even numerically, there is no firmly based conjecture for the sub-dominant term, unlike for ordinary self-avoiding walks, for which the sub-dominant term is widely believed to be $O(\log n)$.

As usual one can define a polygon version of the walk model by requiring the walk to return to the origin. So a three-choice polygon[10] is simply a three-choice self-avoiding walk which returns to the origin, but has no other self-intersections. There are two distinct classes of three-choice polygons. The three-choice rule either leads to staircase polygons or *imperfect staircase polygons*[3] as illustrated in figure 2.1. In the case of staircase polygons any perimeter vertex can act as the origin of the three-choice walk (which then proceeds counter-clockwise), while for imperfect staircase polygons there is only one possible origin but the polygon could be rotated by 180 degrees. If we denote by t_n the number of three-choice polygons with perimeter $2n$ then, $t_n = 2nc_n + 2p_n$, where c_n is the number of staircase polygons and p_n is the number of imperfect staircase polygons with perimeter $2n$. Note that t_n, p_n and c_n all grow like 4^n and in particular we recall the well-known result that $c_{n+1} = C_n = \frac{1}{n+1}\binom{2n}{n}$ where C_n are the Catalan numbers.

In this paper we report on recent work[7] which has led to an exact Fuchsian[11] linear differential equation of order 8 apparently satisfied by the perimeter generating function, $\mathcal{T}(x) = \sum_{n>0} t_n x^n$, for three-choice polygons (that is $\mathcal{T}(x)$ is conjectured to be one of the solutions of the

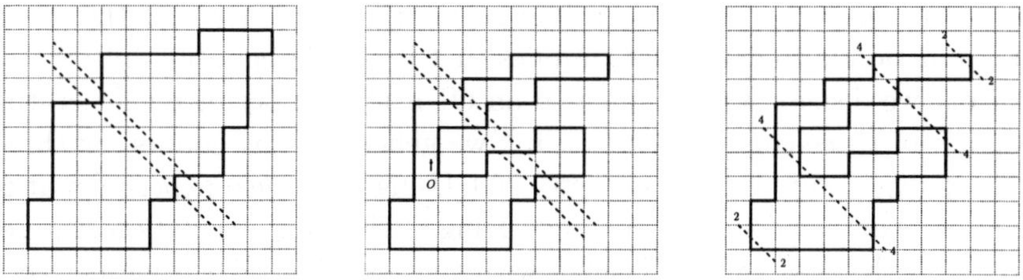

Staircase **Imperfect staircase**

Fig. 2.1. Examples of the two types of three-choice polygons. In the middle panel we indicate the origin (O) and the direction of the first step (note that rotation by 180 degrees also leads to a valid three-choice polygon). The right panle shows the decomposition of an imperfect staircase polygon into a sequence of 2-4-2 non-intersecting walkers, each expressible as a Gessel-Viennot determinant.

ODE, expanded around the origin). The first few terms are

$$\mathcal{T}(x) = 4x^2 + 12x^3 + 42x^4 + 152x^5 + 562x^6 + \cdots .$$

(The generating function for the coefficients p_n is no simpler.)

If we distinguish between steps in the x and y direction, and let $t_{m,n}$ denote the number of three-choice polygons with $2m$ horizontal steps and $2n$ vertical steps, then the anisotropic generating function for $\mathcal{T}(x, y)$ is

$$\mathcal{T}(x,y) = \sum_{m,n} t_{m,n} x^m y^n = \sum_{n} H_n(x) y^n,$$

where $H_n(x) = \frac{R_n(x)}{S_n(x)}$ is the (rational[16]) generating function for three-choice polygons with $2n$ vertical steps. In earlier, unpublished, numerical work, we found that, for imperfect staircase polygons, the denominators are:

$$S_n(x) = (1-x)^{2n-1}(1+x)^{(2n-7)_+} \qquad n \quad \text{even},$$

and

$$S_n(x) = (1-x)^{2n-1}(1+x)^{(2n-8)_+} \qquad n \quad \text{odd}.$$

This was subsequently proved by Bousquet-Mélou[2]. Unfortunately, we still do not have enough information to identify the numerators.

It is also possible to express the generating function $\mathcal{T}(x)$ as a five-fold sum, with one constraint[2], of 4×4 Gessel-Viennot determinants[4]. This is clear from the right panel of figure 2.1, where the enumeration of the lattice paths between the dotted lines is just the classical problem of 4 non-intersecting walkers, and these must be joined to two non-intersecting walkers to the left, and to two non-intersecting walkers to the right. Then one must sum over different possible geometries. The fact that the generating function is so expressible implies that it is differentiably finite[12].

Next we discuss work leading to an ODE for the perimeter generating function of three-choice polygons. In[7] we generated the counts for three-choice polygons up to half-perimeter 260. Using numerical experimentation we found what we believe is the underlying ODE. This calculation required the use of the first 206 coefficients with the resulting ODE then correctly predicting the next 54 coefficients. The possibility that this ODE is incorrect is extraordinarily small, but this does not of course constitute a proof. Unfortunately we cannot usefully bound the size of the underlying ODE, otherwise we could use the knowledge of D-finiteness to provide a proof. Bounds following from closure theorems[12] are too large to be useful.

The algorithm used to enumerate imperfect polygons is a slightly modified version of the algorithm of Conway *et al.*[3], and is described fully in[7].

2.1. *The Fuchsian differential equation*

Recently Zenine *et al.*[18–20] obtained linear differential equations whose solutions give the 3- and 4-particle contributions $\chi^{(3)}$ and $\chi^{(4)}$ to the Ising model susceptibility. In[7] we used their method to find an ODE which has as a solution the generating function $\mathcal{T}(x)$ for three-choice polygons. This involves a systematic search for a differential equation of the form:

$$\sum_{k=0}^{m} P_k(x) \frac{\mathrm{d}^k}{\mathrm{d}x^k} \mathcal{T}(x) = 0, \qquad (2.1)$$

such that $\mathcal{T}(x)$ is a solution to this differential equation, where the $P_k(x)$ are polynomials. To make it as simple as possible we started by searching for a Fuchsian[11] equation. Such equations have only regular singular points.

We searched systematically for solutions by varying m and q_m, the degree of the polynomials $P_m(x)$. In this way a solution with $m = 10$ and $q_m = 12$ was first found, which required the determination of $L = 206$ unknown coefficients. With 260 terms in the half-perimeter series, there are more than 50 additional terms with which to check the correctness of this solution. Having found this conjectured solution the ODE was then turned into a recurrence relation and used to generate more series terms in order to search for a lower order Fuchsian equation. The lowest order equation found was eighth order and with $q_m = 30$, which requires the determination of $L = 321$ unknown coefficients. Thus from the original 260 term series this 8th order solution could not have been found. This raises the question as to whether perhaps there is an ODE of lower order than 8 that generates the coefficients? The short answer to this is no.

So the (half)-perimeter generating function $\mathcal{T}(x)$ for three-choice polygons is conjectured to be a solution of the linear ODE of order 8

$$\sum_{k=0}^{8} P_k(x)\frac{\mathrm{d}^k}{\mathrm{d}x^k}F(x) = 0 \tag{2.2}$$

with

$$P_8(x) = x^3(1-4x)^4(1+4x)(1+4x^2)(1+x+7x^2)Q_8(x), \tag{2.3}$$

where $Q_8(x)$ is a polynomial of degree 25, which together with the remaining polynomials $P_k(x)$ are given in[7].

The singular points of the differential equation are given by the roots of $P_8(x)$. One can easily check that all the singularities (including $x=\infty$) are *regular singular points* so equation (2.2) is indeed of the Fuchsian type. Using the method of Frobenius one can obtain from the indicial equation the critical exponents at the singular points. These are listed in Table 2.1.

Table 2.1. Critical exponents for the regular singular points of the Fuchsian differential equation satisfied by $\mathcal{T}(x)$.

Singularity	Exponents
$x=0$	−1, 0, 0, 0, 1, 2, 3, 4
$x=1/4$	−1/2, −1/2, 0, 1/2, 1, 3/2, 2, 3
$x=-1/4$	0, 1, 2, 3, 4, 5, 6, 13/2
$x=\pm\mathrm{i}/2$	0, 1, 2, 3, 4, 5, 6, 13/2
$1+x+7x^2=0$	0, 1, 2, 2, 3, 4, 5, 6
$x=\infty$	−2, −3/2, −1, −1, −1/2, 1/2, 3/2, 5/2
$Q_8(x)=0$	0, 1, 2, 3, 4, 5, 6, 8

A careful local analysis revealed that near the physical critical point $x=x_c=1/4$ the singular behaviour is

$$\mathcal{T}(x) \sim A(x)(1-4x)^{-1/2} + B(x)(1-4x)^{-1/2}\log(1-4x), \tag{2.4}$$

where $A(x)$ and $B(x)$ are analytic in the neighbourhood of x_c. Note that the terms associated with the exponents 1/2 and 3/2 become part of the analytic correction to the $(1-4x)^{-1/2}$ term. Near the singularity on the negative x-axis, $x=x_-=-1/4$ the singular behaviour is

$$\mathcal{T}(x) \sim C(x)(1+4x)^{13/2}, \tag{2.5}$$

where again $C(x)$ is analytic near x_-. Similar behaviour is expected near the pair of singularities $x = \pm i/2$, and finally at the roots of $1 + x + 7x^2$ one expects the behaviour $\mathcal{T}(x) \sim D(x)(1 + x + 7x^2)^2 \log(1 + x + 7x^2)$.

We can simplify the 8th order differential operator found above. We first found three solutions of the ODE, each corresponding to an order one differential operator. Denoting these by $L_i^{(1)}$, with $i = 1,2,3$, we found that the differential operator could be written as $L^{(8)} = L^{(5)} L_1^{(1)} L_2^{(1)} L_3^{(1)}$, where $L^{(5)}$ is a fifth order differential operator, further decomposable as $L^{(5)} = L^{(3)} L^{(2)}$. This then allows us to write down the form of the 8×8 matrix representing the differential Galois group of $L^{(8)}$, in an appropriate global solution basis. To determine the asymptotics one would need to calculate non-local connection matrices between solutions at different points. This is a huge task for such a large differential operator. Instead, we have developed a numerical technique that avoids all these difficulties, described below.

To analyse the asymptotic behaviour of the coefficients, we first transform the coefficients so that the critical point is at 1. The growth constant of staircase and imperfect staircase polygons is 4, so we consider a new series with coefficients $r_n = t_{n+2}/4^n$. Thus the generating function studied is $\mathcal{R}(y) = \sum_{n \geq 0} r_n y^n = 4 + 3y + 2.625y^2 + \cdots$. From equations (2.4) and (2.5) it follows that the asymptotic form of the coefficients is

$$[y^n]\mathcal{R}(y) = r_n = \frac{1}{\sqrt{n}} \sum_{i \geq 0} \left(\frac{a_i \log n + b_i}{n^i} + (-1)^n \left(\frac{c_i}{n^{7+i}} \right) \right) + \mathrm{O}(\lambda^{-n}). \quad (2.6)$$

The last term includes the effect of other singularities, further from the origin than the dominant singularities. These will decay exponentially since $\lambda > 1$ in the scaled variable $y = x/4$.

Using the recurrence relations for t_n (derived from the ODE) it is easy and fast to generate many more terms r_n. In[7] the first 100000 terms were generated and saved as floating point numbers with 500 digit accuracy (this calculation took less than 15 minutes). With such a long series it is possible to obtain accurate numerical estimates of the first 20 amplitudes a_i, b_i, c_i for $i \leq 19$ with a precision of more than 100 digits for the dominant amplitudes, shrinking to 10–20 digits for the the case when $i = 18$, or 19. In making these estimates the exponentially decaying terms were ignored. In this way an earlier conjecture[3] that $a_0 = \frac{3\sqrt{3}}{\pi^{3/2}}$, was confirmed. Other amplitude estimates include $b_0 = 3.17327538458989848178 5\ldots$ and $c_0 = \frac{-24}{\pi^{3/2}}$, though no one has been able to identify b_0. However, further sub-dominant amplitudes have been estimated[7], such as $a_1 = \frac{-89}{8\sqrt{3}\pi^{3/2}}$, $a_2 = \frac{1019}{384\sqrt{3}\pi^{3/2}}$, and $a_3 = \frac{-10484935}{248832\sqrt{3}\pi^{3/2}}$, and $c_1 = \frac{225}{\pi^{3/2}}$, $c_2 = \frac{-16575}{16\pi^{3/2}}$, and $c_3 = \frac{389295}{128\pi^{3/2}}$. It seems likely that the amplitudes $\pi^{3/2}\sqrt{3}a_i$ and $\pi^{3/2}c_i$ are rational.

We have also looked at the area generating function. For staircase polygons the area generating function is given by

$$A(q) = \sum_{n\geq 1} a_n q^n = \frac{J_1(1,1,q)}{J_0(1,1,q)},$$

where $J_i = \sum_{n\geq 0} \frac{(-1)^n q^{(n+i)(n+i+1)/2}}{(q)_n^2(1-q^{n+1})^i}$, $i = 0, 1$. Based on a 500 term series, our analysis suggests that the area generating function is of the form $\frac{F(q)+G(q)/\sqrt{1-q\eta}}{[J_0(1,1,q)]^2}$. That is to say, the leading singularity occurs at $q = 1/\eta$, where η is the first zero of $J_0(1,1,q)$, and F and G are regular in the neighbourhood of $q = 1/\eta$. The coefficients thus behave asymptotically as

$$a_n = [q^n]A(q) \sim const.\eta^{-n} n^{3/2}.$$

The solution is not, however, of the simple product form as found for staircase polygons. We can see this by constructing Padé approximants of steadily increasing order, which do not stabilise.

3. Punctured staircase polygons

Punctured staircase polygons[6] are staircase polygons with internal holes which are also staircase polygons (the polygons are mutually- as well as self-avoiding). In[6] it was proved that the connective constant μ of k-punctured polygons (polygons with k holes) is the same as the connective constant of un-punctured polygons. Here we discuss only the case with a *single* hole (see figure 3.1). The perimeter length of a punctured staircase polygons is the outer perimeter plus the perimeter of the hole. We denote by p_n the number of punctured staircase polygons of total perimeter $2n$. The results of[6] indicate that the half-perimeter generating function has a simple pole at $x = x_c = 1/\mu = 1/4$, though the analysis[6] clearly indicated a more complicated critical behaviour.

Here we report on recent work[8] which led to an exact Fuchsian linear differential equation of order 8 apparently satisfied by the perimeter generating function, $\mathcal{P}(x) = \sum_{n\geq 0} p_n x^n$, for punctured staircase polygons (that is $\mathcal{P}(x)$ is one of the solutions of the ODE, expanded around the origin). The first few terms in the generating function are

$$\mathcal{P}(x) = x^8 + 12x^9 + 94x^{10} + 604x^{11} + 3463x^{12} + \cdots .$$

The situation is very similar to that of three-choice polygons. This is perhaps not surprising, as one can represent punctured staircase polygons as

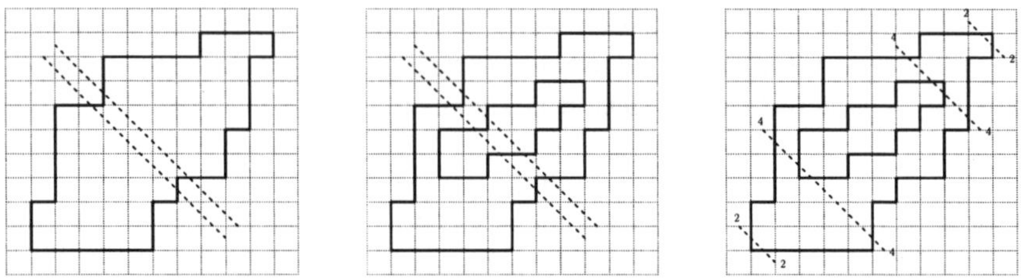

Staircase **Punctured staircase**

Fig. 3.1. Examples of the types of staircase polygons studied in this paper. The right panle shows the decomposition of a punctured staircase polygon into a sequence of 2-4-2 vicious walkers, each expressible as a Gessel-Viennot determinant.

the fusion of two three-choice polygons, with some edges deleted. Again it is possible to express the generating function $\mathcal{P}(x)$ as a sum over 4×4 Gessel-Viennot determinants. This is clear from the right panel of figure 3.1. By arguments similar to those presented above, it follows that the generating function is D-finite. Again we cannot readily bound the size of the underlying ODE, otherwise we could use this observation to provide a proof of our results. However, from the counts of polygons up to half-perimeter 260, the underlying ODE was found experimentally from the first 206 coefficients[8]. The ODE then correctly predicted the next 54 coefficients. While the possibility that the underlying ODE is not the correct one is extraordinarily small, that still does not constitute a proof.

The enumeration algorithm[8] is again a modified version of the algorithm of Conway *et al.*[3] for the enumeration of imperfect staircase polygons.

We identified the ODE in a manner similar to that described above for three-choice polygons, and the (half)-perimeter generating function $\mathcal{P}(x)$ for punctured staircase polygons was found to satisfy an ODE of order 8

$$\sum_{k=0}^{8} P_n(x) \frac{\mathrm{d}^k}{\mathrm{d}x^k} F(x) = 0 \tag{3.1}$$

with

$$P_8(x) = x^4(1-4x)^8(1+4x)(1+4x^2)(1+x+7x^2)Q_8(x), \tag{3.2}$$

where $Q_8(x)$ is a polynomial of degree 22. All polynomials are given in[8]. The singular points as given by the roots of $P_8(x)$ and the associated critical exponents are listed in Table 3.1.

Detailed analysis of the local solutions of the ODE are given in[8]. Near

Table 3.1. Critical exponents for the regular singular points of the Fuchsian differential equation satisfied by $\mathcal{P}(x)$.

Singularity	Exponents
$x = 0$	$-1, 0, 0, 0, 1, 2, 3, 8$
$x = 1/4$	$-1, -1/2, -1/2, 1/2, 1, 3/2, 2, 3$
$x = -1/4$	$0, 1, 2, 3, 4, 5, 6, 13/2$
$x = \pm \mathrm{i}/2$	$0, 1, 2, 3, 4, 5, 6, 13/2$
$1 + x + 7x^2 = 0$	$0, 1, 2, 2, 3, 4, 5, 6$
$1/x = 0$	$-2, -3/2, -1, -1, -1/2, 1/2, 3/2, 5/2$
$Q_8(x) = 0$	$0, 1, 2, 3, 4, 5, 6, 8$

the critical point $x = x_c = 1/4$ the following singular behaviour was found:

$$\mathcal{P}(x) \sim A(x)(1-4x)^{-1} + B(x)(1-4x)^{-1/2} + C(x)(1-4x)^{-1/2}\log(1-4x), \tag{3.3}$$

where $A(x)$, $B(x)$ and $C(x)$ are analytic in a neighbourhood of x_c. Note that the terms associated with the exponents 1/2 and 3/2 become part of the analytic correction to the $(1-4x)^{-1/2}$ term. Near the singularity on the negative x-axis, $x = x_- = -1/4$ the singular behaviour

$$\mathcal{P}(x) \sim D(x)(1+4x)^{13/2}, \tag{3.4}$$

was found, where again $D(x)$ is analytic near x_-. Similar behaviour is expected near the pair of singularities $x = \pm\mathrm{i}/2$, and finally at the roots of $1 + x + 7x^2$ the behaviour $E(x)(1 + x + 7x^2)^2 \log(1 + x + 7x^2)$ is expected.

The asymptotic form of the coefficients was analysed as for three-choice polygons. The growth constant is 4 and we considered the new series with coefficients $r_n = p_{n+8}/4^n$. Using the recurrence relations for p_n (derived from the ODE) we generated many more terms r_n. From equations (3.3) and (3.4) it follows that the asymptotic form of the coefficients is

$$[x^n]\mathcal{R}(y) = r_n = \sum_{i\geq 0}\left(\frac{a_i}{n^i} + \frac{b_i \log n + c_i}{n^{i+1/2}} + (-1)^n\left(\frac{d_i}{n^{15/2+i}}\right)\right). \tag{3.5}$$

Any contributions from the other singularities are exponentially suppressed since their norm (in the scaled variable $y = x/4$) exceeds 1. From the first 100000 terms estimates for the amplitudes were obtained by fitting r_n to the form given above. This led to the refined asymptotic form

$$[x^n]\mathcal{R}(y) = r_n = 1024\left(1 + \frac{1}{\sqrt{n}}\sum_{i\geq 0}\left(\frac{b_i \log n + c_i}{n^i} + (-1)^n\left(\frac{d_i}{n^{7+i}}\right)\right)\right). \tag{3.6}$$

We obtained accurate numerical estimates of many of the amplitudes and found that[8] $b_0 = -\frac{6\sqrt{3}}{\pi^{3/2}}$, $b_1 = \frac{305}{4\sqrt{3}\pi^{3/2}}$, $b_2 = \frac{86123}{192\sqrt{3}\pi^{3/2}}$, $c_0 = 1.5521034004887910537 4\ldots$ and $d_0 = \frac{48}{\pi^{3/2}}$, $d_1 = -\frac{2610}{\pi^{3/2}}$, $d_2 = \frac{640815}{8\pi^{3/2}}$, $d_3 = -\frac{116785575}{64\pi^{3/2}}$, $d_4 = \frac{70325480841}{2048\pi^{3/2}}$, though we have been unable to identify c_0. These amplitudes are known to at least 100 digits accuracy. The excellent convergence is solid evidence (though naturally not a proof) that the assumptions leading to equation (3.5) are correct.

We have also initiated an investigation of the *area* generating function. We find that the area generating function $A(q)$ is of the form

$$A(q) = (G(q) + H(q)\sqrt{1 - q/\eta})/[J_0(1, 1, q)^2],$$

where $J_0(x, y, q)$ is as described above. Here $q = \eta$ is the first zero of $J_0(1, 1, q)$, and G and H are regular in the neighbourhood of $q = \eta$. The coefficients thus behave asymptotically as

$$a_n = [q^n]A(q) \sim const.\eta^{-n}n.$$

Acknowledgments

We would like to thank N Zenine and J-M Maillard for access to their Mathematica routines and useful advice about their use. We would also like to thank M. Bousquet-Mélou for communicating her unpublished results on this problem, which we mentioned in the introduction. We gratefully acknowledge financial support from the Australian Research Council.

References

1. Bousquet-Mélou M 1996 *Disc. Math.* **154** 1
2. Bousquet-Mélou M Private communication.
3. Conway A R, Guttmann A J and Delest M 1997 *Mathl. Comput. Modelling* **26** 51
4. Gessel I and Viennot X G 1989 Determinants, paths and plane partitions *preprint*
5. Guttmann A J and Conway A R 2001 *Ann. Comb.* **5** 319
6. Guttmann A J Jensen I Wong L H and Enting I G 2000 *J. Phys. A* **33** 1735
7. Guttmann A J and Jensen I 2006 Fuchsian differential equation for the perimeter generating function of three-choice polygons. *Séminaire Lotharingien de Combinatoire, in press Preprint:* math.CO/0506317
8. Guttmann A J and Jensen I 2006 Fuchsian differential equation for the perimeter generating function of punctured staircase polygons. *In preparation*
9. Guttmann A J and Jensen I 2006 Properties of Fuchsian differential equations for polygon enumeration problems *In preparation*
10. Guttmann A J, Prellberg T and Owczarek A L 1993 *J. Phys. A* **26** 6615

11. Ince E L 1927 *Ordinary differential equations* (London: Longmans, Green and Co. Ltd.)
12. Lipshitz L 1989 *J. Algebra* **122** 353
13. Manna S S 1984 *J. Phys. A: Math. Gen.* **17** L899
14. Rechnitzer A 2003 *Adv, Appl. Math.* **30** 228
15. Stanley R P 1980 *European J. Combin.* **1** 175
16. Stanley R P 1999 *Enumerative Combinatorics* vol. 2 (Cambridge: Cambridge University Press)
17. Whittington S G 1985 *J. Phys. A* **18** L67
18. Zenine N, Boukraa S, Hassani S and Maillard J M 2004 *J. Phys. A* **37** 9651
19. Zenine N, Boukraa S, Hassani S and Maillard J M 2005 *J. Phys. A* **38** 1875
20. Zenine N, Boukraa S, Hassani S and Maillard J M 2005 *J. Phys. A* **38** 4149

Quasi-exact Solvability of Dirac Equations*

Choon-Lin Ho

Department of Physics, Tamkang University, Tamsui 25137, Taiwan, R.O.C.

We present a general procedure for determining quasi-exact solvability of the Dirac and the Pauli equation with an underlying $sl(2)$ symmetry. This procedure makes full use of the close connection between quasi-exactly solvable systems and supersymmetry. The Dirac-Pauli equation with spherical electric field is taken as an example to illustrate the procedure.

1. In this talk we present a general procedure for determining quasi-exact solvability of the Dirac and the Pauli equation with an underlying $sl(2)$ symmetry. This procedure makes full use of the close connection between quasi-exactly solvable (QES) systems and supersymmetry (SUSY), or equivalently, the factorizability of the equation. Based on this procedure, we have demonstrated that the Pauli and the Dirac equation coupled minimally with a vector potential [1], neutral Dirac particles in external electric fields (which are equivalent to generalized Dirac oscillators) [2], and Dirac equation with a Lorentz scalar potential [3] are physical examples of QES systems.

Here we only give the main ideas of the procedures, and refer the readers to Refs. [1,2,3] for details.

2. For all the cases cited above, one can reduce the corresponding multi-component equations to a set of one-variable equations possessing one-dimensional SUSY after separating the variables in a suitable coordinate

*Talk presented at the XXIII International Conference on Differential Geometric Methods in Theoretical Physics, Aug 20-26, 2005, Nankai, Tianjin, China.

system. Typically the set of equations takes the form

$$\left(\frac{d}{dr} + W(r)\right) f_- = \mathcal{E}^+ f_+ \ , \tag{1}$$

$$\left(-\frac{d}{dr} + W(r)\right) f_+ = \mathcal{E}^- f_- \ , \tag{2}$$

where r is the basic variable, e.g. the radial coordinate, and $f_\pm$ are, say, the two components of the radial part of the Dirac wave function. The superpotential W is related to the external field configuration, and $\mathcal{E}^\pm$ involve the energy and mass of the particle. We can rewrite this set of equations as

$$A^- A^+ f_- = \epsilon f_- \ , \tag{3}$$

$$A^+ A^- f_+ = \epsilon f_+ \ , \tag{4}$$

with

$$A^\pm \equiv \pm \frac{d}{dr} + W \ , \quad \epsilon \equiv \mathcal{E}^+ \mathcal{E}^- \ . \tag{5}$$

Explicitly, the above equations read

$$\left(-\frac{d^2}{dr^2} + W^2 \mp W'\right) f_\mp = \epsilon f_\mp \ . \tag{6}$$

Here and below the prime means differentiation with respect to the basic variable. Eq.(6) clearly exhibits the SUSY structure of the system. The operators acting on $f_\pm$ in Eq.(6) are said to be factorizable, i.e. as products of A^- and A^+. The ground state, with $\epsilon = 0$, is given by one of the following two sets of equations:

$$A^+ f_-^{(0)}(r) = 0 \ , \quad f_+^{(0)}(r) = 0 \ ; \tag{7}$$

$$A^- f_+^{(0)}(r) = 0 \ , \quad f_-^{(0)}(r) = 0 \ , \tag{8}$$

depending on which solution is normalizable.

One can determine the forms of the external field that admit exact solutions of the problem by comparing the forms of the superpotential W with those listed in Table (4.1) of Ref. [4].

Similarly, from Turbiner's classification of the $sl(2)$ QES systems [5], one can determine the forms of W, and hence the forms of external fields admitting QES solutions based on $sl(2)$ algebra. The main ideas of the procedures are outlined below.

3. We shall concentrate only on solution of the upper component f_-, which is assumed to have a normalizable zero energy state.

Eq.(6) shows that f_- satisfies the Schrödinger equation $H_- f_- = \epsilon f_-$, with

$$\begin{aligned} H_- &= A^- A^+ \\ &= -\frac{d^2}{dr^2} + V(r) \ , \end{aligned} \tag{9}$$

with

$$V(r) = W(r)^2 - W'(r) \ . \tag{10}$$

We shall look for $V(r)$ such that the system is QES. According to the theory of QES models, one first makes an "imaginary gauge transformation" on the function f_-

$$f_-(r) = \phi(r) e^{-g(r)} \ , \tag{11}$$

where $g(r)$ is called the gauge function. The function $\phi(r)$ satisfies

$$-\frac{d^2\phi(r)}{dr^2} + 2g'\frac{d\phi(r)}{dr} + \left[V(r) + g'' - g'^2\right]\phi(r) = \epsilon\phi(r) \ . \tag{12}$$

For physical systems which we are interested in, the phase factor $\exp(-g(r))$ is responsible for the asymptotic behaviors of the wave function so as to ensure normalizability. The function $\phi(r)$ satisfies a Schrödinger equation with a gauge transformed Hamiltonian

$$H_G = -\frac{d^2}{dr^2} + 2W_0(r)\frac{d}{dr} + \left[V(r) + W_0' - W_0^2\right] \ , \tag{13}$$

where $W_0(r) = g'(r)$. Now if $V(r)$ is such that the quantal system is QES, that means the gauge transformed Hamiltonian H_G can be written as a quadratic combination of the generators J^a of some Lie algebra with a finite dimensional representation. Within this finite dimensional Hilbert space the Hamiltonian H_G can be diagonalized, and therefore a finite number of eigenstates are solvable. For one-dimensional QES systems the most general Lie algebra is $sl(2)$. Hence if Eq.(13) is QES then it can be expressed as

$$H_G = \sum C_{ab} J^a J^b + \sum C_a J^a + \text{constant} \ , \tag{14}$$

where C_{ab}, C_a are constant coefficients, and the J^a are the generators of the Lie algebra $sl(2)$ given by

$$J^+ = z^2\frac{d}{dz} - Nz \ ,$$

$$J^0 = z\frac{d}{dz} - \frac{N}{2} , \qquad N = 0, 1, 2 \ldots$$
$$J^- = \frac{d}{dz} . \tag{15}$$

Here the variables r and z are related by $z = h(r)$, where $h(\cdot)$ is some (explicit or implicit) function . The value $j = N/2$ is called the weight of the differential representation of $sl(2)$ algebra, and N is the degree of the eigenfunctions ϕ, which are polynomials in a $(N+1)$-dimensional Hilbert space with the basis $\langle 1, z, z^2, \ldots, z^N \rangle$:

$$\phi = (z - z_1)(z - z_2) \cdots (z - z_N) . \tag{16}$$

The requirement in Eq.(14) fixes $V(r)$ and $W_0(r)$, and H_G will have an algebraic sector with $N+1$ eigenvalues and eigenfunctions. For definiteness, we shall denote the potential V admitting $N+1$ QES states by V_N. From Eqs.(11) and (16), the function f_- in this sector has the general form

$$f_- = (z - z_1)(z - z_2) \cdots (z - z_N) \exp\left(- \int^z W_0(r) dr \right) , \tag{17}$$

where z_i $(i = 1, 2, \ldots, N)$ are N parameters that can be determined by plugging Eq.(16) into Eq.(12). The algebraic equations so obtained are called the Bethe ansatz equations corresponding to the QES problem [6,1,2] . Now one can rewrite Eq.(17) as

$$f_- = \exp\left(- \int^z W_N(r, \{z_i\}) dr \right) , \tag{18}$$

with

$$W_N(r, \{z_i\}) = W_0(r) - \sum_{i=1}^{N} \frac{h'(r)}{h(r) - z_i} . \tag{19}$$

There are $N+1$ possible functions $W_N(r, \{z_i\})$ for the $N+1$ sets of eigenfunctions ϕ. Inserting Eq.(18) into $H_- f_- = \epsilon f_-$, one sees that W_N satisfies the Ricatti equation

$$W_N^2 - W_N' = V_N - \epsilon_N , \tag{20}$$

where ϵ_N is the energy parameter corresponding to the eigenfunction f_- given in Eq.(17) for a particular set of N parameters $\{z_i\}$.

From Eqs.(9), (10) and (20) it is clear how one should proceed to determine the external fields so that the Dirac equation becomes QES based on $sl(2)$: one needs only to determine the superpotentials $W(r)$ according to Eq.(20) from the QES potentials $V(r)$ classified in Ref. [5]. This is easily done by observing that the superpotential W_0 corresponding to $N = 0$ is related to the gauge function $g(r)$ associated with a particular class of QES potential $V(r)$ by $g'(r) = W_0(r)$. This superpotential gives the field configuration that allows the weight zero ($j = N = 0$) state, i.e. the ground state, to be known in that class. The more interesting task is to obtain higher weight states (i.e. $j > 0$), which will include excited states. For weight j ($N = 2j$) states, this is achieved by forming the superpotential $W_N(r, \{z_i\})$ according to Eq.(19). Of the $N + 1$ possible sets of solutions of the Bethe ansatz equations, the set of roots $\{z_1, z_2, \ldots, z_N\}$ to be used in Eq.(19) is chosen to be the set for which the energy parameter of the corresponding state is the lowest.

4. Let us illustrate the above procedure by an example. We consider the motion of a neutral fermion of spin-1/2 with mass m coupled non-minimally with an external electromagnetic field with an anomalous magnetic moment μ. The relevant equation describing such particle is the Dirac-Pauli equation [7]. This equation is useful in describing the celebrated Aharonov-Casher effect [8], and is also of some interest in quantum chromodynamics in connection with the problem of quark confinement[9].

We shall consider the situation in which only electric field $\mathbf{E}$ is present. In this case, the Dirac-Pauli equation $H\psi = \mathcal{E}\psi$ is described by the Hamiltonian

$$H = \boldsymbol{\alpha} \cdot \mathbf{p} + i\mu\boldsymbol{\gamma} \cdot \mathbf{E} + \beta m \quad , \tag{21}$$

with $\mathbf{p} = -i\nabla$. We choose the Dirac matrices in the standard representation

$$\boldsymbol{\alpha} = \begin{pmatrix} 0 & \boldsymbol{\sigma} \\ \boldsymbol{\sigma} & 0 \end{pmatrix} , \quad \beta = \begin{pmatrix} 1 & 0 \\ 0 & -1 \end{pmatrix} , \tag{22}$$

where $\boldsymbol{\sigma}$ are the Pauli matrices. We also define $\psi = (\chi, \varphi)^t$, where t denotes transpose, and both χ and φ are two-component spinors. Then the Dirac–Pauli equation becomes

$$\begin{aligned} \boldsymbol{\sigma} \cdot (\mathbf{p} - i\mu\mathbf{E})\chi &= (\mathcal{E} + m)\varphi \ , \\ \boldsymbol{\sigma} \cdot (\mathbf{p} + i\mu\mathbf{E})\varphi &= (\mathcal{E} - m)\chi \ . \end{aligned} \tag{23}$$

We now consider central electric field $\mathbf{E} = E_r\hat{\mathbf{r}}$. In this case, one can choose a complete set of observables to be $(H, \mathbf{J}^2, J_z, \mathbf{S}^2 = 3/4, K)$. Here $\mathbf{J}$ is the total angular momentum $\mathbf{J} = \mathbf{L} + \mathbf{S}$, where $\mathbf{L}$ is the orbital angular momentum, and $\mathbf{S} = \frac{1}{2}\boldsymbol{\Sigma}$ is the spin operator. The operator K is defined as $K = \beta(\mathbf{\Sigma} \cdot \mathbf{L} + 1)$, which commutes with both H and $\mathbf{J}$. Explicitly, we have

$$K = \text{diag}\left(\hat{k}, -\hat{k}\right) ,$$
$$\hat{k} = \boldsymbol{\sigma} \cdot \mathbf{L} + 1 . \tag{24}$$

The common eigenstates can be written as

$$\psi = \frac{1}{r}\begin{pmatrix} f_-(r)\mathcal{Y}^k_{jm_j} \\ if_+(r)\mathcal{Y}^{-k}_{jm_j} \end{pmatrix} , \tag{25}$$

here $\mathcal{Y}^k_{jm_j}(\theta, \phi)$ are the spin harmonics satisfying

$$\mathbf{J}^2\mathcal{Y}^k_{jm_j} = j(j+1)\mathcal{Y}^k_{jm_j} , \quad j = \frac{1}{2}, \frac{3}{2}, \cdots , \tag{26}$$
$$J_z\mathcal{Y}^k_{jm_j} = m_j\mathcal{Y}^k_{jm_j} , \qquad |m_j| \leq j , \tag{27}$$
$$\hat{k}\mathcal{Y}^k_{jm_j} = -k\mathcal{Y}^k_{jm_j} , \qquad k = \pm(j + \frac{1}{2}) , \tag{28}$$

and

$$(\boldsymbol{\sigma} \cdot \hat{\mathbf{r}})\mathcal{Y}^k_{jm_j} = -\mathcal{Y}^{-k}_{jm_j} , \tag{29}$$

where $\hat{\mathbf{r}}$ is the unit radial vector. Eq.(23) then reduces to

$$\left(\frac{d}{dr} + \frac{k}{r} + \mu E_r\right) f_- = (\mathcal{E} + m)\, f_+ , \tag{30}$$
$$\left(-\frac{d}{dr} + \frac{k}{r} + \mu E_r\right) f_+ = (\mathcal{E} - m)\, f_- . \tag{31}$$

This shows that f_- and f_+ forms a one-dimensional SUSY pairs with the superpotential W given by

$$W = \frac{k}{r} + \mu E_r , \tag{32}$$

and the energy parameter $\epsilon = \mathcal{E}^2 - m^2$.

We can now classify the forms of the electric field $E_r(r)$ which allow exact and quasi-exact solutions. To be specific, we consider the situation where $k < 0$ and $\int dr \mu E_r > 0$, so that $f_-^{(0)}$ is normalizable, and $f_+^{(0)} = 0$. The other situation can be discussed similarly. In this case, Eq.(32) becomes

$$W = -\frac{|k|}{r} + \mu E_r . \tag{33}$$

We determine the forms of E_r that give exact/quasi-exact energy $\mathcal{E}$ and the corresponding function f_-. The corresponding function f_+ is obtained using Eq.(30).

5. Comparing the forms of the superpotential W in Eq.(33) with Table (4.1) in [4], one concludes that there are three forms of E_r giving exact solutions of the problem :

i) oscillator-like : $\mu E_r(r) \propto r$;
ii) Coulomb potential-like : $\mu E_r(r) \propto$ constant ;
iii) zero field-like : $\mu E_r(r) \propto 1/r$.

Case (i) and (ii) had been considered in Ref. [10] and [11], and case (iii) in [10].

We mention here that the case with oscillator-like field, i.e. case (i), is none other than the spherical Dirac oscillator [9].

6. The form of the superpotential W in Eq.(33) fits into three classes, namely, Classes VII, VIII and IX of $sl(2)$-based QES systems in [5]. Below we shall illustrate our construction of QES electric fields in Class VII QES systems.

The general potential in Class VII has the form

$$V_N(r) = a^2r^6 + 2abr^4 + \left[b^2 - a\,(4N + 2\gamma + 3)\right] r^2 + \gamma\,(\gamma - 1)\,r^{-2} - b\,(2\gamma + 1)\ , \tag{34}$$

where a, b and γ are constants. The gauge function is

$$g(r) = \frac{a}{4}r^4 + \frac{b}{2}r^2 - \gamma \ln r\ . \tag{35}$$

We must have $a, \gamma > 0$ to ensure normalizability of the wave function. Eqs.(35) and (33), together with the relation $W_0(r) = g'(r)$, give us the electric field $E_r^{(0)}$:

$$\mu E_r^{(0)}(r) = ar^3 + br\ . \tag{36}$$

The Dirac-Pauli equation with this field configuration admits a QES ground state with energy $\mathcal{E}^2 = m^2$ ($\epsilon = 0$) and ground state function $f_- \propto \exp(-g_0(r))$. Also, here we have $\gamma = |k|$.

To determine electric field configurations admitting QES potentials V_N with higher weight, we need to obtain the Bethe ansatz equations for ϕ. Letting $z = h(r) = r^2$, Eq.(12) becomes

$$\left[-4z\frac{d^2}{dz^2} + \left(4az^2 + 4bz - 2\,(2\gamma + 1)\right)\frac{d}{dz} - (4aNz + \epsilon)\right]\phi(z) = 0\ . \tag{37}$$

For $N = 0$, the value of the ϵ is $\epsilon = 0$. For higher $N > 0$ and $\phi(r) =$

$\prod_{i=1}^{N}(z - z_i)$, the electric field $E_r^{(N)}(r)$ is obtained from Eq.(19):

$$\mu E_r^{(N)}(r) = \mu E_r^{(0)}(r) - \sum_{i=1}^{N} \frac{h'(r)}{h(r) - z_i} \ . \tag{38}$$

For the present case, the roots z_i's are found from the Bethe ansatz equations

$$2az_i^2 + 2bz_i - (2\gamma + 1) - \sum_{l \neq i} \frac{z_i}{z_i - z_l} = 0 \ , \qquad i = 1, \ldots, N \ , \tag{39}$$

and ϵ in terms of the roots z_i's is

$$\epsilon = 2\,(2\gamma + 1) \sum_{i=1}^{N} \frac{1}{z_i} \ . \tag{40}$$

For $N = 1$ the roots z_1 are

$$z_1^{\pm} = \frac{-b \pm \sqrt{b^2 + 2a(2\gamma + 1)}}{2a} \ , \tag{41}$$

and the values of ϵ are

$$\epsilon^{\pm} = 2\left(b \pm \sqrt{b^2 + 2a(2\gamma + 1)}\right) \ . \tag{42}$$

For $a > 0$, the root $z_1^- = -|z_1^-| < 0$ gives the ground state. With this root, one gets the superpotential

$$W_1(r) = ar^3 + br - \frac{2r}{r^2 + |z_1^-|} - \frac{\gamma}{r} \ . \tag{43}$$

From Eq.(38), the corresponding electric field is

$$\mu E_r^{(1)}(r) = ar^3 + br - \frac{2r}{r^2 + |z_1^-|} \ . \tag{44}$$

The QES potential appropriate for the problem is

$$\begin{aligned} V(x) &= W_1^2 - W_1' \ , \\ &= V_1 - \epsilon \ . \end{aligned} \tag{45}$$

The one-dimensional SUSY sets the energy parameter of ground state at $\epsilon = 0$. Hence, the ground state and the excited state have energy parameter $\epsilon = 0$ and $\epsilon = \epsilon^+ - \epsilon^- = 4\sqrt{b^2 + 2a(2\gamma + 1)}$, and wave function

$$f_- \propto e^{-g_0(r)}\left(r^2 - z_1^-\right) \tag{46}$$

and

$$f_- \propto e^{-g_0(r)}\left(r^2 - z_1^+\right) \ , \tag{47}$$

respectively.

QES potentials and electric fields for higher degree N can be constructed in the same manner.

This work was supported in part by the National Science Council of the Republic of China through Grant No. NSC 94-2112-M-032-007.

References

1. C.L. Ho and P. Roy, J. Phys. **A36**, 4617 (2003).
2. C.L. Ho and P. Roy, Ann. Phys. **312**, 161 (2004).
3. C.L. Ho, Quasi-exact solvability of Dirac equation with Lorentz scalar potential, Tamkang preprint, Nov 2005. hep-th/0511204.
4. E. Cooper, A. Khare and U. Sukhatme, Phys. Rep. **251**, 267 (1995).
5. A. Turbiner, Comm. Math. Phys. **118**, 467 (1988).
6. A.G. Ushveridze, *Quasi-exactly solvable models in quantum mechanics* (IOP, Bristol, 1994).
7. W. Pauli, Rev. Mod. Phys. **13**, 203 (1941).
8. Y. Aharonov and A. Casher, Phys. Rev. Lett. **53**, 319 (1984).
9. M. Moshinsky and A. Szczepanaik, J. Phys. **A22**, L817 (1989).
10. G.V. Shishkin and V.M. Villalba, J. Math. Phys. **34**, 5037 (1993).
11. Q.-L. Lin, Phys. Rev. **A61**, 022101 (2000).

Exotic Galilean Symmetry, Non-commutativity & the Hall Effect

P. A. Horváthy*
Laboratoire de Mathématiques et de Physique Théorique
Université de Tours
Parc de Grandmont
F-37200 TOURS (France).
e-mail: horvathy@lmpt.univ-tours.fr.

The "exotic" particle model associated with the two-parameter central extension of the planar Galilei group can be used to derive the ground states of the Fractional Quantum Hall Effect. Similar equations arise for a semiclassical Bloch electron. Exotic Galilean symmetry can also be shared by Chern-Simons field theory of the Moyal type.

1. Introduction

Recent interest in non-commuting structures stems, as it often happens, from far remote fields. In high-energy physics, it comes from the theory of strings and membranes [1], or from studying galilean symmetry in the plane[2–5]. Independently and around the same time, very similar structures were considered in condensed matter physics, namely for the semiclassical dynamics of a Bloch electron [6]. Recent devolopments include the Anomalous [7], the Spin [8] and the Optical [9] Hall effects.

Below we first review the exotic point-particle model of [4], followed by a brief outline of the semiclassical Bloch electron.

Our present understanding of the Fractional Quantum Hall Effect is based on the motion of charged vortices in a magnetic field [10,11]. Such vortices arise as exact solutions in a field theory of matter coupled to an abelian gauge field A_ν, whose dynamics is governed by the Chern-Simons term [12,13]. Such theories can be either relativistic or nonrelativistic. For the latter, boosts commute, but exotic Galilean symmetry can be found in a Moyal-version of Chern-Simons field-theory [5], presented in Section 4.

*Partially supported by Nankai Institue of Mathematics, Tianjin, China.

2. "Exotic" mechanics in the plane

It has been known for (at least) 33 years that the planar Galilei group admits an "exotic" two-parameter central extension [3]: unlike in $D \geq 3$ spatial dimensions, the commutator of galilean boosts yields a new central charge,

$$[G_1, G_2] = \kappa. \tag{2.1}$$

This has long remained a sort of mathematical curiosity, though. It has been around 1995 that people started to inquire about the physical consequences of such an extended symmetry. In [2,4], in particular, Souriau's "orbit method" [14] was used to construct a classical system with such an exotic symmetry. The latter is realized by the usual galilean generators, except for the boost and the angular momentum,

$$\begin{aligned} j &= \epsilon_{ij} x_i p_j + \tfrac{\theta}{2}\, p_i p^i, \\ G_i &= m x_i - p_i t + m\theta\, \epsilon_{ij} p_j. \end{aligned} \tag{2.2}$$

The resulting free model moves, however, exactly as in the standard case. The "exotic" structure behaves hence roughly as spin: it contributes to some conserved quantities, but the new terms are separately conserved. The new structure does not seem to lead to any new physics.

The situation changes dramatically if the particle is coupled to a gauge field. The resulting equations of motion read

$$\begin{aligned} m^* \dot{x}_i &= p_i - em\theta\, \varepsilon_{ij} E_j, \\ \dot{p}_i &= eE_i + eB\, \varepsilon_{ij} \dot{x}_j, \end{aligned} \tag{2.3}$$

where $\theta = k/m^2$ is the non-commutative parameter and we have introduced the *effective mass*

$$m^* = m(1 - e\theta B). \tag{2.4}$$

The changes, crucial for physical applications, are two-fold: Firstly, the relation between velocity and momentum, (3.1), contains an "anomalous" term so that $\dot{x}_i$ and p_i are not parallel. The second novelty is the interplay between the exotic structure and the magnetic field, yielding the effective mass m^* in (3.2).

The equations (2.3) come from the Lagrangian

$$\int (\mathbf{p} - \mathbf{A}) \cdot d\mathbf{x} - \frac{p^2}{2}\, dt + \frac{\theta}{2}\, \mathbf{p} \times d\mathbf{p}. \tag{2.5}$$

When $m^* \neq 0$, 2.3 is also a Hamiltonian system, $\dot{\xi} = \{h, \xi^\alpha\}$, with $\xi = (p_i, x^j)$ and Poisson brackets

$$\begin{aligned} \{x_1, x_2\} &= \frac{m}{m^*}\,\theta, \\ \{x_i, p_j\} &= \frac{m}{m^*}\,\delta_{ij}, \\ \{p_1, p_2\} &= \frac{m}{m^*}\,eB. \end{aligned} \tag{2.6}$$

A most remarkable property is that for vanishing effective mass $m^* = 0$ i.e. when the magnetic field takes the critical value

$$B = \frac{1}{e\theta}, \tag{2.7}$$

then the system becomes singular. Then "Faddeev-Jackiw" (alias symplectic) reduction yields an essentially two-dimensional, simple system, similar to "Chern-Simons mechanics" [15]. The symplectic plane plays, simultaneously, the role of both configuration and phase space. The only motions are those which follow a generalized Hall law; quantization of the reduced system yields the "Laughlin wave functions" [10], which are in fact the ground states in the Fractional Quantum Hall Effect (FQHE).

The relations (2.6) diverge as $m^* \to 0$, but after reduction we have

$$\{x_1, x_2\} = \theta. \tag{2.8}$$

3. Semiclassical Bloch electron

Quite remarkably, around the same time and with no relation to the above developments, a very similar theory has arisen in solid state physics [6]. Applying a Berry-phase argument to a Bloch electron in a lattice, a semiclassical model can be derived. The equations of motion in the nth band read

$$\dot{\mathbf{r}} = \frac{\partial \epsilon_n(\mathbf{p})}{\partial \mathbf{p}} - \dot{\mathbf{p}} \times \vec{\Omega}(\mathbf{p}), \tag{3.1}$$

$$\dot{\mathbf{p}} = -e\mathbf{E} - e\dot{\mathbf{r}} \times \mathbf{B}(\mathbf{r}), \tag{3.2}$$

where $\mathbf{r} = (x^i)$ and $\mathbf{p} = (p_j)$ denote the electron's three-dimensional intracell position and quasimomentum, respectively, $\epsilon_n(\mathbf{p})$ is the band energy. The purely momentum-dependent $\vec{\Omega}$ is the Berry curvature of the electronic Bloch states, $\Omega_i(\mathbf{p}) = \epsilon_{ijl}\partial_{\mathbf{p}_j} a_l(\mathbf{p})$, where a_i is the Berry connection.

Recent applications of the model, based on the anomalous velocity term in (3.1), include the Anomalous [7] and the Spin [8] Hall Effects.

Eqns. (3.1-3.2) derive from the Lagrangian

$$L^{Bloch} = \big(p_i - eA_i(\mathbf{r},t)\big)\dot{x}^i - \big(\epsilon_n(\mathbf{p}) - eV(\mathbf{r},t)\big) + a^i(\mathbf{p})\dot{p}_i, \tag{3.3}$$

and are also consistent with the Hamiltonian structure [17,16]

$$\{x^i, x^j\}^{Bloch} = \frac{\epsilon^{ijk}\Omega_k}{1 + e\mathbf{B}\cdot\vec{\Omega}}, \tag{3.4}$$

$$\{x^i, p_j\}^{Bloch} = \frac{\delta^i{}_j + eB^i\Omega_j}{1 + e\mathbf{B}\cdot\vec{\Omega}}, \tag{3.5}$$

$$\{p_i, p_j\}^{Bloch} = -\frac{\epsilon_{ijk}eB^k}{1 + e\mathbf{B}\cdot\vec{\Omega}} \tag{3.6}$$

and Hamiltonian $h = \epsilon_n - eV$.

Restricted to the plane, these equations reduce, furthermore, to the exotic equations (2.3) provided $\Omega_i = \theta\delta_{i3}$. For $\epsilon_n(\mathbf{p}) = \mathbf{p}^2/2m$ and chosing $A_i = -(\theta/2)\epsilon_{ij}p_j$, the semiclassical Bloch Lagrangian (3.3) becomes the "exotic" expression (2.5).

The exotic galilean symmetry is lost if θ is not constant.

4. Non-commutative Chern-Simons theory

Field theory coupled to an abelian gauge field A_ν, whose dynamics is governed by the Chern-Simons (C-S) term admits exact vortex solutions [12,13]. Such theories can be either relativistic or nonrelativistic. In the latter case [13],

$$L = L_{matter} + L_{field} = i\bar{\psi}D_t\psi - \frac{1}{2}\big|\mathbf{D}\psi\big|^2 + \mu\left(\frac{1}{2}\epsilon_{ij}\partial_t A_i A_j + A_t B\right), \tag{4.1}$$

[plus a potential $U(\psi)$], where $D_\nu = \partial_\nu - ieA_\nu$, $\nu = t, i$. Infinitesimal galilean boosts, implemented conventionally as

$$\delta^0\psi = i\mathbf{b}\cdot\mathbf{x}\,\psi - t\mathbf{b}\cdot\vec{\nabla}\psi, \tag{4.2}$$

$$\delta^0 A_i = -t\mathbf{b}\cdot\vec{\nabla}A_i, \tag{4.3}$$

$$\delta^0 A_t = -\mathbf{b}\cdot\mathbf{A} - t\mathbf{b}\cdot\vec{\nabla}A_t, \tag{4.4}$$

are generated by the constants of the motion

$$G_i^0 = tP_i - \int x_i\,|\psi|^2\,d^2\mathbf{x}, \tag{4.5}$$

$$P_i = \int \frac{1}{2i}\big(\bar{\psi}\partial_i\psi - (\overline{\partial_i\psi})\psi\big)d^2\mathbf{x} - \frac{\mu}{2}\int \epsilon_{jk}A_k\partial_i A_j d^2\mathbf{x}. \tag{4.6}$$

The galilean symmetry extends in fact into a Schrödinger symmetry [13]; there is no sign of "exotic" galilean symmetry, however, since $\{G_1^0, G_2^0\} = 0$. Replacing ordinary products with the Moyal star-product,

$$(f \star g)(x_1, x_2) = \exp\left(i\frac{\theta}{2}(\partial_{x_1}\partial_{y_2} - \partial_{x_2}\partial_{y_1})\right) f(x_1, x_2) g(y_1, y_2)\Big|_{\mathbf{x}=\vec{y}} \tag{4.7}$$

where θ is a real parameter, a non-commutative version of the theory can be constructed, though. The classical Lagrangian is formally still (4.1), but the covariant derivative, the field strength, and the Chern-Simons term,

$$D_\mu \psi = \partial_\mu - ieA_\mu \star \psi, \tag{4.8}$$

$$F_{\mu\nu} = \partial_\mu A_\nu - \partial_\nu A_\mu - ie(A_\mu \star A_\nu - A_\nu \star A_\mu), \tag{4.9}$$

$$\text{C-S term} = \frac{\mu}{2}\epsilon_{\mu\nu\sigma}\left(A_\mu \star \partial_\nu A_\sigma - \frac{2ie}{3}A_\mu \star A_\nu \star A_\sigma\right), \tag{4.10}$$

respectively, all involve the Moyal form. The variational equations read

$$iD_t\psi + \frac{1}{2}\mathbf{D}^2\psi = 0, \tag{4.11}$$

$$\kappa E_i - e\epsilon_{ik}{\jmath^l}_k = 0, \tag{4.12}$$

$$\kappa B + e\rho^l = 0, \tag{4.13}$$

where $B = \epsilon_{ij}F_{ij}$, $E_i = F_{i0}$, and ρ^l and $\jmath^{\,l}$ denote the *left density* and *left current*, respectively,

$$\rho^l = \psi \star \bar{\psi}, \qquad \jmath^{\,l} = \frac{1}{2i}\left(\mathbf{D}\psi \star \bar{\psi} - \psi \star (\overline{\mathbf{D}\psi})\right). \tag{4.14}$$

These equations admit, just like their ordinary counterparts, exact vortex solutions [18].

The modified theory is *not* invariant w. r. t. boosts implemented as above. Galilean invariance can be restored, however, by implementing boosts rather as

$$\delta\psi = \psi \star (i\mathbf{b}\cdot\mathbf{x}) - t\mathbf{b}\cdot\vec{\nabla}\psi = (i\mathbf{b}\cdot\mathbf{x})\psi + \frac{\theta}{2}\mathbf{b}\times\vec{\nabla}\psi - t\mathbf{b}\cdot\vec{\nabla}\psi, \tag{4.15}$$

supplemented by (4.3)-(4.4). Then the generators,

$$G_i = tP_i - \int x_i\bar{\psi} \star \psi \, d^2\mathbf{x}, \tag{4.16}$$

do satisfy the "exotic" relation (2.1)

$$[G_1, G_2] = \kappa \qquad \text{with} \qquad \kappa = -\theta\int |\psi|\, d^2\mathbf{x}. \tag{4.17}$$

Acknowledgments

This review is based on joint research with C. Duval, Z. Horváth, L. Martina, M. Plyushchay and P. Stichel to whom I express my indebtedness. I would like to thank Prof. Mo-lin Ge for warm hospitality at Nankai Institute of Mathematics at Tianjin (China).

References

1. N. Seiberg and E. Witten, *JHEP* **9909** 032 (1999).
2. C. Duval, *Exotic Galilei group, IQHE and Chern-Simons electrodynamics.* Unpublished draft (1995); D. R. Grigore, *Journ. Math. Phys.* **37**, 240 (1996); *ibid.* **37** 460 (1996);
3. Y. Brihaye, C. Gonera, S. Giller and P. Kosiński, `hep-th/9503046` (unpublished); J. Lukierski, P. C. Stichel, W. J. Zakrzewski, *Annals of Physics* (N. Y.) **260**, 224 (1997). The "exotic" two-parameter extension of the planar Galilei goup has been first noticed by J.-M. Lévy-Leblond, in *Group Theory and Applications* (Loebl Ed.), **II**, Acad. Press, New York, p. 222 (1972). See also A. Ballesteros, M. Gadella, and M. del Olmo, *Journ. Math. Phys.* **33**, 3379 (1992).
4. C. Duval and P. A. Horváthy, *Phys. Lett.* **B 479**, 284 (2000) [`hep-th/0002233`] C. Duval and P. A. Horváthy, *Journ. Phys.* **A 34**, 10097 (2001) [`hep-th/0106089`]; P. A. Horváthy, *Ann. Phys.* **299**, 128 (2002) [`hep-th/0201007`].
5. P. A. Horváthy, L. Martina and P. Stichel, *Nucl. Phys.* **B 673**, 301-318 (2003).
6. M. C. Chang and Q. Niu, *Phys. Rev. Lett.* **75**, 1348 (1995); *Phys. Rev.* **B 53**, 7010 (1996); G. Sundaram and Q. Niu, *Phys. Rev.* **B59**, 14915 (1999). See also A. Bohm, A. Mostafazadeh, H. Koizumi, Q. Niu and J. Zwanziger, *The Geometric Phase in Quantum Systems.* Chapter 12. Springer Verlag (2003).
7. T. Jungwirth, Q. Niu, and A. H. MacDonald, *Phys. Rev. Lett.* **90**, 207208 (2002); D. Culcer, A. H. MacDonald, and Q. Niu, *Phys. Rev.* **B 68**, 045327 (2003); Fang et al. *Science* **302**, 92 (2003); Similar theories were initiated by R. Karplus and J. M. Luttinger, *Phys. Rev.* **95**, 1154 (1954).
8. S. Murakami, N. Nagaosa, and S.-C. Zhang, *Science* **301**, 1348 (2003) [`cond-mat/0308167`]; Y.-S. Wu, talk given at this Meeting.
9. K. Yu. Bliokh and Yu. P. Bliokh, *Phys. Lett.* **A333**, 181 (2004) [`physics/0402110`]; M. Onoda, S. Murakami, N. Nagaosa, *Phys. Rev. Lett.* **93**, 083901 (2004).
10. R. B. Laughlin, *Phys. Rev. Lett.* **50**, 1395 (1983). On the Quantum Hall Effect see, e.g., *Quantum Hall Effect*, Ed. M. Stone, World Scientific Singapore (1992).
11. S.-C. Zhang, *Intl. Journ. Mod. Phys.* **6**, 25 (1992).
12. H. J. De Vega and F. A. Schaposnik, *Phys. Rev. Lett.* **56**, 2564 (1986); S. K. Paul and A. Khare, *Phys. Lett.* **B174**, 420 (1986); S.-C. Zhang, T. H. Hansson and S. Kivelson, *Phys. Rev. Lett.* **62**, 82 (1989); J. Hong, Y. Kim and P. Y.

Pac, *Phys. Rev. Lett.* **64**, 2230 (1990); R. Jackiw and E. Weinberg, *Phys. Rev. Lett.* **64**, 2234 (1990).
13. R. Jackiw and S. Y. Pi, *Phys. Rev. Lett.* **64**, 2969 (1990); *Phys. Rev.* **D42**, 3500 (1990). For reviews see, e. g., R. Jackiw and S. Y. Pi, *Self-dual Chern-Simons Solitons.* Prog. Theor. Phys. Suppl. **107**, 1 (1992); G. Dunne, *Self-Dual Chern-Simons Theories.* Springer Lecture Notes in Physics. New Series: Monograph 36. (1995); C. Duval and P. A. Horváthy, *Chern-Simons vortices.* In Proc. Clifford Conf., Madeira'95, H. Nencka and J.-P. Bourguignon (eds). *Contemporary Mathematics*, **203**, p. 271 (1997) [hep-th/0307025].
14. J.-M. Souriau, *Structure des systèmes dynamiques*, Dunod: Paris (1970); *Structure of Dynamical Systems: a Symplectic View of Physics.* Birkhäuser: Dordrecht (1997).
15. G. Dunne R. Jackiw and C. A. Trugenberger, *Phys. Rev.* **D41**, 661 (1990); G. Dunne and R. Jackiw, *Nucl. Phys.* **B** (Proc. Suppl.) **33C** 114 (1993). The model goes back to G. Kirchhoff, *Vorlesungen über mathematischen Physik. Mechanik.* 3rd. ed., pp. 251-272. G. B. Teubner, Leipzig (1883), see P. Horváthy, talk given at the *COSLAB-VORTEX-BEC 2000+ Workshop.* Bilbao'03. [hep-th/0307175].
16. P. A. Horváthy, L. Martina and P. Stichel, *Phys. Lett.* **B 615**, 87 (2005). [hep-th/0412090]. C. Duval, Z. Horváth, P. A. Horváthy, L. Martina and P. Stichel, to be published in *Mod. Phys. Lett.* **B**. [cond-mat/0506051].
17. A. Bérard and H. Mohrbach, *Phys. Rev.* **D 69**, 127701 (2004) [hep-th/0310167].
18. G. S. Lozano, E. F. Moreno, F. A. Schaposnik, *Journ. High Energy Phys.* **02** 036 (2001) [hep-th/0012266]; D. Bak, S. K. Kim, K.-S. Soh, and J. H. Yee, *Phys. Rev.* **D64**, 025018 (2001) [hep-th/0102137].

The Energy-momentum and Related Topics in Gravitational Radiation

Wen-ling Huang

Fachbereich Mathematik, Schwerpunkt GD,
Universität Hamburg,
Bundesstr. 55, D-20146 Hamburg, Germany
E-mail: huang@math.uni-hamburg.de

Xiao Zhang*

Institute of Mathematics,
Academy of Mathematics and System Sciences,
Chinese Academy of Sciences,
Beijing 100080, P.R. China
E-mail: xzhang@amss.ac.cn

We report recent works on the conformal character of mappings preserving null geodesics in the Robertson-Walker spacetime and in the Schwarzscild spacetime. They should relate to the coordinate transformation in radiating spacetimes. We also discuss the recent progress on the complete and rigorous proof on the positivity of the Bondi mass, and relations between the ADM mass and the Bondi mass in gravitational radiation.

1. Introduction

Gravitational waves are wave-like solutions of the Einstein field equations which radiate energy. Gravitational waves are predicted by general relativity. However, they are not detected yet. An indirect proof of the existence of gravitational waves comes from observations of the pulsar PSR 1913+16. This binary system rotate rapidly, therefore should emit appreciable amounts of gravitational quadrupole radiation, hence lose energy and rotate faster. The observed relative change in period of $-2.422(\pm 0.006)\cdot 10^{-12}$ is in agreement with the theoretical value remarkably.

*Work partially supported by National Science Foundation of China under grant No. 10231050 and the Innovation Project of Chinese Academy of Sciences.

The theory of gravitational radiation was studied by Bondi, van der Burg, Metzner and Sachs systematically [1–3]. They assumed the vacuum spacetime $\left(\mathbb{L}^{3,1}, \mathbf{g}_{Bondi}\right)$ (possible with black holes) takes the following Bondi's radiating metric

$$\begin{aligned}\mathbf{g}_{Bondi} = & -\Big(-\frac{V}{r}e^{2\beta} + r^2 e^{2\gamma}U^2\cosh 2\delta + r^2 e^{-2\gamma}W^2\cosh 2\delta \\ & +2r^2 UW\sinh 2\delta\Big)du^2 - 2e^{2\beta}dudr \\ & -2r^2\Big(e^{2\gamma}U\cosh 2\delta + W\sinh 2\delta\Big)dud\theta \\ & -2r^2\Big(e^{-2\gamma}W\cosh 2\delta + U\sinh 2\delta\Big)\sin\theta dud\psi \\ & +r^2\Big(e^{2\gamma}\cosh 2\delta d\theta^2 + e^{-2\gamma}\cosh 2\delta\sin^2\theta d\psi^2 \\ & +2\sinh 2\delta\sin\theta d\theta d\psi\Big) \end{aligned} \tag{1.1}$$

where β, γ, δ, U, V and W are functions of $x^0 = u$, $x^1 = r$, $x^2 = \theta$, $x^3 = \psi$ which are smooth for $r \geq r_0 > 0$, u is a retarded coordinate, r is a luminosity distance, θ and ψ are spherical coordinates, $0 \leq \theta \leq \pi$, $0 \leq \psi \leq 2\pi$. The outgoing radiation condition implies that the following asymptotic behaviors hold for r sufficiently large

$$\gamma = \frac{c(u,\theta,\psi)}{r} + O\Big(\frac{1}{r^3}\Big), \tag{1.2}$$

$$\delta = \frac{d(u,\theta,\psi)}{r} + O\Big(\frac{1}{r^3}\Big), \tag{1.3}$$

$$\beta = -\frac{c^2+d^2}{4r^2} + O\Big(\frac{1}{r^4}\Big), \tag{1.4}$$

$$U = -\frac{c_{,2} + 2c\cot\theta + d_{,3}\csc\theta}{r^2} + O\Big(\frac{1}{r^3}\Big), \tag{1.5}$$

$$W = -\frac{d_{,2} + 2d\cot\theta - c_{,3}\csc\theta}{r^2} + O\Big(\frac{1}{r^3}\Big), \tag{1.6}$$

$$V = -r + 2M(u,\theta,\psi) + O\Big(\frac{1}{r}\Big). \tag{1.7}$$

(We denote $f_{,i} = \frac{\partial f}{\partial x^i}$ for $i = 0, 1, 2, 3$ throughout the paper.) The following conditions are assumed:

Condition A: Each of the six functions β, γ, δ, U, V, W together with its derivatives up to the second orders are equal at $\psi = 0$ and 2π.

Condition B: For all u,

$$\int_0^{2\pi} c(u,0,\psi)d\psi = 0, \quad \int_0^{2\pi} c(u,\pi,\psi)d\psi = 0.$$

Furthermore, the physics requires that the retarded time $u = constant$ is a null hypersurface, and this is the cases in the Minkowski spacetime (where $u = t - r$) and in the Schwarzschild spacetime (where $u = t - r - 2m \ln |r - 2m|$).

2. Mappings preserving null geodesics

In general relativity, it is important to study the (not necessary smooth) coordinate transformation in order to remove the coordinate singularity. Therefore it is natural to study the mappings between spacetimes which preserve null geodesics in the theory of gravitation radiation. The first result was due to Brinkmann [4] and subsequently rediscovered by several authors (e.g.,[5]) that a vacuum field can be mapped conformally on another vacuum field if and only if both admit a covariant constant vector. Such a vacuum spacetime is called the plane-fronted gravitational waves with parallel rays (i.e.,pp-waves) and covariant constant null vector field is called parallel ray. The related work refers to the Alexandrov theorem. When Einstein developed his special relativity, he studied the affine transformations between two inertial frames preserving light speed and proved that they are Lorentzian transformations up to a dilatation. Motivated by the understanding the Einstein's assumption of linearity is superfluous, in 1950, Alexandrov proved that any bijective transformation f of n-dimensional Minkowski spacetime ($n \geq 3$) to itself which preserves the distance zero in both directions must be a conformal mapping. That is, f is of the form

$$f(x) = \lambda^* x L + t$$

where L is a Lorentz matrix, $\lambda^* \in \mathbb{R} \setminus \{0\}$ is a scalar and $t \in \mathbb{R}^n$ is a vector. Therefore, distance zero preserving mappings must be affine. It should be emphasized that no regularity conditions such as affinity, differentiability, or even continuity are needed in Alexandrov's theorem. Along this line, in 1982, Lester [6,7] found that if there exists a conformal diffeomorphism from the Robertson-Walker spacetime to a domain of Minkowski spacetime, then any injective mapping of a Robertson-Walker spacetime to itself preserving pair of points jointed by null geodesics in both directions must be conformal. In general, the analogue of the Alexandrov theorem in general relativity does not always hold and the Einstein's cylinder universe $\mathcal{M}^{3,1} = \mathbb{S}^3 \times \mathbb{R}$ provides a counterexample. When two maximal null geodesic lines meet in a point $(x, t) \in \mathcal{M}^{3,1}$, then they will also meet in any point $((-1)^k x, t + k\pi)$, $k \in \mathbb{Z}$. So a bijection which takes (images of) maximal null geodesic lines to the (images of) maximal null geodesic lines need not be continuous, and, therefore, need not to be conformal.

In 1999, the first author [8] proved the Alexandrov type theorem for the Schwarzschild spacetime - the most fundamental curved spacetime which is not conformal to Minkowski spacetime: Let f be a bijective mapping from Schwarzschild spacetime to itself such that f and f^{-1} preserve inextendible null geodesic curves (as point sets). Then f is an isometry. The first author [9] also studied the strongly causal spacetimes including the Minkowski spacetime, the Schwartzschild spacetime, the Einstein cylinder universe, the de-Sitter spacetime, and proved that if a bijection f of a strongly causal space-time $\mathcal{M}$ satisfies the condition: For any null geodesic curve $\gamma \in \mathcal{M}$, $f(\gamma)$ and $f^{-1}(\gamma)$ are null geodesic curves, then f is a conformal transformation. This theorem essentially removes the "homeomorphy" condition in Hawking's theorem [10].

From the point of view of gravitational radiation, if the Alexandrov type theorem holds in a spacetime, then the spacetime will be too "rigid" to radiating coordinate transformation. Therefore, as it is well-known that no gravitational radiation occurs in the Schwarzschild spacetime, it will be interesting to study the Alexandrov type theorem for general radiating spacetimes.

3. The ADM mass at spatial infinity

At spatial infinity of an asymptotically flat spacetime, Arnowitt, Deser and Misner defined the total energy-momentum as follows: Let (M^3, g_{ij}, h_{ij}) be an asymptotically flat spacelike hypersurface that, outside a compact subset, M is diffoomorphic to $\mathbb{R}^3$ minus a ball with the metric g and the symmetric 2-tensor h satisfying the following asymptotic conditions

$$g_{ij} = \delta_{ij} + O(\frac{1}{r}), \quad \partial_k g_{ij} = O(\frac{1}{r^2}), \quad \partial_k \partial_l g_{ij} = O(\frac{1}{r^3}),$$

$$h_{ij} = O(\frac{1}{r^2}), \quad \partial_k h_{ij} = O(\frac{1}{r^3})$$

The ADM total energy $\mathbb{E}$ and the ADM total linear momentum $\mathbb{P}_k$ are

$$\mathbb{E} = \frac{1}{16\pi} \lim_{r\to\infty} \int_{S_r} (\partial_j g_{ij} - \partial_i g_{jj}) * dx^i,$$

$$\mathbb{P}_k = \frac{1}{8\pi} \lim_{r\to\infty} \int_{S_r} (h_{ki} - g_{ki} h_{jj}) * dx^i,$$

where S_r is the sphere of radius r in $\mathbb{R}^3$.

In 1979, Schoen-Yau [11] proved that if a spacetime satisfies the dominant energy condition, i.e., for any timelike vector W, $T_{uv}W^uW^v \geq 0$, and $T^{uv}W_u$ is a non-spacelike vector, then, for asymptotically flat initial data set (M^3, g_{ij}, h_{ij}),

$$\mathbb{E} \geq \sqrt{\sum_k \mathbb{P}_k^2}.$$

That $\mathbb{E} = 0$ implies that the spacetime is flat over M. This solved the long-term positive mass conjecture in general relativity. In 1981, Witten [12] found a new proof by using spinors and the Dirac operator. In 1999 [13], the second author generalized the positive mass theorem to the spacetimes including the total angular momentum.

4. The Bondi mass at null infinity

Now let us go back to the Bondi's radiating vacuum spacetime. The null hypersurface $\{u = u_0\}$ gives null infinity as $r \to \infty$ where the Bondi energy-momentum is defined as [1–3]:

$$m_\nu(u_0) = \frac{1}{4\pi} \int_{S^2} M(u_0, \theta, \psi) n^\nu dS$$

for $\nu = 0, 1, 2, 3$, where $n^0 = 1$, $n^1 = \sin\theta\cos\psi$, $n^2 = \sin\theta\sin\psi$, $n^3 = \cos\theta$. The Bondi energy-momentum is the total energy-momentum measured after the loss due to the gravitational radiation up to that time. In the paper, Bondi proved that the Bondi mass is non-increasing w.r.t. u, i.e., more and more energy is radiated away.

Most physical systems cannot radiate away more energy than they have initially. This is usually a trivial consequence of a conserved stress-energy tensor with a positive timelike component. However, the gravitational field does not have a well-defined stress-energy tensor. It is possible that a finite gravitational system might be able to radiate arbitrarily large amounts of energy. That it is impossible is known as the positive mass conjecture at null infinity. There is no mathematical setting available of this conjecture in general spacetimes. In Bondi's radiating vacuum spacetimes, the conjecture says that the Bondi mass must be nonnegative.

The outlines of the proof that the Bondi mass is nonnegative were given by Schoen-Yau[14] by solving the Jang's equation and physicists (Israel-Nester, Horowitz-Perry, Ashtekar-Horowitz, Ludvigsen-Vickers, Renla-Tod, etc.[15]) by using Witten's spinor argument. However, no mathematical detail was provided in any those proofs. The idea is to choose certain spacelike hypersurfaces approaching to null infinity. These spacelike hypersurfaces are asymptotically hyperbolic with the nontrivial second fundamental forms in the Bondi's radiating spacetimes. Therefore, it requires to establish the positive mass theorem for these spacelike hypersurfaces. In 2002, by using Witten's method, the second author was able to find a complete and rigorous proof of this positive mass theorem near null infinity [16,17]. Recently, together with Yau, the authors were able to find suitable asymptotically null, spacelike hypersurface in vacuum Bondi's radiating spacetimes. Then the positive mass theorem in [16,17] indicates that if there exists u_0 in Bondi's radiating vacuum spacetimes such that $c\big|_{u=u_0} = d\big|_{u=u_0} = 0$ for r sufficiently large, then

$$m_0(u) \geq \sqrt{\sum_{i=1,2,3} m_i^2(u)}$$

for all $u \leq u_0$ [18].

We are still working on Schoen-Yau's method whether the above conditions can be removed, i.e., whether the Bondi mass is always nonnegative.

5. The ADM energy-momentum and the Bondi energy-momentum

Finally, we would like to discuss one of the main problems in gravitational radiation on the relation between the total energy-momentum at spatial infinity and that at null infinity.

In 1979, assuming that the spacetime can be conformally compactified, and asymptotically empty and flat at null and spatial infinity in certain sense, Ashtekar and Magnon-Ashtekar [19] demonstrated the mass at spatial infinity is the past limit of the Bondi mass. Here, the "past limit" means $\lim_{u\to-\infty} m_\nu(u)$. (In 2003, Hayward [20] proved this theorem in a new framework for spacetime asymptotics, replacing the Penrose conformal factor by a product of advanced and retarded conformal factors.) In 1993, Christodoulou and Klainerman [21] proved the global existence of globally hyperbolic, strongly asymptotically flat, maximal foliated vacuum solutions of the Einstein field equations. They also proved rigorously the ADM mass at spatial infinity is the past limit of the Bondi mass in these spacetimes.

In 2004, the second author [22] studied this problem in the Bondi's radiating vacuum spacetime. He defined the spatial infinity as the t-slices where the "real" time t is defined as $t = u + r$. Denote $\mathbb{E}(t_0)$ by $\mathbb{P}_0(t_0)$. Then he verified that

$$\mathbb{P}_\nu(t_0) = m_\nu(-\infty)$$

for $\nu = 0, 1, 2, 3$ under the asymptotic flatness assumptions at spatial infinity which ensure the Schoen-Yau's positive mass theorem. In this case, the ADM total energy, the ADM total linear momentum of (spatial) t_0-slice and the Bondi energy-momentum of (null) u_0-slice satisfy

$$\mathbb{P}_\nu(t_0) = m_\nu(u_0) + \frac{1}{4\pi}\int_{-\infty}^{u_0}\int_{S^2}\Big((c_{,0})^2 + (d_{,0})^2\Big)n^\nu dSdu$$

In particular, if there is *news* $c_{,0}$, $d_{,0}$, then the ADM total energy is always greater than the Bondi mass.

Unfortunately, the asymptotic flatness conditions at spatial infinity in all above works preclude gravitational radiation. The second author therefore assumes certain weaker asymptotic flatness conditions at spatial infinity in order to include gravitational radiation: Roughly speaking, we assume that, as $u \to -\infty$, $\big\{M, c, d, M_{,0}, c_{,0}, d_{,0}, M_{,A}, c_{,A}, d_{,A}\big\} = O\big(1\big)$ where $A, B = 2, 3$. Under these conditions the ADM total energy of any t_0-slice and the past limit of the Bondi mass satisfy:

$$\mathbb{E}(t_0) = m_0(-\infty) + \frac{1}{4\pi}\lim_{u\to-\infty}\int_{S^2}\left(c^2 + d^2\right)_{,0} dS.$$

This formula indicates that, in radiative fields, infinite energy is needed for any t_0-slice goes to a u_0-slice. Very recently, the authors were able to find the relations between the ADM linear momentum of any t_0-slice and the past limit of the Bondi momentum:

$$\mathbb{P}_k(t_0) = m_k(-\infty) + \frac{1}{8\pi}\lim_{u\to-\infty}\int_0^\pi\int_0^{2\pi}\mathcal{P}_k d\psi d\theta$$

for $k = 1, 2, 3$, where $\mathcal{P}_k$ have long expressions and are given in the appendix of [23]. In particular, in axi-symmetric spacetimes where $c = c(u, \theta)$, $d = 0$[1],

$$\mathbb{P}_1(t_0) = m_1(-\infty), \quad \mathbb{P}_2(t_0) = m_2(-\infty).$$

However, one cannot expect the "real" time $t = u + r$ in general (e.g., the Schwarzschild spacetime). That the case t approaches $u + r$ asymptotically is studying by the authors [24].

Acknowledgments

This work was partially done when Wenling Huang visited the Morningside Center of Mathematics, Chinese Academy of Sciences, and she would like to thank the center for its hospitality.

References

1. H. Bondi, M. van der Burg and A. Metzner, *Proc. Roy. Soc. London* **A 269**, 21 (1962.
2. R. Sachs, *Proc. Roy. Soc. London* **A 270**, 103 (1962).
3. M. van der Burg, *Proc. Roy. Soc. London* **A 294**, 112 (1966).
4. H.W. Brinkmann, *Math. Ann.* **18**, 119 (1925).
5. J. Ehlers and W. Kundt, in Gravitation: an introduction to current research, ed. L. Witten (Wiley, New York and London), 49 (1962).
6. J.A. Lester and C. R. Math, *Rep. Acad. Sci. Canada*, **IV**, 175 (1982).
7. J.A. Lester, *Aequationes Math.*, **25**, 216 (1982).
8. W.-l. Huang, *Commun. Math. Phys.*, **201**, 471 (1999).
9. W.-l. Huang, *J. Math. Phys.*, **39(3)**, 1637 (1998).
10. S.W. Hawking, A.R. King and P.J. McCarthy, *J. Math. Phys.*, **17(2)**, 174 (1976).
11. R. Schoen and S.T. Yau, *Commun. Math. Phys.*, **65**, 45 (1979), **79**, 47 (1981), **79**, 231(1981).
12. E. Witten, *Commun. Math. Phys.*, **80**, 381 (1981).
13. X. Zhang, *Commun. Math. Phys.*, **206**, 137 (1999).
14. R. Schoen, S.T. Yau, *Phys. Rev. Lett.*, **48**, 369 (1982).
15. W. Israel and J. Nester, *Phys. Lett.*, **85A**, 259 (1981); G. Horowitz, M. Perry, *Phys. Rev. Lett.*, **48**, 371 (1982); A. Ashtekar and G. Horowitz, *Phys. Lett.*, **89A**, 181 (1982); G. Horowitz and P. Tod, *Commun. Math. Phys.*, **85**, 429 (1982); M. Ludvigsen and J. Vickers, *J. Phys. A: Math. Gen.*, **15**, L67 (1982); O. Reula and K. Tod, *J. Math. Phys.*, **25**, 1004 (1984).
16. X. Zhang, *Commun. Math. Phys.*, **249**, 529 (2004).
17. X. Zhang, *The positive mass theorem near null infinity*, Proceedings of ICCM 2004, December 17-22, Hong Kong (eds. S.T. Yau, etc.), AMS/International Press, Boston, to appear.
18. W.-l. Huang, S.T. Yau, X. Zhang, *Positivity of the Bondi mass in Bondi's radiating spacetimes*, in preparation.
19. A. Ashtekar, A. Magnon-Ashtekar, Phys. Rev. Lett. 43(1979)181
20. S. Hayward, Phys. Rev. D68(2003)104015.
21. D. Christodoulou, S. Klainerman, *The global nonlinear stablity of Minkowski space*, Princeton Math. Series 41(1993), Princeton Univ. Press.
22. X. Zhang, *On the relation between ADM and Bondi energy-momenta*, preprint, 2003, gr-qc/0511036.
23. W.-l. Huang, X. Zhang, *On the relation between ADM and Bondi energy-momenta II - radiative spatial infinity*, preprint, 2003, gr-qc/0511036.
24. W-l. Huang, X. Zhang, in preparation.

Quantum Operators and Hermitian Vector Fields

Josef Janyška*

Department of Mathematics, Masaryk University
Janáčkovo nám 2a, 602 00 Brno, Czech Republic
E-mail: janyska@math.muni.cz

Marco Modugno†

Department of Applied Mathematics, Florence University
Via S. Marta 3, 50139 Florence, Italy
E-mail: marco.modugno@unifi.it

We classify the Lie algebra of Hermitian vector fields of a Hermitian line bundle, by means of a generic Hermitian connection. Then, we specify the base space of the above Hermitian bundle by considering a Galilei, or an Einstein spacetime. In these cases, the geometric structure of the base space yields a distinguished choice for the Hermitian connection. Then, we can prove that the Lie algebra of Hermitian vector fields turns out to be naturally isomorphic to the Lie algebra of special phase functions.

Keywords: Hermitian vector fields, quantum bundle, special phase functions, Galilei spacetime, Lorentz spacetime.

2000 *Mathematics Subject Classification*: 17B66, 17B81, 53B35, 53C07, 53C50, 55R10, 58A10, 81R20, 81S10, 83C99, 83E99.

Introduction

Covariant Quantum Mechanics is a formulation of quantum mechanics on a curved spacetime with absolute time, which is manifestly independent of coordinates and accelerated observers[1,2]. One of the main aspects of this theory deals with the covariant achievement of the Schrödinger equation[3].

*Supported by the Ministry of Education under the project MSM0021622409 and by the Grant Agency under the project GA 201/05/0523 (Czech Republic)
†Supported by the University of Florence, the PRIN 2003 "Sistemi integrabili, teorie classiche e quantistiche" (MIUR) and by the GNFM of INDAM (Italy).

Another complementary aspect deals with the achievement of quantum operators: in this paper, we sketch the most recent approach to this second topic.

The idea is the following. A covariant family of 1st order quantum operators can be obtained through the Lie derivatives of a covariant family of distinguished vector fields of the quantum bundle. A natural canditate for this family consists of the Hermitian vector fields. So, we classify these vector fields and see that they constitute a Lie algebra naturally isomorphic to a distinguished Lie algebra of "special functions" of the classical phase space (in general, the special bracket does not coincide with the Poisson bracket). We assume this result as the correspondence principle of Covariant Quantum Mechanics. We stress that every classical observable and the corresponding quantum operator depend on an observer. But the families of the classical observables and of the associated quantum operators, respectively, as a whole, and the correspondence principle are covariant. This approach allows us to treat position, momentum and energy observables on the same footing. Indeed, we have no ordering problems concerning energy, because it is not deduced from momentum. On the other hand, the quantum operator arising for energy is a 1st order operator. But, combining this operator with the 2nd order Schrödinger operator (which is achieved independently of the energy viewpoint) yields the physically correct quantum operator on the Hilbert bundle. However, this last development is beyond the scope of the present paper, for reasons of space[2].

It is wellknown that quantum mechanics cannot be formulated in an Einstein framework. On the other hand, we show that the above results concerning the covariant formulation of pre–quantum operators in the Galilei framework can be successfully repeated in the Einstein framework with some necessary changes. This fact seems to be interesting by itself and to be useful for a deeper understanding of the Galilei case. For this reason, we discuss also the Einstein case in the present paper.

If $\boldsymbol{M}$ and $\boldsymbol{N}$ are manifolds, then the sheaf of local smooth maps $\boldsymbol{M} \to \boldsymbol{N}$ is denoted by $\mathrm{map}(\boldsymbol{M}, \boldsymbol{N})$. If $\boldsymbol{F} \to \boldsymbol{B}$ is a fibred manifold, then the sheaf of local sections $\boldsymbol{B} \to \boldsymbol{F}$ is denoted by $\mathrm{sec}(\boldsymbol{B}, \boldsymbol{F})$ and the vertical restriction of forms is be denoted by $^{\vee}$.

We assume the following basic spaces of scales: the space of *time intervals* $\mathbb{T}$, the space of *lengths* $\mathbb{L}$, the space of *masses* $\mathbb{M}$. We assume the following "universal scales": the *Planck's constant* $\hbar \in \mathbb{T}^{-1} \otimes \mathbb{L}^2 \otimes \mathbb{M}$ and the *speed of light* $c \in \mathbb{T}^{-1} \otimes \mathbb{L}$. Moreover, we will consider a *particle* of *mass* $m \in \mathbb{M}$ and *charge* $q \in \mathbb{T}^{-1} \otimes \mathbb{L}^{3/2} \otimes \mathbb{M}^{1/2}$.

1. Hermitian vector fields

We start by analysing the Lie algebra of Hermitian vector fields of a Hermitian line bundle over a generic base manifold. Thus, we consider a manifold $\boldsymbol{E}$, which will be specified in the next sections. We denote the charts of $\boldsymbol{E}$ by (x^λ) and the associated local bases of vector fields of $T\boldsymbol{E}$ and forms of $T^*\boldsymbol{E}$ by ∂_λ and d^λ, respectively.

We consider a *complex line bundle* $\pi : \boldsymbol{Q} \to \boldsymbol{E}$ equipped with a scaled Hermitian product $Uh : \boldsymbol{E} \to (\mathbb{L}^{-3} \otimes \mathbb{C}) \otimes (\boldsymbol{Q}^* \otimes \boldsymbol{Q}^*)$.

We shall refer to a normalised *quantum basis* $Ub \in \sec(\boldsymbol{E}, \mathbb{L}^{3/2} \otimes \boldsymbol{Q})$. For each $\Phi, \Psi \in \sec(\boldsymbol{E}, \boldsymbol{Q})$, we write $\Psi = \psi\, Ub$, with $\psi \in \text{map}(\boldsymbol{E}, \mathbb{L}^{-3/2} \otimes \mathbb{C})$, and $\mathbf{E}h(\Phi, \Psi) = \bar{\phi}\,\psi$.

Let $\mathbb{I} : \boldsymbol{Q} \to V\boldsymbol{Q} : q \mapsto (q, q)$ be the *Liouville* vector field.

A Hermitian connection c of the quantum bundle can be locally written as $c = \chi[Ub] + \mathrm{i}\, A[Ub] \otimes \mathbb{I}$, with $A[Ub] \in \sec(\boldsymbol{E}, T^*\boldsymbol{E})$. The curvature of c is $R[c] = -\mathrm{i}\, \Phi[c] \otimes \mathbb{I}$, where $\Phi[c] = 2\, dA[Ub]$.

Each quantum basis Ub yields (locally) the flat connection $\chi[Ub] : \boldsymbol{Q} \to T^*\boldsymbol{E} \otimes T\boldsymbol{Q}$, with expression $\chi[Ub] = d^\lambda \otimes \partial_\lambda$.

A vector field $Y \in \sec(\boldsymbol{Q}, T\boldsymbol{Q})$ is said to be *Hermitian* if it is projectable over an $X \in \sec(\boldsymbol{E}, T\boldsymbol{E})$, is $I\!R$–linear over X and $L[X]\big(Uh(\Psi, \Phi)\big) = Uh\big(Y.\Psi, \Phi\big) + Uh\big(\Psi, Y.\Phi\big)$, for each $\Psi, \Phi \in \sec(\boldsymbol{E}, \boldsymbol{Q})$. The Hermitian vector fields are locally characterised by an expression of the type $Y = X^\lambda \partial_\lambda + \mathrm{i}\, \check{Y}\, \mathbb{I}$, with $X^\lambda, \check{Y} \in \text{map}(\boldsymbol{E}, I\!R)$.

The Hermitian vector fields constitute a subsheaf $\text{her}(\boldsymbol{Q}, T\boldsymbol{Q}) \subset \sec(\boldsymbol{Q}, T\boldsymbol{Q})$, which is closed with respect to the Lie bracket.

In order to classify the Hermitian vector fields globally, we consider a Hermitian connection c and obtain the linear isomorphism $\mathfrak{j}[c] : \sec(\boldsymbol{E}, T\boldsymbol{E}) \times \text{map}(\boldsymbol{E}, I\!R) \to \text{her}(\boldsymbol{Q}, T\boldsymbol{Q})$, with expression $\mathfrak{j}[c](X, \check{Y}) = X^\lambda\, \partial_\lambda + \mathrm{i}\, (A_\lambda\, X^\lambda + \check{Y}) \otimes \mathbb{I}$.

If Φ is a closed 2–form of $\boldsymbol{E}$, then we have the Lie bracket of $\sec(\boldsymbol{E}, T\boldsymbol{E}) \times \text{map}(\boldsymbol{E}, I\!R)$

$$\big[(X_1, \check{Y}_1),\ (X_2, \check{Y}_2)\big]_\Phi =: \big([X_1, X_2],\quad \Phi(X_1, X_2) + X_1.\check{Y}_2 - X_2.\check{Y}_1\big)$$

Now, let us refer to the 2–form $\Phi[c]$ associated with the curvature of c.

Theorem 1.1. *The map* $\mathfrak{j}[c]$ *is a Lie algebra isomorphisms.* □

In the next sections we equip the base manifold $\boldsymbol{E}$ with a geometric structure describing the Galilei, or Einstein spacetime, and obtain a distinguished choice for the Hermitian connection c and a classification of the Hermitian vector fields via the Lie algebra of special phase functions.

2. Galilei case

2.1. *Classical setting*

We consider the absolute *time*, consisting of an affine 1–dimensional space $\boldsymbol{T}$ associated with the vector space $\bar{\mathbb{T}} =: \mathbb{T} \otimes I\!R$ and assume spacetime $\boldsymbol{E}$ to be oriented and equipped with a *time fibring* $t : \boldsymbol{E} \to \boldsymbol{T}$. A *motion* is defined to be a section $s : \boldsymbol{T} \to \boldsymbol{E}$.

We shall refer to a *time unit* $u_0 \in \mathbb{T}$, or, equivalently, to its dual $u^0 \in \mathbb{T}^*$, and to a *spacetime chart* $(x^\lambda) \equiv (x^0, x^i)$ adapted to the orientation, the fibring, the affine structure of $\boldsymbol{T}$ and the time unit u_0. Greek indices will span all spacetime coordinates and Latin indices the fibre coordinates. We have the scaled form $dt : \boldsymbol{E} \to \mathbb{T} \otimes T^*\boldsymbol{E}$, with expression $dt = u_0 \otimes d^0$.

We assume as *metric* a scaled spacelike Riemannian metric $g : \boldsymbol{E} \to \mathbb{L}^2 \otimes (V^*\boldsymbol{E} \otimes V^*\boldsymbol{E})$. With reference to a mass $m \in \mathbb{M}$, it is convenient to introduce the *rescaled metric* $G =: \frac{m}{\hbar}\, g : \boldsymbol{E} \to \mathbb{T} \otimes (V^*\boldsymbol{E} \otimes V^*\boldsymbol{E})$, with expression $G = G^0_{ij}\, u_0 \otimes \check{d}^i \otimes \check{d}^j$.

We assume as *gravitational field* a torsion free linear spacetime connection $K^\natural : T\boldsymbol{E} \to T^*\boldsymbol{E} \otimes TT\boldsymbol{E}$, which fulfills the identities $\nabla^\natural dt = 0$, $\nabla^\natural g = 0$, $R^\natural{}_{\lambda i \mu j} = R^\natural{}_{\mu j \lambda i}$. We observe that $K^\natural$ is determined by dt and g up to a local closed 2–form.

We assume as *electromagnetic field* a closed scaled 2–form $F : \boldsymbol{E} \to (\mathbb{L}^{1/2} \otimes \mathbb{M}^{1/2}) \otimes \Lambda^2 T^*\boldsymbol{E}$.

With reference to a particle with mass m and charge q, we obtain the *joined connection* $K =: K^\natural + K^e = K^\natural - \frac{q}{2m}\,(dt \otimes \widehat{F} + \widehat{F} \otimes dt)$, with $\widehat{F} = g^{\sharp 2}(F)$, which fulfills the same identities of the gravitational connection. Thus, from now on, we shall refer to this joined connection, which incoroporates both the gravitational and the electromagnetic fields.

We assume as classical *phase space* the 1st jet space $J_1\boldsymbol{E}$ of motions $s \in \sec(\boldsymbol{T}, \boldsymbol{E})$. A space time chart (x^λ) induces a chart (x^λ, x^i_0) on $J_1\boldsymbol{E}$. We have the *contact map* $д : J_1\boldsymbol{E} \to \mathbb{T}^* \otimes T\boldsymbol{E}$ and the *complementary contact map* $\theta =: 1 - д \circ dt : J_1\boldsymbol{E} \to T^*\boldsymbol{E} \otimes V\boldsymbol{E}$, with expressions $д = u^0 \otimes (\partial_0 + x^i_0\, \partial_i)$ and $\theta = (d^i - x^i_0\, d^0) \otimes \partial_i$. Moreover, we have a linear isomorphism $\nu : V_0 J_1\boldsymbol{E} \to \mathbb{T}^* \otimes V\boldsymbol{E}$.

An *observer* is defined to be a section $o \in \sec(\boldsymbol{E}, J_1\boldsymbol{E})$. Each observer o yields the affine fibred isomorphism $\nabla[o] =: \mathrm{id} - o : J_1\boldsymbol{E} \to \mathbb{T}^* \otimes V\boldsymbol{E}$ and the linear fibred projection $\nu[o] : T\boldsymbol{E} \to V\boldsymbol{E}$. For each observer o, we define the *kinetic energy* and the *kinetic momentum* as $\mathcal{K}[o] = \frac{1}{2}\, G\,(\nabla[o], \nabla[o])$ and $\mathcal{Q}[o] = \nu[o] \lrcorner \big(G^\flat(\nabla[o])\big)$, with expressions $\mathcal{K}[o] = \frac{1}{2}\, G^0_{ij}\, x^i_0\, x^j_0\, d^0$ and $\mathcal{Q}[o] = G^0_{ij}\, x^j_0\, d^j$. We define the *kinetic Poincaré–Cartan form* $\Theta[o] =: -\mathcal{K}[o] + \mathcal{Q}[o]$

and obtain $\mathcal{K}[o] = -\text{д}[o] \lrcorner \Theta[o]$ and $\mathcal{Q}[o] = \theta[o] \lrcorner \Theta[o]$.

We have a natural bijective map χ between time preserving linear spacetime connections K and affine phase connections $\Gamma : J_1\boldsymbol{E} \to T^*\boldsymbol{E} \otimes TJ_1\boldsymbol{E}$, with expression $\Gamma = d^\lambda \otimes (\partial_\lambda + \Gamma_\lambda{}^i_0 \, \partial^0_i)$, where $\Gamma_\lambda{}^i_0 = \Gamma_\lambda{}^i{}_0{}^0_0 + \Gamma_\lambda{}^i{}_0{}^0_j \, x^j_0$, with $\Gamma_\lambda{}^i{}_0{}^0_\mu = K_\lambda{}^i{}_\mu$.

Then, K yields the *phase connection*, $\Gamma =: \chi(K) : J_1\boldsymbol{E} \to T^*\boldsymbol{E} \otimes TJ_1\boldsymbol{E}$, which splits as $\Gamma = \Gamma^\natural + \Gamma^\mathfrak{e}$, where $\Gamma^\mathfrak{e} = J_1\boldsymbol{E} \to \mathbb{T}^* \otimes (T^*\boldsymbol{E} \otimes V\boldsymbol{E})$ and $\Gamma^\natural = \chi(K^\natural)$. We have $\Gamma^\mathfrak{e} = +(F_{jh} \, x^j_0 + 2 \, F_{0h}) \, d^0) \otimes \partial^0_i$.

Hence, Γ yields the *dynamical phase connection*, $\gamma =: \text{д} \lrcorner \Gamma : J_1\boldsymbol{E} \to \mathbb{T}^* \otimes TJ_1\boldsymbol{E}$, with expression $\gamma = u^0 \otimes (\partial_0 + x^i_0 \, \partial_i + \gamma_0{}^i_0 \, \partial^0_i)$, with $\gamma_0{}^i_0 = K_\lambda{}^i{}_\mu \, \breve{\delta}^\lambda_0 \, \breve{\delta}^\mu_0$, where $\breve{\delta}^\alpha_0 =: \delta^\alpha_0 + \delta^\alpha_h \, x^h_0$. Indeed, γ splits as $\gamma = \gamma^\natural + \gamma^\mathfrak{e}$, where $\gamma^\natural = \text{д} \lrcorner \Gamma^\natural$ and $\gamma^\mathfrak{e} = -\frac{q}{m} \, \text{д} \lrcorner \widehat{F} : J_1\boldsymbol{E} \to (\mathbb{T}^* \otimes \mathbb{T}^*) \otimes V\boldsymbol{E}$.

Moreover, Γ and G yield the *phase 2–form*, $\Omega =: G \lrcorner \big(\nu[\Gamma] \wedge \theta\big) : J_1\boldsymbol{E} \to \Lambda^2 T^* J_1\boldsymbol{E}$, with expression $\Omega = G^0_{ij} \, (d^i_0 - \Gamma_\lambda{}^i_0 \, d^\lambda) \wedge (d^j - x^j_0 \, d^0)$. Indeed, Ω splits as $\Omega = \Omega^\natural + \Omega^\mathfrak{e}$, where $\Omega^\natural = G \lrcorner \big(\nu[\Gamma^\natural] \wedge \theta\big)$ and $\Omega^\mathfrak{e} = \frac{q}{2\hbar} \, F$.

The pair (dt, Ω) is *cosymplectic*, i.e. $d\Omega = 0$ and $dt \wedge \Omega \wedge \Omega \wedge \Omega \not\equiv 0$.

Ω admits *horizontal* potentials $A^\uparrow$, which are defined up to a spacetime 1–form. For each o, we can write $A^\uparrow = \Theta[o] + A[o]$, where $A[o] = o^* A^\uparrow$, and obtain the closed spacetime 2–form $\Phi[o] =: 2 \, o^* \Omega = 2 \, dA[o]$.

Γ and G yield the *phase 2–vector* $\Lambda =: \bar{G} \lrcorner (\Gamma \wedge \nu) : J_1\boldsymbol{E} \to \Lambda^2 V J_1\boldsymbol{E}$, with expression $\Lambda = G^{ij}_0 \, \big(\partial_i + \Gamma_i{}^h_0 \, \partial^0_h\big) \wedge \partial^0_j$. Indeed, Λ splits as $\Lambda = \Lambda^\natural + \Lambda^\mathfrak{e}$, where $\Lambda^\natural = \bar{G} \lrcorner (\Gamma^\natural \wedge \nu)$ and $\Lambda^\mathfrak{e} = \frac{q}{2\hbar} \, G^\sharp(F) : J_1\boldsymbol{E} \to (\mathbb{T}^* \otimes \mathbb{T}^*) \otimes \Lambda^2 V\boldsymbol{E}$. We have the expression $\Lambda^\mathfrak{e} = \frac{q}{2\hbar} \, G^{ih}_0 \, G^{jk}_0 \, F_{hk} \, \partial^0_i \wedge \partial^0_j$.

$f \in \operatorname{map}(J_1\boldsymbol{E}, I\!\!R)$ is said to be a *special phase function* if $D^2 f = f'' \otimes G$, with $f'' \in \operatorname{map}(\boldsymbol{E}, \bar{\mathbb{T}})$. These functions constitute a subsheaf $\mathit{spec}(J_1\boldsymbol{E}, I\!\!R) \subset \operatorname{map}(J_1\boldsymbol{E}, I\!\!R)$.

$f \in \mathit{spec}(J_1\boldsymbol{E}, I\!\!R)$ if and only if $f = f^0 \, \frac{1}{2} \, G^0_{ij} \, x^i_0 \, x^j_0 + f^i \, G^0_{ij} \, x^j_0 + \breve{f}$, with $f^0, f^i, \breve{f} \in \operatorname{map}(\boldsymbol{E}, I\!\!R)$.

For each $f \in \mathit{spec}(J_1\boldsymbol{E}, I\!\!R)$, the map $f'' \lrcorner \text{д} - G^\sharp(Df)$ factorises through a spacetime vector field, $X[f] \in \operatorname{sec}(\boldsymbol{E}, T\boldsymbol{E})$, with expression $X[f] = f^0 \, \partial_0 - f^i \, \partial_i$.

For each observer o, we have the linear isomorphism

$$\mathfrak{s}[o] : \mathit{spec}(J_1\boldsymbol{E}, I\!\!R) \to \operatorname{sec}(\boldsymbol{E}, T\boldsymbol{E}) \times \operatorname{map}(\boldsymbol{E}, I\!\!R) : f \mapsto \big(X[f], \, f \circ o\big).$$

We define the *special Lie bracket* of $\mathit{spec}(J_1\boldsymbol{E}, I\!\!R)$ by

$$[\![f_1, f_2]\!] =: \Lambda(df_1, df_2) + \gamma(f''_1).f_2 - \gamma(f''_2).f_1 .$$

Indeed, for each observer o, we obtain

$$[\![f_1, f_2]\!] = -\big[X[f_1], \, X[f_2]\big] \lrcorner \Theta[o] + \big[(X[f_1], \breve{f}_1), \, (X[f_2], \breve{f}_2)\big]_{\Phi[o]}$$

and $\mathfrak{s}[o]$ turns out to be an isomorphism of Lie algebras.

For example, let us consider a potential $A^\uparrow$ and an observer o. Then, we define the observed *Hamiltonian* and *momentum* to be, respectively, $\mathcal{H}[o] =: -\, д[o] \lrcorner A^\uparrow$ and $\mathcal{P}[o] =: \nu[o] \lrcorner A^\uparrow$, with expressions $\mathcal{H}[o] = (\frac{1}{2}\, G^0_{ij}\, x^i_0\, x^j_0 - A_0)\, d^0$ and $\mathcal{P}[o] = (G^0_{ij}\, x^j_0 + A_i)\, d^i$ where $A^i_0 =: G^{ij}_0\, A_j$. Indeed, $x^\lambda, \mathcal{H}_0, \mathcal{P}_i \in spec(J_1\boldsymbol{E}, I\!R)$. Moreover, we have $X[x^\lambda] = 0$, $X[\mathcal{H}_0] = \partial_0$, $X[\mathcal{P}_i] = -\partial_i$ and $[\![x^\lambda, x^\mu]\!] = 0$, $[\![x^\lambda, \mathcal{H}_0]\!] = -\delta^\lambda_0$, $[\![x^\lambda, \mathcal{P}_i]\!] = \delta^\lambda_i$, $[\![\mathcal{H}_0, \mathcal{P}_i]\!] = 0$, $[\![\mathcal{P}_i, \mathcal{P}_j]\!] = 0$.

2.2. *Quantum setting*

We assume the line bundle $\pi : \boldsymbol{Q} \to \boldsymbol{E}$ as *quantum bundle* over the Galilei spacetime and define the *phase quantum bundle* as $\pi^\uparrow : \boldsymbol{Q}^\uparrow =: J_1\boldsymbol{E} \times_{\boldsymbol{E}} \boldsymbol{Q} \to J_1\boldsymbol{E}$.

We suppose that the cohomolgy class of Ω be integer and assume a connection $Ч^\uparrow : \boldsymbol{Q}^\uparrow \to T^*J_1\boldsymbol{E} \otimes T\boldsymbol{Q}^\uparrow$, which is Hermitian, "universal"[3] and whose curvature is given by $R[Ч^\uparrow] = -2\,\mathrm{i}\,\Omega \otimes \mathbb{I}^\uparrow$.

With reference to a basis Ub and an observer o, the expression of $Ч^\uparrow$ is of the type $Ч^\uparrow = \chi^\uparrow[\mathbf{E}b] + \mathrm{i}\,(\Theta[o] + A[Ub, o]) \otimes \mathbb{I}^\uparrow$, where $A[Ub, o]$ is a potential of $\Phi[o]$ selected by $Ч^\uparrow$ and Ub. Hence, we have $Ч^\uparrow = d^\lambda \otimes \partial_\lambda + d^i_0 \otimes \partial^0_i + \mathrm{i}\,((-\frac{1}{2}\, G^0_{ij}\, x^i_0\, x^j_0 + A_0)\, d^0 + (G^0_{ij}\, x^j_0 + A_i)\, d^i) \otimes \mathbb{I}^\uparrow$.

For each observer o, the expression of $Ч[o] =: o^*Ч^\uparrow$ is $Ч[o] = \chi[Ub] + \mathrm{i}\, A[Ub, o] \otimes \mathbb{I}$, i.e. $Ч[o] = d^\lambda \otimes \partial_\lambda + \mathrm{i}\, A_\lambda\, d^\lambda \otimes \mathbb{I}$. If Ub is a quantum basis and $o, \acute{o} = o + v$ are two observers, then we obtain the transition law $A[Ub, \acute{o}] = A[Ub, o] - \frac{1}{2}\, G(v, v) + \nu[o] \lrcorner G^\flat(v)$.

Eventually, we apply to the Galilei framework the classification of Hermitian vector fields achieved in Theorem 1.1. For this purpose, we choose any observed quantum connection $Ч[o]$ as auxiliary connection c and use the observed representation $\mathfrak{s}[o]$.

Theorem 2.1. *We have the Lie algebra isomorphism*

$$\mathfrak{F} =: \mathfrak{j}\big[Ч[o]\big] \circ \mathfrak{s}[o] : spec(J_1\boldsymbol{E},\, I\!R) \to \mathrm{her}\,(\boldsymbol{Q},\, T\boldsymbol{Q}),$$

with expression $\mathfrak{F}(f) = f^0\,\partial_0 - f^i\,\partial_i + \mathrm{i}\,(f^0\, A_0 - f^i\, A_i + \breve{f}) \otimes \mathbb{I}$, *which turns out to be observer independent.* □

For instance, we have $\mathfrak{F}(x^\lambda) = \mathrm{i}\, x^\lambda\, \mathbb{I}$, $\mathfrak{F}(\mathcal{H}_0[o]) = \partial_0$, $\mathfrak{F}(\mathcal{P}_i[o]) = -\partial_i$.

3. Einstein case

3.1. *Classical setting*

We assume *spacetime* to be an oriented and time oriented 4–dimensional manifold $\boldsymbol{E}$ equipped with a scaled Lorentzian metric $g : \boldsymbol{E} \to \mathbb{L}^2 \otimes (T^*\boldsymbol{E} \otimes T^*\boldsymbol{E})$ with signature $(- + + +)$. With reference to a mass $m \in \mathbb{M}$, it is convenient to introduce the *rescaled metric* $G =: \frac{m}{\hbar}\, g : \boldsymbol{E} \to \mathbb{T} \otimes (T^*\boldsymbol{E} \otimes T^*\boldsymbol{E})$. A *motion* is defined to be a 1–dimensional timelike submanifold $s : \boldsymbol{T} \subset \boldsymbol{E}$.

We shall refer to a *spacetime chart* $(x^\lambda) \equiv (x^0, x^i)$ adapted to the spacetime orientation and such that the vector ∂_0 is timelike and time oriented and the vectors $\partial_1, \partial_2, \partial_3$ are spacelike. Greek indices will span all spacetime coordinates and Latin indices will span the spacelike coordinates. We shall also refer to a time unit $u_0 \in \mathbb{T}$ and its dual $u^0 \in \mathbb{T}^*$. We have the expression $G = G^0_{\lambda\mu}\, u_0 \otimes d^\lambda \otimes d^\mu$.

We assume as *gravitational connection* the Levi–Civita connection $K^\natural : T\boldsymbol{E} \to T^*\boldsymbol{E} \otimes TT\boldsymbol{E}$ induced by G.

We assume as *electromagnetic field* a closed scaled 2–form $F : \boldsymbol{E} \to (\mathbb{L}^{1/2} \otimes \mathbb{M}^{1/2}) \otimes \Lambda^2 T^*\boldsymbol{E}$.

In the Einstein framework there is no way to merge the electromagnetic field into the gravitational connection, hence we have no joined spacetime connection.

We assume as *phase space* the subspace of 1st jets of motions $\mathcal{J}_1\boldsymbol{E} \subset J_1(\boldsymbol{E}, 1)$.

Each spacetime chart (x^0, x^i) induces a fibred chart (x^0, x^i, x^i_0) of $\mathcal{J}_1\boldsymbol{E}$. It is convenient to set $\breve{g}_{0\lambda} =: g(b_0, \partial_\lambda) = g_{0\lambda} + g_{i\lambda}\, x^i_0$, $\breve{\delta}^\lambda_0 =: \delta^\lambda_0 + \delta^\lambda_i\, x^i_0$, $\breve{\delta}^i_\lambda =: \delta^i_\lambda - \delta^0_\lambda\, x^i_0$.

We have the *contact map* $д : \mathcal{J}_1\boldsymbol{E} \to \mathbb{T}^* \otimes T\boldsymbol{E}$, with expression $д = c_0\, \alpha^0\, u^0 \otimes (\partial_0 + x^i_0\, \partial_i)$, where $\alpha^0 =: 1/\sqrt{|g_{00} + 2\, g_{0j}\, x^j_0 + g_{ij}\, x^i_0\, x^j_0|}$.

We define the *time form* as the map $\tau =: -\frac{1}{c^2}\, g^\flat(д) : \mathcal{J}_1\boldsymbol{E} \to \mathbb{T} \otimes T^*\boldsymbol{E}$, with expression $\tau = \tau_\lambda\, d^\lambda$, where $\tau_\lambda = -\frac{\alpha^0}{c_0}\, \breve{g}_{0\lambda}\, u_0$. We have $\tau(д) = 1$ and $g\,(д, д) = -c^2$.

We have the *complementary contact map* $\theta =: 1 - д \otimes \tau : \mathcal{J}_1\boldsymbol{E} \times_{\boldsymbol{E}} T\boldsymbol{E} \to T\boldsymbol{E}$, with expression $\theta = d^\lambda \otimes \partial_\lambda + (\alpha^0)^2\, \breve{g}_{0\lambda}\, d^\lambda \otimes (\partial_0 + x^j_0\, \partial_j)$.

We define the 1–form $\Theta =: -\frac{mc^2}{\hbar}\, \tau$, with expression $\Theta = \alpha^0\, c_0\, \breve{G}^0_{0\lambda}\, d^\lambda$.

We have an isomorphism $\nu_\tau : \mathbb{T}^* \otimes V_\tau\boldsymbol{E} \to V_0\mathcal{J}_1\boldsymbol{E}$, where $V_\tau\boldsymbol{E}$ is the subbundle of $\mathcal{J}_1\boldsymbol{E} \times_{\boldsymbol{E}} T\boldsymbol{E}$ consisting of vector fields killed by τ.

An *observer* is defined to be a section $o \in \sec(\boldsymbol{E}, \mathcal{J}_1\boldsymbol{E})$. An *observing frame* is defined to be a pair (o,ζ), where o is an observer and $\zeta \in \sec(\boldsymbol{E}, \mathbb{T}\otimes T^*\boldsymbol{E})$ is timelike and positively time oriented. In particular, each observer o determines the observing frame $(o, o^*\tau)$. An observing frame is said to be *integrable* if ζ is closed. In this case, there exists locally a scaled function $t \in \operatorname{map}(\boldsymbol{E}, \bar{\mathbb{T}})$, called the *observed time function*, such that $\zeta = dt$. For each observing frame (o,ζ), by splitting Θ into the horizontal and vertical components, we define the observed *kinetic energy* and *kinetic momentum* as $\mathcal{K}[o,\zeta] = -(1/\varsigma)\,\zeta(д[o]\lrcorner\Theta)$ and $\mathcal{Q}[o,\zeta] = \theta[o,\zeta]\lrcorner\Theta$. Thus, we have $\Theta = -\mathcal{K}[o,\zeta] + \mathcal{Q}[o,\zeta]$. For an integrable observing frame we obtain $\mathcal{K}[o] = -c_0\,\alpha^0\,\breve{G}^0_{00}\,d^0$ and $\mathcal{Q}[o] = c_0\,\alpha^0\,\breve{G}^0_{0i}\,d^i$.

We have a natural injective map χ between linear spacetime connections K and phase connections $\Gamma : \mathcal{J}_1\boldsymbol{E} \to T^*\boldsymbol{E}\otimes T\mathcal{J}_1\boldsymbol{E}$, with expression $\Gamma = d^\lambda \otimes (\partial_\lambda + \Gamma_{\lambda}{}^{i}_{0}\,\partial^0_i)$, where $\Gamma_{\lambda}{}^{i}_{0} = \breve{\delta}^i_\nu\,K_\lambda{}^\nu{}_\rho\,\breve{\delta}^\rho_0$.

$K^\natural$ yields the connection $\Gamma^\natural =: \chi(K^\natural) : \mathcal{J}_1\boldsymbol{E} \to T^*\boldsymbol{E}\otimes T\mathcal{J}_1\boldsymbol{E}$.

$\Gamma^\natural$ yields the 2nd order connection $\gamma^\natural =: д\lrcorner\Gamma^\natural : \mathcal{J}_1\boldsymbol{E} \to \mathbb{T}^*\otimes T\mathcal{J}_1\boldsymbol{E}$, with expression $\gamma^\natural = c_0\,\alpha^0\,u^0\otimes(\partial_0 + x^i_0\,\partial_i + \gamma^\natural{}_{0}{}^{i}_{0}\,\partial^0_i)$, where $\gamma^\natural{}_{0}{}^{i}_{0} = \breve{\delta}^i_\nu\,K_\lambda{}^\nu{}_\mu\,\breve{\delta}^\lambda_0\,\breve{\delta}^\mu_0$.

$\Gamma^\natural$ and G yield the 2–form $\Omega^\natural =: G\lrcorner\big((\nu_\tau^{-1}\circ\nu[\Gamma^\natural])\wedge\theta\big) : \mathcal{J}_1\boldsymbol{E} \to \Lambda^2T^*\mathcal{J}_1\boldsymbol{E}$, with expression $\Omega^\natural = c_0\,\alpha^0\,\breve{G}^0_{i\mu}\,\big(d^i_0 - \breve{\delta}^i_\nu\,K^\natural{}_\lambda{}^\nu{}_\rho\,\breve{\delta}^\rho_0)\,d^\lambda\big)\wedge d^\mu$.

The pair $(\Theta,\Omega^\natural)$ is "contact", i.e. $\Omega = d\Theta$ and $\Theta\wedge\Omega^\natural\wedge\Omega^\natural\wedge\Omega^\natural \not\equiv 0$.

$\Gamma^\natural$ and G yield the vertical 2–vector $\Lambda^\natural =: \bar{G}\lrcorner(\Gamma^\natural\wedge\nu_\tau) : \mathcal{J}_1\boldsymbol{E} \to \Lambda^2V\mathcal{J}_1\boldsymbol{E}$, with expression $\Lambda^\natural = \frac{1}{c_0\,\alpha^0}\,\breve{G}^{j\lambda}_0\,(\partial_\lambda + \breve{\delta}^i_\mu\,K^\natural{}_\lambda{}^\mu{}_\rho\,\breve{\delta}^\rho_0\,\partial^0_i)\wedge\partial^0_j$.

Now, we are looking for *joined* phase objects, obtained by merging the electromagnetic field into the above gravitational phase objects, in such a way to preserve the above relations.

We define the connection $\Gamma =: \Gamma^\natural + \Gamma^e$, where $\Gamma^e =: \circ\big(F + 2\tau\wedge(д\lrcorner F)\big)$, i.e., in coordinates, $\Gamma^e = (F_{\lambda\mu} - (\alpha^0)^2\breve{g}_{0\lambda}\,F_{\rho\mu}\,\breve{\delta}^\rho_0)\,d^\lambda\otimes\partial^0_i$.

Γ yields the 2nd order connection $\gamma =: д\lrcorner\Gamma : \mathcal{J}_1\boldsymbol{E} \to \mathbb{T}^*\otimes T\mathcal{J}_1\boldsymbol{E}$, which splits as $\gamma = \gamma^\natural + \gamma^e$, where $\gamma^e = -\frac{q}{m}\,\nu_\tau\circ G^\sharp\circ(д\lrcorner F)$, i.e., in coordinates, $\gamma^e = (F_{0\mu} + F_{j\mu}\,x^j_0)\,u^0\otimes\partial^0_i$.

Γ an G yield the 2–form $\Omega =: G\lrcorner\big(\nu_\tau[\Gamma]\wedge\theta\big)$, which splits as $\Omega = \Omega^\natural + \Omega^e$, where $\Omega^e = \frac{q}{2\hbar}\,F$, i.e., in coordinates, $\Omega^e = \frac{q}{2\hbar}\,F_{\lambda\mu}\,d^\lambda\wedge d^\mu$. The pair $(\Theta,\,\Omega)$ is "cosymplectic" i.e, $d\Omega = d\Omega^\natural + \frac{q}{2\hbar}\,dF = 0$ and $\Theta\wedge\Omega\wedge\Omega\wedge\Omega = \Theta\wedge\Omega^\natural\wedge\Omega^\natural\wedge\Omega^\natural \not\equiv 0$.

Ω admits horizontal potentials $A^\uparrow$, which are defined up to a spacetime 1–form. Indeed, we have $A^\uparrow = \Theta + \frac{q}{\hbar}\,A^e$, with expression $A^\uparrow = (c_0\,\alpha^0\,\breve{G}^0_{0\lambda} + \frac{q}{\hbar}\,A^e{}_\lambda)\,d^\lambda$.

Indeed, γ is the unique 2nd order connection such that $i(\gamma)\tau = 1$ and $i(\gamma)\Omega = 0$.

Γ and G yield the 2–vector $\Lambda =: \bar{G}\lrcorner(\Gamma\wedge\nu^\natural)$, which splits as $\Lambda =$

$\Lambda^{\natural} + \Lambda^{\mathfrak{e}}$, where $\Lambda^{\mathfrak{e}} = \frac{q}{2\hbar}\,(\nu_\tau \wedge \nu_\tau)(G^\sharp(\theta^*(F)))$, i.e., in coordinates, $\Lambda^{\mathfrak{e}} = \frac{q}{2\hbar}\,\frac{1}{(c_0\,\alpha^0)^2}\,\breve{G}^{i\lambda}_0\,\breve{G}^{j\mu}_0\,F_{\lambda\mu}\,\partial^0_i \wedge \partial^0_j$.

$f \in \text{map}(\mathcal{J}_1\boldsymbol{E},\,I\!R)$ is said to be a *special phase function* if $f = -G(\text{д}, X) + \breve{f}$, with $X \in \sec(\boldsymbol{E}, T\boldsymbol{E})$ and $\breve{f} \in \text{map}(\boldsymbol{E},\,I\!R)$, i.e., in coordinates,

$$f = -c_0\,\alpha^0\,(G^0_{\lambda 0} + G^0_{\lambda i}\,x^i_0)\,f^\lambda + \breve{f} = -c_0\,\alpha^0\,(f^0_0 + f^0_i\,x^i_0) + \breve{f}\,,$$

with $f^\lambda =: X^\lambda$ and $f^0_\lambda =: G^0_{\lambda\mu}\,X^\mu$. These functions constitute a subsheaf $spec(\mathcal{J}_1\boldsymbol{E},\,I\!R) \subset \text{map}(\mathcal{J}_1\boldsymbol{E},\,I\!R)$. Thus, we have the linear maps $X : spec(\mathcal{J}_1\boldsymbol{E},\,I\!R) \to \sec(\boldsymbol{E}, T\boldsymbol{E}) : f \mapsto X[f]$ and $\breve{}\; : spec(\mathcal{J}_1\boldsymbol{E},\,I\!R) \to \text{map}(\boldsymbol{E},\,I\!R) : f \mapsto \breve{f}$.

We have the linear isomorphism

$$\mathfrak{s} : spec(\mathcal{J}_1\boldsymbol{E},\,I\!R) \to \sec(\boldsymbol{E}, T\boldsymbol{E}) \times \text{map}(\boldsymbol{E},\,I\!R) : f \mapsto \big(X[f],\,\breve{f}\big)\,.$$

We define the *special Lie bracket* of $spec(\mathcal{J}_1\boldsymbol{E},\,I\!R)$ by

$$[\![\, f_1, f_2 \,]\!] =: \Lambda(df_1, df_2) + (\tau(X[f_1])\,(\gamma.f_2) - (\tau(X[f_2])\,(\gamma.f_1)\,.$$

Indeed, we obtain

$$[\![\, f_1,\, f_2 \,]\!] = X[f_1].\breve{f}_2 - X[f_2].\breve{f}_1 + \tfrac{q}{\hbar}\,F\big(X[f_1], X[f_2]\big)$$

and $\mathfrak{s}$ turns out to be an isomorphism of Lie algebras.

For any spacetime chart (x^λ), the functions x^λ are special phase functions and we obtain $X[x^\lambda] = 0$. Moreover, with reference to a potential $A^\uparrow$ and to an observing frame (o, ζ), we define the observed *Hamiltonian* and *momentum* as $\mathcal{H}[o,\zeta] =: -\,(1/\varsigma)\,(\text{д}[o] \lrcorner A^\uparrow)\,\zeta$ and $\mathcal{P}[o] =: \theta[o,\zeta]\,A^\uparrow$. If the observing frame is integrable, then we have the expressions $\mathcal{H}[o,\zeta] = (-c_0\,\alpha^0\,\breve{G}^0_{00} - \frac{q}{\hbar}\,A^{\mathfrak{e}}{}_0)\,d^0$ and $\mathcal{P}[o,\zeta] = (c_0\,\alpha^0\,\breve{G}^0_{0i} + \frac{q}{\hbar}\,A^{\mathfrak{e}}{}_i)\,d^i$. In this case, $\mathcal{H}_0$ and $\mathcal{P}_i$ are special phase functions and we obtain $X\big[\mathcal{H}_0\big] = \partial_0$ and $X\big[\mathcal{P}_i\big] = -\partial_i$.

We have $[\![\, x^\lambda, x^\mu \,]\!] = 0$ and, with reference to an integrable observing frame, $[\![\, x^\lambda, \mathcal{H}_0 \,]\!] = \delta^\lambda_0$, $[\![\, x^\lambda, \mathcal{P}_i \,]\!] = \delta^\lambda_i$, $[\![\, \mathcal{H}_0, \mathcal{P}_i \,]\!] = 0$.

3.2. *Quantum setting*

We assume the line bundle $\pi : \boldsymbol{Q} \to \boldsymbol{E}$ as *quantum bundle* over the Einstein spacetime. Moreover, we define the *phase quantum bundle* as $\pi^\uparrow : \boldsymbol{Q}^\uparrow =: \mathcal{J}_1\boldsymbol{E} \times_{\boldsymbol{E}} \boldsymbol{Q} \to \mathcal{J}_1\boldsymbol{E}$.

We can refrase the notion of Hermitian systems of connections and associated universal connection that we have discussed in the Galilei case, by replacing $J_1\boldsymbol{E}$ with $\mathcal{J}_1\boldsymbol{E}$.

We suppose that the cohomology class of $\frac{q}{\hbar} F$ be integer and assume a connection $\text{Ч}^{\uparrow} : \boldsymbol{Q}^{\uparrow} \to T^*\mathcal{J}_1\boldsymbol{E} \otimes T\boldsymbol{Q}^{\uparrow}$, which is Hermitian, "universal"[3] and whose curvature is given by $R[\text{Ч}^{\uparrow}] = -2\,\mathfrak{i}\,\Omega \otimes \mathbb{I}^{\uparrow}$.

We have the splitting $\text{Ч}^{\uparrow} = \text{Ч}^{\uparrow\mathfrak{e}} + \mathfrak{i}\,\Theta \otimes \mathbb{I}^{\uparrow}$, where $\text{Ч}^{\uparrow\mathfrak{e}} : \boldsymbol{Q}^{\uparrow} \to T^*\mathcal{J}_1\boldsymbol{E} \otimes T\boldsymbol{Q}^{\uparrow}$, is the pull back of a Hermitian connection $\text{Ч}^{\mathfrak{e}} : \boldsymbol{Q} \to T^*\boldsymbol{E} \otimes T\boldsymbol{Q}$, whose curvature is given by the equality $R[\text{Ч}^{\mathfrak{e}}] = -\mathfrak{i}\,\frac{q}{\hbar}\, F \otimes \mathbb{I}$.

With reference to a basis Ub, the expression of $\text{Ч}^{\uparrow}$ is of the type $\text{Ч}^{\uparrow} = \chi^{\uparrow}[Ub] + \mathfrak{i}\,\big(\Theta + \frac{q}{\hbar}\, A^{\mathfrak{e}}[Ub]\big) \otimes \mathbb{I}^{\uparrow}$, where $A^{\mathfrak{e}}[Ub]$ is a potential of F selected by $\text{Ч}^{\uparrow}$ and Ub. Hence, in a chart adapted to Ub, we have $\text{Ч}^{\uparrow} = d^{\lambda} \otimes \partial_{\lambda} + d^i_0 \otimes \partial^0_i + \mathfrak{i}\,(c_0\, \alpha^0\, \breve{G}^0_{0\lambda} + \frac{q}{\hbar}\, A^{\mathfrak{e}}{}_{\lambda})\, d^{\lambda} \otimes \mathbb{I}^{\uparrow}$.

For each o, the expression of $\text{Ч}[o]$, is $\text{Ч}[o] = \mathfrak{i}\,\Theta[o] \otimes \mathbb{I} + \text{Ч}^{\mathfrak{e}}$, and, in a chart adapted to Ub, $\text{Ч}[o] = d^{\lambda} \otimes \partial_{\lambda} + \mathfrak{i}\,(\Theta[o]_{\lambda} + \frac{q}{\hbar}\, A^{\mathfrak{e}}{}_{\lambda})\, d^{\lambda} \otimes \mathbb{I}$.

Eventually, we apply to the Einstein framework the classification of Hermitian vector fields achieved in Theorem 1.1. For this purpose, we choose the electromagnetic quantum connection $\text{Ч}^{\mathfrak{e}}$ as auxiliary connection c and use the classification of special phase functions.

Theorem 3.1. *We have the Lie algebra isomorphism*

$$\mathfrak{F} =: \mathfrak{j}[\text{Ч}^{\mathfrak{e}}] \circ \mathfrak{s} : spec(\mathcal{J}_1\boldsymbol{E},\, I\!\!R) \to \text{her}\,(\boldsymbol{Q},\, T\boldsymbol{Q}),$$

with coordinate expression $\mathfrak{F}(f) = f^{\lambda}\partial_{\lambda} + \mathfrak{i}\,(\frac{q}{\hbar}\, f^{\lambda}\, A^{\mathfrak{e}}{}_{\lambda} + \breve{f})\, \mathbb{I}$. □

The above result could also be obtained via observers in analogy with the Galilei case.

Hence, the Hermitian vector field associated with f by the connection $\text{Ч}[o]$ does not depend on the observer o. For instance, we have $\mathfrak{F}(x^{\lambda}) = \mathfrak{i}\, x^{\lambda}\, \mathbb{I}$ and, with reference to an integrable observing frame and to an adapted chart, $\mathfrak{F}(\mathcal{H}_0) = \partial_0$ and $\mathfrak{F}(\mathcal{P}_i) = \partial_i$.

References

1. A. Jadczyk, M. Modugno, *An outline of a new geometric approach to Galilei general relativistic quantum mechanics*, in "Diff. geom. meth. in theor. phys.", Eds.: C. N. Yang, M. L. Ge and X. W. Zhou, World Scientific, 1992, 543–556.
2. A. Jadczyk, M. Modugno, *Galilei general relativistic quantum mechanics,* Report of Dep. of Appl. Math., University of Florence, 1994, 1–215.
3. J. Janyška, M. Modugno, *Covariant Schrödinger operator J. Phys.: A, Math. Gen.* **35** (2002) 8407–8434.

Electric-magnetic Duality Beyond Four Dimensions and in General Relativity *

Bernard L. JULIA †

Laboratoire de Physique théorique de l'ENS
24 rue Lhomond
75005 Paris, FRANCE
E-mail: bjulia AT lpt.ens.fr

After reviewing briefly the classical examples of duality in four dimensional field theory we present a generalisation to arbitrary dimensions and to p-form fields. Then we explain how U-duality may become part of a larger non abelian V-symmetry in superstring/supergravity theories. And finally we discuss two new results for 4d gravity theory with a cosmological constant: a new exact gravitational instanton equation and a surprizing linearized classical duality around de Sitter space.

1. Electric-magnetic duality and self-duality

1.1. *Gauge fields*

The discrete (Z_4) and continuous ($SO(2,R)$) invariances of the Maxwell equation and of the gauge fixed Maxwell action [1] are a remarkable feature of 4 dimensional electromagnetism in vacuum. The inclusion of matter requires non trivial topology (like a possibly nontrivial U(1) principal bundle) in order to preserve these symmetries. At the quantum level the lattice of electric-magnetic charges breaks the symmetry down to a discrete one. The Dirac-Schwinger quantization condition constrains the possible charges of a pair of dyons D(e,g) and D'(e',g') to satisfy:

$$4\pi(eg' - e'g)/h = integer \tag{1.1}$$

The two helicities of the electromagnetic field correspond to self-dual and antiself-dual field strengths. In euclidean signature the (real) classical field

*This work is supported by CNRS
†Work done partly in collaboration with Y Dolivet, P H Labordère and L Paulot resp. with J Levie and S Ray.

strength can be decomposed locally into the sum of a self-dual part and an anti self-dual part ie

$$dA \equiv F \, , \, F_{\pm} = \pm * F_{\pm} \tag{1.2}$$

where * is the Hodge dualization operator on two forms. For Yang-Mills fields there is a celebrated generalization of the self-duality projection namely the instanton equation. Note that the usual instanton equation is first order and provides only special solutions to the full (vacuum) Yang-Mills equations.

1.2. *Gauge forms*

Pointlike electric charges are minimally coupled to "vector" potentials and the generalization for scalar fields resp higher (p+1)-form potentials is their coupling respectively to instantons and p-branes. Abelian self-duality is possible in even dimensional spacetimes of dimension (2p+4) of the appropriate signature for a single (p+1)-form potential. There is a generalization of the quantization condition (1.1) to this situation as well and interestingly it involves a plus sign rather than a minus sign in (4k+2) dimensions [2].

One key property of these remarkable self-dual solutions is that they minimize the action by saturating a topological charge bound: the so-called BPS bound. It is E. Bogomolny who analyzed systematically this mechanism and applied it to magnetic monopoles and dyons (independently studied by M. Prasad and C. Sommerfield) . The lower bound is typically a characteristic (for instance Pontryagin) number of the principal bundle under study [3].

2. U-duality: selecta

2.1. *Gravity case*

The Einstein action in D dimensions is invariant under diffeomorphisms of the manifold M_D. Upon dimensional reduction by r commuting one parameter isometry groups the effective action on the (D-r) quotient space (of orbits) the set of equations becomes invariant under a group of internal symmetries that grows with r. Part of it is expected for instance $GL(r, R)$ or at least $SL(r, R)$ but other parts of it come as surprises, the first of which is the so-called Ehlers symmetry $SL(2, R)$ that is easy to verify after reducing ordinary Einstein gravity in $D = 4$ by one dimension (r=1). More generally reduction of pure gravity from D to 3 dimensions leads to a generalised Ehlers symmetry $SL(D-2, R)$, see for instance [4]. This is a major mystery and constitutes one of our motivations to concentrate on dualities in general, to discover new ones and to study their properties.

2.2. *The supergravity magic triangles*

If one considers at first the internal symmetries (commuting with the Poincaré group) one encounters often coset spaces, even Riemannian symmetric spaces, on which these symmetries act as real Lie groups. These cosets are the target spaces where scalar fields (ie 0-forms) take their values. The symmetries are called U-dualities for historical reasons [5], approximately half of them act by (Hodge) dualities on the p-forms in their self-duality dimension. A remarkable collection of (pure in D=4) supergravity theories as well as their dimensional reductions down to 3 dimensions and their higher dimensional ancestors fit into a triangle with partial symmetry under the exchange of the space-time dimension with the number of supercharges see [4]. These groups are expected to play an important role in string theory after being broken down to a discrete (arithmetic) subgroup.

In the example of 4 dimensions for instance the U-duality group of maximal supergravity is the split real form of E_7 it contains a parity conserving subgroup $SL(8, R)$ and the other generators are dualities. The maximal compact subgroup of this real form of E_7 is $SU(8)$ sometimes called R-symmetry just to confuse us. The string "gauge group" is expected to be the intersection of the split E_7 with the discrete group $Sp(56, Z)$. E_7 is indeed a subgroup of $Sp(56, R)$. One must double the number of vector potentials from 28 to 56 to realize locally the action of dualities, it turns out that the doubled set of fields obeys first order equations that are now equivalent to the second order original equations. We shall recognize this phenomenon as rather general and this will lead us to V-dualities. The doubled set of fist order equations is nothing but a (twisted) self-duality condition. For an early discussion of doubling see for instance [6].

$$E.F = S * E.F \tag{2.1}$$

In our example F is the 56-plet of field strengths, E is a representative of the scalar fields taking their values in the exceptional group E_7 and written in the **56** representation and S is a pseudo involution of square ±1 that compensates for the square of the Hodge operation $** = \pm1$.

3. V-duality

3.1. *del Pezzo surfaces and Borcherds algebras*

Another mystery of duality is the occurrence not only of the exceptional group E_7 but of the full (extended in fact) E series: E_8, E_7, E_6, $E_5 = D_5$, $E_4 = A_4$, $E_3 = A_2 \times A_1$... both as the U-duality groups of maximal supergravity reduced to 3,4,5,6,7,8... dimensions and as symmetry groups of type II string theories after torus compactifications. The equally mysterious occurrence of the E groups or rather of their Weyl groups acting on the middle cohomology of the socalled del Pezzo complex surfaces may be a related phenomenon. There are in fact two candidates for E_1 so let us choose A_1 which is known to be associated to the trivial bundle $CP^1 \times CP^1$ (one of the two "minimal del Pezzo surfaces"). $SL(2, Z) = A_1$ is known to be also the U-duality group of type IIB superstring theory in 10 dimensions (the top dimension). Besides the information provided by algebraic geometers (Y. Manin...) we used [7] one important remark of C. Vafa and collaboratorsi who stressed the importance of rational cycles within the second cohomology of the del Pezzo complex surfaces. For instance in the case of $CP^1 \times CP^1$ the middle cohomology is quite boringly equal to $Z + Z$, yet one axis of this lattice is selected by the complex geometry to be the root lattice of the above mentioned A_1 and the correspondence between spheres on the del Pezzo surface and D-branes on the string side [8] suggested to us that one should combine the Weyl cone of A_1 and the Mori cone of the cohomology into a Borcherds cone associated to the simple (positive) roots of a generalized Cartan matrix obtained from that of A_2 by replacing one of the diagonal elements (2) by a zero! The correspondence is best understood in this case but more generally it is still useful [7]. The intersection form on the surface is in this case the metric on the Cartan subalgebra of a Borcherds algebra.

3.2. *Truncated Borcherds algebras and V-duality*

On the string/supergravity side we have known for a while [9] that there is a natural generalization of the Borel subgroup of U-duality (isomorphic to the corresponding non-compact symmetric space and target of the scalar fields) to a solvable group encompassing all the p-forms and encoding their non linear couplings but not the graviton field yet. The question was to give a name to this solvable group despite the absence of any reasonable classification of non semi-simple Lie algebras. It generalizes the encoding of nonlinear sigma model fields' couplings within the structure constants of a group, to that of higher forms' couplings in the (super)group structure of this solvable algebra. A (p+1)-form will have degree (p+1) and the Z-graded solvable superalgebra reduces in degree zero precisely the U-duality algebra or if one prefers its Borel subalgebra. There is a remarkable correspondence between the del Pezzo data and the string/M-theory data [7]. Two steps are left to ascend: firstly one should include gravity which only trickles down into this formalism after dimensional reduction, and secondly one must incorporate the fermions (this will require the enlargment of the Borel algebras to full V-duality symmetry groups in order to allow for their "maximal compact subgroups" whatever this means to act on the fermions, but we have lots of experience even in the infinite dimensional case of spacetime dimension 2).

4. Λ-Instantons

4.1. *Gravitational instantons*

Let us consider now a 4 dimensional Riemannian manifold and its Riemann curvature 4-tensor R. It is well known [3] that one may impose (Hodge) self-duality on the first (or second) pair of indices, this defines the usual gravitational instantons which are necessarily Ricci flat and provide a nice subset of solutions of the second order Einstein equations. One may also require to have double self-duality exactly as in (2.1)

$$R = S * R \tag{4.1}$$

where S is the dualization on the first pair of indices if * is the dualization on the second pair. This is equivalent to the Einstein space condition (with unspecified cosmological constant). There is the conformal self-duality equation too that guarantees the existence of a twistor space see for instance [10].

4.2. *Λ-instantons*

It seems to have gone unnoticed that there is yet another equation for any given value Λ of the cosmological constant that provides what we call Λ-instantons [11]. It is obtained by adding in the ordinary gravitational instanton equation to the Riemann curvature tensor the combination

$$-\Lambda/3(g_{\mu\rho}g_{\nu\sigma} - g_{\nu\rho}g_{\mu\sigma}) \tag{4.2}$$

the resulting tensor $Z_{\mu\nu\rho\sigma}$ turns out to be equal to the MacDowell Mansouri tensor associated to a de Sitter bundle [12]. The Λ-instanton equation reads simply

$$Z = *Z. \tag{4.3}$$

It implies the Einstein equation for that particular value of the cosmological constant but it is not equivqlent to it.

5. Duality in the gravitational sector

5.1. *Near flat space*

In a nice paper [13] the dual form of 4d linearized Einstein gravity was found to be again of the same type. The authors introduced 2 prepotentials and their associated pregauge invariances beyond diffeomorphism symmetry and showed they were interchangeable by a continuous duality rotation on shell. Even off shell the non-covariant action is invariant under duality exactly as in the Maxwell case. Such a duality exists at the nonlinear level in the presence of one Killing vector field it is precisely the Ehlers symmetry, whereas such an isometry is not assumed anymore here. The prepotentials are defined by solving the hamiltonian and momentum constraints.

5.2. *Near de Sitter space*

It maybe encouraging to go beyond this linear truncation to linearize around a different background and to try and see whether such a duality symmetry persists. Around de Sitter space (but the sign of the cosmological constant is not really important for local questions) indeed the duality rotation exchanges the relevant components of the modified curvature tensor Z, the electric part is Z_{0m0n} and the magnetic part $1/2\, Z^{pq}_{0m}\epsilon^{pqn}$. When the cosmological constant tends to zero the near flat space result is recovered smoothly.

6. Conclusion

We must now go nonlinear and it seems natural to expect from M-theory considerations that the dual theory does exist and that it is worth our efforts. More specifically the dual diffeomorphism invariance is suggestive of a doubling of spacetime, allowing for some self-duality condition that reduces the effective dimension to 4. This doubling is very familiar in string theory. We had no time to review quantum effects like quantum anomaly or NUT charge quantization.

References

1. S. Deser and C. Teitelboim, *Phys. Rev.* **D13**, 1592 (1976).
2. S. Deser, A. Gomberoff, M. Henneaux and C. Teitelboim , *Nucl. Phys.* **B520**, 179 (1998).
3. T. Eguchi, P. Gilkey and A. Hanson , *Phys. Rep.* **66** , 213 (1980).
4. E. Cremmer, H. Lu, B. Julia and C. Pope, *Preprint hep-th/9909099*, HIGHER DIMENSIONAL ORIGIN OF D = 3 COSET SYMMETRIES.
5. C. Hull and P. Townsend, *Nucl. Phys.* **B438** , 109 (1995).
6. D. Zwanziger, *Phys. Rev.* **D3** , 880 (1970).
7. P. Henry-Labordère, B. Julia and L. Paulot, *JHEP* **0204** , 049 (2002). see also Y. Dolivet, P. Henry-Labordère, B. Julia and L. Paulot, *Preprint*, SUPERALGEBRAS OF OXIDATION CHAINS (2005)
8. A. Iqbal, A. Neitzke and C. Vafa, *Adv. Theor. Math. Phys.* **5**, 769 (2002).
9. E. Cremmer, B. Julia, H. Lu and C. Pope *Nucl. Phys.* **B535**, 242 (1998).
10. R. Ward, *Com. Math. Phys.* **78**, 1 (1980).
11. B. Julia, J.Levie and S. Ray, *Preprint hep-th/0507262*, GRAVITATIONAL DUALITY NEAR DE SITTER SPACE.
12. S. MacDowell and F. Mansouri *Phys. Rev. Lett.* **38** , 739 (1977).
13. M. Henneaux and C. Teitelboim *Phys. Rev.* **D71** , 024018 (2005).

Topology and Quantum Information

Louis H. Kauffman

Department of Mathematics
University of Illinois at Chicago
851 South Morgan Street
Chicago, Illinois 60607-7045
E-mail: kauffman@uic.edu

This paper is a short survey of relationships among topology, quantum topology and quantum information theory.

1. Introduction

This paper is a summary of recent research of the author, much of it in collaboration with Sam Lomonaco, and more recently with Mo Lin Ge and Yong Zhang. The main thrust of this research has been an exploration of the relationship between quantum topology and quantum computing. This has included an exploration of how a quantum computer could compute the Jones polynomial, theorems establishing that generic 4×4 solutions to the Yang-Baxter equation are universal quantum gates, relationships between topological linking and quantum entanglement, new universal gates via solutions to the Yang-Baxter equation that include the spectral parameter [31,32], new ways to understand teleportation using the categorical formalism of quantum topology and a new theory of unitary braid group representations based on the bracket model of the Jones polynomial. These representations include the Fibonacci model of Kitaev, and promise to yield new insights into anyonic topological quantum computation.

2. Quantum entanglement and Topological Entanglement

It is natural to ask whether there are relationships between topological entanglement and quantum entanglement. Topology studies global relationships in spaces, and how one space can be placed within another (e.g.

knotting and linking of curves in three-dimensional space.) Link diagrams can be used as graphical devices and holders of information. In this vein, Aravind [1] proposed that the entanglement of a link should correspond to the entanglement of a quantum state. We discussed this approach in [18,19]. Observation at the link level is modeled by cutting one component of the link. A key example is the Borommean rings. See Figure 1.

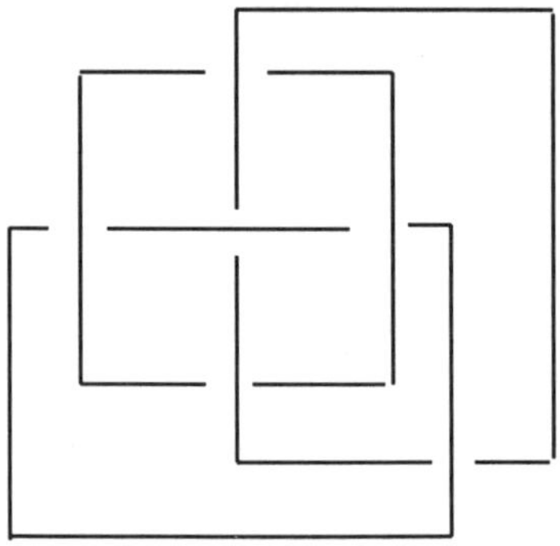

Figure 1 Borommean Rings

Cutting any component of this link yields a remaining pair of unlinked rings: The Borommean rings are entangled (viz., the link is not split), but any two of them are unentangled. In this sense, the Borommean rings are analogous to the GHZ state $|GHZ\rangle = (1/\sqrt{2})(|000\rangle + |111\rangle)$. Observation of any factor (qubit) of the GHZ yields an unentangled state. Aravind points out that this property is basis dependent, and we further point out that *there are states whose entanglement after an observation is probabilistic.* Consider, for example, the state $(1/2)(|000\rangle + |001\rangle + |101\rangle + |110\rangle)$. Observation in any coordinate yields an entangled or an unentangled state with equal probability. New ways to use link diagrams must be invented to map the properties of such states. See [30].

Our analysis of the Aravind analogy places it as an important question to which no definitive answer has yet been given. Our work shows that the analogy, taken literally, requires that a given quantum state would have to be correlated with a multiplicity of topological configurations. We are nevertheless convinced that the classification of quantum states according to their correspondence to topological entanglement will be of practical importance to quantum computing, distributed quantum computing and relations with quantum information protocols.

3. Entanglement, Universality and Unitary R-matrices

Another way to approach the analysis of quantum entanglement and topological entanglement is to look at solutions to the Yang-Baxter equation (see below) and examine their capacity to entangle quantum states. A solution to the Yang-Baxter equation is a mathematical structure that lives in two domains. It can be used to measure the complexity of braids, links and tangles, and it can (if unitary) be used as a gate in a quantum computer. We decided to investigate the quantum entangling properties of unitary solutions to the Yang-Baxter equation.

We consider unitary gates R that are both universal for quantum computation and are also solutions to the condition for topological braiding. A *Yang-Baxter operator* or R-matrix [3] is an invertible linear operator $R\colon\ V \otimes V \longrightarrow V \otimes V$, where V is a vector space, so that R satisfies the *Yang-Baxter equation*:

$$(R \otimes I)(I \otimes R)(R \otimes I) = (I \otimes R)(R \otimes I)(I \otimes R),$$

where I is the identity map of V. This concept generalizes the permutation of the factors (i.e., it generalizes a swap gate when V represents one qubit.)

Topological quantum link invariants are constructed by the association of an R-matrix R to each elementary crossing in a link diagram, so that an R-matrix R is regarded as representing an elementary bit of braiding given by one string crossing over another. In Figure 2 below, we have illustrated the braiding identity that corresponds to the Yang-Baxter equation. There is no room in this brief description to give the full translation from the topological picture into the algebraic one. Suffices it to say that each braiding picture with its three input lines (below) and output lines (above) corresponds to a mapping of the three fold tensor product of the vector space V to itself as required by the algebraic equation quoted above, and the pattern of placement of the crossings in the diagram correspond to the factors $R \otimes I$ and $I \otimes R$. The point is that this crucial topological move has an algebraic expression in terms of the R-matrix R.

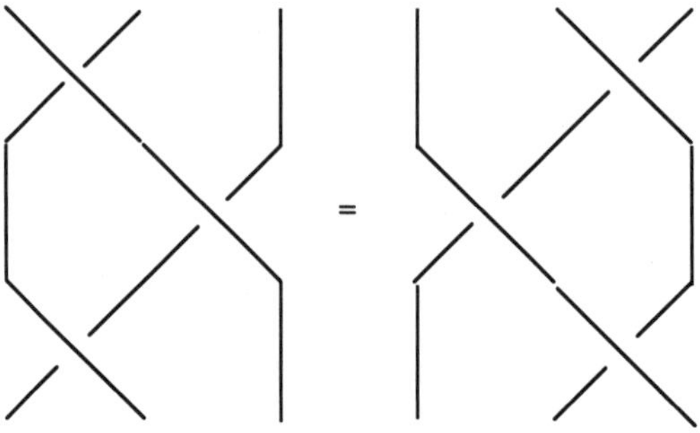

Figure 2 The Yang-Baxter Equation at the braid level

We worked on relating *topology, quantum computing, and quantum entanglement* through the use of R-matrices. In order to accomplish this aim, we have the following studied unitary R-matrices, interpreting them as *both* braidings *and* quantum gates.

The problem of finding unitary R-matrices turns out to be surprisingly difficult. Dye [6] has classified all such matrices of size 4×4, and we are still working on a general theory for the classification and of unitary R-matrices in other dimensions.

A key question about unitary R-matrices is to understand their capability of entangling quantum states. We use the criterion that $\phi = a|00\rangle + b|01\rangle + c|10\rangle + d|11\rangle$ is entangled if and only if $ad - bc \neq 0$. This criterion is generalized to higher dimensional pure states in the papers [18,19] by Kauffman and Lomonaco. We discovered families of R-matrices that detect topological linking if and only if they can entangle quantum states. A recent example in [29] is a unitary R-matrix that is highly entangling for quantum states. It takes the standard basis for the tensor product of two single-qubit spaces onto the Bell basis. On the topological side, R generates a non-trivial invariant of knots and links that is a specialization of the well-known link invariant, the Homflypt polynomial.

Entanglement and quantum computing are related in a myriad of ways, not the least of which is the fact that one can replace the $CNOT$ gate by another gate R and maintain universality (as described above) just so long as R can entangle quantum states. That is, R can be applied to some unentangled state to produce an entangled state. It is of interest to examine other sets of universal primitives that are obtained by replacing $CNOT$ by such an R.

We proved that certain solutions R to the Yang-Baxter equation together with local unitary two dimensional operators form a universal set of quantum gates. Results of this kind follow from general results of the Brylinskis [4] about universal quantum gates. The Brylinskis show that a gate R is universal in this sense, if and only if it can entangle a state that is initially unentangled. We show that *generically, the* 4×4 *solutions to the Yang-Baxter equation are universal quantum gates.*

For example, the following solutions to the Yang-Baxter equation are universal quantum gates (in the presence of local unitary transformations):

$$R = \begin{pmatrix} 1/\sqrt{2} & 0 & 0 & 1/\sqrt{2} \\ 0 & 1/\sqrt{2} & -1/\sqrt{2} & 0 \\ 0 & 1/\sqrt{2} & 1/\sqrt{2} & 0 \\ -1/\sqrt{2} & 0 & 0 & 1/\sqrt{2} \end{pmatrix}$$

$$R' = \begin{pmatrix} 1 & 0 & 0 & 0 \\ 0 & 0 & 1 & 0 \\ 0 & 1 & 0 & 0 \\ 0 & 0 & 0 & -1 \end{pmatrix}$$

$$R'' = \begin{pmatrix} a & 0 & 0 & 0 \\ 0 & 0 & b & 0 \\ 0 & b & 0 & 0 \\ 0 & 0 & 0 & a \end{pmatrix}$$

where a,b are unit complex numbers with $a^2 \neq b^2$.

R is the Bell-Basis change matrix, alluded to above. R' is a close relative to the swap-gate (which is not universal). R'' is both a universal gate and a useful matrix for topological purposes (it detects linking numbers). In this last example, we see a solution to the Yang-Baxter equation that detects topological linking exactly when it entangles quantum states.

These results about R-matrices are fundamental for understanding topological relationships with quantum computing, but they are only a first step in the direction of topological quantum computing. In topological quantum computing one wants to have all gates and compositions of gates intepreted as part of a single representation of the Artin Braid Group. By taking only a topological operator as a replacement for $CNOT$, we leave open the question of the topological interpretation of local unitary operators.

One must go on and examine braiding at the level of local unitary transformations and the problem of making fully topological models. The first step in this process (although made only recently and by us [23]) is to classify representations of the three-strand braid group into $SU(2)$. To go further involves finding brading representations into $U(2)$ that extend to dense representations in $U(N)$ for larger values of N. This is where topological quantum field theory comes into play. In the next section we outline our appproach to full topological quantum computation.

4. Topological Quantum Field Theory and Topological Quantum Computation

As described above, one comes to a barrier if one only attempts to construct individual topological gates for quantum computing. In order to go further, one must find ways to make global unitary representations of the Artin Braid Group. One way to accomplish this aim is via topological quantum field theory. Topological quantum field theory originated in the work of Witten [26] with important input from Atiyah [2]. This work opened up quantum field theoretic intepretations of the Jones polynomial (an invariant on knots and links, new at that time) and gave rise to new representations of the braid groups. The basic ideas of topological quantum field theory generalize concepts of angular momentum recombination in classical quantum physics. In [22,23] we use generalizations (so-called q-deformations) of the Penrose [24] formalism of spin networks to make models of topological quantum field theories that are finite dimensional, unitary and that produce dense representations of the braid group into the unitary group. These representations can be used to do quantum computing. In this way, we recover a version of the results of Freedman [7–11] and his collaborators and, by making very concrete representations, open the way for many applications of these ideas. Our methods are part of the approach to Witten's invariants that is constructed in the book of Kauffman and Lins [22]. This work is directly based on the combinatorial knot theory associated with the Jones polynomial. Thus our work provides a direct and fundamental relationship between quantum computing and the Jones polynomial.

Here is a very condensed presentation of how unitary representations of the braid group are constructed via topological quantum field theoretic methods. The structure described here is sometimes called the *Fibonacci model* [23,25,12]. One has a mathematical particle with label P that can interact with itself to produce either itself labeled P or itself with the null label $*$. When $*$ interacts with P the result is always P. When $*$ interacts with $*$ the result is always $*$. One considers process spaces where a row of particles labeled P can successively interact subject to the restriction that the end result is P. For example the space $V[(ab)c]$ denotes the space of interactions of three particles labeled P. The particles are placed in the positions a, b, c. Thus we begin with $(PP)P$. In a typical sequence of interactions, the first two P's interact to produce a $*$, and the $*$ interacts with P to produce P.

$$(PP)P \longrightarrow (*)P \longrightarrow P.$$

In another possibility, the first two P's interact to produce a P, and the P interacts with P to produce P.

$$(PP)P \longrightarrow (P)P \longrightarrow P.$$

It follows from this analysis that the space of linear combinations of processes $V[(ab)c]$ is two dimensional. The two processes we have just described can be taken to be the the qubit basis for this space. One obtains a representation of the three strand Artin braid group on $V[(ab)c]$ by assigning appropriate phase changes to each of the generating processes. One can think of these phases as corresponding to the interchange of the particles labeled a and b in the association $(ab)c$. The other operator for this representation corresponds to the interchange of b and c. This interchange is accomplished by a *unitary change of basis mapping*

$$F : V[(ab)c] \longrightarrow V[a(bc)].$$

If

$$A : V[(ab)c] \longrightarrow V[(ba)c]$$

is the first braiding operator (corresponding to an interchange of the first two particles in the association) then the second operator

$$B : V[(ab)c] \longrightarrow V[(ac)b]$$

is accomplished via the formula $B = F^{-1}AF$ where the A in this formula acts in the second vector space $V[a(bc)]$ to apply the phases for the interchange of b and c.

In this scheme, vector spaces corresponding to associated strings of particle interactions are interrelated by *recoupling transformations* that generalize the mapping F indicated above. A full representation of the Artin braid group on each space is defined in terms of the local intechange phase gates and the recoupling transfomations. These gates and transformations have to satisfy a number of identities in order to produce a well-defined representation of the braid group. These identities were discovered originally in relation to topological quantum field theory. In our approach [23] the structure of phase gates and recoupling transformations arise naturally from the structure of the bracket model for the Jones polynomial. Thus we obtain a knot-theoretic basis for topological quantum computing.

Many questions arise from this approach to quantum computing. The deepest question is whether there are physical realizations for the mathematical particle interactions that constitute such models. It is possible that such realizations may come about by way of the fractional quantum Hall effect or by other means. We are working on the physical basis for such models by addressing the problem of finding a global Hamiltonian for them, in analogy to the local Hamiltonians that can be constructed for solutions to the Yang-Baxter equation. We are also investigating specific ways to create and approximate gates in these models, and we are working on the form of quantum computers based on recoupling and braiding transformations.

These models are based on the structure of the Jones polynomial [13,15–17,21]. They lead naturally to the question of whether or not there exists a polynomial time quantum algorithm for computing the the Jones polynomial. The problem of computing the Jones polynomial is known to be classically P#-hard, and hence, classically computationally harder than NP-complete problems. Should such a polynomial time quantum algorithm exist, then it would be possible to create polynomial time quantum algorithms for any NP-complete problem, such as for example, the traveling salesman problem. This would indeed be a major breakthrough of greater magnitude than that arising from Shor's and Simon's quantum algorithms. The problem of determining the quantum computational hardness of the Jones polynomial would indeed shed some light on the very fundamental limits of quantum computation.

A polynomial time quantum algorithm (called the AJL algorithm) for approximating the value of the Jones polynomial $L(t)$ at primitive roots of unity can be found in [14]. We are currently writing a paper [23] that shows that this algorithm can not sucessfully be extended by polynomial interpolation to a polynomial time quantum algorithm for computing the Jones polynomial. However, there is a loop hole. It may well still be possible to modify the AJL algorithm in such a way that it can be used to create a polynomial time algorithm for $L(t)$. We propose to investigate why this is or is not the case. Our objective is to come to a better understanding of the exact divide between classical and quantum algorithms.

References

1. P.K. Aravind, Borromean entanglement of the GHZ state. in *Potentiality, Entanglement and Passion-at-a-Distance*, R. S. Cohen et al, (eds.) Kluwer, 1997, pp. 53–59.
2. M.F. Atiyah, *The Geometry and Physics of Knots*, Cambridge University Press, 1990.
3. R.J. Baxter *Exactly Solved Models in Statistical Mechanics*, Academic Press, 1982.
4. J-L Brylinski and R. Brylinski, Universal Quantum Gates, in [5] Chapman & Hall/CRC, Boca Raton, Florida, 2002, 101–116.
5. R. Brylinski and G. Chen, *Mathematics of Quantum Computation*, Chapman & Hall/CRC Press, Boca Raton, Florida, 2002.
6. H. Dye, Unitary solutions to the Yang-Baxter equation in dimension four, quant-ph/0211050, v3 1, August 2003.
7. M. Freedman, A magnetic model with a possible Chern-Simons phase, quant-ph/0110060v1 9 Oct 2001, (2001), preprint
8. M. Freedman, *Topological Views on Computational Complexity*, Documenta Mathematica - Extra Volume ICM, 1998, pp. 453–464.
9. M. Freedman, M. Larsen, and Z. Wang, A modular functor which is universal for quantum computation, quant-ph/0001108v2, 1 Feb 2000.
10. M. H. Freedman, A. Kitaev, Z. Wang, Simulation of topological field theories by quantum computers, quant-ph/0001071.
11. M. Freedman, Quantum computation and the localization of modular functors, quant-ph/0003128.
12. A. Kitaev, Anyons in an exactly solved model and beyond. arxiv:cond-mat/0506438.
13. V.F.R. Jones, A polynomial invariant for links via von Neumann algebras, Bull. Amer. Math. Soc. **129** (1985), 103–112.
14. D. Aharonov, V. Jones, and Z. Landau, On the quantum algorithm for approximating the Jones polynomial, preprint.
15. L.H. Kauffman, State models and the Jones polynomial, Topology **26** (1987), 395–407.
16. L.H. Kauffman, Statistical mechanics and the Jones polynomial, AMS Contemp. Math. Series **78** (1989), 263–297.

17. L.H. Kauffman, *Knots and Physics*, World Scientific Pub. Co. 1991, 1994, 2001.
18. L.H. Kauffman and S.J. Lomonaco Jr., Quantum entanglement and topological entanglement, New Journal of Physics **4** (2002), 73.1–73.18 (http://www.njp.org/).
19. L.H. Kauffman and S.J. Lomonaco Jr., Entanglement Criteria - Quantum and Topological, in *Quantum Information and Computation - Spie Proceedings, 21-22 April, 2003, Orlando, FL*, Donkor, Pinch and Brandt (eds.), Volume 5105, pp. 51–58.
20. L.H. Kauffman, Quantum topology and quantum computing, in *Quantum Computation*, S. Lomonaco (ed.), AMS PSAPM/58, 2002, pp. 273–303.
21. L.H. Kauffman, Quantum computation and the Jones polynomial , in *Quantum Computation and Information*, S. Lomonaco, Jr. (ed.), AMS CONM/305, 2002, pp. 101–137.
22. L.H. Kauffman, *Temperley-Lieb Recoupling Theory and Invariants of Three-Manifolds*, Princeton University Press, Annals Studies **114** (1994).
23. L.H. Kauffman and S. Lomonaco, q-Deformed Spin Networks, Temperley Lieb Recoupling Theory and Anyonic Topological Quantum Computing, (paper in preparation).
24. R. Penrose, Angular momentum: An approach to Combinatorial Spacetime, In *Quantum Theory and Beyond*, edited by T. Bastin, Cambridge University Press (1969).
25. J. Preskill, Topological computing for beginners, (slide presentation), Lecture Notes for Chapter 9 - Physics 219 - Quantum Computation. *http://www.iqi.caltech.edu/ preskill/ph219*
26. E. Witten, Quantum field theory and the Jones polynomial, Commun. Math. Phys. **121** (1989), 351–399.
27. Louis H. Kauffman and Samuel J. Lomonaco, Jr. Quantum entanglement and topological entanglement. New Journal of Physics 4 (2002), pp. 73.1 - 73.18.
28. L.H. Kauffman and S.J. Lomonaco Jr.,Entanglement Criteria - Quantum and Topological, in *Quantum Information and Computation - Spie Proceedings, 21-22 April, 2003, Orlando, FL*, Donkor, Pirich and Brandt (eds.), Volume 5105, pp. 51–58.
29. Louis H. Kauffman and Samuel J. Lomonaco, Braiding Operators are Universal Quantum Gates, New Journal of Physics 6 (2004) 134, pp. 1-39.
30. Louis H. Kauffman and Samuel J. Lomonaco Jr., Quantum knots, in *Quantum Information and Computation II, Proceedings of Spie, 12 -14 April 2004* (2004), ed. by Donkor Pirich and Brandt, pp. 268-284.
31. Louis H. Kauffman, Yong Zhang and Mo Lin Ge), Yang–Baxterizations, Universal Quantum Gates and Hamiltonians, *Quantum Information Processing*, Vol 4. No. 3, August 2005, pp. 159 - 197. quant-ph/0502015.
32. Louis H. Kauffman, Yong Zhang and Mo Lin Ge, Universal Quantum Gates, Yang–Baxterization and Hamiltonian, quant-ph/0412095, (to appear in IJQI - World Scientific)
33. Samuel J. Lomonaco and Louis H. Kauffman, On quantum algorithms for the Jones polynomial. (in preparation).

Generalized Cohomologies and Differential Forms of Higher Order

Richard Kerner

Laboratoire LPTMC, Université Paris-VI - CNRS UMR-7600
Tour 24, 4-ème étage, Boîte 121
4 Place Jussieu, 75252 Paris Cedex 5, France
e-mail: rk@ccr.jussieu.fr

We show how the condition $d^N = 0$ replacing the usual exterior differential's property $d^2 = 0$ leads to a natural generalization of cohomology. The case of $N = 3$ is analyzed in more detail, and simple algebraic realizations are constructed. The notion of a connection 1-form and the corresponding covariant differential are generalized, too. A Z_N-graded differential calculus is introduced and the corresponding gauge invariants of higher order are defined.

1. Introduction

This paper presents the results obtained in a series of papers ([2]), ([4]), ([5]), ([7]) in which a framework for the $d^N = 0$, $N \geq 2$ generalization of classical exterior differential calculus (with $d^2 = 0$) has been introduced and developed.

Our starting point consists in introducing a Z_N-graded algebra A of generalized exterior forms with an associative mutliplication rule. There are no particular conditions imposed on this product, except for the case when the result attains the highest degree, i.e. N: if ω is a form of degree p and θ is a form of degree $N-p$, we must have

$$\omega\,\theta = q^{|\omega|}\,\theta\,\omega, \tag{1.1}$$

where $|\omega|$ denotes the degree of the form ω, and q is an N-th primitive root of unity satisfying $1 + q + q^2 + ... + q^{N-1} = 0$. Next, we introduce a differential operator on the algebra of abstract $p-forms$ satisfying the following q-Leibniz rule:

$$d\,(\omega\,\theta) = (d\omega)\,\theta + q^{|\omega|}\,d\theta. \tag{1.2}$$

We also impose the condition $d^N = 0$. The immediate consequence of this assumption is the appearance of an entirely new set of generalized differentials $d^2 f$, $d^3 f$,..., up to $d^{(N-1)} f$, where f is a 0-form. In the simplest case of $N = 3$ we have to add the second-order forms $d^2 x^i$ to the usual set of first-order differentials dx^i.

The following obvious questions should be answered:

1) How the notion of cohomology can be generalized ?
2) How to find simple realizations of such calculus ?
3) How Stokes' theorem should be generalized ?
4) What are the analogs of covariant derivation, linear connection and curvature ?
5) How a gauge theory can be developed within this formalism ?

These questions have been partly or fully answered in the aforementioned series of papers co-authored with M. Dubois-Violette, V. Abramov and B. Niemeyer ([6]), ([2]), ([3]), ([4]), ([5]), ([7]). In what follows, we shall display shortly the main results in this novel field yet to be explored.

2. Generalized Cohomology

The usual definition of cohomology is related to the definition of quotient spaces $Ker(d)/Im(d)$ i.e. the forms whose exterior differential vanishes ("closed forms") but which are not differentials. Now, assuming that the higher-order differentials do not vanish, we can define new cohomology spaces. For example, for $N = 3$ we have not only first-order differentials df, but also the second-order forms $d^2 f$, whereas $d^3 f = 0$ identically. Obviously, $Im(d) \subset Ker(d^2)$, $Im(d^2) \subset Ker(d)$, and we define two space of forms that are "2-closed" but not "2-exact": $H^{(2)} = Ker(d^2)/Im(d)$, and the more usual $H^{(1)} = Ker(d)/Im(d^2)$,

Of course, each type of cohomology spaces contains a whole series of subspaces labeled by another index related to the degree of forms we are considering; thus, we should have $H^{(1)}_\alpha$ and $H^{(2)}_\alpha$ with the subscript α covering the range of values including all possible degrees.

In the case of arbitrary $N > 2$ one has $H^{(k)} = Ker(d^k)/Im(d^{N-k})$ Let l and m such that $l + m \leq N$. One obviously has $H^{(0)} = \{0\}$, $H^{(N)} = \{0\}$. The natural inclusion

$$i^l : Ker(d^m) \subset Ker(d^{(l+m)}$$

induces a linear mapping $[i^l] : H^{(m)} \rightarrow H^{(l+m)}$, since $Im(d^{N-m}) \subset Im(d^{N-(l+m)})$ On the other hand, one has $d^m\,(Ker(d^{(l+m)})) \subset Ker(d^l)$ and $d^m\, Im(d^{N-(l+m)}) \subset Im(d^{N-l})$ therefore the operator d^m induces a linear mapping $[d^m] : H^{(l+m)} \to H^l$.

Due to the above identities and definitions the following "magic hexagon" can be drawn, in which all sequences are exact:

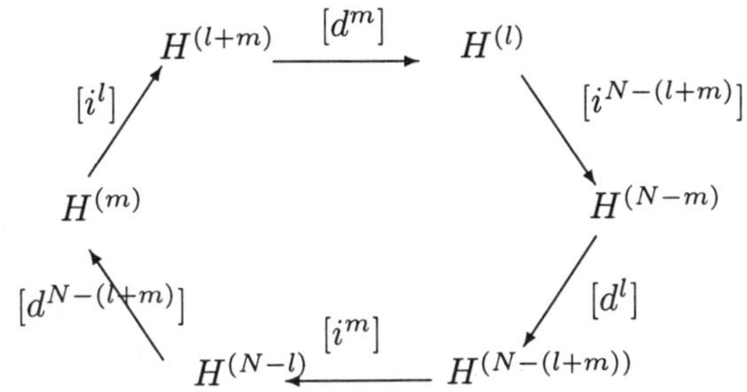

The simplest example is provided by the non-commutative geometry of the 3×3 complex matrices. Let $q = e^{\frac{2\pi i}{3}}$ so that $q^3 = 1$.

An atrbitrary matrix $B \in A = \mathbf{M}_3(\mathbf{C})$ can be decomposed into three parts with corresponding grades 0, 1 and 2 as follows: $A = A_0 + A_1 + A_2$

$$\left\{\begin{pmatrix} x & 0 & 0 \\ 0 & y & 0 \\ 0 & 0 & z \end{pmatrix}\right\} \subset A_0, \quad \left\{\begin{pmatrix} 0 & \alpha & 0 \\ 0 & 0 & \beta \\ \gamma & 0 & 0 \end{pmatrix}\right\} \subset A_1, \quad \left\{\begin{pmatrix} 0 & 0 & \xi \\ \lambda & 0 & 0 \\ 0 & \rho & 0 \end{pmatrix}\right\} \subset A_2. \tag{2.1}$$

Under ordinary matrix multplication the grades add up modulo 3. In non-commutative geometry of matrices the infinite-dimensional commutative algebra of smooth functions on a manifold is replaced by the non-commutative algebra A. The Z_3-graded differential is defined as a q-deformed commutator: for any $B \in A = \mathbf{M}_3(\mathbf{C})$, we define

$$d_q B = \eta\, B - q^{|b|} B\, \eta, \quad \text{with} \quad \eta = \begin{pmatrix} 0 & 1 & 0 \\ 0 & 0 & 1 \\ 1 & 0 & 0 \end{pmatrix} \tag{2.2}$$

where $\mid b \mid = grade(B)$. One checks easily that $d_q^3 B = 0$ for any B, and that

$$Ker(d_q) = \left\{ \begin{pmatrix} 1 & 0 & 0 \\ 0 & 1 & 0 \\ 0 & 0 & 1 \end{pmatrix}, \begin{pmatrix} 0 & q & 0 \\ 0 & 0 & q^2 \\ 1 & 0 & 0 \end{pmatrix}, \begin{pmatrix} 0 & 0 & 1 \\ q^2 & 0 & 0 \\ 0 & q & 0 \end{pmatrix} \right\}$$

whereas

$$Ker(d_q^2) = Ker(d_q) \oplus \left\{ \begin{pmatrix} 1 & 0 & 0 \\ 0 & q & 0 \\ 0 & 0 & q^2 \end{pmatrix}, \begin{pmatrix} 0 & q^2 & 0 \\ 0 & 0 & q \\ 1 & 0 & 0 \end{pmatrix}, \begin{pmatrix} 0 & 0 & 1 \\ 1 & 0 & 0 \\ 0 & 1 & 0 \end{pmatrix} \right\}$$

It is easy to check that

$$Im(d_q^2) = Ker(d_q), \quad \text{and} \quad Im(d_q) = Ker(d_q^2),$$

so that all cohomology spaces are trivial (contain just the zero element).

3. Stokes' theorem

In classical differential geometry the Stokes theorem can be written in a compact form as follows:

$$\int_C d\omega = \int_{\partial C} \omega, \tag{3.1}$$

where ω is a p-form and C is a $(p+1)$-chain, so that ∂C is a p-chain and $d\omega$ is a $(p+1)$-form, and we have $\partial^2 C = 0$ and $d^2\omega = 0$.

In the case when $d^2 \neq 0$ but $d^N = 0$ instead, the straightforward generalization of Stokes' formula (3.1) should read

$$\int_C d^{N-1}\omega = \int_{\partial C} d^{N-2}\omega = ... = \int_{\partial^2 C} d^{N-3}\omega = ... \int_{\partial^{N-1}C} \omega, \tag{3.2}$$

where ω is a p-form and C is an $(N+p-1)$-chain.

Let us show the realization of this formula on integration defined over the Z_3-graded matrix geometry introduced in the previous section. Because the integration over a p-chain can be considered as linear functional acting in the space of p-forms, in the case of matrix geometry the definition of a p-chain becomes unique: it is a matrix from $A = \mathbf{M}_N(\mathbf{C})$ of degree p, too. Then the integral of a p-form ω over a p-chain C is defined as a trace of $Tr(C^T\omega)$. The grade of C^T is equal to $(2p)_{\text{mod}\,3}$, so that the matrix $C^T\omega$ is diagonal (of grade 0). The matrices belonging to the grade 1 and grade 2 subspaces are traceless. Let us prove the formula (3.2) with the integration of matrix forms over matrix chains. What remains to be defined is the boundary of a p-chain. We postulate the following:

$$\partial C = \eta^T C - q^{2_m idc|} C\eta^T, \tag{3.3}$$

It is enough to prove that $Tr(C^T\, d\omega) = Tr((\partial C)^T\, \omega)$. We have:

$$Tr(C^T\, d\omega) = Tr(C^T\eta\omega - q^{|\omega|}C^T\omega\eta) =)Tr(C\eta\omega) - q^{|\omega|}\,Tr(C\omega\eta),$$

$$Tr(\partial C\omega) = Tr((\eta^T C - q^{2|c|}\, C\eta^T)^T\omega), = Tr(C^T\eta\omega) - q^{2|c|}\,\eta C^T\omega) \tag{3.4}$$

The first terms coincide, while the second terms in both expressions are also identical because the trace of any poduct of matrices is invariant under cyclic permutations. This completes the proof, and from here, the generalized Stokes' formula 3.2 is obtained by iteration.

4. Z_3-graded differentials calculus on manifolds

Consider a manifold described locally by a set of real coordinates $\{\xi^i\}$, $i = 1, 2, ..., n$. We postulate that the differential df of a function f coincides by definition with the usual one:

$$df = \frac{\partial f}{\partial \xi^k} d\xi^k = (\partial_k f)\, d\xi^k \tag{4.1}$$

When computing formally higher-order differentials, we shall suppose that our exterior differential operator d obeys the Z_N-graded Leibniz rule:

$$d\,(\omega\,\phi) = d\,\omega\,\phi + q^{\deg(\omega)}\,\omega\,d\,\phi, \tag{4.2}$$

where we suppose that q is an N-th order root of unity, instead of -1 in the Z_2-graded case, and that the grades add up modulo N under the associative multiplication of forms; the functions are of grade 0, and the operator d raises the grade of a form by 1, which means that the linear operator d applied to ξ^k produces a 1-form whose Z_N-grade is 1 by definition; when applied two times, by iteration, it will produce a new entity, which we shall call a *1-form of grade* 2, denoted by $d^2\xi^k$. Finally, we require that $d^N = 0$.

Let F denote the algebra of functions of n variables $C^\infty(\xi^k)$, over which the Z_N-graded algebra generated by the forms $d\xi^i$, $d^2\xi^k$, $d^3\xi^k$, etc., behaves as a *left* module. In other words, we shall multiply the forms $d\xi^i$, $d^2\xi^k$, $d\xi^i d\xi^k$.., by the functions *on the left* only; right multiplication will just not be considered here. We shall write by definition

$$d(\xi^i\xi^k) := \xi^i d\xi^k + \xi^k d\xi^i. \tag{4.3}$$

This amounts to suppose that the coordinates (functions) commute with the 1 forms, but do not necessarily commute with the forms of higher order.

From now on, we shall consider the simplest example of such structure when $N = 3$ and $q = e^{\frac{2\pi i}{3}}$.

With the Z_3-graded Leibniz rule established in 4.2 the postulate $d^3 = 0$ imposes certain ternary and binary commutation rules on the differentials $d\xi^i$ and $d^2\xi^k$. Consider the differentials of a function of the coordinates ξ^k:

$$\begin{aligned} df &:= (\partial_i f) d\xi^i \;\; ; \;\; d^2 f := (\partial_k\partial_i f) d\xi^k d\xi^i + (\partial_i f) d^2\xi^i \; ; \\ d^3 f &= (\partial_m\partial_k\partial_i f) d\xi^m d\xi^k d\xi^i + (\partial_k\partial_i f) d^2\xi^k d\xi^i \\ &\quad + q(\partial_k\partial_i f) d\xi^i d^2\xi^k + (\partial_k\partial_i f) d\xi^k d^2\xi^i \; ; \end{aligned}$$

(we remind that the last part of the differential, $(\partial_i f)d^3\xi^i$, vanishes by virtue of the postulate $d^3\xi^i = 0$). Supposing that partial derivatives commute, exchanging the summation indices i et k in the last expression and replacing $1+q$ by $-q^2$, we arrive at the following two conditions that lead to the vanishing of $d^3 f$:

$$d\xi^m d\xi^k d\xi^i + d\xi^k d\xi^i d\xi^m + d\xi^i d\xi^m d\xi^k = 0 \, d^2\xi^k d\xi^i - q^2 d\xi^i d^2\xi^k = 0. \quad (4.4)$$

which leads in turn to the following choice of relations:

$$d\xi^i d\xi^k d\xi^m = q d\xi^k d\xi^m d\xi^i, \quad \text{and} \quad d\xi^i d^2\xi^k = q d^2\xi^k d\xi^i. \qquad (4.5)$$

Strictly speaking, the above formulae hold only for the *symmetric* part of the above expression; we choose to impose stronger relations in order to make the resulting space of forms finite-dimensional.

Extending these rules to *all* the expressions with a well-defined grade, and applying the associativity of the Z_3-exterior product, we see that all products of the type $d\xi^i d\xi^k d\xi^m d\xi^n$ and $d\xi^i d\xi^k d^2\xi^m$ must vanish, and along with them, also the monomials of higher order containing these as factors. The proof is straightforward: consider the algebra of forms spanned by the basis of n forms of degree 1, θ^a, $a, b, ... = 1, 2, ...n$. Let us form a product 4 such forms, $\theta^a\theta^b\theta^c\theta^d$. As we have now $\theta^a\theta^b\theta^c = q\,\theta^b\theta^c\theta^a$, we can use this formula to evaluate several permutations:

$$(\theta^a\theta^b\theta^c)\theta^d = q\,\theta^b(\theta^c\theta^a\theta^d) = q^2\,\theta^b(\theta^a\theta^d\theta^c) = q^3\,\theta^a(\theta^d\theta^b\theta^c) = q^4\,\theta^a\theta^b\theta^c\theta^d$$

and as $q^4 = q \neq 1$, the four-product $\theta^a\theta^b\theta^c\theta^d$ must be zero.

Still, this is not sufficient in order to satisfy the rule $d^3 = 0$ on all the forms spanned by the generators $d\xi^1$ and $d^2\xi^k$. It can be proved without much pain that the expressions containing $d^2\xi^i d^2\xi^k$ must vanish, too; so we set forward the additional rule declaring that any expression containing *four or more* operators d must identically vanish. With this set of rules we can check that $d^3 = 0$ on all forms, whatever their grade or degree.

The dimension D of this module is

$$D = \frac{n^3 + 6n^2 + 5n}{3},$$

As a matter of fact, we have n first order differentials dx^i. There are n^2+n monomials spanning the module of 2-forms because we have n^2 independent products $dx^i dx^j$ and n second order differentials d^2x^i. The number of monomials spanning the module of 3-forms is $(n^3-n)/3+n^2$ since there are $(n^3-n)/3$ independent monomials $dx^i dx^j dx^k$ and n^2 independent monomials $dx^i d^2x^j$. Summing all these one finally obtains the dimension of the module $\Omega(U)$

Although we have described the construction of the algebra $\Omega(U)$ only in the case $N = 3$ it can be extended to any integer

$N > 3$. In this case our algebra is generated by the differentials $d\xi^1, \ldots, d\xi^n, \ldots, d^{N-1}\xi^1, \ldots, d^{N-1}\xi^n$.

The operator d satisfies the q-graded Leibniz rule and if we require that $d^N = 0$, we should impose the following minimal set of generalized commutation rules on the products of forms of order N:

$$d\xi^{k_1} d\xi^{k_2} \ldots d\xi^{k_N} = q\, d\xi^{k_2} \ldots d\xi^{k_N} d\xi^{k_1} = q^2\, d\xi^{k_3} \ldots d\xi^{k_N} d\xi^{k_1} d\xi^{k_2} .., \tag{4.6}$$

In the above formula (4.6) one can insert a higher-order differential $d^\alpha \xi^k$, $\alpha = 2, .., N-1$ replacing a product of α 1-forms, and the formula should still hold, e.g. $d\xi^j d^{N-1}\xi^k = q\, d\xi^j d^{N-1}\xi^k$ As a corollary, one can conjecture that for $N \geq 3$ any product of more than N such 1-forms must vanish, and the proof is similar as in the casee $N = 3$.

5. Linear connection and curvature

Let $\{\mathbf{e}_k\}$ denote the set of N independent vectors defined at any point of our space (which we suppose locally isomorphic with R^N), forming a basis. We define the *covariant differential* of $\mathbf{e}_k$ by means of the *covariant derivatives* of the $\mathbf{e}_k$ which define the *connection coefficients* Γ^l_{ik}:

$$\nabla\, \mathbf{e}_k = \nabla_i\, \mathbf{e}_k\, d\xi^k \;=\; \Gamma^l_{ik}\, \mathbf{e}_l\, d\xi^i \tag{5.1}$$

Now, when applying this operation second time, we get:

$$\nabla^2\, \mathbf{e}_k = \partial_m\, \Gamma^l_{ik} \mathbf{e}_l\, d\xi^m d\xi^i + \Gamma^l_{ik} (\nabla_m\, \mathbf{e}_k)\, d\xi^m d\xi^i + \Gamma^l_{ik}\, \mathbf{e}_l\, d^2\, \xi^i \tag{5.2}$$

which in view of the definition of $\nabla_m \mathbf{e}_k$ can be written as:

$$\nabla^2\, \mathbf{e}_k = \left(\partial_m\, \Gamma^l_{ik} + \Gamma^l_{mj}\, \Gamma^j_{ik} \right) \mathbf{e}_l\, d\xi^m d\xi^i + \Gamma^l_{ik}\, \mathbf{e}_l\, d^2 \xi^i \tag{5.3}$$

In the usual differential geometry we would set by definition $d^2\xi^k = 0$, and $d\xi^i\, d\xi^k \;=\; -\, d\xi^k\, d\xi^i$, which automatically leads to the well-known expression

$$\nabla^2\, \mathbf{e}_k = R^l_{mi\,k}\, \mathbf{e}_l\, d\xi^m \wedge d\xi^m \tag{5.4}$$

with

$$R^l_{mi\,k} = \partial_m \Gamma^l_{ik} - \partial_i \Gamma^l_{mk} + \Gamma^l_{mj}\, \Gamma^j_{ik} - \Gamma^l_{ij}\, \Gamma^j_{mk},$$

Here we do not assume $d^2\, \xi^k = 0$ anymore, neither a particular symmetry of the tensorial product of the differenitals $d\,\xi^k \otimes d\,\xi^m$. Therefore we must write instead:

$$\nabla^2\, \mathbf{e}_k = \left(\frac{1}{2}\, R^l_{mi\,k} + \frac{1}{2}\, P^l_{mi\,k} \right) \mathbf{e}_l\, d\xi^m\, d\xi^i + \Gamma^l_{ik}\, \mathbf{e}_l\, d^2\, \xi^i \tag{5.5}$$

with a new entity

$$P^{l}_{mi\,k} = \partial_m \Gamma^{l}_{ik} + \partial_i \Gamma^{l}_{mk} + \Gamma^{l}_{mj} \Gamma^{j}_{ik} + \Gamma^{l}_{ij} \Gamma^{j}_{mk}$$

Note that $P^{l}_{mi\,k}$ does not transform as a tensor under a change of coordinates, but obeys instead a non-homogeneous transformation law, like the connection coefficients.

However, if we compute the *third* covariant derivative of $\mathbf{e}_k$, $\nabla^3\,\mathbf{e}_k$, supposing the differentials obey the Z_3-graded exterior algebra defined in previous section, we get the following expression:

$$\nabla^3\,\mathbf{e}_k = R^{l}_{mi\,k}\,\mathbf{e}_l\,d^2\xi^m\,d\xi^i + \frac{1}{2}\,[\nabla_n\,R^{l}_{im\,k} - \nabla_m\,R^{l}_{in\,k}]\,\mathbf{e}_l\,d\xi^n d\xi^i d\xi^m$$

$$+ \frac{i\sqrt{3}}{2}\,[\nabla_n\,R^{l}_{im\,k} + \nabla_m\,R^{l}_{in\,k}]\,\mathbf{e}_l\,d\xi^n d\xi^i d\xi^m. \qquad (5.6)$$

It is interesting to note that only *two* combinations of the covariant derivative of $R^{l}_{ik\,m}$ appear here; as a matter of fact, the third one, $\nabla_i\,R^{l}_{mn\,k}$ is linearly dependent by virtue of Bianchi identity.

The expression for $\nabla^3\,\mathbf{e}_k$ in the case of Z_3-graded differential calculus defined above has a clear geometrical meaning. In the usual (Z_2-graded) case, the vanishing of the expression $\nabla^2\mathbf{e}_k$ was equivalent with a zero-curvature condition, $R^{l}_{im\,k} = 0$; here, the vanishing of $\nabla^3\,\mathbf{e}_k$ also implies vanishing curvature, however, another invariant and interesting condition can be formulated, i.e.

$$\nabla^3\,\mathbf{e}_k = R^{l}_{im\,k}\,\mathbf{e}_l$$

implying *constant* curvature condition satisfied in symmetric spaces.

6. A Z_3-graded Gauge Field Model

Let A be the above associative Z_3-graded matrix algebra with unit element, and let H be a free left module over this algebra. Let A be a grade 1 matrix belonging to our algebra.

We shall introduce the *covariant differential* of a form Φ as usual:

$$D\Phi := d\Phi + A\Phi; \qquad (6.1)$$

If the module is a free one, any of its elements Φ can be represented by an appropriate element of the algebra acting on a fixed element of H, so that one can always write $\Phi = B\Phi_o$; then the action of the group of automorphisms of H can be translated as the action of the same group on the algebra A.

Let U be a function defined on M with its values in the group of the automorphisms of H. The definition of a covariant differential is equivalent with the requirement $DU^{-1}B = U^{-1}DB$; as in the usual case, this leads to the following well-known transformation for the connection 1-form A :

$$A \Rightarrow U^{-1}AU + U^{-1}dU; \tag{6.2}$$

But here, unlike in the usual theory, the second covariant differential $D^2\Phi$ is not an automorphism: as a matter of fact, we have:

$$D^2\Phi = d(d\Phi + A\Phi) + A(d\Phi + A\Phi) = d^2\Phi + dA\Phi + (1+q)\,Ad\Phi + A^2\Phi; \tag{6.3}$$

the expression containing $d^2\Phi$ and $d\Phi$; whereas $D^3\Phi$ is an automorphism indeed, because it contains only Φ multiplied on the left by an algebra-valued 3-form:

$$D^3\Phi = d(D^2\Phi) + A(D^2\Phi), \tag{6.4}$$

which gives explicitly:

$$d(d^2\Phi + dA\Phi + qAd\Phi + A^2\Phi) + A(d^2\Phi + dA\Phi + qAd\Phi + Ad\Phi + A^2\Phi) \tag{6.5}$$

With a direct calculus one observes that all terms containing $d\Phi$ or $d^2\Phi$ simplify because of the identity $1 + q + q^2 = 0$, leaving only

$$D^3\Phi = (d^2A + d(A^2) + AdA + A^3)\Phi = (D^2A)\Phi := \Omega\Phi; \tag{6.6}$$

Obviously, because $D(U^{-1}\Phi) = U^{-1}(D\Phi)$, one also has:

$$D^3(U^{-1}\Phi) = U^{-1}(D^3\Phi) = U^{-1}\Omega\Phi = U^{-1}\Omega UU^{-1}\Phi,$$

which proves that the 3-form Ω transforms as usual, $\Omega \Rightarrow U^{-1}\Omega U$ when the connection transforms according to the law: $A \Rightarrow U^{-1}AU + U^{-1}dU$.

It can be also proved by a direct calculus that the curvature 3-form Ω does vanish identically for $A = U^{-1}dU$ (see [3], [7])).

Now let show how such a Z_3-graded gauge theory can be realized with our Z_3-graded differential forms on a manifold.

The curvature 3-form $\Omega = d^2A + d(A^2) + AdA + A^3$ is of grade 0; therefore it must be decomposed along the elements $d\xi^i d\xi^k d\xi^m$ and $d^2\xi^i d\xi^k$. Here is how we can compute its components in a local coordinate system. By definition, $A = A_i d\xi^i$, so we have:

$$dA = \partial_i A_k d\xi^i d\xi^k + A_k d^2\xi^k; \tag{6.7}$$

After replacing $1 + q$ by $-q^2$, and taking into account the relation $d\xi^k d^2\xi^i = q d^2\xi^i d\xi^k$, we get:

$$d^2A = (\partial_m \partial_i A_k) d\xi^m d\xi^i d\xi^k + (\partial_i A_k - \partial_k A_i) d^2\xi^i d\xi^k; \tag{6.8}$$

$$\text{Then,}\ d(A^2) + AdA = dAA + qAdA + AdA = dAA - q^2 AdA, \qquad (6.9)$$

due to the relations

$$d\xi^m d^2\xi^k = qd^2\xi^k d\xi^m \quad \text{and} \quad d\xi^m d\xi^i d\xi^k = qd\xi^i d\xi^k d\xi^m,$$

it can be shown by simple calculus that the curvature 3-form can be written in local coordinates as follows:

$$\Omega = d^2A + d(A^2) + AdA + A^3 = \Omega_{ikm} d\xi^i d\xi^k d\xi^m + F_{ik} d^2\xi^i d\xi^k \qquad (6.10)$$

$$\text{where}\ \Omega_{ikm} := \partial_i \partial_k A_m + A_i \partial_k A_m - \partial_k A_m A_i + A_i A_k A_m, \qquad (6.11)$$

$$\text{and}\ F_{ik} := \partial_i A_k - \partial_k A_i + A_i A_k - A_k A_i; \qquad (6.12)$$

F_{ik} is the 2-form of curvature (the field tensor) of usual gauge theories. We know that the expression F_{ik} is covariant with respect to the gauge transformations; on the other hand, the 3-form Ω is also covariant; therefore, the local expression Ω_{ijk} must be covariant, too. Indeed, it can be expressed as a combination of covariant derivatives of the 2-form F_{ik}:

$$\Omega_{ikm} = -\frac{1}{6}[D_i F_{mk} + D_k F_{mi}] + \frac{i\sqrt{3}}{6}[D_i F_{mk} - D_k F_{mi}] \qquad (6.13)$$

It is interesting that only two independent combinations appear here, the third one being determined automatically by the Bianchi identity.

References

1. M. Kapranov, *On the q-analog of homological algebra*, preprint, Cornell University (1992)
2. R. Kerner, *Z_3-graded differential calculus and Gauge Thories of Higher Order*, Lett. Math. Phys., **36**, p. 441-454 (1996)
3. R. Kerner, *The cubic chessboard*, Class.and Quant.Gravity, **14** 1(A), p. A203-A225 (1997)
4. M. Dubois-Violette and R. Kerner, *Universal q-differential calculus and the q-analog of homological algebra*, Acta Uniuversitatis Comenianae, (1996)
5. M. Dubois-Violette and R. Kerner, *Universal Z_N-graded differential calculus*, Journal of Geometry and Physics, **23** p. 235-246 (1997)
6. R. Kerner, V. Abramov, Reports on Math. Phys., 43 (1/2), 179-194, (1999).
7. R. Kerner, B. Niemeyer, Lett. in Math. Phys., 45,161-176, (1998).
8. M. Dubois-Violette, $d^N = 0$: Generalized homology, K-Theory, **14**, p. 371-404 (1998)
9. R. Kerner, *Z_3-graded algebras and the cubic root of Dirac equation*, Journal of Math. Phys., **33** (2), p. 403-411 (1992)
10. V. Abramov, R. Kerner, Journal of Math. Phys., **41** (3), (2000)

Periodic Cellular Automata and Bethe Ansatz

Atsuo Kuniba

Institute of Physics, University of Tokyo, Tokyo 153-8902, Japan
E-mail: atsuo@gokutan.c.u-tokyo.ac.jp

Akira Takenouchi

Institute of Physics, University of Tokyo, Tokyo 153-8902, Japan
E-mail: takenouchi@gokutan.c.u-tokyo.ac.jp

We review and generalize the recent progress in a soliton cellular automaton known as the periodic box-ball system. It has the extended affine Weyl group symmetry and admits the commuting transfer matrix method and the Bethe ansatz at $q = 0$. Explicit formulas are proposed for the dynamical period and the number of states characterized by conserved quantities.

1. Introduction

In [13], a class of periodic soliton cellular automata is introduced associated with non-exceptional quantum affine algebras. The dynamical period and a state counting formula are proposed by the Bethe ansatz at $q = 0$ [11]. In this paper we review and generalize the results on the $A_n^{(1)}$ case, where the associated automaton is known as the periodic box-ball system [14, 19]. The box-ball system was originally introduced on the infinite lattice without boundary [18, 17]. Here is a collision of two solitons with amplitudes 3 and 1 interchanging internal degrees of freedom with a phase shift:

$$\cdots 1114221111131111111111111111 \cdots$$
$$\cdots 1111114221113111111111111111 \cdots$$
$$\cdots 1111111114223111111111111111 \cdots$$
$$\cdots 1111111111114232111111111111 \cdots$$
$$\cdots 1111111111111121432111111111 \cdots$$
$$\cdots 1111111111111112111432111111 \cdots$$
$$\cdots 1111111111111111211111432111 \cdots$$

The system was identified [4, 3] with a solvable lattice model [1] at $q = 0$, which led to a direct formulation by the crystal base theory [5] and generalizations to the soliton cellular automata with quantum group symmetry [7, 6]. Here we develop the approach to the periodic case in [13] further by combining the commuting transfer matrix method [1] and the Bethe ansatz [2] at $q = 0$.

In section 2 we formulate the most general periodic automaton for $\mathfrak{g}_n = A^{(1)}$ in terms of the crystal theory. A commuting family of time evolutions $\{T^{(a)}_j\}$ is introduced as commuting transfer matrices under periodic boundary condition. The associated conserved quantities form an n-tuple of Young diagrams $m = (m^{(1)}, \ldots, m^{(n)})$, which we call the *soliton content.*

In section 3 we invoke the Bethe ansatz at $q = 0$ [11] to study the Bethe eigenvalue $\Lambda^{(r)}_l$ relevant to $T^{(r)}_l$. The Bethe equation is linearized into the string center equation and the $\Lambda^{(r)}_l$ is shown to be a root of unity. We also recall an explicit weight multiplicity formula obtained by counting the off-diagonal solutions to the string center equation [11]. It is a version of the fermionic formula called the combinatorial completeness of the string hypothesis at $q = 0$. These results are parameterized with the number of strings, which we call the *string content.*

In section 4 two applications of the results in section 3 are presented under the identification of the soliton and the string contents. First we relate the root of unity in the Bethe eigenvalue $\Lambda^{(r)}_l$ to the dynamical period of the periodic $A^{(1)}_n$ automaton under the time evolution $T^{(r)}_l$. Second we connect each summand in the weight multiplicity formula [11] to the number of states characterized by conserved quantities.

In [13], similar results have been announced concerning the highest states. Our approach here is based on conserved quantities and covers a wider class of states without recourse to the combinatorial Bethe ansatz at $q = 1$ [10]. We expect parallel results in general $\mathfrak{g}_n$. In fact all the essential claims in this paper make sense also for $\mathfrak{g}_n = D^{(1)}_n$ and $E^{(1)}_{6,7,8}$. Our formulas (3.9) and (3.11) include the results in [19] proved by a different approach as the case $\mathfrak{g}_n = A^{(1)}_1$ with $B = (B^{1,1})^{\otimes L}$ and $l = \infty$. For the standard notation and facts in the crystal theory, we refer to [5, 9, 6].

2. Periodic $A^{(1)}_n$ automaton

Let $B^{a,j}$ $(1 \le a \le n, j \in \mathbb{Z}_{\ge 1})$ be the crystal [9] of the Kirillov-Reshetikhin module $W^{(a)}_j$ over $U_q(A^{(1)}_n)$. Elements of $B^{a,j}$ are labeled with semistandard tableaux on an $a \times j$ rectangular Young diagram with letters $\{1, 2, \ldots, n+1\}$. For example when $n = 2$, one has $B^{1,1} =$

$\{1,2,3\}, B^{1,2} = \{11,12,13,22,23,33\}$, $B^{2,2} = \left\{ {\scriptstyle 11 \atop 22}, {\scriptstyle 11 \atop 23}, {\scriptstyle 11 \atop 33}, {\scriptstyle 12 \atop 23}, {\scriptstyle 12 \atop 33}, {\scriptstyle 22 \atop 33} \right\}$ as sets. $\mathrm{Aff}(B^{a,j}) = \{\zeta^d b \mid b \in B^{a,j}, d \in \mathbb{Z}\}$ denotes the affine crystal. The combinatorial R is the isomorphism of affine crystals $\mathrm{Aff}(B^{a,j}) \otimes \mathrm{Aff}(B^{b,k}) \xrightarrow{\sim} \mathrm{Aff}(B^{b,k}) \otimes \mathrm{Aff}(B^{a,j})$ [16]. It has the form $R(\zeta^d b \otimes \zeta^e c) = \zeta^{e+H}\tilde{c} \otimes \zeta^{d-H}\tilde{b}$, where $H = H(b \otimes c)$ is the energy function. We normalize it so as to attain the maximum at $H(u^{a,j} \otimes u^{b,k}) = 0$, where $u^{a,j} \in B^{a,j}$ denotes the classically highest element. We set $B = B^{r_1,l_1} \otimes B^{r_2,l_2} \otimes \cdots \otimes B^{r_L,l_L}$ and write $\mathrm{Aff}(B^{r_1,l_1}) \otimes \cdots \otimes \mathrm{Aff}(B^{r_L,l_L})$ simply as $\mathrm{Aff}(B)$. An element of B is called a state. Given a state $p = b_1 \otimes \cdots \otimes b_L \in B$, regard it as the element $\zeta^0 b_1 \otimes \cdots \otimes \zeta^0 b_L \in \mathrm{Aff}(B)$ and seek an element $v \in B^{r,l}$ such that $\zeta^0 v \otimes p \simeq (\zeta^{d_1} b'_1 \otimes \cdots \otimes \zeta^{d_L} b'_L) \otimes \zeta^e v$ under the isomorphism $\mathrm{Aff}(B^{r,l}) \otimes \mathrm{Aff}(B) \simeq \mathrm{Aff}(B) \otimes \mathrm{Aff}(B^{r,l})$. If such a v exists and $\zeta^{d_1} b'_1 \otimes \cdots \otimes \zeta^{d_L} b'_L$ is unique even if v is not unique, we say that p is (r,l)-*evolvable* and write $T^{(r)}_l(p) = b'_1 \otimes \cdots \otimes b'_L \in B$ and $E^{(r)}_l(p) = e = -d_1 - \cdots - d_L$. Otherwise we say that p is not (r,l)-evolvable or $T^{(r)}_l(p) = 0$. We formally set $T^{(r)}_l(0) = 0$. $E^{(r)}_l \in \mathbb{Z}_{\geq 0}$ holds under this normalization. Our $A^{(1)}_n$ automaton is a dynamical system on $B \sqcup \{0\}$ equipped with the family of time evolutions $\{T^{(r)}_l \mid 1 \leq r \leq n, l \in \mathbb{Z}_{\geq 1}\}$. $T^{(r)}_l$ is the $q = 0$ analogue of the transfer matrix in solvable vertex models. It is invertible and weight preserving on B. Using the Yang-Baxter equation of the combinatorial R, one can show (cf. [3, 6])

Theorem 2.1. *The commutativity* $T^{(a)}_j T^{(b)}_k(p) = T^{(b)}_k T^{(a)}_j(p)$ *is valid for any* $(a,j), (b,k)$ *and* $p \in B$, *where the both sides are either in* B *or* 0. *In the former case* $E^{(a)}_j(T^{(b)}_k(p)) = E^{(a)}_j(p)$ *and* $E^{(b)}_k(T^{(a)}_j(p)) = E^{(b)}_k(p)$ *hold.*

Thus, $\{E^{(a)}_j \mid 1 \leq a \leq n, j \in \mathbb{Z}_{\geq 1}\}$ is a family of conserved quantities.

Conjecture 2.1. *For any* $1 \leq a \leq n$ *and* $p \in B$, *there exists* $i \geq 1$ *such that* $T^{(a)}_k(p) \neq 0$ *if and only if* $k \geq i$. *The limit* $\lim_{k\to\infty} T^{(a)}_k(p) \in B$ *exists and* $E^{(a)}_i(p) < E^{(a)}_{i+1}(p) < \cdots < E^{(a)}_j(p) = E^{(a)}_{j+1}(p) = \cdots$ *holds for some* $j \geq i$.

Let $S_0, S_1, \ldots, S_n$ be the Weyl group operators [5] and pr be the promotion operator [16] acting on B component-wise. For instance for $A^{(1)}_3$, $\mathrm{pr}\left({\scriptstyle 223 \atop 334} \otimes 1344\right) = {\scriptstyle 133 \atop 444} \otimes 1124 \in B^{2,3} \otimes B^{1,4}$. They act on B as the extended affine Weyl group $\widetilde{W}(\mathfrak{g}_n) = \widetilde{W}(A^{(1)}_n) = \langle \mathrm{pr}, S_0, S_1, \ldots, S_n \rangle$.

Theorem 2.2. *If* $T^{(a)}_j(p) \neq 0$, *then for any* $w \in \widetilde{W}(A^{(1)}_n)$, *the relations* $wT^{(a)}_j(p) = T^{(a)}_j(w(p))$ *and* $E^{(a)}_j(w(p)) = E^{(a)}_j(p)$ *are valid.*

A state $p \in B$ is called *evolvable* if it is (a,j)-evolvable for any (a,j). In Conjecture 2.1, we expect that the convergent limit $T^{(a)}_\infty$ equals a translation in $\widetilde{W}(A^{(1)}_n)$. Compared with T_l in [13], the family $\{T^{(a)}_j\}$ here is more general and enjoys a larger symmetry $\widetilde{W}(A^{(1)}_n)$. Define the subset of B by

$$P(m) = \{p \in B \mid p : \text{evolvable}, E^{(a)}_j(p) = \sum_{k\geq 1} \min(j,k) m^{(a)}_k\}. \qquad (2.1)$$

Pictorially, $m = (m^{(1)}, \ldots, m^{(n)})$ is the n-tuple of Young diagrams and $E^{(a)}_k$ (resp. $m^{(a)}_k$) is the number of nodes in the first k columns (resp. number of length k rows) in $m^{(a)}$. We call m the *soliton content*.

Remark 2.1. $P(m)$ is $\widetilde{W}(A^{(1)}_n)$-invariant due to Theorem 2.2.

Given $p \in P(m)$, $T^{(a)}_j(p) \in P(m)$ is not necessarily valid. For instance $p = 112233 \in P(((22),(2))) \subset B = (B^{1,1})^{\otimes 6}$ but $T^{(2)}_1(p) = 213213$ is not evolvable since $(T^{(2)}_1)^2(p) = 0$. On the other hand one can show

Proposition 2.1. *If $p \in P(m)$ and $(T^{(a)}_j)^t(p) \neq 0$ for any t, then $(T^{(a)}_j)^t(p) \in P(m)$ for any t.*

Let W, Λ_a, α_a be the Weyl group, the fundamental weights and the simple roots of A_n, respectively. We specify $p^{(a)}_j = p^{(a)}_j(m)$ by (3.4) and set

$$\lambda(m) = \sum_{a=1}^{n} p^{(a)}_\infty \Lambda_a, \quad H = \{(a,j) \mid 1 \leq a \leq n, j \in \mathbb{Z}_{\geq 1}, m^{(a)}_j > 0\}. \quad (2.2)$$

Conjecture 2.2. *$P(m) \neq \emptyset$ if and only if $p^{(a)}_j \geq 0$ for all $(a,j) \in H$. $\{\mathrm{wt}p \mid p \in P(m)\} = W\lambda(m)$.*

The claim on the weights is consistent with the $\widetilde{W}(A^{(1)}_n)$-invariance of $P(m)$.

Here is an example of time evolutions in $B = B^{1,1} \otimes B^{1,1} \otimes B^{1,3} \otimes B^{1,1} \otimes B^{1,1} \otimes B^{1,1} \otimes B^{1,2}$ with $\widetilde{W}(A^{(1)}_3)$ symmetry. The leftmost column is $p_0, T^{(1)}_2(p_0), T^{(2)}_1 T^{(1)}_2(p_0)$ and $T^{(3)}_2 T^{(2)}_1 T^{(1)}_2(p_0)$ from the top to the bottom. At each time step, the states connected by the Weyl group actions S_0 and S_1 are shown, forming commutative diagrams. ($\cdot$ signifies $\otimes$.) All these states belong to $P(((3111),(21),(1)))$.

$2\cdot1\cdot233\cdot4\cdot1\cdot2\cdot12$	$\overset{S_0}{\mapsto}$	$2\cdot4\cdot233\cdot4\cdot1\cdot2\cdot24$	$\overset{S_1}{\mapsto}$	$1\cdot4\cdot133\cdot4\cdot1\cdot2\cdot14$
$1\cdot2\cdot123\cdot3\cdot4\cdot1\cdot22$		$4\cdot2\cdot234\cdot3\cdot4\cdot1\cdot22$		$4\cdot1\cdot134\cdot3\cdot4\cdot1\cdot12$
$1\cdot2\cdot112\cdot3\cdot2\cdot3\cdot24$		$1\cdot2\cdot244\cdot3\cdot2\cdot3\cdot24$		$1\cdot2\cdot144\cdot3\cdot1\cdot3\cdot14$
$2\cdot3\cdot112\cdot4\cdot2\cdot3\cdot12$		$2\cdot3\cdot244\cdot4\cdot2\cdot3\cdot12$		$2\cdot3\cdot144\cdot4\cdot1\cdot3\cdot11$

Remark 2.2. Let $R_j(1 \le j \le L-1)$ be the combinatorial R that exchanges the j-th and $(j+1)$-st components in $B = B^{r_1,l_1} \otimes \cdots \otimes B^{r_L,l_L}$ and $\pi(b_1 \otimes \cdots \otimes b_L) = b_L \otimes b_1 \otimes \cdots \otimes b_{L-1}$. Together with $R_0 = \pi^{-1}R_1\pi$, they act on $\cup_{s\in\mathfrak{S}_L} B^{r_{s_1},l_{s_1}} \otimes \cdots \otimes B^{r_{s_L},l_{s_L}}$ as the extended affine Weyl group $\widetilde{W}(A^{(1)}_{L-1}) = \langle \pi, R_0, \dots, R_{L-1}\rangle$. Theorem 2.2 is actually valid for $w \in \widetilde{W}(\mathfrak{g}_n = A^{(1)}_n) \times \widetilde{W}(A^{(1)}_{L-1})$ as in [8]. In the homogeneous case $(r_1, l_1) = \cdots = (r_L, l_L)$, the $\widetilde{W}(A^{(1)}_{L-1})$ symmetry shrinks down to the π-symmetry, which is the origin of the adjective "periodic".

3. Bethe ansatz at $q = 0$

Eigenvalues of row transfer matrices in trigonometric vertex models are given by the analytic Bethe ansatz [15, 12]. Let $Q_r(u) = \prod_k \sinh \pi(u - \sqrt{-1}u^{(r)}_k)$ be Baxter's Q-function where $\{u^{(a)}_j\}$ satisfy the Bethe equation eq.(2.1) in [11]. We set $q = e^{-2\pi\hbar}$ and $\zeta = e^{2\pi u}$. For the string solution ([11] Definition 2.3), the relevant quantity to our $T^{(r)}_l$ is the top term of the eigenvalue $\Lambda^{(r)}_l(u)$ (cf.[12] (2.12)):

$$\lim_{q\to 0} \frac{Q_r(u - l\hbar)}{Q_r(u + l\hbar)} = \zeta^{-E^{(r)}_l} \Lambda^{(r)}_l, \ \Lambda^{(r)}_l := \prod_{j\alpha} (-z^{(r)}_{j\alpha})^{\min(j,l)}. \tag{3.1}$$

Here $z^{(a)}_{j\alpha}$ is the center of the α-th string having color a and length j. Denote by $m^{(a)}_j$ the number of such strings. We call the data $m = (m^{(a)}_j)$ the *string content*. The product in (3.1) is taken over $j \in \mathbb{Z}_{\ge 1}$ and $1 \le \alpha \le m^{(r)}_j$. $E^{(r)}_l$ is given by the same expression as in (2.1) as the function of m. At $q = 0$ the Bethe equation becomes the string center equation ([11] (2.36)):

$$\prod_{(b,k)\in H} \prod_{\beta=1}^{m^{(b)}_k} (z^{(b)}_{k\beta})^{A_{aj\alpha,bk\beta}} = (-1)^{p^{(a)}_j + m^{(a)}_j + 1}, \tag{3.2}$$

$$A_{aj\alpha,bk\beta} = \delta_{ab}\delta_{jk}\delta_{\alpha\beta}(p^{(a)}_j + m^{(a)}_j) + C_{ab}\min(j,k) - \delta_{ab}\delta_{jk}, \tag{3.3}$$

$$p^{(a)}_j = \sum_{i=1}^{L} \min(j, l_i)\delta_{ar_i} - \sum_{(b,k)\in H} C_{ab}\min(j,k)m^{(b)}_k, \tag{3.4}$$

where $(C_{ab})_{1\le a,b\le n}$ is the Cartan matrix of A_n. To avoid a notational complexity we temporally abbreviate the triple indices $aj\alpha$ to j, $bk\beta$ to k and accordingly $z^{(b)}_{k\beta}$ to z_k etc. Then (3.1) and (3.2) read

$$\Lambda^{(r)}_l = \prod_k (-z_k)^{\rho_k}, \quad \prod_k (-z_k)^{A_{j,k}} = (-1)^{s_j}, \tag{3.5}$$

where ρ_k is given by $\rho_k = \delta_{br}\min(k,l)$ for k corresponding to $bk\beta$, and s_j is an integer. Note that $A_{j,k} = A_{k,j}$. Suppose that the $q=0$ eigenvalue satisfies $(\Lambda_l^{(r)})^{\mathcal{P}_l^{(r)}} = \pm 1$ for generic solutions to the string center equation *. It means that there exist integers ξ_j such that $\sum_j \xi_j A_{j,k} = \mathcal{P}_l^{(r)}\rho_k$, or equivalently $\xi_j = \mathcal{P}_l^{(r)}\frac{\det A[j]}{\det A}$, where $A[j]$ denotes the matrix $A = (A_{j,k})$ with its j-th column replaced by ${}^t(\rho_1, \rho_2, \ldots)$. In view of the condition $\forall \xi_j \in \mathbb{Z}$, the minimum integer allowed for $\mathcal{P}_l^{(r)}$ is $\mathcal{P}_l^{(r)} = \mathrm{LCM}\Big(1, \bigcup_k{}' \frac{\det A}{\det A[k]}\Big)$, where LCM stands for the least common multiple and $\cup_k'$ means the union over those k such that $A[k] \neq 0$. Back in the original indices, the determinants here can be simplified (cf. [11] (3.9)) to those of matrices indexed with H:

$$\mathcal{P}_l^{(r)} = \mathrm{LCM}\Big(1, \bigcup_{(b,k)\in H}{}' \frac{\det F}{\det F[b,k]}\Big), \tag{3.6}$$

where the matrix $F = (F_{aj,bk})_{(a,j),(b,k)\in H}$ is defined by

$$F_{aj,bk} = \delta_{ab}\delta_{jk}p_j^{(a)} + C_{ab}\min(j,k)m_k^{(b)}. \tag{3.7}$$

The matrix $F[b,k]$ is obtained from F by replacing its (b,k)-th column as

$$F[b,k]_{aj,cm} = \begin{cases} F_{aj,cm} & (c,m) \neq (b,k), \\ \delta_{ar}\min(j,l) & (c,m) = (b,k). \end{cases} \tag{3.8}$$

The union in (3.6) is taken over those (b,k) such that $\det F[b,k] \neq 0$.

The LCM (3.6) can further be simplified for $A_1^{(1)}, r=1, \forall r_i = \forall l_i = 1$. We write $p_j^{(1)}$ just as p_j and parameterize the set $H = \{j \in \mathbb{Z}_{\geq 1} \mid m_j^{(1)} > 0\}$ as $H = \{(0<)J_1 < \cdots < J_s\}$. Setting $i_k = \min(J_k, l)$ and $i_0 = 0$, one has

$$\mathcal{P}_l^{(1)} = \mathrm{LCM}\left(1, \bigcup_{k=0}^{t}{}' \frac{p_{i_{k+1}}p_{i_k}}{(i_{k+1}-i_k)p_{i_s}}\right), \tag{3.9}$$

where $0 \leq t \leq s-1$ is the maximum integer such that $i_{t+1} > i_t$.

Let us turn to another Bethe ansatz result, the character formula called combinatorial completeness of the string hypothesis at $q=0$ [11]:

$$\prod_{i=1}^{L} \mathrm{ch}B^{r_i,l_i} = \sum_m \Omega(m)e^{\lambda(m)}, \tag{3.10}$$

$$\Omega(m) = \det F \prod_{(a,j)\in H} \frac{1}{m_j^{(a)}}\binom{p_j^{(a)} + m_j^{(a)} - 1}{m_j^{(a)} - 1} \in \mathbb{Z}, \tag{3.11}$$

---

*$\mathcal{P}_l^{(r)}$ here should not be confused with the symbol in (3.4).

where $\binom{s}{t} = s(s-1)\cdots(s-t+1)/t!$ and $\mathrm{ch}B^{r,l}$ is the character of $B^{r,l}$. $\lambda(m)$, $p_j^{(a)}$ and F are defined by (2.2), (3.4) and (3.7). The sum in (3.10) extends over all $m_j^{(a)} \in \mathbb{Z}_{\geq 0}$ canceling out exactly leaving the character of B. (3.10) and (3.11) are the special cases of eq.(5.13) and eq.(4.1) in [11], respectively. $\Omega(m)$ (denoted by $R(\nu, N)$ therein) is the number of off-diagonal solutions to the string center equation with string content m. It is known ([11] Lemma 3.7) that $\Omega(m) \in \mathbb{Z}_{\geq 1}$ provided that $p_j^{(a)} \geq 0$ for all $(a,j) \in H$.

4. Dynamical period and state counting

In (2.1), the soliton content $m = (m_j^{(a)})$ is introduced as the conserved quantity associated with the commuting transfer matrices. One the other hand, the $m_j^{(a)}$ in the string content $m = (m_j^{(a)})$ is the number of strings of color a and length j in the Bethe ansatz in section 3. From now on we identify them motivated by the factor $\zeta^{-E_l^{(r)}}$ in (3.1) and some investigation of Bethe vectors at $q=0$. In view of Conjecture 2.2, the data of the form $m = (m_j^{(a)})$ is defined to be a *content* if and only if $p_j^{(a)} \geq 0$ for all $(a,j) \in H$. Thus $\det F > 0$ and $\lambda(m)$ in (2.2) is a dominant weight for any content m.

Conjecture 4.1. *If $p \in P(m)$ and $(T_l^{(r)})^t(p) \neq 0$ for any t, the dynamical period of p under $T_l^{(r)}$ (minimum positive integer t such that $(T_l^{(r)})^t(p) = p$) is equal to $\mathcal{P}_l^{(r)}$* (3.6) *generically and its divisor otherwise.*

In the situation under consideration, the whole $T_l^{(r)}$ orbit of p belongs to $P(m)$ due to Proposition 2.1. Naturally we expect $(\Lambda_l^{(r)})^{\mathcal{P}_l^{(r)}} = 1$, which can indeed be verified for $A_1^{(1)}$. Conjecture 4.1 has been checked, for example in $A_3^{(1)}$ case, for $B = (B^{1,1})^{\otimes 3} \otimes B^{2,2}$ and $B^{2,1} \otimes B^{2,1} \otimes B^{3,1} \otimes B^{3,2}$.

Let us present more evidence of Conjecture 4.1. To save the space, $\cdot = \otimes$ is dropped when $B = (B^{1,1})^{\otimes L}$. In each table, the period under $T_l^{(r)}$ with maximum l is equal to that under $T_\infty^{(r)}$.

$A_1^{(1)}$, state $= 1221121122221$, content $= ((321))$

(r,l)	LCM of				$=$ period
(1,1)	1,	13,	13,	13	13
(1,2)	1,	$\frac{91}{3}$,	$\frac{91}{16}$,	$\frac{91}{16}$	91
(1,3)	1,	91,	$\frac{273}{16}$,	$\frac{273}{107}$	273

$A_1^{(1)}$, state $= 122 \cdot 112 \cdot 12 \cdot 1222 \cdot 2 \cdot 11111 \cdot 1122 \cdot 111$, content $= ((4321))$

(r,l)	LCM of					= period
(1,1)	1,	2				2
(1,2)	1,	7,	$\frac{7}{2}$,	21,	42	42
(1,3)	1,	14,	7,	$\frac{21}{4}$,	$\frac{21}{2}$	42
(1,4)	1,	21,	$\frac{21}{2}$,	$\frac{63}{8}$,	$\frac{126}{29}$	126

$A_3^{(1)}$, state $= 134 \cdot 34 \cdot 1 \cdot 134 \cdot 23 \cdot 1 \cdot 13$, content $= ((432), (31), (1))$

(r,l)	LCM of							= period
(1,1)	1,	$\frac{380}{39}$,	$\frac{95}{6}$,	$\frac{95}{6}$,	$\frac{380}{31}$,	$\frac{380}{27}$,	$\frac{380}{29}$	380
(1,2)	1,	$\frac{190}{39}$,	$\frac{95}{12}$,	$\frac{95}{12}$,	$\frac{190}{31}$,	$\frac{190}{27}$,	$\frac{190}{29}$	190
(2,1)	1,	$\frac{190}{13}$,	$\frac{95}{4}$,	$\frac{95}{4}$,	$\frac{190}{137}$,	$\frac{190}{9}$,	$\frac{190}{73}$	190
(2,2)	1,	$\frac{76}{5}$,	$\frac{38}{3}$,	$\frac{38}{3}$,	$\frac{76}{41}$,	$\frac{76}{21}$,	$\frac{76}{31}$	76
(2,3)	1,	$\frac{95}{6}$,	$\frac{95}{11}$,	$\frac{95}{11}$,	$\frac{95}{34}$,	$\frac{95}{48}$,	$\frac{95}{41}$	95
(3,1)	1,	$\frac{380}{13}$,	$\frac{95}{2}$,	$\frac{95}{2}$,	$\frac{380}{137}$,	$\frac{380}{9}$,	$\frac{380}{263}$	380

$A_3^{(1)}$, state $= 233 \cdot \begin{matrix}12\\23\\34\end{matrix} \cdot \begin{matrix}11\\34\end{matrix} \cdot 1$, content $= ((3), (3), (2))$

(r,l)	LCM of				= period
(1,1)	1,	$\frac{11}{2}$,	11,	22	22
(1,2)	1,	$\frac{11}{4}$,	$\frac{11}{2}$,	11	11
(1,3)	1,	$\frac{11}{6}$,	$\frac{11}{3}$,	$\frac{22}{3}$	22
(2,1)	1,	11,	$\frac{33}{7}$,	$\frac{66}{7}$	66
(2,2)	1,	$\frac{11}{2}$,	$\frac{33}{14}$,	$\frac{33}{7}$	33
(2,3)	1,	$\frac{11}{3}$,	$\frac{11}{7}$,	$\frac{22}{7}$	22
(3,2)	1,	11,	$\frac{33}{7}$,	$\frac{33}{20}$	33

Here, content=((3111), (44), (2)) for example means that $m_1^{(1)} = 3, m_3^{(1)} = m_2^{(3)} = 1, m_4^{(2)} = 2$ and the other $m_j^{(a)}$'s are 0.

Let us turn to another application of the Bethe ansatz results (3.10) and (3.11). We introduce $\mathcal{T}(P(m)) = \bigcup_{a=1}^{n} \bigcup_{j \geq 1} \{T_j^{(a)}(p) \mid p \in P(m)\}$, which is the subset of B consisting of all kinds of one step time evolutions of $P(m)$. Under Conjecture 2.1, any state $p \in P(m)$ is (a, j)-evolvable for j sufficiently large. Thus from Proposition 2.1, p is expressed as $p = (T_j^{(a)})^k(p)$ for some k, showing that $\mathcal{T}(P(m)) \supseteq P(m)$. In general $\mathcal{T}(P(m))$

can contain non-evolvable states which do not belong to $P(m)$.

Conjecture 4.2. *For any content m such that $\mathcal{T}(P(m)) = P(m)$, the following relation holds:*

$$\Omega(m) = \frac{|P(m)|}{|W\lambda(m)|}. \tag{4.1}$$

In view of Remark 2.1 and Conjecture 2.2, the right hand side is the number of states in the periodic $A_n^{(1)}$ automaton having the content m and a fixed weight. Thus it is equal to $\sharp\{\mathrm{wt}p = \lambda(m) \mid p \in P(m)\}$. In case $\mathcal{T}(P(m)) \supsetneq P(m)$, we expect that $|P(m)|/|W\lambda(m)|$ is a divisor of $\Omega(m)$.

Let us present two examples of Conjecture 4.2. In the periodic $A_3^{(1)}$ automaton with $B = B^{1,2} \otimes B^{1,1} \otimes B^{1,2} \otimes B^{1,1}$, there are 1600 states among which 824 are evolvable. They are classified according to the contents m in the following table.

m	$\lambda(m)$	$\lvert W\lambda(m)\rvert$	$\lvert P(m)\rvert$	$\Omega(m)$
$(\emptyset, \emptyset, \emptyset)$	$(6,0,0,0)$	4	4	1
$((1), \emptyset, \emptyset)$	$(5,1,0,0)$	12	48	4
$((11), \emptyset, \emptyset)$	$(4,2,0,0)$	12	24	2
$((2), \emptyset, \emptyset)$	$(4,2,0,0)$	12	72	6
$((21), \emptyset, \emptyset)$	$(3,3,0,0)$	6	24	4
$((3), \emptyset, \emptyset)$	$(3,3,0,0)$	6	36	6
$((11), (1), \emptyset)$	$(4,1,1,0)$	12	96	8
$((22), (2), \emptyset)^*$	$(2,2,2,0)$	4	24	12
$((21), (1), \emptyset)$	$(3,2,1,0)$	24	432	18
$((111), (11), (1))$	$(3,1,1,1)$	4	16	4
$((211), (11), (1))$	$(2,2,1,1)$	6	48	8

In the second column, $(\lambda_1, \lambda_2, \lambda_3, \lambda_4)$ means $\lambda(m) = (\lambda_1 - \lambda_2)\Lambda_1 + (\lambda_2 - \lambda_3)\Lambda_2 + (\lambda_3 - \lambda_4)\Lambda_3$. In the last two cases, the subsets of $P(m)$ having the dominant weight $\lambda(m)$ are given by

$$\{11{\cdot}2{\cdot}13{\cdot}4,\ 12{\cdot}3{\cdot}14{\cdot}1,\ 13{\cdot}4{\cdot}11{\cdot}2,\ 14{\cdot}1{\cdot}12{\cdot}3\} \text{ for } m = ((111),(11),(1)),$$

$$\{11{\cdot}2{\cdot}23{\cdot}4,\ 12{\cdot}2{\cdot}13{\cdot}4,\ 12{\cdot}3{\cdot}24{\cdot}1,\ 13{\cdot}4{\cdot}12{\cdot}2,$$
$$14{\cdot}1{\cdot}22{\cdot}3,\ 22{\cdot}3{\cdot}14{\cdot}1,\ 23{\cdot}4{\cdot}11{\cdot}2,\ 24{\cdot}1{\cdot}12{\cdot}3\} \text{ for } m = ((211),(11),(1)).$$

In the case of $B = B^{2,1} \otimes B^{2,1} \otimes B^{2,2}$, there are 720 states among which 518 are evolvable.

m	$\lambda(m)$	$\lvert W\lambda(m)\rvert$	$\lvert P(m)\rvert$	$\Omega(m)$
$(\emptyset,\emptyset,\emptyset)$	$(4,4,0,0)$	6	6	1
$(\emptyset,(1),\emptyset)$	$(4,3,1,0)$	24	72	3
$(\emptyset,(11),(1))$	$(4,2,1,1)$	12	36	3
$(\emptyset,(2),\emptyset)$	$(4,2,2,0)$	12	48	4
$((1),(11),\emptyset)$	$(3,3,2,0)$	12	36	3
$((1),(11),(1))$	$(3,3,1,1)$	6	72	12
$((1),(21),(1))$	$(3,2,2,1)$	12	240	20
$((2),(22),(2))^*$	$(2,2,2,2)$	1	8	32

The assumption $\mathcal{T}(P(m)) = P(m)$ of the conjecture is valid for all the contents except $((22),(2),\emptyset)$ and $((2),(22),(2))$ marked with $*$.

References

1. R. J. Baxter, *Exactly solved models in statistical mechanics*, Academic Press, London (1982).
2. H. A. Bethe, *Z. Physik* **71** 205 (1931).
3. K. Fukuda, M. Okado, Y. Yamada, *Int. J. Mod. Phys.* **A15** 1379 (2000).
4. G. Hatayama, K. Hikami, R. Inoue, A. Kuniba, T. Takagi and T. Tokihiro, *J. Math. Phys.* **42** 274 (2001).
5. M. Kashiwara, *Duke Math. J.* **71** 839 (1993).
6. G. Hatayama, A. Kuniba, M. Okado, T. Takagi and Y. Yamada, *Contemporary Math.* **297** 151 (2002).
7. G. Hatayama, A. Kuniba, and T. Takagi, *Nucl. Phys.* **B577**[PM] 619 (2000), *J. Stat. Phys.* **102** 843 (2001), *J. Phys. A: Math. Gen.* **34** 10697 (2001).
8. K. Kajiwara, M. Noumi and Y. Yamada, *Lett. Math. Phys.* **60** 211 (2002).
9. S-J. Kang, M. Kashiwara, K. C. Misra, T. Miwa, T. Nakashima and A. Nakayashiki, *Duke Math. J.* **68** 499 (1992).
10. S.V.Kerov, A.N.Kirillov and N.Yu.Reshetikhin, *Zap. Nauch. Semin. LOMI.* **155** 50 (1986).
11. A. Kuniba and T. Nakanishi, *J. Alg.* **251** 577 (2002).
12. A. Kuniba and J. Suzuki, *Commun. Math. Phys.* **173** 225 (1995).
13. A. Kuniba and A. Takenouchi, Bethe ansatz at $q = 0$ and periodic box-ball systems, preprint (nlin.SI/0509001).
14. J. Mada, M. Idzumi and T. Tokihiro, *J. Math. Phys.* **46** 022701 (2005).
15. N. Yu. Reshetikhin, *Sov. Phys. JETP* **57** 691 (1983).
16. M. Shimozono, *J. Algebraic Combin.* **15** 151 (2002).
17. D. Takahashi, *Proceedings of the International Symposium on Nonlinear Theory and Its Applications (NOLTA '93)* 555 (1993).
18. D. Takahashi and J. Satsuma, *J. Phys. Soc. Jpn.* **59** 3514 (1990).
19. D. Yoshihara, F. Yura and T. Tokihiro, *J. Phys. A: Math. Gen.* **36** 99 (2003).

An L^2–Alexander–Conway Invariant for Knots and the Volume Conjecture *

Weiping Li

Department of Mathematics, Oklahoma State University
Stillwater, Oklahoma 74078-0613
U. S. A
E-mail address: wli@math.okstate.edu

Weiping Zhang

Chern Institute of Mathematics & LPMC
Nankai University
Tianjin 300071, P. R. China
E-mail address: weiping@nankai.edu.cn

Dedicated to the memory of Professor Shiing–Shen Chern

1. Introduction

In this paper, we focus on the L^2–Alexander invariant defined in [13, 14] from the twisted Alexander invariant point of view. The Alexander polynomial is a knot invariant discovered by J. W. Alexander [1] in 1928. The Alexander polynomial remained the only known knot polynomial until the Jones polynomial was discovered by V. Jones [8] in 1984. It is well–known that the Alexander polynomial plays an important role in the theory of knots.

The paper is organized as follows. In §1, we review the twisted Alexander polynomials. The necessary background on the L^2–invariant is given in §2. An L^2–analogue of the Alexander–Conway invariant for knots is pre-

*2000 *Mathematics Subject Classification.* Primary 57M25, 57M27; Secondary 58J52, 46L

Key words and Phases. Braid, Knot, Fuglede–Kadison determinant, Alexander–Conway invariant, L^2–Reidemeister torsion, volume conjecture.

W.L is partially supported by NSF and W.Z is partially supported by MOEC and the 973 project.

sented in §3. A possible relation between our L^2–Alexander invariant and the volume conjecture is discussed in the last section.

Let L be a link in S^3 with $\mu(L)$–components and exterior $X = S^3 \setminus L$. Let P be a base point of X and $p : \tilde{X} \to X$ be the maximal Abelian covering space with $\pi_1(X) \xrightarrow{\alpha} H_1(X) \cong \mathbb{Z}^{\mu(L)}$. The module $H_1(\tilde{X}, \mathbb{Z})$ depends only on the fundamental group of X. Any generator of the ideal order $H_1(\tilde{X}, \mathbb{Z})$ is called the *Alexander polynomial* $\Delta_L(t)$ of $\pi_1(X)$ (see [1]).

A twisted version of the Alexander polynomial has been introduced and studied first by Lin [15] from the Seifert surface point of view. Wada defined twisted Alexander polynomial via the free calculus method for Wirtinger presentations of knots in [22]. Using the twisted homology of the maximal Abelian covering space, Kirk and Livingston [11] defined a version of twisted Alexander polynomial via the ideal order in certain module.

Let ρ be a representation of $\pi_1(X)$ on a finitely generated free module V over some unique factorization domain R. Choosing a basis for V with $\dim_R V = N$, ρ can be realized as a homomorphism $\rho : \pi_1(X) \to \mathrm{Aut}(V) = GL_N(R)$. The associated ring homomorphism is

$$\overline{\rho} : \mathbb{Z}\pi_1(X) \to \mathbb{Z}GL_N(R) = M_N(R),$$

where $M_N(R)$ is the matrix algebra.

Let $\{x_1, \cdots, x_n | r_1, \cdots, r_m\}$ be a presentation of $\pi_1(X)$. The twisted version of Alexander polynomials defined in [22] is by working on the following group ring homomorphism

$$\mathbb{Z}F_n \xrightarrow{\psi} \mathbb{Z}\pi_1(X) \xrightarrow{\overline{\rho}\otimes\alpha} M_N(R) \otimes \mathbb{Z}G \cong M_N(R[t_1^{\pm 1}, \cdots, t_{\mu(L)}^{\pm 1}]). \tag{1.1}$$

Denote $\Phi = (\overline{\rho}\otimes\alpha)\circ\psi$ and $R[G] = R[t_1^{\pm 1}, \cdots, t_{\mu(L)}^{\pm 1}]$. The matrix $\Phi(\frac{\partial r_j}{\partial x_i})(1 \leq i \leq n, 1 \leq j \leq m)$ is called the Alexander matrix of $\pi_1(X)$ associated to the representation ρ. The matrix $\Phi(\frac{\partial r_j}{\partial x_i})$ is a presentation matrix of $H_1(\tilde{X}, \tilde{P})$ as $M_N(R[G])$–module. The twisted Alexander module of L associated to ρ is the $R[G]$–module $A(L, \rho) = H_1(\tilde{X}, \tilde{P}; R[G]^N)$.

For a Wirtinger presentation of $\pi_1(X)$ of the knot complement in S^3, one has $\pi_1(X) = \{x_1, \cdots, x_n | r_1, \cdots, r_{n-1}\}$ and hence each matrix M_j is a square matrix. So

$$\Delta_{L,\rho}(t_1, \cdots, t_{\mu(L)}) = \frac{\det M_j}{\det(\Phi(x_j) - \mathrm{Id})}, \tag{1.2}$$

where the matrix M_j is a $(n-1) \times (n-1)$ minor of the Jacobian $\Phi(\frac{\partial r_j}{\partial x_i})_{n\times n}$ for the Wirtinger presentation of a knot group.

The twisted Alexander polynomial $\Delta_{L,\rho}$ is independent of the choice of the presentation of $\pi_1(X)$ by Theorem 1 and Theorem 2 of [22]. The definition works for any finitely presentable group (see [22]). In general, the twisted Alexander polynomial is a rational function.

Note that both the Kinoshita–Terasaka knot and the Conway's 11 crossing knot have the same trivial Alexander polynomial and different twisted Alexander polynomial by [22]. Kitano [12] interprets these twisted invariants in terms of Reidemeister torsions along the lines in [18].

2. L^2–invariants

Let Γ be a finitely generated discrete (infinite) group. Let $l^2(\Gamma)$ be the standard Hilbert space of squared summable formal sums over Γ with complex coefficients. An element in $l^2(\Gamma)$ can be written as $a = \sum_{\gamma\in\Gamma} a_\gamma \gamma$ with $a_\gamma \in \mathbf{C}$ and $\sum_{\gamma\in\Gamma} |a_\gamma|^2 < +\infty$. If $a = \sum_{\gamma\in\Gamma} a_\gamma\gamma$ and $b = \sum_{\gamma\in\Gamma} b_\gamma\gamma$ are two elements in $l^2(\Gamma)$, then their inner product is given by $\langle a, b\rangle = \sum_{\gamma\in\Gamma} a_\gamma \bar{b}_\gamma$.

The left multiplication with elements in Γ defines a natural unitary action of Γ on $l^2(\Gamma)$. The group von Neumann algebra $\mathcal{N}(\Gamma)$ is the algebra of Γ–equivariant bounded linear operators from $l^2(\Gamma)$ to $l^2(\Gamma)$. The von Neumann trace on $\mathcal{N}(\Gamma)$ is defined by

$$\mathrm{Tr}_\tau : \mathcal{N}(\Gamma) \to \mathbf{C}, \quad f \mapsto \langle f(e), e\rangle, \tag{2.1}$$

where $e \in \Gamma \subset l^2(\Gamma)$ is the unit element. The right multiplication induces a natural action of Γ on $l^2(\Gamma)$ commuting with the left multiplication of Γ. Thus $\Gamma \subset \mathcal{N}(\Gamma)$. Moreover, for any $\gamma \in \Gamma \subset \mathcal{N}(\Gamma)$, $\mathrm{Tr}_\tau[\gamma] = 1$ if $\gamma = e$ and $\mathrm{Tr}_\tau[\gamma] = 0$ if $\gamma \neq e$. For any positive integer n, set $l^2(\Gamma)^{[n]} = \underbrace{l^2(\Gamma) \oplus \cdots \oplus l^2(\Gamma)}_{n}$.

We call $l^2(\Gamma)^{[n]}$ a free $\mathcal{N}(\Gamma)$–Hilbert module of rank n. A morphism between two free $\mathcal{N}(\Gamma)$–Hilbert modules is a Γ–equivariant bounded linear map between them. Let $f : l^2(\Gamma)^{[n]} \to l^2(\Gamma)^{[n]}$ be such a morphism. Let e_i $(i = 1, \cdots, n)$ be the unit element in the i–th copy of $l^2(\Gamma)$ in $l^2(\Gamma)^{[n]}$. Then we can extend the von Neumann trace in (2.1) to define

$$\mathrm{Tr}_\tau[f] = \sum_{i=1}^{n} \langle f(e_i), e_i\rangle. \tag{2.2}$$

The Fuglede–Kadison determinant $\mathrm{Det}_\tau(f)$ of f can be defined as follows:

(i) If f is invertible and f^* is the adjoint of f, then define (cf. [4, Definition] and [16, Lemma 3.15 (2)])

$$\mathrm{Det}_\tau(f) = \exp\left(\frac{1}{2}\mathrm{Tr}_\tau\left[\log\left(f^*f\right)\right]\right); \tag{2.3}$$

(ii) If f is injective, then define (cf. [4, Lemma 5] and [16, Lemma 3.15 (4), (5)])

$$\mathrm{Det}_\tau(f) = \lim_{\varepsilon\to 0+}\sqrt{\mathrm{Det}_\tau\left(f^*f+\varepsilon\right)} = \sqrt{\mathrm{Det}_\tau\left(f^*f\right)}. \tag{2.4}$$

(iii) If $f : l^2(\Gamma)^{[n]} \to l^2(\Gamma)^{[n]}$ is an invertible morphism, then there exists a C^1 path f_u, $u \in [0,1]$, of invertible morphisms such that $f_0 = f$, $f_1 = \mathrm{Id}$, and (cf. [4, Theorem 1 and Lemma 2]),

$$\log\left(\mathrm{Det}_\tau\left(f\right)\right) = -\mathrm{Re}\left(\int_0^1 \mathrm{Tr}_\tau\left[f_u^{-1}\frac{df_u}{du}\right]du\right). \tag{2.5}$$

Example: Let $\gamma \in \Gamma$ be of *infinite order*, and $|t| < 1$. It is clear that $\mathrm{Id} - t\gamma \in \mathcal{N}(\Gamma)$ is invertible and $\mathrm{Det}_\tau(\mathrm{Id} - t\gamma) = 1$ by (2.5) (cf. [13]).

Let (C_*, ∂) be a finite length $\mathcal{N}(\Gamma)$–chain complex

$$(C_*, \partial) : 0 \to C_n \xrightarrow{\partial_n} C_{n-1} \xrightarrow{\partial_{n-1}} \cdots \xrightarrow{\partial_1} C_0 \to 0, \tag{2.6}$$

where each C_i $(0 \le i \le n)$ is a (finite rank) $\mathcal{N}(\Gamma)$ free Hilbert module. Assume that (C_*, ∂) is weakly acyclic: $\ker(\partial_i) = \overline{\mathrm{Im}(\partial_{i-1})}$, $0 \le i \le n$. Let $\partial_i^* : C_{i-1} \to C_i$ be the adjoint of $\partial_i : C_i \to C_{i-1}$. Then $\partial_i\partial_i^* : \overline{\mathrm{Im}(\partial_i)} \to \overline{\mathrm{Im}(\partial_i)}$ is injective $(0 \le i \le n)$.

We call (C_*, ∂) is of *determinant class* if $\partial_i\partial_i^* : \overline{\mathrm{Im}(\partial_i)} \to \overline{\mathrm{Im}(\partial_i)}$ $(0 \le i \le n)$ is of *determinant class* (i.e. $\mathrm{Det}_\tau(\partial_i\partial_i^*|_{\overline{\mathrm{Im}(\partial_i)}}) > 0$). In this case, **the L^2–Reidemeister torsion** of (C_*, ∂) is defined to be a real number $T^{(2)}(C_*, \partial)$ given by (cf. [16, Definition 3.29])

$$\log T^{(2)}(C_*, \partial) = -\frac{1}{2}\sum_{i=0}^{n}(-1)^i \log \mathrm{Det}_\tau\left(\partial_i\partial_i^*|_{\overline{\mathrm{Im}(\partial_i)}}\right). \tag{2.7}$$

Let $\rho : \pi_1(X) \to GL(H)$ be an $\mathcal{N}(\Gamma)$–linear representation of $\Gamma = \pi_1(X)$ on a (finite rank) free $\mathcal{N}(\Gamma)$ Hilbert module, where X is a finite cell complex. Let $\widetilde{X}$ be the universal covering of X. Thus the chain complex $(C_*(\widetilde{X}) \otimes H, \widetilde{\partial})$ induces canonically a chain complex $(C_*(X, H_\rho), \partial_\rho)$ in the sense of (2.6) with $C_*(X, H_\rho) = (C_*(\widetilde{X}) \otimes_{\pi_1(X),\rho} H)$.

If $(C_*(X, H_\rho), \partial_\rho)$ is weakly acyclic and of *determinant class*, then its L^2–Reidemeister torsion $T^{(2)}(C_*(X, H_\rho), \partial_\rho)$ as in (2.7) is defined. If ρ :

$\pi_1(X) \to GL(H)$ is unitary, then $T^{(2)}(C_*(X, H_\rho), \partial_\rho)$ is a well–defined piecewise linear invariant.

Note that the L^2–Reidemeister torsion detects the unknot by [16, Theorem 4.7 (2)].

3. An L^2–Alexander–Conway invariant for knots

Combining the methods in §1 and §2, we provide the construction of an L^2–Alexander–Conway invariant for knots in this section. See [13,14] for more details.

Let $K \subset S^3$ be a knot. Let $P(r) = \{x_1, \cdots, x_k | r_1, \cdots, r_{k-1}\}$ be a Wirtinger presentation of $\Gamma = \pi_1(S^3 \setminus K)$. Let $\phi : F_k = \{x_1, \cdots, x_k\} \to \Gamma$ be the canonical map from the free group F_k to Γ.

Define α to be the canonical abelianization $\alpha : \Gamma \to U(1)$ with $\alpha(x_i) = t$ for $1 \leq i \leq k$. Let $GL(l^2(\Gamma))$ denote the set of invertible elements in $\mathcal{N}(\Gamma)$. Let $\rho_\Gamma : \Gamma \to GL(l^2(\Gamma))$ denote the fundamental representation of Γ, which is given by the right multiplication of the elements in Γ. The tensor product representation $\rho \otimes \alpha$ induces a ring homomorphism of the integral group rings

$$\widetilde{\rho_\Gamma \otimes \alpha} : \mathbf{Z}[\Gamma] \to \mathcal{N}(\Gamma) \otimes \mathbf{Z}[t^{\pm 1}] \subset \mathcal{N}(\Gamma). \tag{3.1}$$

Let $\Psi = (\widetilde{\rho_\Gamma \otimes \alpha}) \circ \widetilde{\phi} : \mathbf{Z}[F_k] \to \mathcal{N}(\Gamma)$ be the composition of the ring homomorphisms. Consider the morphism

$$A_{\rho_\Gamma \otimes \alpha} : l^2(\Gamma)^{[k-1]} \to l^2(\Gamma)^{[k]} \tag{3.2}$$

which when written in the $(k-1) \times k$ matrix form, the (i,j)–component is given by

$$A_{\rho_\Gamma \otimes \alpha, (i,j)} = \Psi \left(\frac{\partial r_i}{\partial x_j} \right) \in \mathcal{N}(\Gamma) \otimes \mathbf{Z}[t^{\pm 1}] \subset \mathcal{N}(\Gamma), \tag{3.3}$$

where $\frac{\partial r_i}{\partial x_j}$ is the standard Fox derivative.

We call $A_{\rho_\Gamma \otimes \alpha}$ the L^2–Alexander matrix of the presentation $P(\Gamma)$ associated to the fundamental representation ρ_Γ and the representation α. In [13], we proved the following proposition.

Proposition 3.1. *(1) $\Psi(x_j - 1) \in \mathcal{N}(\Gamma)$ is injective and has dense image for any $1 \leq j \leq k$.*

(2) If one of the $A^j_{\rho_\Gamma \otimes \alpha}$'s, $1 \leq j \leq k$, is injective, then every $A^j_{\rho_\Gamma \otimes \alpha}$, $1 \leq j \leq k$, is injective.

(3) For any $1 \leq j < j' \leq k$, *one has*

$$\mathrm{Det}_\tau\left(A^j_{\rho_\Gamma\otimes\alpha}\right)\mathrm{Det}_\tau\left(\Psi(x_{j'}-1)\right) = \mathrm{Det}_\tau\left(A^{j'}_{\rho_\Gamma\otimes\alpha}\right)\mathrm{Det}_\tau\left(\Psi(x_j-1)\right). \tag{3.4}$$

(4) $\mathrm{Det}_\tau\left(\Psi(x_j-1)\right) = 1$ *for* $1 \leq j \leq k$

(5) $\Delta^{(2)}_K(t) = \mathrm{Det}_\tau\left(A^1_{\rho_\Gamma\otimes\alpha}\right)$ *is independent of the choice of the Wirtinger presentation of the knot* K.

Thus we define $\Delta^{(2)}_K(t)$ to be the L^2*–Alexander invariant* of the knot K in S^3.

When $t = 1$, $\Delta^{(2)}_K(t)$ has been studied by Lück (see [16, Theorem 4.9]), who shows that $\Delta^{(2)}_K(1)$ is equivalent to the L^2–Reidemeister torsion of $S^3 \setminus K$. In [13], we identify $\Delta^{(2)}_K(t)$ with $t \in U(1)$ as certain twisted L^2–Reidemeister torsion of $S^3 \setminus K$ (see [13, Proposition 5.1]). In view of [22, Section 5], the above construction can also be applied to links. We also proved a rigidity result for the $U(1)$ twisted L^2–torsion on a knot complement in [13, Theorem 6.1].

By considering $\alpha : H_1\left(S^3 \setminus K\right) \longrightarrow \mathbf{C}^*$ with $\alpha(h) = t$, we can prove that $\mathrm{Det}_\tau\left(A^1_{\rho_\Gamma\otimes\alpha}\right)$ is well–defined up to the multiplicative group $\{|t|^p\}_{p\in\mathbf{Z}}$ (see [13]). However, one can resolve this $\{|t|^p\}$ ambiguity through the following theorem.

Theorem 3.2 (Li–Zhang 2005 [14]). *The quantity*

$$\Delta^{(2)}_K(t) = \sqrt{\frac{\mathrm{Det}_\tau\left(A^1_{\rho_\Gamma\otimes\alpha}\right)}{\max\{1,|t|\}}\cdot\frac{\mathrm{Det}_\tau\left(A^1_{\rho_\Gamma\otimes\alpha^{-1}}\right)}{\max\{1,|t|^{-1}\}}}$$

does not depend on the choice of the Wirtinger presentation of Γ. *Moreover, it depends only on* $|t|$.

Definition 3.3. The term $\Delta^{(2)}_K(t)$ in the above theorem is called an L^2–Alexander–Conway invariant of the knot K.

By the rigidity result in [14], this definition coincides with [13, Definition 3.5] for $t \in U(1)$.

Example. Let $K = 4_1$ be the figure eight knot with its Wirtinger presentation $P(\Gamma) = \left\langle x, y | zxz^{-1}y^{-1}\right\rangle$, where $z = x^{-1}yxy^{-1}x^{-1}$. Then one has

(1) If $|t| > 4$, then $\Delta^{(2)}_{4_1}(t) = \sqrt{t}$;

(ii) If $|t| = 1$, then $\Delta_{4_1}^{(2)}(t) = \exp\left(\frac{\mathrm{vol}(S^3\setminus 4_1)}{6\pi}\right) \sim \exp\left(\frac{1}{6\pi}\cdot 2.029\right) \neq 1$.

Thus $\Delta_{4_1}^{(2)}(t)$ is a non–trivial deformation of the hyperbolic volume of 4_1. It would be interesting to study the behavior of $\Delta_K^{(2)}(t)$ on $\mathbf{R}^*$.

Now let $\beta \in \mathbf{B}_k$ be a braid representative of the knot K. By Artin's theorem [3], the knot group Γ admits a presentation

$$\left\langle x_1, \ldots, x_k | \beta(x_1)x_1^{-1} = \cdots = \beta(x_{k-1})x_{k-1}^{-1} = 1\right\rangle .$$

By proceeding similarly as in the Wirtinger presentation case, one can define the L^2–Alexander matrix denoted now by $\widetilde{A}_{\rho_\Gamma\otimes\alpha}$, and define an L^2–Alexander–Conway invariant by, for $t \in \mathbf{C}^*$,

$$\widetilde{\Delta}_K^{(2)}(t) = \sqrt{\frac{\mathrm{Det}_\tau\left(\widetilde{A}^1_{\rho_\Gamma\otimes\alpha}\right)}{\max\{1,|t|\}}\cdot\frac{\mathrm{Det}_\tau\left(\widetilde{A}^1_{\rho_\Gamma\otimes\alpha^{-1}}\right)}{\max\{1,|t|^{-1}\}}}.$$

Theorem 3.4 (Li–Zhang 2005 [14]). *(i) The L^2–Alexander–Conway invariant $\widetilde{\Delta}_K^{(2)}(t)$ does not depend on the braid representative β for the knot K. So it defines an invariant for K.*

(ii) For $t \in U(1)$, $\widetilde{\Delta}_K^{(2)}(t) = \Delta_K^{(2)}(t)$ $\left(= \Delta_K^{(2)}(1)\right)$.

Theorem 3.4 indicates the interactive relation of our L^2–invariant on the braid representatives of the knot K. It can be viewed as an L^2–analogue of the Burau theorem [3, Theorem 3.11]. It is an interesting problem to answer our expectation $\widetilde{\Delta}_K^{(2)}(t) = \Delta_K^{(2)}(t)$. Note that $\widetilde{\Delta}_{4_1}^{(2)}(t) = \Delta_{4_1}^{(2)}(t) = t^{1/2}$, $\widetilde{\Delta}_{5_1}^{(2)}(t) = \Delta_{5_1}^{(2)}(t) = t^{3/2}$ for $|t| > 4$ and $\widetilde{\Delta}_{5_1}^{(2)}(1) = \Delta_{5_1}^{(2)}(1) = 1$.

4. The volume conjecture

The volume conjecture given by Kashaev [9] is derived from the theory of quantum dilogarithm to build a possible relation between the combinatorial TQFT to quantum $2+1$ dimensional gravity. H. and J. Murakami in [19] reinterpreted the Kashaev invariant in [9] as a special case of the colored Jones polynomial associated with the quantum group $SU_q(2)$ evaluated at $q = e^{2\pi i/N}$. The volume conjecture for any knot K in S^3 can be stated as the following,

$$\lim_{N\to\infty}\frac{\log|J_N(K,q)|}{N} = \frac{1}{2\pi}\mathrm{Vol}(S^3\setminus K),$$

where the volume is the simplicial volume. The volume conjecture is true for torus knots [10] and the figure eight knot [20]. See also [5,7] for related topics.

By [13, Proposition 5.1 and Theorem 6.1], the volume conjecture can be restated as follows (cf. [16, Conjecture 4.8]),

$$\lim_{N\to+\infty}\left|J_N\left(K,\exp\left(\frac{2\pi\sqrt{-1}}{N}\right)\right)\right|^{\frac{1}{3N}}=\Delta_K^{(2)}(1). \tag{4.1}$$

Using the 3–dimensional Chern–Simons theory with complex gauge groups $SL_2(\mathbb{C})$, Gukov [6] derived a generalized volume conjecture

$$\lim_{N\to\infty}\frac{\log J_N(K,q)}{N}=\frac{1}{2\pi}(\mathrm{Vol}(M)+i2\pi^2\mathrm{CS}(M)).$$

By comparing (4.1) with the Melvin–Morton conjecture (the Melvin–Morton conjecture was proved formally in [21] and rigorously in [2]), it seems plausible to view the volume conjecture as a kind of L^2–analogue of the Melvin–Morton conjecture. This fits with the picture outlined by Gukov in [6]. In particular, the rigidity property in [13, Theorem 6.1] fits with the form of the generalized Melvin–Morton conjecture stated in [6], where the hyperbolic torsion in the right hand side of [6, (6.30)] (which should play a role of the L^2–torsion, or the L^2–Alexander invariant here) does not contain a (unitary) deformed parameter.

We would like to end our article by listing some natural questions.

(Q1) Note that the generalized volume conjecture in (5.12) of [6] can be thought as a parametrized volume conjecture via the zero locus of the A-polynomial. Is our invariant $\Delta_K^{(2)}(t)$ related to the volume $\mathrm{Vol}(\rho)$ for $\rho:\Gamma\to SL_2(C)$ in the zero locus ?

(Q2) $\Delta_K^{(2)}(t)$ is upper semi–continuous with respect to $t\in C^*$. Whether it is a continuous function or with only first kind of discontinuity ? Whether $\Delta_K^{(2)}(t)\neq 0$ for all knots ?

(Q3) It would be interesting to give a topological proof of Lück–Schick's result in [17], identifying $\Delta_K^{(2)}(1)$ with the simplicial volume of $S^3\setminus K$, up to a constant scalar. Is there a direct proof by passing Lück–Schick's result ?

(Q4) Whether there is a knot polynomial whose Mähler measure equals to the L^2–Alexander invariant $\Delta_K^{(2)}(1)$ (or equivalently, the L^2–torsion of the knot complement) ? This is Question 8.1 of [13].

References

1. J. W. Alexander, *Topological Invariants of Knots and Links*, Trans. Amer. Math. Soc. 30, 275-306, 1928.
2. D. Bar-Natan and S. Garoufalidis, *On the Melvin-Morton-Rozansky conjecture*, Invent. Math. **1** (1996), 103-133.
3. J. S. Birman, *Braids, links and mapping class groups*, Ann. Math. Studies, No. **82**, Princeton University Press (1974).
4. B. Fuglede and R. V. Kadison, *Determinant theory in finite factors*, Ann. of Math. **55** (1952), 520-530.
5. S. Garoufalidis and T.Q. Le, *Asymptotics of the colored Jones function of a knot*, Preprint, arXiv: math.GT/0508100.
6. S. Gukov, *Three-dimensional quantum gravity, Chern-Simons theory, and the A-polynomial*, Commun. Math. Phys. **255** (2005), 577-627.
7. V. Huynh and T.Q. Le, *On the colored Jones polynomial and the Kashaev invariant*, Preprint, arXiv: math.GT/0503296.
8. V. F. R. Jones, *Hecke algebra representations of braid groups and link polynomials*, Annals of Math, **126** (1987), 335-388.
9. R. M. Kashaev, *The hyperbolic volume of knots from the quantum dilogarithm*, Lett. Math. Phys. **39** (1997), 269-275.
10. R. M. Kashaev and O. Tirkkonen, *A proof of the volume conjecture for torus knots*, translation in J. Math. Sci. (N. Y.) **115** (2003), No. 1, 2033–2036.
11. P. Kirk and C. Livingston, *Twisted Alexander invariants, Reidemeister torsion, and Casson-Gordon invariants*, Topology **38** (1999), 635-661.
12. T. Kitano, *Twisted Alexander polynomial and Reidemeister torsion*, Pacific J. Math., **174** (1996), 431-442.
13. W. Li and W. Zhang, *An L^2-Alexander invariants for knots*, to appear in Commun. Contemp. Math.
14. W. Li and W. Zhang, *An L^2-Alexander–Conway invariants for knots*, in preparation.
15. X.-S. Lin, *Representations of knot groups and twisted Alexander polynomials*, Acta Math. Sinica (English Series) **17** (2001), 361-380.
16. W. Lück, *L^2-Invariants: Theory and Applications to Geometry and K-Theory*. Springer-Verlag, 2002.
17. W. Lück and T. Schick, *L^2-torsion of hyperbolic manifolds of finite volume*, Geom. Funct. Anal. **9** (1999), 518-567.
18. J. Milnor, *A duality theorem for Reidemeister torsion*, Ann. of Math., **76** (1962), 137-147.
19. H. Murakami and J. Murakami, *The colored Jones polynomials and the simplicial volume of a knot*, Acta. Math., **186** (2001), 85-104.
20. H. Murakami, *The colored Jones polynomials of the figure-eight knot and the volumes of three-manifolds obtained by Dehn surgeries*, Fund. Math. **184** (2004), 269–289.

21. L. Rozansky, *A contribution of the trivial connection to the Jones polynomial and Witten's invariant of 3-manifolds*, Commun. Math. Phys., **175** (1996), 275-318.
22. M. Wada, *Twisted Alexander polynomial for finitely presented groups*, Topology **33** (1994), 241-256.

Faddeev Knots, Skyrme Solitons, and Concentration-Compactness*

Fanghua Lin

Courant Institute of Mathematical Sciences
New York University
New York, New York 10021

Yisong Yang

Department of Mathematics
Polytechnic University
Brooklyn, New York 11201

In this lecture, we present a series of existence theorems for the locally concentrated static solutions arising as the energy minimizers in the Faddeev model and the Skyrme model in relativistic quantum field theory.

1. The Faddeev knots

Brief Review. For mathematicians, knot theory has long been a theory of classification of knots by means of combinatorics and topology (Tait, Alexander, Jones, Witten, Vassiliev). Recently, there is considerable interest in realizing knots as the solution configurations of suitable quantum field theory models. Of these, the most interesting one that promises to provide a broad spectrum of knot phenomena is the Faddeev quantum field theory model [11] in which the knots are energy-minimizing solitons and characterized by the Hopf charge which is a topological index. Using computer simulation, Faddeev and Niemi [13,12] first produced a ring-shaped (unknotted) Hopf charge one soliton. Shortly after the seminal work of Faddeev and Niemi, a more extensive computer investigation was conducted

*Lecture at the 23rd International Conference on Differential Geometry Methods in Theoretical Physics, Nankai Institute of Mathematics, Tianjin, China, August 20 – 26, 2005

by Battye and Sutcliffe [2–4] who performed fully three-dimensional, highly convincing, computations for the solution configurations of the Hopf charge Q from $Q = 1$ up to $Q = 8$ and found that, for $Q = 1, 2, 3, 4, 5$, the energy-minimizing solitons are ring-shaped and higher charges cause greater distortion, and for $Q = 6, 7, 8$, the solitons become knotted or linked. In particular, the trefoil knot appears at $Q = 7$. The main aim of this talk is to present a series of existence theorems for such knotted solitons.

The Faddeev Model. Recall that, in normalized form, the action density of the Faddeev model [13,11,2–4] over the standard $(3 + 1)$-dimensional Minkowski space of signature $(+ - - -)$ reads $\partial_\mu \mathbf{n} \cdot \partial^\mu \mathbf{n} - \frac{1}{2} F_{\mu\nu}(\mathbf{n}) F^{\mu\nu}(\mathbf{n})$, where the field $\mathbf{n} = (n_1, n_2, n_3)$ assumes its values in the unit 2-sphere, i.e., $\mathbf{n}^2 = n_1^2 + n_2^2 + n_3^2 = 1$, and $F_{\mu\nu}(\mathbf{n}) = \mathbf{n} \cdot (\partial_\mu \mathbf{n} \wedge \partial_\nu \mathbf{n})$. Since $\mathbf{n}$ is parallel to $\partial_\mu \mathbf{n} \wedge \partial_\nu \mathbf{n}$, it is seen that $F_{\mu\nu}(\mathbf{n}) F^{\mu\nu}(\mathbf{n}) = (\partial_\mu \mathbf{n} \wedge \partial_\nu \mathbf{n}) \cdot (\partial^\mu \mathbf{n} \wedge \partial^\nu \mathbf{n})$, which may be identified with the well-known Skyrme term [21–24,28] when one embeds S^2 into $S^3 \approx SU(2)$. Hence, as observed by Cho [7], the Faddeev model may be viewed as a refined Skyrme model and the solution configurations of the former are the solution configurations of the latter with a restrained range. In what follows, we shall only be interested in static fields which make the Faddeev energy

$$E(\mathbf{n}) = \int_{\mathbb{R}^3} \left\{ \sum_{1 \le k \le 3} |\partial_k \mathbf{n}|^2 + \sum_{1 \le k < \ell \le 3} F_{k\ell}^2(\mathbf{n}) \right\} \mathrm{d}x \tag{1.1}$$

finite. The finite-energy condition implies that $\mathbf{n}$ approaches a constant vector $\mathbf{n}_\infty$ at spatial infinity (of $\mathbb{R}^3$). Hence we may compactify $\mathbb{R}^3$ into S^3 and view the fields as maps from S^3 to S^2. As a consequence, we see that each finite-energy field configuration $\mathbf{n}$ is associated with an integer, $Q(\mathbf{n})$, in $\pi_3(S^2) = \mathbb{Z}$. In fact, such an integer $Q(\mathbf{n})$ is known as the Hopf invariant.

The Faddeev Minimization Problem. The Faddeev knots are the solutions to the following topologically constrained minimization problem

$$E_m = \inf\{E(\mathbf{n}) \,|\, E(\mathbf{n}) < \infty, \; Q(\mathbf{n}) = m\}, \tag{1.2}$$

where m is an integer. The computer simulations in [13,2–4] are for $m = 1, 2, \cdots, 8$ for which the problem (1.2) is truncated over a finite large box which "approximates" the full space $\mathbb{R}^3$.

Main Difficulty. In his ICM 2002 address in Beijing, Faddeev [12] proposed the above problem to the mathematicians and noted that the main difficulty involved is the lack of compactness. In fact, this difficulty is not isolated and arises also in the general existence problem for topological solitons in other quantum field theory models. For example, recall that Belavin and Polyakov [5] were able to construct all static solitons characterized by an arbitrary topological charge (the Brouwer degree) for the σ-model modeling the spin vector orientation for a planar ferromagnet. The four-dimensional extension of this construction is of course the well-known resolution [19] of the classical Yang–Mills instantons realizing again any prescribed topological charge (the second Chern number). The common feature of these two soluble models is that they are both conformally invariant field theories. When conformal invariance becomes invalid, the above-described complete solvability may not be available. For example, except in the critical phase [14] between two types of superconductivity, people have not been able to establish for the Ginzburg–Landau theory on $\mathbb{R}^2$ the existence of an energy minimizer realizing a given quantized flux (the first Chern number), and a similar situation happens for the Chern–Simons theory [8,27]; except in the BPS limit [6,18,25], people have not been able to establish the existence of a Yang–Mills–Higgs monopole of any monopole number (the winding number); although there have been some works on the existence of energy-minimizing unit-charge Skyrme solitons [9,10,20,17], the proofs are problematic unfortunately.

Existence Theory. Now we state our main existence theorems for the fundamental minimization problem (1.2). First recall that the lower bound

$$E(\mathbf{n}) \geq C|m|^{3/4} \tag{1.3}$$

was derived a long time ago by Vakulenko and Kapitanski [26]. This lower bound ensures the existence of an integer $m_0 \neq 0$ such that

$$E_{m_0} = \min\{E_m \,|\, m \in \mathbb{Z} \setminus \{0\}\}.$$

For such m_0, we have

Theorem 1.1. *The problem (1.2) with $m = m_0$ has a solution.*

Furthermore, we have

Theorem 1.2. *There exists an infinite subset $\mathbb{S}$ of the set $\mathbb{Z}$ of all integers so that for any $m \in \mathbb{S}$ the problem (1.2) has a solution.*

Although we do not know how big the set $\mathbb{S}$ is or whether $\mathbb{S} \neq \mathbb{Z}$, we can make the above statement more precise as follows.

Theorem 1.3. *For any $m \in \mathbb{Z}$, there is a decomposition*

$$m = m_1 + \cdots + m_\ell, \quad m_s \in \mathbb{Z}, \quad s = 1, \cdots, \ell, \tag{1.4}$$

so that the following sub-additivity relation

$$E_m \geq E_{m_1} + \cdots + E_{m_\ell} \tag{1.5}$$

holds. In fact, all the integers $m_1, \cdots, m_\ell$ in (1.4) may be chosen to be members of the set $\mathbb{S}$. Besides, the sublinear growth (upper) bound

$$E_m \leq C|m|^{3/4}. \tag{1.6}$$

is valid. Here, in (1.6), C is a universal positive constant.

The above result may be interpreted physically as follows: If E_m is viewed as the mass of a particle of charge m and $E_{m_1}, \cdots, E_{m_\ell}$ are the masses of constituent particles or substances, then (1.4) is a charge conservation law and (1.5) says that the mass of the composite particle is greater than or equal to the sum of the masses of its constituents or substances because possible extra energy may be needed for the constituents or substances to form a bound state and, as a result, the composite particle may look "heavier". For this reason, we may call (1.5) "the Substantial Inequality", which will be seen to be a crucial technical ingredient of our method.

Comparing (1.3) and (1.6), we see that the sharp sublinear growth estimate $E_m \sim |m|^{3/4}$ holds asymptotically for a large Hopf charge $|m|$. It will be seen that the upper bound (1.6) is another crucial technical ingredient.

Significant difficulties arise when we attempt to gain further knowledge about the set $\mathbb{S}$ stated in Theorem 1.2 because a minimizing sequence of the problem (1.2) may fail to "concentrate" in $\mathbb{R}^3$. On the other hand, when we consider the problem over a bounded contractible domain, a more satisfactory result is valid because, technically, a bounded domain prohibits the minimizing sequence to "float" away and "concentration" is trivially guaranteed:

Theorem 1.4. *Let Ω be a bounded contractible domain in $\mathbb{R}^3$ and consider the admissible set of all the field configurations which assume a constant value on the boundary of and outside Ω. Then, over such an admissible set, the problem (1.2) has a solution for any $m \in \mathbb{Z}$.*

This theorem ensures the existence of the knotted solutions of respective Hopf charges obtained in [13,2–4] where the full space $\mathbb{R}^3$ is replaced by a large box in order to carry out computer simulations.

The Concentration-Compactness and Substantial Inequality. When one considers the minimization problem (1.2), one naturally encounters the three alternatives in the concentration-compactness principle due to P.-L. Lions for the minimizing sequence, namely,

(i) compactness (concentration of energy up to translations);
(ii) vanishing (energy density is flattened to zero everywhere);
(iii) dichotomy (energy splitting into floating chunks).

In order to achieve convergence (compactness), one usually needs to rule out (ii) and (iii) to arrive at (i). For our problem, however, it is impossible to rule out (iii) completely. Indeed, we are in a situation where we have to accept (iii) (splitting) and achieve something less than (i) (concentration). More precisely, we show that the energy at the worst would split into "topologically concentrated" floating chunks characterized by (1.4) and (1.5), which we referred to as the Substantial Inequality. We shall see later that this important inequality will allow us to obtain existence (hence convergence) indirectly.

The 3/4-Power Upper Bound and Knotted Solitons. We note that a profound implication of (1.6) is indeed the existence of knotted solitons at sufficiently high Hopf charges. To see this, we show that, for a large value of Q, a Faddeev energy minimizer prefers to appear as a clustered configuration (a knotted soliton) realizing the topology designated by Q than appear as a field configuration with widely separated energy lumps of a simpler topology (a multisoliton of a sum of unknotted solitons) realizing the same topology. Such a result may be illustrated most easily by showing that, if $m > 0$ is sufficiently large, a Faddeev energy minimizer with the Hopf invariant $Q = m$ can never be represented as a multisoliton of the sum of m widely separated solitons, each of a Hopf charge $Q = 1$ (an unknot). If the above described multisolitons were allowed, then, away from the local concentration regions of these unknots, the field configurations gave negligible contributions to the total energy. Hence, approximately, we would have $E_m \approx mE_1$, which contradicts (1.6) for large m. Therefore, unlike vortices, monopoles, instantons, and cosmic strings, which do not mind to stay apart at least at the BPS limit, the Faddeev knots prefer to stay together in a clustered structure. In other words, the Faddeev knots like to stay knotted.

Existence Theorems Obtained by the Substantial Inequality and the 3/4-Power Upper Bound. To prove Theorem 1.1, we write the decomposition $m_0 = m_1 + \cdots + m_\ell$, $m_s \in \mathbb{Z}$, $s = 1, \cdots, \ell$ so that $E_{m_0} \geq E_{m_1} + \cdots + E_{m_\ell}$ (splitting into ℓ floating chunks) $\geq \ell E_{m_0}$. Hence $\ell = 1$ (no splitting) and concentration-compactness is achieved. In other words, Theorem 1.1 is proved. In order to prove Theorem 1.2, we suppose otherwise that $\mathbb{S}$ is finite. Set $m^0 = \max\{m \in \mathbb{S}\}$. Again, since for any $m \in \mathbb{N}$, there is a decomposition $m = m_1 + \cdots + m_\ell$, $m_s \in \mathbb{S}$, $s = 1, \cdots, \ell$, we have $m \leq \ell m^0$. On the other hand, we also have $E_m \geq E_{m_1} + \cdots + E_{m_\ell} \geq \ell E_{m_0}$. Combining these two inequalities, we have $E_m \geq (E_{m_0}/m^0)m$, which contradicts the upper bound (1.6). This proves Theorem 1.2.

Proof of the 3/4-Power Upper Bound. We first recall the following fact: if $u \in C^1(\mathbb{R}^3, S^2)$ is such that $u(x)$ =constant for $|x|$ sufficiently large and that $v : S^2 \to S^2$ is a smooth map of degree $\deg(v)$, then the Hopf invariant of $\tilde{u} = v \circ u : \mathbb{R}^3 \to S^2$ satisfies $Q(\tilde{u}) = (\deg(v))^2 Q(u)$.

We begin by considering the case $m = n^2$, for a positive integer n. We decompose the upper hemisphere S^2_+ as $S^2_+ = \cup_{i=1}^n B(i) \cup D$. Here $B(i)$'s are mutually disjoint geodesic balls of radius $r \approx 1/\sqrt{n}$ inside S^2_+. We define a Lipschitz map $v : S^2 \to S^2$ as follows: $v(x) = (0, 0, 1)$ for all $x \in S^2 \backslash \cup_{i=1}^n B(i)$, and on each $B(i)$, v is such that $v|_{\partial B(i)} = (0, 0, 1)$, $v(B(i))$ covers S^2 exactly once, and $v : B(i) \to S^2$ is orientation-preserving. In other words, the degree of the map from $B(i)$ onto S^2 is exactly 1. We can further require that $\|\nabla v\|_{L^2(S^2)} \leq c\sqrt{n}$ for a positive constant c independent of n.

We then construct a map $h : \mathbb{R}^3 \to S^2$ such that h is a constant outside the ball $B_{\sqrt{n}}$, $\|\nabla h\|_{L^\infty(\mathbb{R}^3)} \leq c/\sqrt{n}$ for a constant c independent of n, and that $Q(h) = 1$.

Let $u = v \circ h \in B$. Then $Q(u) = n^2 = (\deg v)^2 = m$. On the other hand, $\|\nabla u\|_{L^\infty(\mathbb{R}^3)} \leq c^2$ and $u(x)$ is a constant for x outside the ball $B_{\sqrt{n}}$. Hence $E(u) \leq C(\sqrt{n})^3 = C|m|^{3/4}$.

For the general case, we have $n^2 \leq m < (n+1)^2$ for some positive integer n. We observe that $k = m - n^2 < (n+1)^2 - n^2 = 2n + 1$. Let $h_0 : B_1 \to S^2$ be a smooth map with $h_0|_{\partial B_1} = (0, 0, 1)$ and $Q(h_0) = 1$. Take k points $x_1, \cdots, x_k \in \mathbb{R}^3$ such that $|x_i| >> \sqrt{n}$ and that $|x_i - x_j| >> 1 + \sqrt{n}$ for all $i, j = 1, \cdots, k$, $i \neq j$. We then define $\tilde{u} : \mathbb{R}^3 \to S^2$ as follows: $\tilde{u}(x) = u(x) = (v \circ h)(x)$ for $x \in B_{\sqrt{n}}(0)$, $\tilde{u}(x) = h_0(x - x_i)$ for $x \in B_1(x_i)$, $i = 1, \cdots, k$, and $\tilde{u}(x) = (0, 0, 1)$ otherwise. Here u is constructed as in the case $m = n^2$ before. It is obvious that $\tilde{u}$ is a Lipschitz map from $\mathbb{R}^3$ into S^2 with $\tilde{u}(x) = (0, 0, 1)$ for $|x|$ large. Besides, $Q(\tilde{u}) = Q(u) + k = n^2 + k = m$. Moreover, $E(\tilde{u}) = E(u) + kE(h_0) \leq C_1(\sqrt{n})^3 \leq C_2 m^{3/4}$. We have thus proved (1.6) in the case that m is positive. For negative m, one simply needs to change orientation.

2. The Skyrme solitons

Brief Review. The Skyrme model [21–24] is a quantum field theory for baryons (including subatomic particles such as proton and neutron and hyperons). The static Skyrme energy has two terms which are similar to the Faddeev model and governs a map from $\mathbb{R}^3$ into $S^3 \approx SU(2)$. The Skyrme solitons are the energy minimizers among the topological class defined by a Brouwer degree, which is physically the baryon number of the system. Therefore, as will be seen below, the technical structure of the Skyrme model is similar to that of the Faddeev model. In particular, the existence of an absolute energy minimizer of the Skyrme model of unit baryon number cannot be proved via a direct application of the concentration-compactness principle as was originally conceived in [9,10] but can only be proved indirectly via a use of the Substantial Inequality. The second aim of this talk is to present a correct proof for the existence of a Skyrme soliton in the class of unit baryon number, which is in fact a by-product of our method for the Faddeev knot problem.

The Skyrme Model and Existence Theorem. The static Skyrme energy has the form [21–24]

$$E(\mathbf{n}) = \int_{\mathbb{R}^3} \left\{ \sum_{1\leq k\leq 3} |\partial_k \mathbf{n}|^2 + \sum_{1\leq k<\ell\leq 3} |\partial_k \mathbf{n} \wedge \partial_\ell \mathbf{n}|^2 \right\} \mathrm{d}x, \tag{2.1}$$

where the field configuration $\mathbf{n}$ maps $\mathbb{R}^3$ into S^3. Hence, similar to the Faddeev model, the relevant topological invariant is the Brouwer degree (when $\mathbf{n}$ is viewed as a map from $S^3 = \mathbb{R}^3 \cup \{\infty\}$ into itself) which may be represented as an integral as well, $\deg(\mathbf{n}) = \frac{1}{2\pi^2}\int_{\mathbb{R}^3} \det(\mathbf{n}, \nabla \mathbf{n})\,\mathrm{d}x$.

For any integer m, one is interested in the solvability of the optimization problem

$$E_m = \inf\{E(\mathbf{n}) \,|\, E(\mathbf{n}) < \infty,\ \deg(\mathbf{n}) = m\}. \tag{2.2}$$

It can be shown that all the existence results (except the (3/4)-growth law) parallel to those in Theorems 1.3 and 1.4 hold for the problem (2.2) and we skip their corresponding statements. Instead, we will only indicate that the following result holds for the Skyrme model as a corollary from our analysis:

Theorem 2.1. *For $m = \pm 1$, the problem (2.2) has a solution.*

Proof of Theorem by Substantial Inequality. It suffices to consider the case $m = 1$. Recall that there is a decomposition $1 = m_1 + \cdots + m_\ell, m_s \in \mathbb{Z}, s = 1, \cdots, \ell$, so that $E_1 \geq E_{m_1} + \cdots + E_{m_\ell}$. We assert that this decomposition must be trivial. That is, $\ell = 1$ and $m_1 = \pm 1$. In fact, if this is not trivial, then there is an m_s $(1 \leq s \leq \ell)$ so that $|m_s| \geq 2$.

Hence $E_1 > E_{m_s}$. On the other hand, however, Esteban [9] has obtained the estimates $E_m \geq 6|S^3||m|, \forall m \in \mathbb{Z}; E_1 \leq 6\sqrt{2}|S^3|$, where $|S^3|$ is the volume of the unit 3-sphere. Thus we arrive at a contradiction.

3. The two-dimensional Skyrme solitons

Brief Review. Recently, there has been some interest in formulating a Skyrme theory over a $(2+1)$-dimensional spacetime, following the original idea of Skyrme. In such a lower-dimensional field theory, in addition to the usual Skyrme term, one must impose a potential term in order to stabilize the solitons. Hence, parallel to the classical minimization problem of the static Skyrme energy over the spatial domain $\mathbb{R}^3$, we encounter the minimization problem over the spatial domain $\mathbb{R}^2$, for which the static Skyrme energy now contains an additional potential term. This is the problem of the existence of two-dimensional (2D) Skyrmions. In this context, interestingly, the technical difficulties (in 3D) in the Esteban paper [9] may all be overcome to yield a complete proof of an existence theorem for 2D Skyrmions following the ideas given in [9] by directly using the method of concentration-compactness. We emphasize that such an approach works only for the 2D Skyrme model, which is the third aim of this talk.

The Static Energy Functional and Existence Theorem. In normalized form, the two-dimensional Skyrme energy functional governing a configuration map $\mathbf{n} : \mathbb{R}^3 \to S^2$ is defined by (cf. [1] and references therein)

$$E(\mathbf{n}) = \frac{1}{2}\int_{\mathbb{R}^2}\left\{|\nabla\mathbf{n}|^2 + \frac{\lambda}{2}|\partial_1\mathbf{n}\wedge\partial_2\mathbf{n}|^2 + \frac{\mu}{2}(1-\mathbf{k}\cdot\mathbf{n})^2\right\}\mathrm{d}x, \tag{3.1}$$

where $\mathbf{k} = (0,0,1)$ is the north pole of S^2 in $\mathbb{R}^3$, and λ, μ are positive coupling constants. Note that, sometimes in literature, the potential term in (3.1) is chosen to be of a lower power, $\mu(1-\mathbf{k}\cdot\mathbf{n})$, which makes the potential energy of a stereographic projection take infinite value. In order to maintain a finite value for the potential of a stereographic projection, we observe the above (common) convention for the choice of the potential density. However, our general analysis is not affected by such a convenient, definitive, choice.

Finite-energy condition implies that $\mathbf{n}$ tends to $\mathbf{k}$ as $|x| \to \infty$. Therefore $\mathbf{n}$ may be viewed as a map from S^2 to itself which defines a homotopy class in $\pi_2(S^2) = \mathbb{Z}$, whose integer representative is the Brouwer degree of $\mathbf{n}$ with the integral representation $\deg(\mathbf{n}) = \frac{1}{4\pi}\int_{\mathbb{R}^2}\mathbf{n}\cdot(\partial_1\mathbf{n}\wedge\partial_2\mathbf{n})\,\mathrm{d}x$.

Like before, we are interested in the basic minimization problem

$$E_m = \inf\{E(\mathbf{n}) \mid E(\mathbf{n}) < \infty, \deg(\mathbf{n}) = m\}, \tag{3.2}$$

where $m \in \mathbb{Z}$. Below is our main existence result for 2D Skyrmions.

Theorem 3.1. *If the coupling constants λ and μ satisfy*

$$\lambda\mu \leq 48, \tag{3.3}$$

then the minimization problem (3.2) has a solution for $m = \pm 1$. Moreover, $E_1 < E_m$ for all $|m| \geq 2$ if $\lambda\mu \leq 12$.

Note that the condition (3.3) guarantees $E_1 < E_m$ for all $|m| \geq 3$.

Recent Development. It is interesting to mention that the Substantial Inequality method may be exploited further to obtain some new existence results. For example, with the help of a sharpened estimate of the universal constant in (1.3) and a suitable estimate of E_1, we can show that E_1 is in fact attainable for the Faddeev model. That is, $1 \in \mathbb{S}$. Moreover, for the 2D Skyrme model, the Substantial Inequality is also valid for the full parameter regime which allows us to prove that the least-positive energy of the functional (3.1) is always attainable and that E_1 is actually attainable under the condition $\lambda\mu \leq 192$, instead of (3.3). These results will be published elsewhere.

In conclusion, we have presented a series of existence theorems for the Faddeev knots (in 3D) and Skyrme solitons (in 3D and 2D) characterized by their respective topological invariants. The full details are in our papers [15,16].

Acknowledgments. FL was supported in part by NSF grant DMS–9896391. YY was supported in part by NSF grants DMS–9972300 and DMS–9729992 through IAS. We thank Weiping Zhang for kindly inviting us to present our work at the conference. We thank Chengming Bai for all the help he gave us.

References

1. K. Arthur, G. Roche, D. H. Tchrakian, and Y. Yang, J. Math. Phys. **37** (1996) 2569.
2. R. A. Battye and P. M. Sutcliffe, Phys. Rev. Lett. **81** (1998) 4798.
3. R. A. Battye and P. M. Sutcliffe, Phys. Rev. Lett. **81** (1998) 4798.

4. R. A. Battye and P. M. Sutcliffe, Solitons, Proc. Roy. Soc. A **455** (1999) 4305.
5. A. A. Belavin and A. M. Polyakov, JETP Lett. **22** (1975) 245.
6. E. B. Bogomol'nyi, Sov. J. Nucl. Phys. **24** (1976) 449.
7. Y. M. Cho, Phys. Rev. Lett. **87** (2001) 252001.
8. G. Dunne, *Self-Dual Chern–Simons Theories*, Lecture Notes in Phys., vol. m **36**, Springer, Berlin, 1995.
9. M. Esteban, Commun. Math. Phys. **105** (1986) 571.
10. M. J. Esteban, in *Variational Methods*, Birkhäuser, Boston, 1988. p. 77.
11. L. Faddeev, in *Relativity, Quanta, and Cosmology*, vol. 1 (ed. M. Pantaleo and F. de Finis), 1979, p. 247.
12. L. Faddeev, in *Proc. Internat. Congress Mathematicians*, Vol. I, Higher Ed. Press, Beijing, 2002, p. 235.
13. L. Faddeev and A. J. Niemi, Nature **387** (1997) 58.
14. A. Jaffe and C. H. Taubes, *Vortices and Monopoles*, Birkhäuser, Boston, 1980.
15. F. Lin and Y. Yang, Commun. Math. Phys. **249** (2004) 273.
16. F. Lin and Y. Yang, Comm. Pure Appl. Math. **LVII** (2004) 1332.
17. V. G. Makhankov, Y. P. Rybakov, and V. I. Sanyuk, *The Skyrme Model*, Springer, Berlin and Heidelberg, 1993.
18. M. K. Prasad and C. M. Sommerfield, Phys. Rev. Lett. **35** (1975) 760.
19. R. Rajaraman, *Solitons and Instantons*, North-Holland, Amsterdam, 1982.
20. Y. P. Rybakov and V. I. Sanyuk, Internat. J. Mod. Phys. A **7** (1992) 3235.
21. T. H. R. Skyrme, Proc. Roy. Soc. A **260** (1961) 127.
22. T. H. R. Skyrme, Proc. Roy. Soc. A **262** (1961) 237.
23. T. H. R. Skyrme, Nucl. Phys. **31** (1962) 556.
24. T. H. R. Skyrme, Internat. J. Mod. Phys. A **3** (1988) 2745.
25. C. H. Taubes, Commun. Math. Phys. **86** (1982) 257.
26. A. F. Vakulenko and L. V. Kapitanski, Sov. Phys. Dokl. **24** (1979) 433.
27. Y. Yang, *Solitons in Field Theory and Nonlinear Analysis*, Springer, New York, 2001.
28. I. Zahed and G. E. Brown, Phys. Reports **142** (1986) 1.

Dynamics of Bose-Einstein Condensates

Wu-Ming Liu

Beijing National Laboratory for Condensed Matter Physics, Institute of Physics, Chinese Academy of Sciences, Beijing 100080, China
E-mail: wmliu@aphy.iphy.ac.cn

We obtain exact solutions of the nonlinear Schrodinger equation, the discrete nonlinear Schrodinger equation, the two and three coupled nonlinear Schrodinger equations which describe the dynamics of one component, two component and spinor Bose-Einstein condensates with the short-range on-site interactions, the long-range dipole-dipole interactions, the time-dependent interatomic interaction near Feshbach resonance in an external potential. We find one-, two-, and three-component solitons of the polar and ferromagnetic types in spinor Bose-Einstein condensates. We study the magnetic soliton dynamics of spinor Bose-Einstein condensates in an optical lattice.

1. Introduction

The dynamics of ultracold atoms including atoms near Feshbach resonance in an external potential can be described by the nonlinear Schrodinger equation in the mean field approximation. It gives a link between the dynamics of condensed matter and the physics of nonlinear media. This will allow us to get a better understanding, both at the classical and at the quantum level, the interplay between on-site – intersite interactions as well as integrability – nonintegrability and discrete – continuum properties of condensed matter in an external potential such as an optical lattice. From the other side, we show that the system of BEC in an external potential such as an optical lattice give us a new tool to study the different solitary excitations as the physical parameters of the system of condensed matter in external potential varied. Such a highly controllable system may be crucial in answering some unresolved questions in the theory of quantum nonlinear dynamics. One of major developments in BEC was the study of spinor condensates. Spinor BEC feature an intrinsic three-component structure, due to the distinction between different hyperfine spin states of the atoms. When spinor BEC are trapped in the magnetic potential, the spin degree of freedom is frozen. However, in the condensate held by an optical potential, the spin is free, making it possible to observe a rich variety of phenomena, such as spin domains and textures. Recently, properties of BEC with this degree of freedom were investigated in detail, experimentally and theoretically. An important result demonstrated that, under special constraints imposed on parameters, the matrix nonlinear Schrödinger equation, which is a model of the one-dimensional spinor BEC in the free space, may be integrable by means of the inverse scattering transform. For that case, exact single-soliton solutions, as well as solutions describing collisions between two solitons, were found. Now we give a brief review of our original works for dynamics of BEC [1–5].

2. Dynamics of BEC near Feshbach Resonance

Our starting point is based on the well-established concept that at low enough temperatures, the nonlinear Schrodinger equation governs the evolution of the macroscopic wave function of a three dimensional BEC [1]. In the physically important case of the cigar-shaped BEC, it is reasonable to reduce three dimensional nonlinear Schrodinger equation into one dimensional Schrodinger equation,

$$i\frac{\partial\psi(x,t)}{\partial t}+\frac{\partial^2\psi(x,t)}{\partial x^2}+2a(t)|\psi(x,t)|^2\psi(x,t)+\frac{1}{4}\lambda^2x^2\psi(x,t)=0, \quad (2.1)$$

where time t and coordinate x are measured in units $2/\omega_\perp$ and $a_\perp$, $a_\perp = (\hbar/m\omega_\perp)^{1/2}$ and $a_0 = (\hbar/m\omega_0)^{1/2}$ are linear oscillator lengths in the transverse and cigar-axis directions, respectively. $\omega_\perp$ and ω_0 are respective harmonic oscillator frequencies, m is the atomic mass and $\lambda = 2|\omega_0|/\omega_\perp \ll 1$. The Feshbach-managed nonlinear coefficient reads $a(t) = |a_s(t)|/a_B = g_0\exp(\lambda t)$ (a_B is the Bohr radius).

The so-called "seed" solution of Eq. (2.1) can be chosen as $\psi_0(x,t) = A_c\exp[\frac{\lambda t}{2}+i\varphi_c]$, where $\varphi_c = k_0x\exp(\lambda t) - \frac{\lambda x^2}{4} + \frac{(2g_0A_c^2-k_0^2)(\exp(2\lambda t)-1)}{2\lambda}$ and A_c and k_0 are the arbitrary real constants. We perform the Darboux transformation $\psi_1 = \psi_0 + \frac{2}{\sqrt{g_0}}\frac{(\zeta+\bar{\zeta})\phi_1\overline{\phi_2}}{\phi^T\overline{\phi}}\exp(-\lambda t/2 - i\lambda x^2/4)$ to obtain the new solution of Eq. (2.1) by taking ψ_0 as the seed. Then we obtain the exact solution of Eq. (2.1) as follows:

$$\psi = [A_c+A_s\frac{(\gamma\cosh\theta+\cos\varphi)+i(\alpha\sinh\theta+\beta\sin\varphi)}{\cosh\theta+\gamma\cos\varphi}]\exp(\frac{\lambda t}{2}+i\varphi_c), \quad (2.2)$$

where $\theta = -\frac{[(k_0+k_s)\Delta_R-\sqrt{g_0}A_s\Delta_I][\exp(2\lambda t)-1]}{2\lambda} + \Delta_R x\exp(\lambda t)$, $\varphi = -\frac{[(k_0+k_s)\Delta_I+\sqrt{g_0}A_s\Delta_R][\exp(2\lambda t)-1]}{2\lambda} + \Delta_I x\exp(\lambda t)$, $\alpha = \frac{\sqrt{g_0}A_c(k_0-k_s+\Delta_I)}{\Lambda}$, $\beta =$

$1-\frac{2g_0A_c^2}{\Lambda}$, $\gamma = \frac{\sqrt{g_0}A_c(\Delta_R-\sqrt{g_0}A_s)}{\Lambda}$, $\Delta = \sqrt{[-\sqrt{g_0}A_s + i(k_s-k_0)]^2 - 4g_0A_c^2} \equiv \Delta_R + i\Delta_I$, $\Lambda = g_0A_c^2 + \frac{(\Delta_R-\sqrt{g_0}A_s^2)}{4} + \frac{(k_s-k_0+\Delta_I)^2}{4}$, where k_s is the arbitrary real constant. On the one hand, when $A_c = k_0 = 0$, Eq. (2.2) reduces to the well-known one soliton solution, $\psi_s = A_s \text{sech}\theta_s \exp i\varphi_s$, where $\theta_s = -\sqrt{g_0}\exp(\lambda t)A_s x + 2\sqrt{g_0}k_s A_s[\exp(2\lambda t)-1]/2\lambda$ and $\varphi_s = \varphi_c - g_0A_c^2[\exp(2\lambda t)-1]/2\lambda$.

3. Dynamics of Dipolar BEC in Optical Lattice

We consider a dilute gas of bosons in the optical lattice with the following Hamiltonian [2],

$$H = \int d\mathbf{r}\Psi^\dagger(\mathbf{r})[-\frac{\hbar^2}{2m}\nabla^2 + V_{opt}]\Psi(\mathbf{r}) + \int d\mathbf{r}d\mathbf{r}'\Psi^\dagger(\mathbf{r})\Psi^\dagger(\mathbf{r}')V_{int}\Psi(\mathbf{r}')\Psi(\mathbf{r}) + \int d\mathbf{r}\Psi^\dagger(\mathbf{r})V_{ext}\Psi(\mathbf{r}), \quad (3.1)$$

where $\Psi^\dagger(\mathbf{r})$ and $\Psi(\mathbf{r})$ are the boson field operators that annihilate and create a particle at the position $\mathbf{r}$, $V_{opt} = V_0\sin^2(2\pi z/\lambda)$ is the optical lattice potential, λ is the light wave length, V_{ext} is an external potential such as the gravity in the Yale experiment or magnetic traps, V_{int} includes on-site and nearest-neighbor interactions. In the case of polarized dipoles the interaction potential is $V_{int} = \frac{d^2(1-3\cos^2\theta)}{(\mathbf{r}-\mathbf{r}')^3} + \frac{4\pi\hbar^2 a}{m}\delta(\mathbf{r}-\mathbf{r}') = V_{dd} + U_0\delta(\mathbf{r}-\mathbf{r}')$, where the first term V_{dd} is the dipole-dipole interaction characterized by the dipole d and the angle θ between the dipole direction and the vector $\mathbf{r}-\mathbf{r}'$, and the second term is the short-range interaction given by the s wave scattering length a.

The boson field operators of $\Psi(\mathbf{r})$ and $\Psi^+(\mathbf{r})$ can be expanded over Wannier functions $w(\mathbf{r}-\mathbf{r}_n)$ of the lowest energy band, localized on this site. This implies that the energies involved in the system are small compared to the excitation energies of the second band, $\Psi(\mathbf{r}) = \sum_n C_n w(\mathbf{r}-\mathbf{r}_n)$, $\Psi^\dagger(\mathbf{r}) = \sum_n C_n^\dagger w^*(\mathbf{r}-\mathbf{r}_n)$. If we only consider nearest-neighbor sites of n, which is a good approximation for the BEC in 1D optical lattice as the large lattice constant, we will get an effective Hamiltonian. We now introduce a coherent state $|\alpha(t)\rangle$ of the atomic matter field in a potential well. Evaluating the atomic field operator C_n for such a state, we find then the macroscopic matter wave field, $\psi_n = \langle\alpha(t)|C_n|\alpha(t)\rangle$. Using the time-dependent variation principle, we can get the equation of motion for BEC in the optical lattice,

$$i\frac{\partial\psi_n}{\partial t} + J\psi_{n+1} + J\psi_{n-1} - \varepsilon_n\psi_n - U_0|\psi_n|^2\psi_n - U_{dd}(\psi_{n+1}+\psi_{n-1})|\psi_n|^2 = 0, \quad (3.2)$$

where $\varepsilon_n = \varepsilon_0 + \varepsilon_{ext}$ is the total energy of each lattice site, $U_{dd} = U_2$ is the coefficient which denotes the dipole-dipole interaction.

Dynamics of BEC with repulsive on-site interaction Using the substitution $\psi_n \longrightarrow (2J/(U_0 + 2U_{dd}))^{1/2}\varphi_n \exp[-i(\varepsilon_0 - 2J + J\lambda^2)]t$, λ^2 is the background amplitude, we can obtain the following general discrete nonlinear Schrödinger equation, $i\frac{\partial \varphi_n}{\partial t} + (\varphi_{n+1} + \varphi_{n-1} - 2\varphi_n) - \epsilon(\varphi_{n+1} + \varphi_{n-1})|\varphi_n|^2 + 2(\epsilon - 1)|\varphi_n|^2\varphi_n + 2\rho^2\varphi_n = 0$, where $\epsilon = 2U_{dd}/(U_0 + 2U_{dd})$, $\rho^2 = (\lambda^2 - \varepsilon_{ext}/J)/2$, $t \longrightarrow Jt$.

When $\epsilon = 1$, the previous equation is reduced to the integrable Ablowitz-Ladik model which can be solved by the inverse scattering technique, and it leads to the so-called dark soliton solution with Bloch oscillations in a constant electric field. When $0 < \epsilon < 1$, this equation is non-integrable and only the approximate solution can be obtained by the multiple scale expansion method. There are singular points in this equation when $(U_0 + 2U_{dd})/2U_{dd} = (\lambda^2 - \varepsilon_{ext}/J)/2$. At the singular points, the dispersive term becomes zero and the given site is decoupled from its neighbors. Near these singular points or far from them, the dynamic behaviors of φ_n are quite different.

When the excitations are in the vicinity of the singular points, there are soliton solutions which can be described by the *Toda lattice* model and the solution of this equation in the small-amplitude limit is $\varphi_n = \kappa^n\epsilon^{-1/2}(1 - \gamma^2\mu a_n)\exp(-i\gamma\chi_n + i\omega_0 t)$, where $\epsilon \leq 1$, and $\gamma \ll 1$ is a small parameter, $a_n = a_n(\tau)$ and $\chi_n = \chi_n(\tau)$ are two real functions of the time $\tau = 2\gamma t\sqrt{2(\epsilon^{-1} - 1 + \kappa)}$, $\omega_0 = 2((\lambda^2 - \varepsilon_{ext}/J)/2 - \epsilon^{-1})$, $\mu = \kappa\,\mathrm{sgn}(\epsilon - 1 + \kappa)$.

When the excitations are far from the singular points, the dynamics of BEC in the optical lattice can be described by the small-amplitude limit. The solution can be sought using the multiscale expansion technique. We can find the soliton solution of the previous equation in the small amplitude approximation, $\varphi_n = [\rho - (12\nu^2 G_2/G_1)\mathrm{sech}^2(\nu(z - V(\tau)))]e^{-i\Phi_n(t)}$, where ν is an arbitrary parameter, $G_1 = -\frac{1}{3}(1 - \epsilon\rho^2)[3 - (3\epsilon + 1)\rho^2]$, $V = -2\nu^2 G_2/C$, $G_2 = -8\rho^2(3 - 4\epsilon\rho^2)$, $C = -4\rho\sqrt{1 - \epsilon\rho^2}$.

Dynamics of BEC with attractive on-site interaction Without any external potential and for the small kinetic energy, Eq. (3.2) is reduced to $i\frac{\partial \psi_n}{\partial t} + (\psi_{n+1} + \psi_{n-1}) - \frac{U_0}{J}\psi_n^\dagger\psi_n\psi_n - \frac{U_{dd}}{J}(\psi_{n+1} + \psi_{n-1})\psi_n^\dagger\psi_n = 0$. This equation is non-integrable, and a first order adiabatic approximation solution can be obtained by perturbation method. Treating the term $\nu\psi_n|\psi_n|^2$ as a perturbation, where $U_0 < 0$, $\nu = -U_0/J > 0$, and using the adiabatic approximation, a soliton retains its functional form in the presence of perturbation, the solution to the first order of ν can be written as

$\psi_n = \frac{1}{\sqrt{\mu}} \sinh\beta \text{sech}[\beta(n-x)]e^{i\alpha(n-x)+i\sigma}$, where $dx/dt = (2\sinh\beta\sin\alpha)/\beta$, $d\beta/dt = 0$, $\frac{\partial\alpha}{\partial t} = \nu\frac{\partial}{\partial x}\sum_{s=1}^{\infty}\frac{4\pi^2 s\sinh^2\beta}{\beta^3\sinh(\pi^2 s/\beta)}\cos(2\pi s x)$, $\frac{d\sigma}{dt} = 2\cos\alpha\cos\beta + \frac{2\alpha}{\beta}\sin\alpha\sinh\beta$, where $U_{dd} < 0$, $\mu = -U_{dd}/J > 0$. This solution is a bright soliton but the role of the dipole-dipole interaction may follow from a refinement of the general consideration described above. We conclude that for the attractive interaction, the equation of motion of dipolar BEC can be treated by perturbation methods and the bright soliton solution can be found.

4. Solutions of Two-Species BEC in an Optical Lattice

The two-species BECs in a 1D periodic potential can be described by the coupled nonlinear Schrödinger equations [3],

$$\begin{aligned} i\hbar\frac{\partial\psi_1}{\partial t} &= -\frac{\hbar^2}{2m_1}\frac{\partial^2\psi_1}{\partial x^2} + \frac{2\hbar^2 a_1}{m_1 l_1^2}|\psi_1|^2\psi_1 + \frac{2\hbar^2 a_{12}}{\sqrt{m_1 m_2}l_1 l_2}|\psi_2|^2\psi_1 + V_1(x)\psi_1, \\ i\hbar\frac{\partial\psi_2}{\partial t} &= -\frac{\hbar^2}{2m_2}\frac{\partial^2\psi_2}{\partial x^2} + \frac{2\hbar^2 a_{12}}{\sqrt{m_1 m_2}l_1 l_2}|\psi_1|^2\psi_2 + \frac{2\hbar^2 a_2}{m_2 l_2^2}|\psi_2|^2\psi_2 + V_2(x)\psi_2, \end{aligned} \tag{4.1}$$

where ψ_i, m_i, $l_i = \sqrt{\hbar/m_i\omega_0}$ are the macroscopic wave functions of the condensates, the mass and the harmonic oscillator lengths in the radial direction of the ith species ($i = 1, 2$) respectively. a_1, a_2 and a_{12} denote the s-wave scattering lengths between same-species and interspecies collisions. $V_i(x)$ are the periodic potentials, $V_i(x) = V_{0,i}\text{sn}^2(k_L x, k)$, with $V_{0,i}$ denoting the magnitude of potentials, where $k_L = 2\pi/\lambda$ is the wave vector of the laser light and λ is the wavelength, corresponding to a lattice period $d = \lambda/2$. $\text{sn}(k_L x, k)$ is the Jacobian elliptic sine function with modulus k ($0 \le k \le 1$). In the limit $k = 0$, the Jacobian elliptic sine reduces to sinusoid function and thus $V(x)$ possesses a standard form of the standing light wave. For values of $k < 0.9$ the potential is virtually indistinguishable from a standing light wave. Finally, for $k \to 1$, $V(x)$ becomes an array of well-separated hyperbolic secant potential barriers or wells.

For the case of weakly coupled condensates in an optical lattice, the wave function ψ can be decomposed as a sum of wave functions localized in each well of the periodic potential (tight binding approximation) with the assumption relying on the fact that the height of the interwell barrier is much higher than the chemical potential. We, however, do not restrict ourself on the low energy case and look for the global condensates wave functions of excitations: $\psi_i(x,t) = \phi_i(x)\exp(-i\mu_i t/\hbar)$, where μ_i $(i=1,2)$ are the chemical potentials. With the general form of spatial wave functions $\phi_i(x)$ written as $\phi_i(x) = r_i(x)\exp[i\varphi_i(x)]$, Eq. (4.1) can be separated as real and imaginary parts. We then integrate once for the imaginary part and obtain the first-order differential equations for the phases $\varphi_i(x)$, $\varphi_i'(x) = \frac{\alpha_i}{r_i^2(x)}$, where parameters α_i $(i=1,2)$ are constants of integration to be determined.

We then construct the solutions as $r_i^2(x) = A_i \mathrm{sn}^2(k_L x, k) + B_i$, where the constants B_i $(i=1,2)$ determine the mean amplitudes and act as the dc offsets for the numbers of the condensed atoms and parameters A_i $(i=1,2)$ are to be determined. We find $A_1 = \frac{\frac{\sqrt{m_1}l_1l_2a_{12}}{\sqrt{m_2}}(m_2V_{0,2}-\hbar^2k_L^2k^2)-a_2l_1^2(m_1V_{0,1}-\hbar^2k_L^2k^2)}{2\hbar^2(a_1a_2-a_{12}^2)}$, $A_2 = \frac{\frac{\sqrt{m_2}l_1l_2a_{12}}{\sqrt{m_1}}(m_1V_{0,1}-\hbar^2k_L^2k^2)-a_1l_2^2(m_2V_{0,2}-\hbar^2k_L^2k^2)}{2\hbar^2(a_1a_2-a_{12}^2)}$, where $\alpha_1^2 = B_1k_L^2[\frac{k^2}{A_1}B_1^2 + (1+k^2)B_1 + A_1]$, $\alpha_2^2 = B_2k_L^2[\frac{k^2}{A_2}B_2^2 + (1+k^2)B_2 + A_2]$, $\mu_1 = \frac{\hbar^2k_L^2}{2m_1}(1 + k^2 + \frac{6a_1}{l_1^2k_L^2}B_1 + \frac{4a_{12}\sqrt{m_1}}{l_1l_2k_L^2\sqrt{m_2}}B_2 + \frac{2a_{12}\sqrt{m_1}}{l_1l_2k_L^2\sqrt{m_2}}\frac{A_2}{A_1}B_1 + \frac{m_1V_{0,1}}{\hbar^2k_L^2}\frac{B_1}{A_1})$, $\mu_2 = \frac{\hbar^2k_L^2}{2m_2}(1 + k^2 + \frac{6a_2}{l_2^2k_L^2}B_2 + \frac{4a_{12}\sqrt{m_2}}{l_1l_2k_L^2\sqrt{m_1}}B_1 + \frac{2a_{12}\sqrt{m_2}}{l_1l_2k_L^2\sqrt{m_1}}\frac{A_1}{A_2}B_2 + \frac{m_2V_{0,2}}{\hbar^2k_L^2}\frac{B_2}{A_2})$.

5. Magnetic Soliton of Spinor BEC in an Optical Lattice

The dynamics of spinor BECs trapped in an optical lattice is primarily governed by three types of two-body interactions [4]: spin-dependent collision characterized by the s-wave scattering length, magnetic dipole-dipole interaction (of the order of Bohr magneton μ_B), and light-induced dipole-dipole interaction adjusted by the laser frequency in experiment. Our starting point is the Hamiltonian describing an $F=1$ spinor condensate at zero temperature trapped in an optical lattice, which is subject to the magnetic and the light-induced dipole-dipole interactions and is coupled to an external magnetic field via the magnetic dipole Hamiltonian H_B,

$$H = \sum_\alpha \int d\mathbf{r}\hat{\psi}_\alpha^\dagger(\mathbf{r})[-\frac{\hbar^2\nabla^2}{2m} + U_L(\mathbf{r})]\hat{\psi}_\alpha(\mathbf{r}) + \sum_{\alpha,\beta,\upsilon,\tau}\int d\mathbf{r}d\mathbf{r}' \\ \hat{\psi}_\alpha^\dagger(\mathbf{r})\hat{\psi}_\beta^\dagger(\mathbf{r}')[U_{\alpha\upsilon\beta\tau}^{coll}(\mathbf{r},\mathbf{r}') + U_{\alpha\upsilon\beta\tau}^{d-d}(\mathbf{r},\mathbf{r}')]\hat{\psi}_\tau(\mathbf{r}')\hat{\psi}_\upsilon(\mathbf{r}) + H_B, \quad (5.1)$$

where $\hat{\psi}_\alpha(r)$ is the field annihilation operator for an atom in the hyperfine state $|f=1, m_f=\alpha\rangle$, $U_L(\mathbf{r})$ is the lattice potential, the indices $\alpha,\beta,\upsilon,\tau$ which run through the values $-1,0,1$ denote the Zeeman sublevels of the ground state. The parameter $U_{\alpha\upsilon\beta\tau}^{coll}(\mathbf{r},\mathbf{r}')$ describes the two-body ground-state collisions and $U^{d-d}_{\alpha\upsilon\beta\tau}(\mathbf{r},\mathbf{r}')$ includes the magnetic dipole-dipole interaction and the light-induced dipole-dipole interaction.

When the optical lattice potential is deep enough there is no spatial overlap between the condensates at different lattice sites. We can then expand the atomic field operator as $\hat{\psi}(\mathbf{r}) = \sum_n \sum_{\alpha=0,\pm1} \hat{a}_\alpha(n)\phi_n(\mathbf{r})$, where n labels the lattice sites, $\phi_n(\mathbf{r})$ is the condensate wave function for the nth microtrap and the operators $\hat{a}_\alpha(n)$ satisfy the bosonic commutation relations $[\hat{a}_\alpha(n), \hat{a}^\dagger_\beta(l)] = \delta_{\alpha\beta}\delta_{nl}$. It is assumed that all Zeeman components share the same spatial wave function. If the condensates at each lattice site contain the same number of atoms N, the ground-state wave functions for different sites have the same form $\phi_n(\mathbf{r}) = \phi_n(\mathbf{r}-\mathbf{r}_n)$. The spin operators are defined as $\mathbf{S}_n = \hat{a}^\dagger_\alpha(n)\mathbf{F}_{\alpha\upsilon}\hat{a}_\upsilon(n)$, where $\mathbf{F}$ is the vector operator for the hyperfine spin of an atom, with components represented by 3×3 matrices in the $|f=1, m_f=\alpha\rangle$ subspace. We obtain both the one- and two-soliton solutions denoted by $\mathbf{S}(n)$ with $n = 1, 2$ in the following form: $S^x_n = 1-(\chi_{2,n}+2\chi_{3,n}\sin^2\Phi_n)/\Lambda_n$, $S^y_n = (\chi_{1,n}\eta_n\cosh\Theta_n\sin\Phi_n+\chi_{2,n}\sinh\Theta_n\cos\Phi_n)/\Lambda_n$, $S^z_n = (\chi_{1,n}\cosh\Theta_n\cos\Phi_n+\chi_{2,n}\eta_n\sinh\Theta_n\sin\Phi_n)/\Lambda_n$, where $\Lambda_n = \cosh^2\Theta_n + \chi_{3,n}\sin^2\Phi_n$, $\Theta_n = 2\kappa_{4,n}(z - V_n t - z_n)$, $\Phi_n = 2\kappa_{3,n}z - \Omega_n t + \phi_n$, $V_n = 2(\kappa_{1,n} + \frac{\kappa_{3,n}}{\kappa_{4,n}}\kappa_{2,n})$, $\Omega_n = 4(\kappa_{1,n}\kappa_{3,n} - \kappa_{2,n}\kappa_{4,n})$, $\kappa_{1,n} = \mu_n(1+\rho^2/|\zeta_n|^2)$, $\kappa_{2,n} = \nu_n(1-\rho^2/|\zeta_n|^2)$, $\kappa_{3,n} = \mu_n(1-\rho^2/|\zeta_n|^2)$, $\kappa_{4,n} = \nu_n(1+\rho^2/|\zeta_n|^2)$, $\eta_n = (|\zeta_n|^2+\rho^2)/(|\zeta_n|^2-\rho^2)$, $\chi_{1,n} = (2\mu_n\nu_n)/|\zeta_n|^2$, $\chi_{2,n} = (2\nu_n^2)/|\zeta_n|^2$, and $\chi_{3,n} = (4\rho^2\nu_n^2)/(|\zeta_n|^2-\rho^2)^2$.

6. Nonlinear Modulation Instability in Spinor BEC

We start with an effectively one-dimensional BEC trapped in a pencil-shaped region, which is elongated in x and tightly confined in the transverse directions y, z [5]. Atoms in the $F = 1$ hyperfine state can be described by a 1D vectorial wave function, $\mathbf{\Phi}(x,t) = [\Phi_{+1}(x,t), \Phi_0(x,t), \Phi_{-1}(x,t)]^T$, with the components corresponding to the three values of the vertical spin projection, $m_F = +1, 0, -1$. The wave functions obey a system of coupled nonlinear Schrodinger equations,

$$\begin{aligned} i\hbar\partial_t\Phi_{\pm1} &= -\frac{\hbar^2}{2m}\partial_x^2\Phi_{\pm1} + (c_0+c_2)(|\Phi_{\pm1}|^2+|\Phi_0|^2)\Phi_{\pm1} \\ &\quad +(c_0-c_2)|\Phi_{\mp1}|^2\Phi_{\pm1} + c_2\Phi^*_{\mp1}\Phi_0^2, \\ i\hbar\partial_t\Phi_0 &= -\frac{\hbar^2}{2m}\partial_x^2\Phi_0 + (c_0+c_2)(|\Phi_{+1}|^2+|\Phi_{-1}|^2)\Phi_0 \\ &\quad +c_0|\Phi_0|^2\Phi_0 + 2c_2\Phi_{+1}\Phi_{-1}\Phi_0^*, \end{aligned} \tag{6.1}$$

where $c_0 = (g_0+2g_2)/3$ and $c_2 = (g_2-g_0)/3$ denote effective constants of the mean-field (spin-independent) and spin-exchange interaction, respectively. Here $g_f = 4\hbar^2 a_f/[ma_\perp^2(1-ca_f/a_\perp)]$, with $f = 0, 2$, are effective 1D coupling constants, a_f is the s-wave scattering length in the channel with the total hyperfine spin f, $a_\perp$ is the size of the transverse ground state, m is the atomic mass, and $c = -\zeta(1/2) \approx 1.46$. Redefining the wave function as $\mathbf{\Phi} \to (\phi_{+1}, \sqrt{2}\phi_0, \phi_{-1})^T$ and measuring time and length in units of $\hbar/|c_0|$ and $\sqrt{\hbar^2/2m|c_0|}$, respectively, we cast Eqs. (6.1) in a normalized form $i\partial_t\phi_{\pm1} = -\partial_x^2\phi_{\pm1} - (\nu+a)(|\phi_{\pm1}|^2 + 2|\phi_0|^2)\phi_{\pm1} - (\nu-a)|\phi_{\mp1}|^2\phi_{\pm1} - 2a\phi_{\mp1}^*\phi_0^2$, $i\partial_t\phi_0 = -\partial_x^2\phi_0 - 2\nu|\phi_0|^2\phi_0 - (\nu+a)(|\phi_{+1}|^2 + |\phi_{-1}|^2)\phi_0 - 2a\phi_{+1}\phi_{-1}\phi_0^*$, where $\nu \equiv -\mathrm{sgn}(c_0)$, $a \equiv -c_2/|c_0|$.

6.1. *Exact single-, two-, and three-component soliton*

Single-component ferromagnetic soliton a single-component ferromagnetic soliton is given by a straightforward solution, $\phi_{-1} = \phi_0 = 0$, $\phi_{+1} = \sqrt{\frac{-2\mu}{\nu+a}}\mathrm{sech}(\sqrt{-\mu}x)e^{-i\mu t}$, where the negative chemical potential μ is the intrinsic parameter of the soliton family.

Single-component polar soliton The simplest polar soliton, that has only the ϕ_0 component, can be found for $\nu = +1$, $\phi_0 = \sqrt{-\mu}\mathrm{sech}(\sqrt{-\mu}x)e^{-i\mu t}$, $\phi_{\pm1} = 0$.

Two-component polar soliton In the same case as considered above, $\nu = +1$, a two-component polar soliton can be easily found too, $\phi_0 = 0$, $\phi_{+1} = \pm\phi_{-1} = \sqrt{-\mu}\mathrm{sech}(\sqrt{-\mu}x)e^{-i\mu t}$.

Three-component polar solitons One of three-component solitons of the polar type is $\phi_0 = \sqrt{1-\epsilon^2}\sqrt{-\mu}\mathrm{sech}(\sqrt{-\mu}x)e^{-i\mu t}$, $\phi_{+1} = -\phi_{-1} = \pm\epsilon\sqrt{-\mu}\mathrm{sech}(\sqrt{-\mu}x)e^{-i\mu t}$, where ϵ is an arbitrary parameter taking values $-1 < \epsilon < +1$ (the presence of this parameter resembles the feature typical to solitons in the Manakov's system, and, as well as the one- and two-component polar solitons displayed above, the solution does not explicitly depend on the parameter a. We stress that the phase difference of π between the ϕ_{+1} and ϕ_{-1} components is a necessary ingredient of the solution.

There is another three-component polar solution similar to the above one (i.e., containing the arbitrary parameter ϵ, and independent of a), but with equal phases of the $\phi_{\pm1}$ components and a phase shift of $\pi/2$ in the ϕ_0 component. This solution is $\phi_0 = i\sqrt{1-\epsilon^2}\sqrt{-\mu}\mathrm{sech}(\sqrt{-\mu}x)e^{-i\mu t}$, $\phi_{+1} = \phi_{-1} = \pm\epsilon\sqrt{-\mu}\mathrm{sech}(\sqrt{-\mu}x)e^{-i\mu t}$, where the sign $\pm$ is the same for both components.

In addition, there is a species of three-component polar solitons that explicitly depend on a, $\phi_0 = (\mu_{+1}\mu_{-1})^{1/4}\mathrm{sech}(\sqrt{-\mu}x)e^{-i\mu t}$, $\phi_{\pm1} = \sqrt{-\mu_{\pm1}}\mathrm{sech}(\sqrt{-\mu}x)e^{-i\mu t}$, where $\mu_{\pm1}$ are two arbitrary negative parameters, and the chemical potential is $\mu = -(\nu+a)(\sqrt{-\mu_{+1}} + \sqrt{-\mu_{-1}})^2/2$, which implies that $\nu + a > 0$ (note that this solution admits $v = -1$, i.e., repulsive spin-independent interaction). Each species of the three-component polar soliton depends on two arbitrary parameters: either μ and ϵ, or μ_{-1} and μ_{+1}.

Finite background solitons In special cases, it is possible to find exact solutions for solitons sitting on a nonzero background. Namely, for $\nu = 1$ and $a = -1/2$, one can find a two-component polar soliton with a continuous-wave background attached to it, in the following form: $\phi_0 = 0$, $\phi_{+1} = e^{-i\mu t}\sqrt{-\mu}[\frac{1}{\sqrt{2}} \pm \mathrm{sech}(\sqrt{-\mu}x)]$, $\phi_{-1} = e^{-i\mu t}\sqrt{-\mu}[\frac{1}{\sqrt{2}} \mp \mathrm{sech}(\sqrt{-\mu}x)]$.

For $\nu = a = 1$, a three-component polar solution with the background can be found too, $\phi_{+1} = \phi_{-1} = \frac{1}{2}\sqrt{-\mu}e^{-i\mu t}[\frac{1}{\sqrt{2}} \pm \mathrm{sech}(\sqrt{-\mu}x)]$, $\phi_0 = \frac{1}{2}\sqrt{-\mu}e^{-i\mu t}[\frac{1}{\sqrt{2}} \mp \mathrm{sech}(\sqrt{-\mu}x)]$. In the latter case, the availability of the exact solution is not surprising, as the case of $\nu = a = 1$ is the exactly integrable one.

6.2. *Modulational instability*

Now we focus on the integrable case, with $\nu = a = 1$, which corresponds to the attractive interactions. As explained above, the spinor BEC obeys this condition if a special (but physically possible) constraint is imposed on the scattering lengths which determine collisions between atoms. Then, Eqs. (6.1) can be rewritten as a 2×2 matrix NLS equation, $i\partial_t \mathbf{Q} + \partial_x^2 \mathbf{Q} + 2\mathbf{Q}\mathbf{Q}^\dagger\mathbf{Q} = 0$, $\mathbf{Q} \equiv \begin{pmatrix} \phi_{+1} & \phi_0 \\ \phi_0 & \phi_{-1} \end{pmatrix}$. This equation is a completely integrable system. We obtain its new family of solutions in the form $\mathbf{Q}_1 = [\mathbf{A}_c + 4\xi(\mathbf{I}+\mathbf{A}\mathbf{A}^*)^{-1}\mathbf{A}]e^{i\varphi_c}$, where $\mathbf{A} = (\mathbf{\Pi}e^{\theta - i\varphi} + \kappa^{-1}\mathbf{A}_c)(\kappa^{-1}\mathbf{A}_c\mathbf{\Pi}e^{\theta - i\varphi} + \mathbf{I})^{-1}$, $\theta = M_I x + [2\xi M_R - (k+2\eta)M_I]t$, $\varphi = M_R x - [2\xi M_I + (k+2\eta)M_R]t$, $M = \sqrt{(k+2i\lambda)^2 + 4(\alpha_c^2 + \beta_c^2)} \equiv M_{\mathrm{R}} + iM_{\mathrm{I}}$, $\kappa \equiv \frac{1}{2}(ik - 2\lambda + iM)$, $\lambda = \xi + i\eta$ is the spectral parameter, and $\mathbf{\Pi} = \begin{pmatrix} \beta & \alpha \\ \alpha & \gamma \end{pmatrix}$ is an arbitrary complex symmetric matrix. It is worth noting that the three-component polar soliton considered in the previous section is not a special example of the solution $\mathbf{Q}_1$.

7. Conclusion

In conclusion, our results describe the dynamics of BEC near Feshbach resonance in an expulsive parabolic potential. Furthermore, under the condition of $|a_s(t)| < a_{cr}$, it is possible to squeeze a bright soliton of BEC into the assumed peak matter density, which can provide an experimental tool for investigating the range of validity of the 1D nonlinear Schrodinger equation. In addition we find different regions in which stable bright or dark soliton excitations will exist and on the boundaries of these regions the system becomes effectively dispersionless and the formation of shock waves becomes possible. These different excitations are observable when we modify the wavelength and intensity of the lattice and change the magnitude of the external fields in the experiment. Phase diagram is determined analytically according to the order parameters and persistent currents in an optical lattice ring are obtained explicitly in terms of the exact wave functions which are seen to be travelling matter waves. The magnetic soliton of spinor BECs in an optical lattice is mainly caused by the magnetic and the light-induced dipole-dipole interactions between different lattice sites.

Acknowledgement I express my sincerely thanks to Z.X. Cao, P.B. He, E.I. Kats, L. Li, Z.D. Li, J.Q. Liang, Z.X. Liang, B.A. Malomed, D. Mihalache, Z.W. Xie, Z.D. Zhang, G.P. Zheng for helpful discussions in original papers. This work was supported by the NSF of China under grant 90403034, 90406017, 60525417, and the National Key Basic Research Special Foundation of China under 2005CB724508.

References

1. Z.X. Liang, Z.D. Zhang, W.M. Liu, Phys. Rev. Lett. **94**, 050402 (2005).
2. Z.W. Xie, Z.X. Cao, E.I. Kats, W.M. Liu, Phys. Rev. A **71**, 025601 (2005).
3. G.P. Zheng, J.Q. Liang, W.M. Liu, Phys. Rev. A **71**, 053608 (2005).
4. Z.D. Li, P.B. He, L. Li, J.Q. Liang, W.M. Liu, Phys. Rev. A **71**, 053611 (2005).
5. L. Li, Z.D. Li, B.A. Malomed, D. Mihalache, W.M. Liu, Phys. Rev. A **72**, 033611 (2005).

Twisted Space-Time Symmetry, Non-Commutativity and Particle Dynamics

J. LUKIERSKI and M. WORONOWICZ

Institute of Theoretical Physics
50-205 Wrocław,
pl. Maxa Borna 9, Poland
E-mail: lukier,woronow@ift.uni.wroc.pl

We describe the twisted space-time symmetries which imply the quantum Poincaré covariance of noncommutative Minkowski spaces, with constant, Lie algebraic and quadratic commutators. Further we present the relativistic and nonrelativistic particle models invariant respectively under twisted relativistic and twisted Galilean symmetries.

1. Introduction

Since the work of Doplicher et all. (see e.g.[1,2]) there is a strong indication that due to quantum gravity effects the space-time coordinates are becoming noncommutative. In general case one can write*

$$[\widehat{x}_\mu, \widehat{x}_\nu] = \frac{i}{\kappa^2}\,\theta_{\mu\nu}(\kappa\widehat{x}_\rho) = \frac{i}{\kappa^2}\,\theta^{(0)}_{\mu\nu} + \frac{i}{\kappa}\,\theta^{(1)\rho}_{\mu\nu}\,\widehat{x}_\rho + i\theta^{(2)\rho\tau}_{\mu\nu}\widehat{x}_\rho\widehat{x}_\tau\,, \tag{1.1}$$

where the fundamental mass parameter κ has been introduced in order to exhibit the mass dimensions of respective terms and have the constant tensors $\theta^{(0)}_{\mu\nu}$, $\theta^{(1)\rho}_{\mu\nu}$, $\theta^{(2)\rho\tau}_{\mu\nu}$ as dimensionless. If we link (1.1) with quantum gravity one can put $\kappa = m_{\rm pl}$ ($m_{\rm pl}$ - Planck mass). Further we add that the relation (1.1) describes in D=10 first-quantized open string theory the noncommutative coordinates on D-branes providing the localizations of the ends of the strings [3,4].

*Formula (1.1) is not the most general one. One can assume that the rhs of (1.1) depends also on momenta (or derivative operators) as well as on other operators, e.g. spin variables. In this note we shall not consider such extensions of (1.1). The expansion (1.1) is only up to quadratic term because higher orders do not have classical limit $\kappa \to \infty$.

There are two important problems related with the application of formula (1.1) to physical models:

i) In standard relativistic theory, with classical Poincaré symmetries, the first term on rhs of (1.1) breaks the Lorentz invariance, and further two terms break both Lorentz and translational invariance. One can ask how looks the deformation of classical Poincaré invariance which permits to consider relations (1.1) as covariant under deformed Poincaré transformations, i.e. the same in any deformed Poincaré frame.

ii) There should be given prescriptions how to formulate the classical mechanics and field theory models with noncommutative space-time coordinates (1.1), covariant under the twisted Poincaré symmetries.

If the time coordinate remains classical (i.e. in formula (1.1) $\theta_{0\mu} = 0$) both points i) and ii) can be applied to the nonrelativistic noncommutative theories with classical Galilean invariance broken by relation (1.1).

2. Twisted Space-Time Symmetries

We shall look for the quantum relativistic symmetries implying the covariance of noncommutative Minkowski spaces. In systematic study firstly one should consider all possible quantum relativistic symmetries (quantum Poincaré algebras) in the form of noncommutative Hopf algebras, and then derive corresponding quantum Minkowski spaces as deformed Hopf algebra modules. An example of such a construction which is already more than ten years old is the κ-deformed Minkowski space [5–7]

$$[\widehat{x}_0, \widehat{x}_i] = \frac{i}{\kappa}\widehat{x}_i\,, \qquad [\widehat{x}_i, \widehat{x}_j] = 0\,, \tag{2.1}$$

corresponding in (1.1) to the choice $\theta^{(0)}_{\mu\nu} = \theta^{(2)\rho\tau}_{\mu\nu} = 0$ and $\theta^{(1)\rho}_{\mu\nu} = \eta_{\nu 0}\delta_\mu{}^\rho - \eta_{\mu 0}\delta_\nu{}^\rho$. Using the Hopf-algebraic formulae of κ-deformed Poincaré algebra in bicrossproduct basis one can show [6] that the relations (2.1) are covariant under the Hopf-algebraic action of κ-deformed Poincaré algebra.

It appears that the most effective way of describing the noncommutative space-times covariant under quantum relativistic symmetries is to consider twisted symmetry algebras. In such a case the classical Poincaré-Hopf algebra is modified only in the coalgebraic sector, with all the algebraic relations preserved. We change the classical Poincaré Hopf algebra $\mathcal{H}^{(0)} = (\mathcal{U}(\mathcal{P}_4), m, \Delta_0, S_0, \epsilon)$ into twisted Poincaré Hopf algebra $\mathcal{H} = (\mathcal{U}(\mathcal{P}_4), m, \Delta, S, \epsilon)$ by means of the twist factor $\mathcal{F} \in \mathcal{U}(\mathcal{P}_4) \otimes \mathcal{U}(\mathcal{P}_4)$ as follows ($\mathcal{P}_4 \ni \hat{g} = (P_\mu, M_{\mu\nu})$)

$$\Delta(\hat{g}) = \mathcal{F} \circ \Delta_0(\hat{g}) \circ \mathcal{F}^{-1}\,, \qquad S(\hat{g}) = U S_0(\hat{g}) U^{-1}\,, \tag{2.2}$$

$$\Delta_0(\hat{g}) = \hat{g} \otimes 1 + 1 \otimes \hat{g}\,, \qquad S_0(\hat{g}) = -\hat{g}\,, \qquad \epsilon(\hat{g}) = 0\,, \tag{2.3}$$

where $(a \otimes b) \circ (c \otimes d) = ac \otimes bd$. The twist $\mathcal{F}$ satisfies the cocycle and normalization conditions [8]

$$\mathcal{F}_{12}\,(\Delta_0 \otimes 1)\,\mathcal{F} = \mathcal{F}_{23}\,(1 \otimes \Delta_0)\,\mathcal{F}\,, \qquad (\epsilon \otimes 1)\mathcal{F} = (1 \otimes \epsilon)\mathcal{F} = 1\,, \tag{2.4}$$

where $\mathcal{F}_{12} = \mathrm{f}_{(1)} \otimes \mathrm{f}_{(2)} \otimes 1$ etc. ($\mathcal{F} = \mathrm{f}_{(1)} \otimes \mathrm{f}_{(2)}$) and $U = \mathrm{f}_{(1)} S(\mathrm{f}_{(2)})$.

The advantage of using twisted Poincaré algebra is the explicit formula for the multiplication in twisted Hopf algebra module $\mathcal{A}$ which should satisfy the condition (see e.g. [9], $h \in \mathcal{U}(\mathcal{P}_4)$, $a, b \in \mathcal{A}$)

$$h \triangleright (a \bullet b) = (h_1 a) \bullet (h_2 b)\,, \tag{2.5}$$

where $\Delta(h) = h_1 \otimes h_2$. We see from (2.5) that if $h_1 \neq h_2$ then $a \bullet b \neq b \bullet a$, i.e. from quantum-deformed relativistic symmetry follow necessarily the noncommutative Minkowski space as its Hopf-algebraic module.

One can show that the multiplication in $\mathcal{A}$ for twisted Hopf algebra $\mathcal{H}$ which is consistent with the relation (2.5) ($h \in \mathcal{H}$) provides the formula [10–12]

$$a \bullet b = (\bar{\mathrm{f}}_{(1)} a)(\bar{\mathrm{f}}_{(2)} b)\,, \qquad \mathcal{F}^{-1} = \bar{\mathrm{f}}_{(1)} \otimes \bar{\mathrm{f}}_{(2)}\,. \tag{2.6}$$

In the case of relativistic symmetries one can use the classical space-time representation for the Poincaré generators P_μ, $M_{\mu\nu}$

$$P_\mu = i\partial_\mu\,, \qquad M_{\mu\nu} = i(x_\nu \partial_\mu - x_\mu \partial_\nu)\,. \tag{2.7}$$

Subsequently in the formula (2.6) one can assume that a, b are classical functions on commutative Minkowski space x_μ, and define $\bar{\mathrm{f}}_{(i)}(P_\mu,\, M_{\mu\nu}) \equiv \bar{\mathfrak{f}}_{(i)}(x,\, \partial)$, $i = 1, 2$. One gets the following star product multiplication which is a particular representation of algebraic formula (2.6)

$$\xi(x) \star \zeta(x) = (\bar{\mathfrak{f}}_{(1)}(x,\partial)\xi(x))(\bar{\mathfrak{f}}_{(2)}(x,\partial)\zeta(x))\,. \tag{2.8}$$

The important application of twisted Poincaré algebras to the covariant description of noncommutative Minkowski spaces, namely describing the quantum covariance of (1.1) for the case $\theta_{\mu\nu} = \theta^{(0)}_{\mu\nu}$ is quite recent[†]. The quantum symmetry which leaves invariant the simplest form of (1.1)[‡]

$$[\widehat{x}_\mu, \widehat{x}_\nu]_\bullet = \frac{i}{\kappa^2}\theta^{(0)}_{\mu\nu}\,, \tag{2.9}$$

[†]The twisted Poincaré symmetries corresponding to $\theta_{\mu\nu} = \theta^{(0)}_{\mu\nu}$ were earlier discussed in [13–16], but the full consequences of the twisted description were realized in 2004 (see e.g. [12,17–19]).

[‡]Below, in chapter 2 and 3, we shall use explicitly the *fat dot notation* for the algebra of functions on quantum Minkowski space in order to stress its Hopf algebra module origin.

(where $[a,b]_\bullet = a \bullet b - b \bullet a$) is generated by the following Abelian twist

$$\mathcal{F}_\theta = \exp \frac{i}{2\kappa^2}(\,\theta^{\mu\nu}_{(0)}\, P_\mu \wedge P_\nu\,)\,. \tag{2.10}$$

We obtain the twisted Poincaré-Hopf structure with classical Poincaré algebra relations and modified coproducts of Lorentz generators $M_{\mu\nu}$

$$\Delta_\theta(P_\mu) = \Delta_0(P_\mu), \tag{2.11}$$

$$\begin{aligned}\Delta_\theta(M_{\mu\nu}) &= \mathcal{F}_\theta \circ \Delta_0(M_{\mu\nu}) \circ \mathcal{F}_\theta^{-1} \\ &= \Delta_0(M_{\mu\nu}) - \frac{1}{\kappa^2}\theta^{\rho\sigma}_{(0)}[(\eta_{\rho\mu}P_\nu - \eta_{\rho\nu}\,P_\mu)\otimes P_\sigma \\ &\qquad + P_\rho \otimes (\eta_{\sigma\mu}P_\nu - \eta_{\sigma\nu}P_\mu)]\,.\end{aligned} \tag{2.12}$$

One can consider however also other Abelian twists of Poincaré symmetries, depending on the Lorentz generators $M_{\mu\nu}$ (see [20,13,14,21]). It appears that only subclass of general commutator (1.1) with linear and quadratic terms can be covariantized by twisted Poincaré algebras. In the following section we shall consider the quantum Poincaré symmetries corresponding to the following two twist functions [21]:

i) Lie-algebraic relations for noncommutative Minkowski space

$$\mathcal{F}_{(\alpha\beta)} = \exp \frac{i}{2\kappa}(\zeta^\lambda\, P_\lambda \wedge M_{\alpha\beta})\,, \tag{2.13}$$

where $\alpha,\ \beta = 0,1,2,3$ are fixed and the vector $\zeta^\lambda = \theta^{\lambda\alpha\beta}_{(1)}$ has vanishing components $\zeta^\alpha,\ \zeta^\beta$.

ii) Quadratic deformations of Minkowski space

$$\mathcal{F}_{(\alpha\beta\gamma\delta)} = \exp \frac{i}{2}\zeta\, M_{\alpha\beta} \wedge M_{\gamma\delta}\,, \tag{2.14}$$

where $\zeta = \theta^{\alpha\beta\delta\gamma}_{(2)}$ is a numerical parameter, all the four indices $\alpha,\beta,\gamma,\delta$ are fixed and different.

3. Lie-algebraic and Quadratic Quantum-Covariant Noncommutative Minkowski Spaces

In this Section we shall report on results presented in [21], which we supplement by the proof of quantum translational invariance.

In the formalism of quantum-deformed Hopf-algebraic symmetries the quantum-covariant noncommutative Minkowski space can be introduced in two ways:

i) as the translation sector of quantum Poincaré group,

ii) as the quantum representation space (a Hopf algebra module) for quantum Poincaré algebra with the action of the deformed symmetry generators satisfying suitably deformed Leibnitz rule (2.5).

In the case of constant tensor $\theta_{\mu\nu} = \theta^{(0)}_{\mu\nu}$ the quantum Poincaré group algebra dual to the coproducts (2.11), (2.12) is known[15,19,21], and the quantum translations do not satisfy the relation (2.9). It appears that the relation (2.9) as describing quantum-covariant noncommutative Minkowski space can be obtained only as the Hopf algebra module. To the contrary, in the case of twisted relativistic symmetries generated by the twist factors (2.13, 2.14) it can be shown that both definitions i) and ii) coincide [21].

i) Lie-algebraic noncommutative Minkowski space.
The commutator algebra following from (2.13) and the formula (2.6) has the form [21]

$$[\widehat{x}_\mu, \widehat{x}_\nu]_\bullet = C^\rho{}_{\mu\nu} \widehat{x}_\rho \,, \tag{3.1}$$

where

$$C^\rho{}_{\mu\nu} = \frac{i}{\kappa}\zeta_\mu(\eta_{\beta\nu}\delta^\rho{}_\alpha - \eta_{\alpha\nu}\delta^\rho{}_\beta) + \frac{i}{\kappa}\zeta_\nu(\eta_{\alpha\mu}\delta^\rho{}_\beta - \eta_{\beta\mu}\delta^\rho{}_\alpha) \,. \tag{3.2}$$

The relations (3.1) can be written in more transparent way as follows (α, β are fixed by the choice of twist function)

$$[\widehat{x}_\alpha, \widehat{x}_\lambda]_\bullet = \frac{i}{\kappa}\zeta_\lambda \eta_{\alpha\alpha}\widehat{x}_\beta \,, \qquad [\widehat{x}_\beta, \widehat{x}_\lambda]_\bullet = -\frac{i}{\kappa}\zeta_\lambda \eta_{\beta\beta}\widehat{x}_\alpha \,, \tag{3.3}$$

where $\zeta_\alpha = \zeta_\beta = 0$.
The quantum Lorentz covariance of (3.1) under the Hopf action of the Lorentz generators $M_{\mu\nu}$ has been shown in [21]. We shall show the quantum translational invariance of (3.1) using the differential realization (2.7). The fourmomentum coproduct generated by twist (2.13) has the form [21]

$$\Delta(P_\mu) = \Delta_0(P_\mu) + \frac{1}{2\kappa}\zeta^\lambda P_\lambda \wedge (\eta_{\alpha\mu}P_\beta - \eta_{\beta\mu}P_\alpha) + \mathcal{O}(P^3) \,. \tag{3.4}$$

Putting in (2.5) $h \equiv P_\mu$, $a \equiv x_\rho$, $b \equiv x_\sigma$ and using (3.2) we obtain

$$\begin{aligned} P_\mu \triangleright (x_\rho \bullet x_\sigma) &= i x_{\{\rho}\eta_{\sigma\}\mu} + \frac{1}{2\kappa}(\eta_{\alpha\mu}\zeta_{[\sigma}\eta_{\rho]\beta} - \eta_{\beta\mu}\zeta_{[\sigma}\eta_{\rho]\alpha}) \,, \\ &= i x_{\{\rho}\eta_{\sigma\}\mu} + \frac{1}{2}P_\mu \triangleright C^\lambda{}_{\rho\sigma} x_\lambda \,. \end{aligned} \tag{3.5}$$

Finally we get

$$P_\mu \triangleright [x_\rho, x_\sigma]_\bullet = P_\mu \triangleright C^\lambda{}_{\rho\sigma} x_\lambda \,, \tag{3.6}$$

i.e. the relation (3.1) is covariant.

ii) Quadratic noncommutativity of Minkowski space coordinates.
After using the formula (2.6) with inserted twist (2.14) one gets the following commutation relations of space-time coordinates ($\{a,b\}_\bullet = a\bullet b + b\bullet a$)

$$
\begin{aligned}
[\widehat{x}_\mu, \widehat{x}_\nu]_\bullet = i \sinh\frac{\zeta}{2}\cosh\frac{\zeta}{2}(&\eta_{\alpha[\mu}\eta_{\gamma\nu]}\{\widehat{x}_\beta, \widehat{x}_\delta\}_\bullet - \eta_{\alpha[\mu}\eta_{\delta\nu]}\{\widehat{x}_\beta, \widehat{x}_\gamma\}_\bullet \\
&-\eta_{\beta[\mu}\eta_{\gamma\nu]}\{\widehat{x}_\alpha, \widehat{x}_\delta\}_\bullet + \eta_{\beta[\mu}\eta_{\delta\nu]}\{\widehat{x}_\alpha, \widehat{x}_\gamma\}_\bullet) \\
- \sinh^2\frac{\zeta}{2}(&\sum_{\substack{k=\alpha,\beta \\ l=\gamma,\delta}} \delta^k{}_{[\mu}\delta^l{}_{\nu]}[\widehat{x}_k, \widehat{x}_l]_\bullet)\,,
\end{aligned}
\tag{3.7}
$$

or in more explicit form ($k=\alpha,\beta$ and $l=\gamma,\delta$)

$$
\begin{aligned}
[\widehat{x}_k, \widehat{x}_l]_\bullet = i \tanh\frac{\zeta}{2}(&\eta_{\alpha k}\eta_{\gamma l}\{\widehat{x}_\beta, \widehat{x}_\delta\}_\bullet - \eta_{\alpha k}\eta_{\delta l}\{\widehat{x}_\beta, \widehat{x}_\gamma\}_\bullet \\
&-\eta_{\beta k}\eta_{\gamma l}\{\widehat{x}_\alpha, \widehat{x}_\delta\}_\bullet + \eta_{\beta k}\eta_{\delta l}\{\widehat{x}_\alpha, \widehat{x}_\gamma\}_\bullet)\,,
\end{aligned}
\tag{3.8}
$$

and $[\widehat{x}_\alpha, \widehat{x}_\beta]_\bullet = [\widehat{x}_\gamma, \widehat{x}_\delta]_\bullet = 0$.
We conjecture that the relations (3.7) are covariant under the action of quantum Poincaré symmetries, generated by twist (2.14).
The linear and quadratic relations (3.3) and (3.8) provide special choices of the constant parameters $\theta^{(1)\rho}_{\mu\nu}$, $\theta^{(2)\rho\tau}_{\mu\nu}$ for which the quantum covariance group was found in [21].

4. Particle Dynamics Invariant Under Twisted Relativistic and Galilean Symmetries

The discussion of the noncommutative dynamical theories one begins naturally with the consideration of classical mechanics models. We shall restrict our considerations here to the case $\theta_{\mu\nu} = \theta^{(0)}_{\mu\nu}$, i.e. the noncommutative space-time described by (2.9). One can introduce the Lagrangian models describing free point particles moving in noncommutative space-time in the following two ways:

i) If $\theta_{\mu 0} = 0$, i.e. we have the relations

$$[\widehat{x}_i, \widehat{x}_j] = i\theta_{ij}\,, \tag{4.1}$$

$$[\widehat{x}_0, \widehat{x}_i] = 0\,, \tag{4.2}$$

we deal with classical time variable t, where $\hat{x}_0 = ct$ and noncommutative space coordinates $\hat{x}_i$. In such a case one can look for the non-relativistic Lagrangian models with constraints, which provide the relation (4.1) as the quantized Dirac bracket. Such a first model was constructed in [22] in $D = (2+1)$ dimensions with the following Lagrangian

$$\mathcal{L} = \frac{m\dot{x}_i^2}{2} - k\epsilon_{ij}\dot{x}_i\ddot{x}_j \,. \tag{4.3}$$

The higher order Lagrangian (4.3) can be expressed if first order form in six-dimensional phase space $(x_i, p_i, \tilde{p}_i)$[§] and after introducing the linear transformations

$$\begin{aligned} X_i &= x_i - \frac{2}{m}\tilde{p}_i \,, \\ P_i &= p_i \,, \\ \tilde{P}_i &= \epsilon_{ij}\tilde{p}_j + \frac{k}{m}p_i \,, \end{aligned} \tag{4.4}$$

one obtains the following symplectic structure for the variables $Y_A = (X_i, P_i, \tilde{P}_i)$, $(A = 1...6)$

$$\{Y_A, Y_B\} = \Omega_{AB} \,, \qquad \Omega = \begin{pmatrix} \frac{2k}{m^2}\epsilon & 1_2 & 0 \\ -1_2 & 0 & 0 \\ 0 & 0 & \frac{k}{2}\epsilon \end{pmatrix} . \tag{4.5}$$

One can identify (4.1) with quantized PB for the space variables X_i if we put in (4.1) $\theta_{ij} = \frac{2k}{m^2}\epsilon_{ij}$.
In [22] the dimension $D = 2+1$ was chosen because in two space dimensions one can put $\theta_{ij} = \theta\epsilon_{ij}$, i.e. the relation (4.1) does not break the classical Galilean invariance. However if $k \neq 0$ the Galilean algebra is centrally extended by second *exotic* central charge [23].

ii) For general constant $\theta_{\mu\nu}$ one obtains the noncommutative action describing free particle motion if we introduce in the first order action for classical massive relativistic particle

$$S = \int d\tau [\dot{y}_\mu p^\mu - e(p^2 - m^2)] \,, \tag{4.6}$$

the following change of variables (we recall that $\theta_{\mu\nu} = \theta^{(0)}_{\mu\nu}$)

$$y_\mu = x_\mu + \frac{1}{a}\theta_{\mu\nu}p^\nu \,. \tag{4.7}$$

[§]The momenta $p_i, \tilde{p}_i$ are described by the following formulae

$$p_i = \frac{\partial\mathcal{L}}{\partial\dot{x}_i} - \frac{d}{dt}\frac{\partial\mathcal{L}}{\partial\ddot{x}_i} \,, \qquad \tilde{p}_i = \frac{\partial\mathcal{L}}{\partial\ddot{x}_i} \,.$$

It is easy to check that if we introduce CCR following from (4.6)

$$[y_\mu, y_\nu] = 0\,, \qquad [y_\mu, p^\nu] = i\delta_\mu{}^\nu\,, \qquad [p^\mu, p^\nu] = 0\,, \tag{4.8}$$

then the variables x_μ in (4.7) satisfy the relation (2.9) if we put $a = 2\kappa^2$. Using the relation (4.7) one can rewrite the action (4.6) as follows

$$S = \int d\tau[\dot{x}_\mu p^\mu - e(p^2 - m^2) + \frac{1}{a}\theta^{\mu\nu}\dot{p}_\mu p_\nu]\,. \tag{4.9}$$

The variables y_μ, p_μ in (4.6) are classical, i.e. transform under Lorentz rotations in standard way

$$y'_\mu = \Lambda_\mu{}^\nu y_\nu\,, \qquad p'_\mu = p_\mu\,. \tag{4.10}$$

Using (4.7) and (4.10) one gets however

$$\begin{aligned} x'_\mu &= y'_\mu + \frac{1}{a}\theta_{\mu\nu}\Lambda^\nu{}_\rho p^\rho \\ &= \Lambda_\mu{}^\nu x'_\nu + \frac{1}{a}(\Lambda_\mu{}^\rho\theta_{\rho\nu} + \theta_{\mu\rho}\Lambda^\rho{}_\nu)p^\nu\,. \end{aligned} \tag{4.11}$$

Interestingly enough, the transformations (4.11) describe exactly the twisted Lorentz transformations, generated by the coproduct (2.12), which leave invariant the action (4.9) for the noncommutative relativistic particle.

The model (4.9) has been firstly obtained without reference to twisted Lorentz symmetries by Deriglazov [24] and its non-relativistic version

$$S_{NR} = \int dt[\dot{x}_i\dot{p}_i - \frac{1}{2m}\vec{p}^{\,2} + \frac{1}{a}\theta_{ij}\dot{p}_i p_j]\,, \tag{4.12}$$

in $D = 2+1$, when $\theta_{ij} = \epsilon_{ij}$, it was proposed by Duval and Horvathy [25]. It is well-known however that the model (4.12) can be also derived from the model (4.3). Indeed, the first order formulation of the model (4.3) in Faddeev-Jackiw approach [26] to higher order Lagrangians provides the action [27,28]

$$\mathcal{L} = p_i(\dot{x}_i - y_i) + \frac{\vec{y}^{\,2}}{2m} + \frac{1}{a}\epsilon_{ij}\dot{p}_i p_j\,. \tag{4.13}$$

The Lagrangian (4.13) after introducing the new coordinates

$$X_i = x_i + \frac{1}{a}\epsilon_{ij}(y_j - p_j)\,, \tag{4.14}$$

provides the Lagrangian (4.12) (with x_i replaced by X_i) and additional term which depends on auxiliary internal variables commuting with (X_i, p_i) [27].

The nonrelativistic model (4.12) can be considered in any space dimension d . If $d = 2$ the action (4.12) is, similarly as (4.3), invariant under the transformations of exotic $(2+1)$ - dimensional Galilean group. If $d > 2$ the invariance of the nonrelativistic model (4.12) can be achieved by considering quantum Galilean symmetries, with twisted space rotations generated by the following nonrelativistic twist

$$\mathcal{F}_\theta^{NR} = \exp \frac{i}{a} \theta_{ij} P_i \wedge P_j \,. \tag{4.15}$$

The formulation of twisted quantum mechanics invariant under twisted quantum Galilei group is now under our consideration.

5. Final Remarks

We presented in this paper some selected aspects of the theory of noncommutative space-times, with new results on quantum Poincaré covariance of a class of linearly and quadratically deformed Minkowski spaces. We also considered the non-relativistic and relativistic particle models on noncommutative space-time with numerical value of the noncommutativity function $\theta_{\mu\nu} = \theta_{\mu\nu}^{(0)}$ and have pointed out their twisted quantum covariance. We see that the role of quantum deformations is to introduce in place of broken classical symmetries a modified transformations which imply the quantum covariance. Such a possibility selects only particular class of tensors $\theta_{\mu\nu}^{(1)\rho}$ and $\theta_{\mu\nu}^{(2)\rho\tau}$ in formula (1.1).

Most of the applications of the noncommutative space-times in the literature assume the choice $\theta_{\mu\nu} = \theta_{\mu\nu}^{(0)}$ (see (2.9)). In this talk we presented also the results for linear ($\theta_{\mu\nu}^{(1)\rho} \neq 0$) and quadratic ($\theta_{\mu\nu}^{(2)\rho\tau} \neq 0$) deformations of Minkowski space. The extension of particle models on noncommutative space-times to linearly and quadratically deformed Minkowski spaces is now studied.

Acknowledgements One of the authors (JL) would like to thank Prof. Ge Mo-Lin and Dr. Cheng-Ming Bai for their fantastic hospitality during the conference at Nankai Mathematical Institute in Tianjin. We also acknowledge the support of KBN grant 1P03B 01828 and the EPSRC.

References

1. S. Doplicher, K. Fredenhagen and J.E. Roberts, Phys. Lett. **B331**, 39 (1994); Commun. Math. Phys. 172, 187 (1995); hep-th/0303037.
2. A. Kempf and G. Mangano, Phys. Rev. **D55**, 7909 (1997); hep-th/9612084.

3. N. Seiberg and E. Witten, JHEP **09**, 032 (1999); hep-th/9908142.
4. Chong-Sun Chu, *Noncommutative geometry from strings*; hep-th/0502167.
5. S. Zakrzewski, Journ. of Phys. **A27**, 2075 (1994).
6. S. Majid, H. Ruegg, Phys. Lett **B334**, 338 (1994).
7. J. Lukierski, H. Ruegg, W. Zakrzewski, Ann. Phys. **243**, 90 (1995).
8. V.G. Drinfeld, Leningrad Math. J. 1, 1419 (1990).
9. S. Majid, *Foundation of Quantum Groups*, Cambridge University Press (1994).
10. P.P. Kulish, A.I. Mudrov, Proc. Stekl. Inst. Math. **226**, 97 (1999); q-alg/9901019.
11. C. Blohmann, Jour. Math. Phys. **44**, 4736 (2003).
12. M. Chaichian, P.P Kulish, K. Nishijima and A. Tureanu, Phys. Lett. **B604**, 98 (2004); hep-th/0408069.
13. P. Aschieri and L. Castellani, Int. J. Mod. Phys. **A11**, 4513 (1996); q-alg/9601006.
14. P. Aschieri, L. Castellani and A.M. Scarfone, Eur. Phys. J. **C7**, 159 (1999).
15. R. Oeckl, Nucl. Phys. **B581**, 559 (2000); hep-th/0003018.
16. C. Jambor and A. Sykora, *Realization of algebras with the help of ⋆-products*; hep-th/0405268.
17. J. Wess, *Deformed coordinate spaces: Derivatives*; hep-th/0408080.
18. F. Koch and E. Tsouchnika, Nucl. Phys. **B717**, 387 (2005); hep-th/0409012.
19. P. Kosiński and P. Maślanka, *Lorentz - invariant interpretation of noncommutative space-time: Global version*; hep-th/0408100.
20. J. Lukierski, A. Nowicki, H. Ruegg and V.N. Tolstoy, J. Phys. **A27**, 2389 (1994); hep-th/9312068.
21. J. Lukierski and M. Woronowicz, *New Lie-algebraic and quadratic deformations of Minkowski space from twisted Poincare symmetries*; hep-th/0508083, PLB (in press).
22. J. Lukierski, P. Stichel and W.J. Zakrzewski, Ann. of Phys. **260**, 224 (1997).
23. J.M. Levy-Leblond, *Group Theory and Applications*, vol. 2, ed. Loeble, Acad. Press, New York (1972), p.222.
24. A.A. Deriglazov, *Noncommutative relativistic particle*; hep-th/0207274.
25. C.A. Duval and P. Horvathy, J. Phys. **A34**, 10097 (2001).
26. L. Faddeev and R. Jackiw, Phys. Rev. Lett **60**, 1968 (1988).
27. P. Horvathy and M.S. Plyushchay, JHEP **206** 33 (2002).
28. J. Lukierski, P. Stichel and W.J. Zakrzewski, Ann. Phys. **306**, 78 (2003).

Toeplitz Quantization and Symplectic Reduction

Xiaonan Ma

Centre de Mathématiques Laurent Schwartz,
UMR 7640 du CNRS,
Ecole Polytechnique,
91128 Palaiseau Cedex, France
E-mail: ma@math.polytechnique.fr

Weiping Zhang

Chern Institute of Mathematics & LPMC
Nankai University
Tianjin 300071, P. R. China
E-mail address: weiping@nankai.edu.cn

Dedicated to the memory of Professor Shiing–Shen Chern

In [9], we announced the asymptotic expansion of the G-invariant Bergman kernel of the spin^c Dirac operator associated with high tensor powers of a positive line bundle on a symplectic manifold. In this note, we describe several consequences of our asymptotic expansion of the G-invariant Bergman kernel in the Kähler case, especially, we study the Toeplitz quantization in the framework of the symplectic reduction. The full details can be found in [10].

1. Toeplitz quantization

Let (X, ω) be a compact Kähler manifold with Kähler form ω, and $\dim_{\mathbb{C}} X = n$. Let J be the almost complex structure on the real tangent bundle TX. Let $g^{TX}(v, w) := \omega(v, Jw)$ be the corresponding Riemannian metric on TX.

Let L be a holomorphic line bundle over X with Hermitian metric h^L. Let ∇^L be the holomorphic Hermitian connection on (L, h^L) with curvature $R^L := (\nabla^L)^2$. We suppose that (L, h^L) is a pre-quatum line bundle of (X, ω), i.e.

$$\frac{\sqrt{-1}}{2\pi} R^L = \omega. \tag{1.1}$$

According the geometric quantization introduced by Kostant and

Souriau, the Kähler manifold (X,ω) is the classical phase space and $H^0(X,L)$, the space of holomorphic sections of L on X, is the quantum space. The set of classical observables is the Poisson algebra $\mathscr{C}^\infty(X)$, the quantum observables are the linear operators on $H^0(X,L)$. The semi-classical limit is a way to relate the classical and quantum observables, basically, for any $p\in\mathbb{N}$, we replace L by L^p, then we obtain a sequence of spaces $H^0(X,L^p)$, the semi-classical limit is the process of $p\to\infty$. In this note, we will restrict ourself to a family of quantum observables : Toeplitz operators.

Let $\{\, ,\}$ be the Poisson bracket on $(X,2\pi\omega)$: for $f_1,f_2\in\mathscr{C}^\infty(X)$, if ξ_{f_2} is the Hamiltonian vector field generated by f_2 which is defined by $2\pi i_{\xi_{f_2}}\omega=df_2$, then

$$\{f_1,f_2\}(x)=(\xi_{f_2}(df_1))(x). \tag{1.2}$$

Let dv_X be the Riemannian volume form of (X,g^{TX}), then $dv_X=\omega^n/n!$. We define the L^2-scalar product $\langle\ \ \rangle$ on $\mathscr{C}^\infty(X,L^p)$ by

$$\langle s_1,s_2\rangle=\int_X\langle s_1,s_2\rangle_{L^p}(x)\,dv_X(x)\,. \tag{1.3}$$

Let Π_p denote the orthogonal projection from $(L^2(X,L^p),\langle\ \ \rangle)$, the space of L^2 sections of L^p on X, to $H^0(X,L^p)$, the space of holomorphic sections of L^p on X.

For any $f\in\mathscr{C}^\infty(X)$, consider the Toeplitz operators

$$T_p(f)=\Pi_pf\Pi_p:H^0(X,L^p)\to H^0(X,L^p). \tag{1.4}$$

We denote by $\|T_p(f)\|$ the operator norm of $T_p(f)$ with respect to the scalar product $\langle\ \ \rangle$.

We now state two results of Bordemann-Meinrenken-Schlichenmaier[2], concerning the asymptotic behavior of $T_p(f)$ as $p\to+\infty$.

Theorem 1.1. *As $p\to+\infty$, one has*

$$\lim_{p\to+\infty}\|T_p(f)\|=\|f\|_\infty, \tag{1.5a}$$

$$[T_p(f),T_p(g)]=\frac{1}{\sqrt{-1}p}T_p(\{f,g\})+O(p^{-2}). \tag{1.5b}$$

2. Hamiltonian action and symplectic reduction

Let E be a holomorphic vector bundle on X with Hermitian metric h^E. Let ∇^E be the holomorphic Hermitian connection on (E,h^E). Let G be a compact connected Lie group. Let $\mathfrak{g}$ be the Lie algebra of G.

Suppose that G acts holomorphically on X, and the action of G lifts holomorphically on L, E and preserves the metrics h^L, h^E. Then the action of G preserves ω, the connections ∇^L, ∇^E.

For $K \in \mathfrak{g}$, we denote by K^X the vector field on X generated by K, and by L_K the infinitesimal action induced by K on the corresponding vector bundles. Let $\mu : X \to \mathfrak{g}^*$ be defined by

$$2\pi\sqrt{-1}\mu(K) := \nabla^L_{K^X} - L_K, \ K \in \mathfrak{g}. \tag{2.1}$$

Then μ is the corresponding **moment map**, i.e. for any $K \in \mathfrak{g}$,

$$d\mu(K) = i_{K^X}\omega. \tag{2.2}$$

Definition 2.1. The Marsden-Weinstein **symplectic reduction** space X_G is defined to be

$$X_G = \mu^{-1}(0)/G. \tag{2.3}$$

Basic assumption: $0 \in \mathfrak{g}^*$ is a regular value of the moment map $\mu : X \to \mathfrak{g}^*$.

Then $\mu^{-1}(0)$ is a closed manifold. For simplicity, also assume that G acts on $\mu^{-1}(0)$ freely, then X_G is a compact smooth manifold and carries an induced symplectic form ω_G.

Moreover, J induces a complex structure J_G on TX_G such that $\omega_G(\cdot, J_G\cdot)$ determines a Riemannian metric g^{TX_G} on TX_G. Thus (X_G, ω_G, J_G) is also Kähler.

The line bundle (L, h^L) induces a Hermitian line bundle (L_G, h^{L_G}) on X_G by identifying G-invariant sections of L on $\mu^{-1}(0)$. In fact (L_G, h^{L_G}) is a pre-quantized holomorphic line bundle over (X_G, ω_G), cf. [5].

In the same way, (E, h^E) induces a holomorphic Hermitian vector bundle (E_G, h^{E_G}) on X_G.

3. Toeplitz quantization and symplectic reduction

We now assume that a connected compact Lie group acts on (X, ω, J, L) in a Hamiltonian way as before.

Let $i : \mu^{-1}(0) \hookrightarrow X$ denote the canonical embedding. We assume as before that 0 is a regular value of μ and G acts on $\mu^{-1}(0)$ freely. Then

$$\pi : \mu^{-1}(0) \to X_G$$

is a principal fibration with fiber G.

Let $H^0(X, L^p \otimes E)^G$ be the G-invariant part of $H^0(X, L^p \otimes E)$, the space of holomorphic sections of $L^p \otimes E$ on X. Let $\mathscr{C}^\infty(X, L^p \otimes E)^G$ (resp.

$\mathscr{C}^\infty(\mu^{-1}(0), L^p \otimes E)^G$) be the G-invariant smooth sections of $L^p \otimes E$ on X (resp. $\mu^{-1}(0)$). Let $\pi_G : \mathscr{C}^\infty(\mu^{-1}(0), L^p \otimes E)^G \to \mathscr{C}^\infty(X_G, L^p_G \otimes E_G)$ be the natural identification. By a result of Zhang[13], for p large enough, the map

$$\pi_G \circ i^* : \mathscr{C}^\infty(X, L^p \otimes E)^G \to \mathscr{C}^\infty(X_G, L^p_G \otimes E_G)$$

induces a natural isomorphism

$$\sigma_p = \pi_G \circ i^* : H^0(X, L^p \otimes E)^G \to H^0(X_G, L^p_G \otimes E_G). \tag{3.1}$$

(When $E = \mathbb{C}$, this result was first proved by Guillemin-Sternberg[5].)

Let dv_{X_G} be the Riemannian volume form on (X_G, g^{TX_G}). Let $\Pi_{G,p}$ be the orthogonal projection from $\mathscr{C}^\infty(X_G, L^p_G \otimes E_G)$ (with the scalar product $\langle\ \ \rangle$ induced by h^{L_G}, h^{E_G} and dv_{X_G} as in (1.3)), onto $H^0(X_G, L^p_G \otimes E_G)$.

Definition 3.1. A family of operators $T_p : H^0(X_G, L^p_G \otimes E_G) \to H^0(X_G, L^p_G \otimes E_G)$ is a Toeplitz operator if there exists a sequence of sections $g_l \in \mathscr{C}^\infty(X_G, \mathrm{End}(E_G))$ with an asymptotic expansion $g(\cdot, p)$ of the form $\sum_{l=0}^\infty p^{-l} g_l(x) + \mathscr{O}(p^{-\infty})$ in the $\mathscr{C}^\infty$ topology such that

$$T_p = \Pi_{G,p}\, g(\cdot, p) \Pi_{G,p} + \mathscr{O}(p^{-\infty}). \tag{3.2}$$

We call $g_0(x)$ the principal symbol of T_p.

For any $x \in X_G$, let $\mathrm{vol}(\pi^{-1}(x))$ be the volume of the orbit $\pi^{-1}(x)$ equipped with the metric induced by g^{TX}. We define the potential function

$$h(x) = \sqrt{\mathrm{vol}(\pi^{-1}(x))}. \tag{3.3}$$

For any $p > 0$, let P^G_p denote the orthogonal projection from $(\mathscr{C}^\infty(X, L^p \otimes E), \langle\ \ \rangle)$ to $H^0(X, L^p \otimes E)^G$. Set

$$\sigma^G_p = \sigma_p P^G_p : \mathscr{C}^\infty(X, L^p \otimes E) \to H^0(X_G, L^p_G \otimes E_G). \tag{3.4}$$

Let

$$(\sigma^G_p)^* : H^0(X_G, L^p_G \otimes E_G) \to \mathscr{C}^\infty(X, L^p \otimes E)$$

denote the adjoint of σ_p.

Theorem 3.1. *For any $f \in \mathscr{C}^\infty(X, \mathrm{End}(E))$, let $f^G \in \mathscr{C}^\infty(X_G, \mathrm{End}(E_G))$ denote the associated G-invariant section defined by $f^G(x) = \int_G g f(g^{-1}x) dg$, here dg is a Haar measure on G. Then*

$$\mathcal{T}_p(f) = p^{-\frac{\dim G}{2}} \sigma^G_p f (\sigma^G_p)^* : H^0(X_G, L^p_G \otimes E_G) \to H^0(X_G, L^p_G \otimes E_G) \tag{3.5}$$

is a Toeplitz operator with principal symbol $2^{\frac{\dim G}{2}} \frac{f^G}{h^2}(x)$. *Especially,*

$$\mathcal{T}_p(f) = \Pi_{G,p} 2^{\frac{\dim G}{2}} \frac{f^G}{h^2} \Pi_{G,p} + \mathcal{O}(1/p) \tag{3.6}$$

as $p \to +\infty$. *In particular,* $p^{-\dim G/2} \sigma_p^G (\sigma_p^G)^*$ *is a Toeplitz operator with principal symbol* $2^{\dim G/2}/h^2$.

Corollary 3.1. *For any* $f_1, f_2 \in \mathscr{C}^\infty(X)$, *we identify them as sections of* $\mathrm{End}(E)$ *by multiplications, then one has*

$$[\mathcal{T}_p(f_1), \mathcal{T}_p(f_2)] = \frac{2^{\dim G}}{\sqrt{-1}p} \Pi_{G,p} \left\{ \frac{f_1^G}{h^2}, \frac{f_2^G}{h^2} \right\} \Pi_{G,p} + \mathcal{O}(p^{-2}). \tag{3.7}$$

One can view this corollary as a generalization of the Bordemann-Meinrenken-Schlichenmaier theorem, Theorem 1.1, in the framework of geometric quantization. If $E = \mathbb{C}$ and $G = \{1\}$, Corollary 3.1 is (1.5b). If $G = \{1\}$ and general E, Corollary 3.1 was obtained in [7], [8].

On the other hand, if one defines the unitary operator

$$\Sigma_p = (\sigma_p^G)^* (\sigma_p^G (\sigma_p^G)^*)^{-1/2} : H^0(X_G, L_G^p \otimes E_G) \to \mathscr{C}^\infty(X, L^p \otimes E), \tag{3.8}$$

then one has the following result:

Theorem 3.2. *For any* $f \in \mathscr{C}^\infty(X, \mathrm{End}(E))$,

$$T_p^G(f) = \Sigma_p^* f \Sigma_p : H^0(X_G, L_G^p \otimes E_G) \to H^0(X_G, L_G^p \otimes E_G) \tag{3.9}$$

is a Toeplitz operator on X_G *with principal symbol* f^G.

Remark 3.1. If $E = \mathbb{C}$, Paoletti[11] also claimed that $p^{-\frac{\dim G}{2}} \sigma_p^G (\sigma_p^G)^*$ is a Toeplitz operator. When $G = T^k$ is a torus, and $E = \mathbb{C}$, Theorem 3.2 was first proved by Charles[3].

Let $\langle \, , \rangle_{L_G^p \otimes E_G}$ be the metric on $L_G^p \otimes E_G$ induced by h^{L_G} and h^{E_G}. In view of Tian and Zhang's analytic approach (cf. [12]. (3.54)) of geometric quantization conjecture of Guillemin-Sternberg, the natural Hermitian product on $\mathscr{C}^\infty(X_G, L_G^p \otimes E_G)$ is the following weighted Hermitian product $\langle \, , \rangle_h$:

$$\langle s_1, s_2 \rangle_h = \int_{X_G} \langle s_1, s_2 \rangle_{L_G^p \otimes E_G}(x_0) h^2(x_0) \, dv_{X_G}(x_0). \tag{3.10}$$

Theorem 3.3. *The isomorphism* $(2p)^{-\frac{\dim G}{4}} \sigma_p$ *is an asymptotic isometry from* $(H^0(X, L^p \otimes E)^G, \langle \, , \rangle)$ *onto* $(H^0(X_G, L_G^p \otimes E_G), \langle \, , \rangle_h)$: *i.e. if* $\{s_i^p\}_{i=1}^{d_p}$

is an orthonormal basis of $(H^0(X, L^p \otimes E)^G, \langle\ ,\ \rangle)$, *then*

$$(2p)^{-\frac{\dim G}{2}} \langle \sigma_p s_i^p, \sigma_p s_j^p \rangle_h = \delta_{ij} + \mathscr{O}(\frac{1}{p}). \tag{3.11}$$

4. The asymptotic expansion of the G-invariant Bergman kernel

Definition 4.1. The G-invariant Bergman kernel $P_p^G(x, x')$ with $x,\ x' \in X$ is the smooth kernel of the orthogonal projection $P_p^G : \mathscr{C}^\infty(X, L^p \otimes E) \to H^0(X, L^p \otimes E)^G$ with respect to $dv_X(x')$.

Our proof of the results in Section 3 relies on the asymptotic behavior as $p \to +\infty$ of the G-invariant Bergman kernel $P_p^G(x, x')$. We now describe some behavior of $P_p^G(x, y)$, as $p \to +\infty$.

Let U be an arbitrary (fixed) small open G-invariant neighborhood of $\mu^{-1}(0)$. At first, we have that for any $x,\ x' \in X \setminus U$, as $p \to +\infty$,

$$|P_p^G(x, x')|_{\mathscr{C}^\infty} = \mathscr{O}(p^{-\infty}). \tag{4.1}$$

This result shows that when $p \to +\infty$, $P_p^G(x, x')$ "localizes" near $\mu^{-1}(0)$ (and thus close to X_G). The main technical result of [9]. Theorem 2.2, and [10]. Theorem 0.2 is the asymptotic expansion of $P_p^G(x, x')$ for $x,\ x' \in U$ when $p \to \infty$ whose proofs use techniques adapting from the works of Bismut-Lebeau [1], Dai-Liu-Ma[4] and Ma-Marinescu[6]. One key step is to deform the Laplacian of the spin^c Dirac operator by a Casimir type operator. We refer the readers to [9], [10] for the details.

References

1. J.-M. Bismut and G. Lebeau, *Complex immersions and Quillen metrics*, Inst. Hautes Études Sci. Publ. Math. (1991), no. 74, ii+298 pp. (1992).
2. M. Bordemann and E. Meinrenken and M. Schlichenmaier, *Toeplitz quantization of Kähler manifolds and* gl(N), $N \to \infty$ *limits*, Comm. Math. Phys. **165** (1994), no. 2, 281–296.
3. L. Charles, *Toeplitz operators and hamiltonian torus action*, Preprint (2004), math.SG/0405128.
4. X. Dai, K. Liu, and X. Ma, *On the asymptotic expansion of Bergman kernel*, C. R. Math. Acad. Sci. Paris **339** (2004), no. 3, 193–198. The full version: J. Differential Geom. to appear, math.DG/0404494.
5. V. Guillemin and S. Sternberg, *Geometric quantization and multiplicities of group representations*, Invent. Math. **67** (1982), no. 3, 515–538.
6. X. Ma and G. Marinescu, *Generalized Bergman kernels on symplectic manifolds*, C. R. Math. Acad. Sci. Paris **339** (2004), no. 7, 493–498. The full version: math.DG/0411559.

7. X. Ma and G. Marinescu, *Toeplitz operators on symplectic manifolds*, Preprint.
8. X. Ma and G. Marinescu, *Holomorphic Morse inequalities and Bergman kernels*, book in preparation, (2005).
9. X. Ma and W. Zhang, *Bergman kernels and symplectic reduction*, C. R. Math. Acad. Sci. Paris **341** (2005), 297-302.
10. X. Ma and W. Zhang, *Bergman kernels and symplectic reduction*, Preprint 2005.
11. R. Paoletti, *The Szegö kernel of a symplectic quotient*, Adv. Math. **197** (2005), 523-553.
12. Y. Tian and W. Zhang, *An analytic proof of the geometric quantization conjecture of Guillemin-Sternberg*, Invent. Math. **132** (1998), no. 2, 229–259.
13. W. Zhang, *Holomorphic quantization formula in singular reduction*, Commun. Contemp. Math. **1** (1999), no. 3, 281–293.

On Mysteriously Missing T-duals, H-flux and the T-duality Group *

Varghese Mathai

Department of Mathematics
University of Adelaide
Adelaide 5005, Australia
E-mail: mathai.varghese@adelaide.edu.au

Jonathan Rosenberg

Department of Mathematics
University of Maryland
College Park, MD 20742, USA
E-mail: jmr@math.umd.edu

A general formula for the topology and H-flux of the T-duals of type II string theories with H-flux on toroidal compactifications is presented here. It is known that toroidal compactifications with H-flux do not necessarily have T-duals which are themselves toroidal compactifications. A big puzzle has been to explain these mysterious "missing T-duals", and our paper presents a solution to this problem using noncommutative topology. We also analyze the T-duality group and its action, and illustrate these concepts with examples.

T-duality is a symmetry of type II string theories that involves exchanging a theory compactified on a torus with a theory compactified on the dual torus. The T-dual of a type II string theory compactified on a circle, in the presence of a topologically nontrivial NS 3-form H-flux, was analyzed in special cases in [2,5–7]. There it was observed that T-duality changes not only the H-flux, but also the spacetime topology. A general formalism for dealing with T-duality for compactifications arising from a free circle action was developed in [8]. This formalism was shown to be compatible with two physical constraints: (1) it respects the local Buscher rules [1], and (2) it yields an isomorphism on twisted K-theory, in which the Ramond-Ramond charges and fields take their values [11–13]. It was shown in [8] that T-duality exchanges the first Chern class with the fiberwise integral of the H-flux,

* Dedicated to the memory of Professor S.S. Chern

thus giving a formula for the T-dual spacetime topology. In this note we will present an account for physicists of the results in [16], consisting of a formula for the T-dual of a toroidal compactification, that is a theory compactified via a free torus action, with H-flux. One striking new feature that occurs for higher dimensional tori is that not every toroidal compactification with H-flux has a T-dual; moreover, even if it has a T-dual, then the T-dual need not be another toroidal compactification with H-flux. A big puzzle has been to explain these mysterious "missing T-duals", and our paper presents a solution to this problem using noncommutative topology. A similar phenomenon was noticed in [15] in the special case of the trivial $\mathbb{T}^2$ bundle over $\mathbb{T}$ with non-trivial H-flux. We also show that the generalized T-duality group $GO(n,n;\mathbb{Z})$, n being the rank of the torus, acts to generate the complete list of T-dual pairs related to a given toroidal compactification with H-flux. We will explain these results by providing examples and applications.

In this letter we will consider type II string theories on target d-dimensional manifolds X, which are assumed to admit free, rank n torus actions. While for most physical applications one wants $d = 10$, we do not need to assume this, and in fact X could represent a partial reduction of the original 10-dimensional spacetime after preliminary compactification in $10 - d$ dimensions. The space of orbits of the torus action on X is given by a $(d-n)$-dimensional manifold, which we call Z. The freeness of the action implies that each orbit is a torus and that none of these tori degenerate. As a result X is a principal torus bundle over the base Z, and so its topology is entirely determined by the topology of the base Z together with the first Chern class c of the bundle $X \xrightarrow{p} Z$ in $H^2(Z,\mathbb{Z}^n)$. This viewpoint is useful in that it automatically identifies some gauge equivalent configurations, excludes configurations not satisfying some equations of motion and imposes the Dirac quantization conditions. The Chern class c is represented by a vector valued closed 2-form with integral periods, the curvature F. We will discuss conditions under which the pair $(X \xrightarrow{p} Z, H)$ has a T-dual, either another pair $(X \xrightarrow{p^\#} Z, H^\#)$ with the same base Z (the "classical" case) or a more general non-commutative object (the "nonclassical" case). In both cases, there should be a sense in which string theory on the original space X (with H-flux H) is equivalent to a theory on the T-dual.

Basic setup: *Let* $p\colon X \to Z$ *be a principal* T*-bundle as above, where* $T = (S^1)^n = \mathbb{T}^n$ *is a rank* n *torus. Let* $H \in H^3(X,\mathbb{Z})$ *be an H-flux on* X *satisfying* $\iota^* H = 0$, $\iota^*\colon H^3(X,\mathbb{Z}) \to H^3(T,\mathbb{Z})$, *where* $\iota\colon T \hookrightarrow X$ *is the* *inclusion of a fiber. (This condition is automatically satisfied when $n \leq 2$.)*

The simplest case when the condition $\iota^*(H) = 0$ does *not* apply is $X = \mathbb{T}^3$, when considered as a rank 3, principal torus bundle over a point, with H-flux a non-zero integer multiple of the volume 3-form on $\mathbb{T}^3$. When $\iota^*(H) \neq 0$, there is no T-dual in the sense we are considering, even in what we call the "nonclassical" sense.

It turns out that nontrivial bundles are always T-dual to trivial bundles with non-zero H-flux. Therefore we will need to include the fluxes H and $H^\#$ in our toroidal compactifications, which are then topologically determined by the triples (Z, c, H) and $(Z, c^\#, H^\#)$, where H and $H^\#$ are closed three-forms on the total spaces X and $X^\#$ respectively.

Our results on classical T-duals: *Suppose that we are in the basic setup as above. Choose a basis $\{\mathbb{T}^2_j\}_{j=1}^k$, $k = \binom{n}{2}$ for $H_2(T, \mathbb{Z})$ consisting of 2-tori, and push this forward into $H_2(X, \mathbb{Z})$ via ι_*. We can consider the cohomology classes*

$$\int_{\mathbb{T}^2_j} H = H \cap \iota_*(\mathbb{T}^2_j) \in H^1(X, \mathbb{Z}).$$

These classes restrict to 0 on the fibers, since $\iota^*(H) = 0$. Using the following exact sequence, derived from the spectral sequence of the torus bundle,

$$0 \to H^1(Z, \mathbb{Z}) \xrightarrow{p^*} H^1(X, \mathbb{Z}) \xrightarrow{\iota^*} H^1(T, \mathbb{Z}) \to \cdots, \tag{1}$$

we see that the classes $\int_{\mathbb{T}^2_j} H = H \cap \iota_*(\mathbb{T}^2_j) \in H^1(X, \mathbb{Z})$ come from unique classes $\{\beta_j\}_{j=1}^k$ in $H^1(Z, \mathbb{Z})$. Set

$$p_!(H) = \Big(\beta_1, \dots, \beta_k\Big) \in H^1(Z, \mathbb{Z}^k). \tag{2}$$

If $p_!(H) = 0 \in H^1(Z, \mathbb{Z}^k)$, and in particular if Z is simply connected, then there is a classical T-dual to (p, H), consisting of $p^\#\colon X^\# \to Z$, which is another principal T-bundle over Z, and $H^\# \in H^3(X^\#, \mathbb{Z})$, the T-dual H-flux on $X^\#$. One obtains a commuting diagram of the form

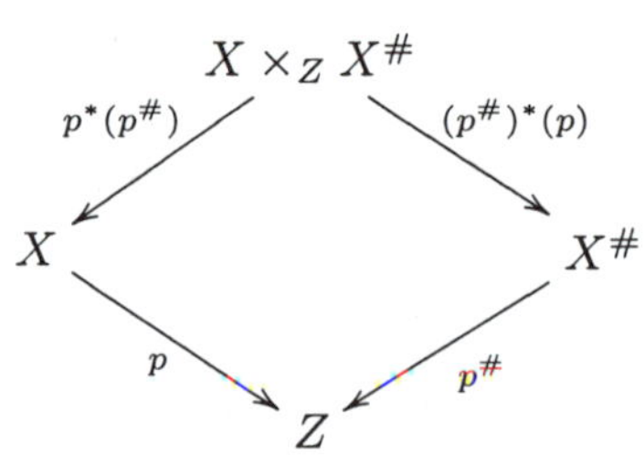

(3)

In this case, the compactifications topologically specified by (Z, c, H) *and* $(Z, c^\#, H^\#)$ *are T-dual if* c, $c^\# \in H^2(Z, \mathbb{Z}^n)$ *are related as follows:*

Let c_j, $j = 1, \cdots, n$, *be the components of* c. *Let* $X_j \xrightarrow{\pi_j} Z$ *be the principal* $\mathbb{T}^{n-1}$ *subbundle of* X *obtained by deleting* c_j, *i.e. the Chern class of* X_j *is*

$$c(\pi_j) = (c_1, \ldots, \hat{c}_j, \ldots, c_n).$$

Then $X \xrightarrow{p_j} X_j$ *is a principal* S^1*-bundle whose Chern class is equal to* $\pi_j^*(c_j)$. *Define* $X_j^\# \xrightarrow{\pi_j^\#} Z$, $X^\# \xrightarrow{p_j^\#} X_j^\#$ *etc. similarly. Then we have*

$$(\pi_j)^*(c_j^\#) = (p_j)_!(H) \quad \textbf{and} \quad (\pi_j^\#)^*(c_j) = (p_j^\#)_!(H^\#).$$

Here the correspondence space $X \times_Z X^\#$ is the submanifold of $X \times X^\#$ consisting of pairs of points (x, y) such that $p(x) = p^\#(y)$, and has the property that it implements the T-duality between (p, H) and $(p^\#, H^\#)$. It also turns out that $p_!^\#(H^\#) = 0 \in H^1(Z, \mathbb{Z}^k)$ and that the T-dual of $(p^\#, H^\#)$ is (p, H). So in this case, T-duality *exchanges the integral of the H-flux* (over a basis of circles in the fibers) *with the first Chern class.* The condition in the result above determines, at the level of cohomology, the curvatures F and $F^\#$. However the NS field strengths are only determined up to the addition of a three-form on the base Z, because the integral of such a form over a basis of circles in the fibers vanishes. This settles a conjecture in [8], and was also considered by [9].

The simplest higher rank example is $X = S^2 \times \mathbb{T}^2$, considered as the trivial $\mathbb{T}^2$ bundle over $Z = S^2$, with H-flux equal to $H = k_1 a \wedge b_1 + k_2 a \wedge b_2$, where we use the Künneth theorem to identify $H^3(S^2 \times \mathbb{T}^2, \mathbb{Z})$ with $H^2(S^2, \mathbb{Z}) \otimes H^1(\mathbb{T}^2, \mathbb{Z})$, and a is the generator of $H^2(S^2, \mathbb{Z}) \cong \mathbb{Z}$, b_1, b_2 are the generators of $H^1(\mathbb{T}^2, \mathbb{Z}) \cong \mathbb{Z}^2$ and $k_1, k_2 \in \mathbb{Z}$. Since S^2 is simply connected, $p_!(H) = 0$ and the T-dual of $(S^2 \times \mathbb{T}^2, H)$ is the nontrivial rank 2 torus bundle P over S^2 with Chern class $c_1(P) = (k_1 a, k_2 a) \in H^2(S^2, \mathbb{Z}) \oplus H^2(S^2, \mathbb{Z}) = H^2(S^2, \mathbb{Z}^2)$, and with H-flux equal to zero. This example generalizes easily by taking the Cartesian product with a manifold M, and pulling back the H-flux to the product and arguing as before, we see that the T-dual of $(M \times S^2 \times \mathbb{T}^2, H)$ is $(M \times P, 0)$.

Our results on nonclassical T-duals: *Suppose that we are in the basic setup as above. If* $p_!(H) \neq 0 \in H^1(Z, \mathbb{Z}^k)$, *then there is* no *classical T-dual to* (p, H); *however, there is a nonclassical T-dual consisting of a continuous field of (stabilized) noncommutative tori* A_f *over* Z, *where the fiber over the* *point $z \in Z$ is equal to the rank n noncommutative torus $A_{f(z)}$ (see Figure 1 below). Here $f\colon Z \to \mathbb{T}^k$ is a continuous map representing $p_!(H)$.*

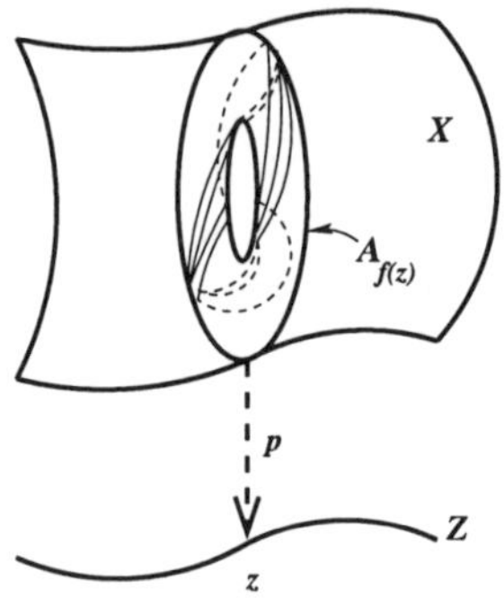

Fig. 1. In the diagram, the fiber over $z \in Z$ is the noncommutative torus $A_{f(z)}$, which is represented by a foliated torus, with foliation angle equal to $f(z)$.

This suggests an unexpected link between classical string theories and the "noncommutative" ones, obtained by "compactifying" matrix theory on tori, as in [4] (cf. also [19]). We now recall the definition of the rank n noncommutative torus A_θ, cf. [18]. This algebra (stabilized by tensoring with the compact operators $\mathcal{K}$) occurs geometrically as the foliation algebra associated to Kronecker foliations on the torus [3]. In [4], the same algebra occurs naturally from studying the field equations of the IKKT (Ishibashi-Kawai-Kitazawa-Tsuchiya) model compactified on n-tori, or from the study of BPS states of the BFSS (Banks-Fisher-Shenker-Susskind) model. (The IKKT and BFSS models are both large-N matrix models in which Poisson brackets in the Lagrangian are replaced by matrix commutators.) For each $\theta \in \mathbb{T}^k$, identified with a hermitian matrix $\theta = (\theta_{ij}),\ i,j = 1,\dots n,\ \theta_{ij} \in S^1$ with 1's down the diagonal, the *noncommutative torus* A_θ is defined abstractly as the C^*-algebra generated by n unitaries $U_j,\ j = 1,\dots,n$ in an infinite dimensional Hilbert space satisfying the commutation relation $U_iU_j = \theta_{ij}U_jU_i,\ i,j = 1,\dots,n$. Elements in A_θ can be represented by infinite power series

$$f = \sum_{m\in\mathbb{Z}^n} a_m U^m, \tag{4}$$

where $a_m \in \mathbb{C}$ and $U^m = U_1^{m_1}\dots U_n^{m_n}$, for all $m = (m_1,\dots,m_n) \in \mathbb{Z}^n$.

A famous example of a principal torus bundle with non-T-dualizable H-flux is provided by $\mathbb{T}^3$, considered as the trivial $\mathbb{T}^2$-bundle over $\mathbb{T}$, with

H given by k times the volume form on $\mathbb{T}^3$. H is non T-dualizable in the classical sense since $p_!(H) \neq 0 \in H^1(\mathbb{T}, \mathbb{Z})$. Alternatively, there are no non-trivial principal $\mathbb{T}^2$-bundles over $\mathbb{T}$, since $H^2(\mathbb{T}, \mathbb{Z}^2) = 0$, that is, there is no way to dualize the H-flux by a (principal) torus bundle over $\mathbb{T}$, cf. [7]. This is an example of a mysteriously missing T-dual. This example is covered by our result on nonclassical T-duals above. The T-dual is realized by a field of stabilized *noncommutative tori* fibered over $\mathbb{T}$. Let $\mathcal{H} = L^2(\mathbb{T})$ and consider the the projective unitary representation $\rho_\theta \colon \mathbb{Z}^2 \to \mathrm{PU}(\mathcal{H})$ in which the generator of the first $\mathbb{Z}$ factor acts by multiplication by z^k (where $\mathbb{T}$ is thought of as the unit circle in $\mathbb{C}$) and the generator of the second $\mathbb{Z}$ factor acts by translation by $\theta \in \mathbb{T}$. Then the Mackey obstruction of ρ_θ is $\theta^k \in \mathbb{T} \cong H^2(\mathbb{Z}^2, \mathbb{T})$. Let $\mathcal{K}(\mathcal{H})$ denote the algebra of compact operators on $\mathcal{H}$ and define an action α of $\mathbb{Z}^2$ on continuous functions on the circle with values in compact operators, $C(\mathbb{T}, \mathcal{K}(\mathcal{H}))$, given at the point θ by ρ_θ. Define the C^*-algebra B, which is obtained by inducing the $\mathbb{Z}^2$ action to an action of $\mathbb{R}^2$ on $B = \mathrm{Ind}_{\mathbb{Z}^2}^{\mathbb{R}^2} (C(\mathbb{T}, \mathcal{K}(\mathcal{H})), \alpha)$, i.e. $B = \{f : \mathbb{R}^2 \to C(\mathbb{T}, \mathcal{K}(\mathcal{H})) : f(t+g) = \alpha(g)(f(t)), \ \ t \in \mathbb{R}^2, g \in \mathbb{Z}^2\}$. Then B is a continuous-trace C^*-algebra having spectrum $\mathbb{T}^3$ and Dixmier-Douady invariant H. B also has an action of $\mathbb{R}^2$ whose induced action on the spectrum of B is the trivial bundle $\mathbb{T}^3 \to \mathbb{T}$. Then our noncommutative T-dual is the crossed product algebra $B \rtimes \mathbb{R}^2 \cong C(\mathbb{T}, \mathcal{K}(\mathcal{H})) \rtimes \mathbb{Z}^2 = A_f$, which has fiber over $\theta \in \mathbb{T}$ given by $\mathcal{K}(\mathcal{H}) \rtimes_{\rho_\theta} \mathbb{Z}^2 \cong A_\theta \otimes \mathcal{K}(\mathcal{H},)$ where A_θ is the noncommutative 2-torus. In fact, the crossed product $B \rtimes \mathbb{R}^2$ is isomorphic to the (stabilized) group C^*-algebra $C^*(H_\mathbb{Z}) \otimes \mathcal{K}$, where $H_\mathbb{Z}$ is the integer Heisenberg-type group,

$$H_\mathbb{Z} = \left\{ \begin{pmatrix} 1 & x & \frac{1}{k}z \\ 0 & 1 & y \\ 0 & 0 & 1 \end{pmatrix} : x, y, z \in \mathbb{Z} \right\}. \tag{5}$$

In summary, the nonclassical T-dual of $(\mathbb{T}^3, H = k)$ is $A_f = C^*(H_\mathbb{Z}) \otimes \mathcal{K}$. As required in order to match up RR charges, the K-theory of this algebra is the same as the K-theory of $\mathbb{T}^3$ with twist given by our H-flux, or k times the volume form.

This example generalizes easily by taking the Cartesian product with a manifold M. Pulling back the H-flux to the product and arguing as before, we see that $(M \times \mathbb{T}^3, H = k)$ is T-dual to $C(M) \otimes C^*(H_\mathbb{Z}) \otimes \mathcal{K}$. For instance, if the dimension of M is seven, then $M \times \mathbb{T}^3$ is ten dimensional, yielding examples of spacetime manifolds that are relevant to type II string theory.

It is important to realize that a fixed space X can sometimes be given the structure of a principal torus bundle over Z in many different ways. For example, given a free action of a torus $T = \mathbb{T}^n$ on X, with quotient space $Z = X/T$, we can for every element $g \in \mathrm{Aut}(\mathbb{T}^n) = GL(n, \mathbb{Z})$ define a new free action of T on X, twisted by g, by the formula $x \cdot_g t = x \cdot g(t)$. (Here $t \in T$, $\cdot$ is the original free right action of T on X, and $\cdot_g$ is the new twisted action.) If $c \in H^2(Z, \mathbb{Z}^n)$ was the Chern class of the original bundle, the Chern class of the g-twisted bundle is $g \cdot c$, with g acting via the action of $GL(n, \mathbb{Z})$ on $\mathbb{Z}^n$.

The group $GL(n, \mathbb{Z})$ embeds in $O(n, n; \mathbb{Z})$, the subgroup of $GL(2n, \mathbb{Z})$ preserving the quadratic form defined by $\begin{pmatrix} 0 & I_n \\ I_n & 0 \end{pmatrix}$, via $a \mapsto \begin{pmatrix} a & 0 \\ 0 & (a^t)^{-1} \end{pmatrix}$ (see [10]). This larger group $O(n, n; \mathbb{Z})$ is often called the *T-duality group*. In fact we will consider the still larger *generalized T-duality group* $GO(n, n; \mathbb{Z}) = O(n, n; \mathbb{Z}) \rtimes (\mathbb{Z}/2)$ of matrices in $GL(2n, \mathbb{Z})$ preserving the form $\begin{pmatrix} 0 & I_n \\ I_n & 0 \end{pmatrix}$ up to sign. Good references for the T-duality group include [10] (for the state of the theory up to 1994) and [14] for more current developments.

Our results on the T-duality group: *Suppose that we are in the basic setup as above, with Z simply connected, so that one is always guaranteed to have a classical T-dual. Then the generalized T-duality group $GO(n, n; \mathbb{Z})$ acts on the set of T-dual pairs (p, H) and $(p^\#, H^\#)$ to generate all related T-dual pairs.* All of these pairs are physically equivalent. *The restriction of the action to $GL(n, \mathbb{Z})$ (as embedded above) corresponds to twisting of the action on the same underlying space as above. When Z is not simply connected and $p_!(H) \neq 0$, it is not clear that one has an action of the full T-duality group. But the action of $GL(n, \mathbb{Z})$ always sends the pair consisting of (p, H) and its nonclassical T-dual to another nonclassical pair, involving continuous fields of (stabilized) noncommutative tori over Z.*

We illustrate the action of the generalized T-duality group in the simplest case of circle bundles with H-flux, in which case the generalized T-duality group reduces to $GO(1, 1; \mathbb{Z})$, a dihedral group of oder 8.

Consider the example of the 3 dimensional lens space $L(1, p) = S^3/\mathbb{Z}_p$, with H-flux $H = q$ times the volume form, cf. [17]. Here $p, q \in \mathbb{Z}$, and initially we take p, $q > 0$. Then $L(1, p)$ is a circle bundle over the 2-dimensional sphere S^2 and has first Chern class equal to p times the volume form of S^2. Then, as shown in [8] , $(L(1, p), H = q)$ and $(L(1, q), H = p)$ are T-dual to each other, and the element $\begin{pmatrix} 0 & 1 \\ 1 & 0 \end{pmatrix}$ of $O(1, 1; \mathbb{Z})$ interchanges them.

The element $\begin{pmatrix} -1 & 0 \\ 0 & -1 \end{pmatrix}$ of the T-duality group $O(1,1;\mathbb{Z})$ lies in the subgroup $GL(1,\mathbb{Z})$, embedded as above, and acts by twisting the S^1 action on $L(1,p)$. This twisted action makes $L(1,p)$ into a circle bundle over S^2 having first Chern class equal to $-p$ times the volume form of S^2. This bundle is denoted $L(1,-p)$, and its total space is diffeomorphic to $L(1,p)$, though by an orientation-reversing diffeomorphism. Therefore the action of $\begin{pmatrix} -1 & 0 \\ 0 & -1 \end{pmatrix}$ on the pair $(L(1,p), H = q)$ and $(L(1,q), H = p)$ gives rise to a new T-dual pair $(L(1,-p), H = -q)$ and $(L(1,-q), H = -p)$. The group $GO(1,1;\mathbb{Z})$ is generated by the two elements of $O(1,1;\mathbb{Z})$ just discussed and by $\begin{pmatrix} 1 & 0 \\ 0 & -1 \end{pmatrix}$, which replaces the original T-dual pair by the pair consisting of $(L(1,p), H = -q)$ and $(L(1,-q), H = p)$. Here we have tacitly assumed p, $q \geq 2$; we can extend things to other values of p and q by making the convention that $L(1,1) = S^3$ and $L(1,0) = S^2 \times S^1$. This refines the T-duality in [8] . Thus in general there are 8 different (bundle, H-flux) pairs with equivalent physics, corresponding to $(\pm p, \pm q)$ and $(\pm q, \pm p)$.

This example generalizes easily by taking the Cartesian product with a manifold M. For instance, if the dimension of M is seven, then we obtain 8 different (bundle, H-flux) pairs in the same $GO(1,1;\mathbb{Z})$-orbit as $M \times L(1,p)$. All of these are ten-dimensional spacetime manifolds relevant to type II string theory.

We end with some open problems. A critical verification of any proposed duality is that the anomalies should match on both sides. This was checked for T-duality involving circle bundles with H-flux in [8], but remains to be analyzed in the general torus bundle case with H-flux. It also remains to be determined whether or not the group $GO(n,n;\mathbb{Z})$ also operates in the nonclassical case. Another problem is to extend our results to non-free torus actions [20], in which case it could be relevant to mirror symmetry.

Acknowledgments:

VM was financially supported by the Australian Research Council and JR was partially supported by NSF Grant DMS-0103647.

References

1. T. Buscher, *Phys. Lett.* **B194**, 59 (1987); *ibid.*, *Phys. Lett.* **B201**, 466 (1988).
2. E. Álvarez, L. Álvarez-Gaumé, J. L. F. Barbón and Y. Lozano, *Nucl. Phys.* **B415**, 71 (1994), hep-th/9309039.

3. A. Connes, *A survey of foliations and operator algebras*, in *Operator algebras and applications, Part I* (Kingston, Ont., 1980), R. V. Kadison, ed., pp. 521–628, Proc. Sympos. Pure Math., 38, Amer. Math. Soc., Providence, R.I., 1982.
4. A. Connes, M. R. Douglas and A. Schwarz, *J. High Energy Phys.* 02 (1998) 003, hep-th/9711162.
5. M. J. Duff, H. Lü and C. N. Pope, *Nucl. Phys.* **B532**, 181 (1998), hep-th/9803061.
6. S. Gurrieri, J. Louis, A. Micu and D. Waldram, *Nucl. Phys.* **B654**, 61 (2003), hep-th/0211102.
7. S. Kachru, M. Schulz, P. Tripathy and S. Trivedi, *J. High Energy Phys.* 03 (2003) 061, hep-th/0211182.
8. P. Bouwknegt, J. Evslin and V. Mathai, *Comm. Math. Phys.* **249**, 383 (2004), hep-th/0306062; *ibid.*, Phys. Rev. Lett. **92**, 181601 (2004).
9. P. Bouwknegt, K. Hannabuss and V. Mathai, *J. High Energy Phys.* 03 (2004) 018, hep-th/0312284.
10. A. Giveon, M. Porrati and E. Rabinovici, *Phys. Rep.* **244**, no. 2–3, 77 (1994), hep-th/9401139.
11. R. Minasian and G. Moore, *J. High Energy Phys.* 11 (1997) 002, hep-th/9710230.
12. E. Witten, *J. High Energy Phys.* 12 (1998) 019, hep-th/9810188.
13. P. Bouwknegt and V. Mathai, *J. High Energy Phys.* 03 (2000) 007, hep-th/0002023.
14. C. M. Hull, hep-th/0406102.
15. D. A. Lowe, H. Nastase and S. Ramgoolam, Massive IIA string theory and matrix theory compactification, Nucl. Phys. **B667** 55 (2003), hep-th/0303173.
16. V. Mathai and J. Rosenberg, *Comm. Math. Phys.* **253** 705 (2005), hep-th/0401168; *ibid.*, hep-th/0508084
17. J. Maldacena, G. Moore and N. Seiberg, *J. High Energy Phys.* 10 (2001) 005, hep-th/0108152.
18. M. Rieffel, *Pacific J. Math.* **93**, 415 (1981).
19. N. Seiberg and E. Witten, *J. High Energy Phys.* 09 (1999) 032, hep-th/9908142.
20. R. Donagi and T. Pantev, math.AG/0306213.

Murphy Operators in Knot Theory

H. R. Morton

Department of Mathematical Sciences
University of Liverpool
Peach Street, Liverpool, L69 7ZL, UK
E-mail: morton@liv.ac.uk

The Murphy operators in the Hecke algebra H_n are commuting elements which arose originally in an algebraic setting in connection with representation theory. They can be represented diagrammatically in a Homfly skein theory version of H_n. Symmetric functions of the Murphy operators are known to lie in the centre of H_n. Diagrammatic views of these are given which demonstrate their algebraic properties readily, and how analogous central elements can be constructed diagrammatically in some related algebras.

Introduction

This article is based closely on a talk given at the meeting on Differential Geometric Methods in Theoretical Physics on the occasion of the opening of the new building for the Nankai Institute of Mathematics. More detailed accounts of the results described during the talk can be found in the references notcd.

I first heard about the Murphy operators on my previous visit to Nankai ten years ago for a statistical mechanics satellite meeting. At that meeting Chakrabarti gave a talk about the properties of what he termed the 'fundamental element' which generated the centre of the Hecke algebra H_n[3].

At that time Aiston and I had been studying geometrically based models for H_n in terms of the group B_n of n-string braids, and I initially expected that his fundamental element must be represented by the well-known generator for the centre of the braid group, namely the full twist braid Δ^2. However it soon became clear that Chakrabarti was referring to a different, and more useful, element of H_n, with the algebraic feature that it had distinct eigenvalues on the different irreducible submodules of H_n.

Chakrabarti then told me that this element was the sum of the Murphy elements (Murphy operators) in H_n. These are elements which have their origin in work of Jucys[2] and subsequently Murphy[8].

Having been introduced to these elements, Aiston and I looked at them in our geometrical model in order to understand them in that context, and to see if their algebraic properties could be readily established there.

While we were able to understand their basic appearance, and establish the eigenvalue property quite quickly[5] it was not until a few years later that I came across a more satisfactory geometric way to represent them, and a particularly striking way to produce their sum as an obviously central element in H_n[4].

This in turn led me to a natural description for other central elements, and similar descriptions of central elements in some natural extensions of the Hecke algebras.

A further consequence of the eigenvalue property led me also to a very helpful way of identifying the elements in a natural combinatorial model constructing 2-variable knot invariants which correspond neatly to the invariants produced by irreducible quantum $SL(N)$ modules.

I shall give here a brief account of the Jucys-Murphy elements in an algebraic context, before describing the geometric models for H_n and for the further construction.

1. Murphy operators in Hecke algebras

The Hecke algebra H_n is a deformed version of the group algebra $\mathbf{C}[S_n]$ of permutations. Jucys[2] and Murphy[8] studied certain sums of transpositions $m(j) \in \mathbf{C}[S_n]$.

$$\begin{aligned} m(2) &= (1\,2) \\ m(3) &= (1\,3) + (2\,3) \\ m(4) &= (1\,4) + (2\,4) + (3\,4) \\ \vdots &= \\ m(j) &= \sum_{i=1}^{j-i} (i\,j) \\ \vdots &= \end{aligned}$$

These elements have the following two properties:

1. The elements $m(j)$ commute.

2. Every symmetric polynomial in them, for example their sum, or the sum of their squares, lies in the *centre* of the algebra.

Dipper and James[1] found corresponding elements $M(j)$ in H_n which they named the *Murphy operators*, having similar properties:

1. The elements $M(j)$ commute.
2. Every symmetric polynomial in them lies in the centre of H_n.

The Hecke algebra H_n can be readily presented as linear combinations of n-string braids subject to a simple linear relation depending on a single parameter z.

The elementary braids $\sigma_i^{\pm 1}$ when composed by placing one below another will generate all n-braids. Here

$\sigma_i =$ 

is the braid on n strings in which string i crosses string $i+1$ once in the positive sense.

They satisfy Artin's braid relations

$$\sigma_i\sigma_j = \sigma_j\sigma_i, \quad |i-j| > 1,$$
$$\sigma_i\sigma_{i+1}\sigma_i = \sigma_{i+1}\sigma_i\sigma_{i+1},$$

Elements of H_n can be regarded as linear combinations of braids on which we impose the further quadratic relations

$$\sigma_i^2 = z\sigma_i + 1.$$

These relations can be visualised in the form $\sigma_i - \sigma_i^{-1} = z$ as

Setting the parameter $z = 0$ gives $\sigma_i = \sigma_i^{-1}$ and reduces each braid to the permutation defined by following its strings, when σ_i becomes the transposition $(i\ i+1)$. The elements $M(j)$ were based on a choice of braids which each reduce to individual transpositions when $z = 0$.

Ram[9] pointed out that these could be combined into a single braid

$T(j) =$ 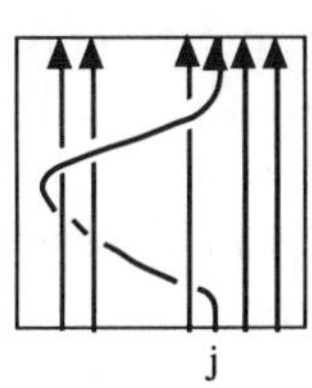

to represent each $M(j)$, up to linear combination with the identity.

Explicitly $T(j) = 1 + zM(j)$. So long as $z \neq 0$ the elements $T(j)$ will do equally well in place of $M(j)$.

The geometric braids $T(j)$ clearly commute. Their product is the full twist braid Δ^2 which commutes with all braids, and so lies in the centre of H_n. It is not immediately clear however that their sum, or any other symmetric function of them is central.

2. A skein theory version

I shall now construct a model of H_n based on more general diagrams which will provide a simple representative for the sum. In this wider context, known as skein theory, we work with pieces of oriented knot diagrams, lying with some prescribed boundary conditions in a fixed surface F. Diagrams consist of arcs respecting the boundary conditions along with further closed curves, and may be altered by sequences of the standard Reidemeister moves R_{II} and R_{III}. The moves can be interpreted as the natural physical moves allowed on pieces of ribbon representing the curves.

The *skein* $\mathcal{S}(F)$ consists of formal linear combinations of diagrams in F (sometimes known as *tangles*) modulo two linear relations

(1) 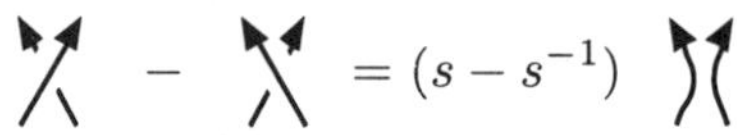

and (2) 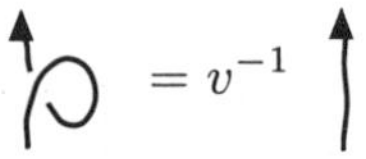

between diagrams which differ only as shown. The coefficient ring can be taken as $\Lambda = \mathbf{Z}[v^{\pm 1}, s^{\pm 1}]$ with powers of $s^k - s^{-k}$ in the denominators.

Theorem 2.1. *(Morton-Traczyk[7]) The skein of the rectangle with n input points at the bottom and n output points at the top is the Hecke algebra H_n, with scalars extended to Λ and $z = s - s^{-1}$.*

Any diagram in the rectangle can be reduced to a Λ-combination of braids by use of relations (1) and (2). For braids, the relation (1) becomes the algebraic relation $\sigma_i = \sigma_i^{-1} = z$.

The algebra composition in the skein version of H_n is given by placing diagrams one below the other, as for braids. We can then exhibit lots of diagrams which belong to the centre of H_n in this model.

For a start the diagram

$T^{(n)} =$ 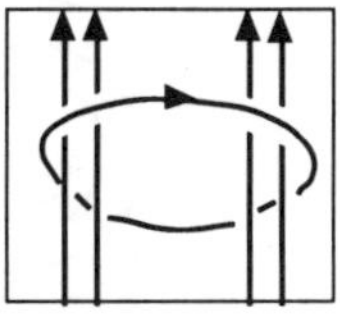

is central.

This can be readily seen, since any diagram A can be passed through using only Reidemeister moves II and III.

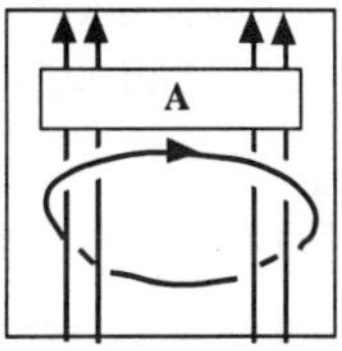

$=$
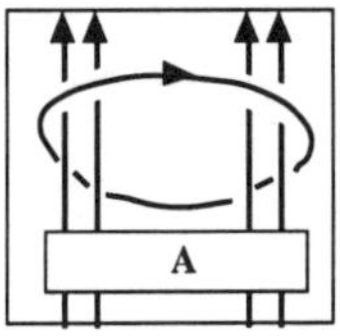

.

Theorem 2.2. *(Morton[4]) $T^{(n)}$ is the sum of the variant Murphy operators $T(j)$, up to linear combination with the identity.*

This result depends essentially on a repeated application of the skein relation (1), leading to the equation

$T^{(n)} -$ 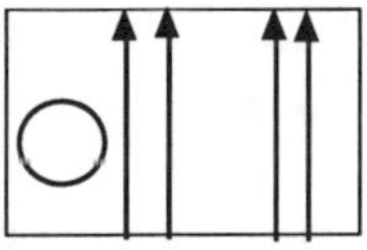 $= v^{-1}z\sum_{j=1}^{n} T(j).$

Replacing the encircling curve in $T^{(n)}$ by a more complicated combination of diagrams

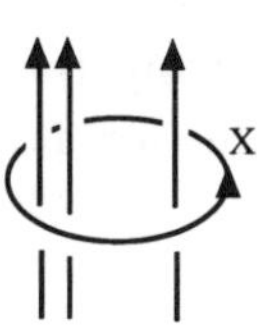

gives a huge range of further central elements.

The choices for X are best thought of as elements in the skein $\mathcal{C}$ of the annulus without prescribed boundary points, for example

$X =$ 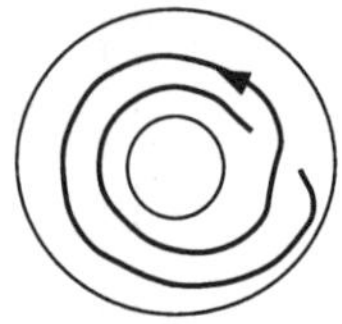 .

There is a nice choice X_m for each m which gives the sum of the mth powers of the Murphy operators $T(j)$ in H_n no matter what n may be[4]. It is then possible to produce any symmetric polynomial in $T(j)$ from a suitable choice of X.

In the same spirit, algebras $H_{n,p}$ can be constructed as the skeins based on the rectangle with inputs and outputs arranged as shown,

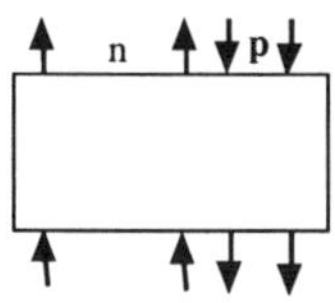

where elements are again composed by placing one below the other. There is again a large choice of similarly constructed elements

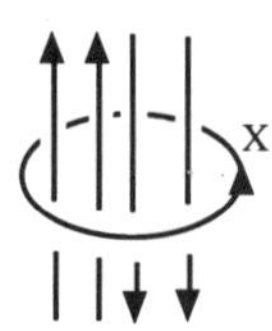

in the centre of the algebra. These can all be expressed as supersymmetric polynomials in two families of commuting elements in the algebra which can be considered as an analogue of Murphy elements in this setting.

Even where the basic skein relation is altered, for example to Kauffman's 4-term relation on non-oriented diagrams, similar diagrams to these will give central elements. In this setting too these central elements may be interpreted as polynomials in some form of Murphy elements.

3. The annulus

Representation theory of Hecke algebras also leads naturally to the skein of the annulus. We are interested in finding trace functions on H_n, namely linear functions tr to a commutative algebra such that $\mathrm{tr}(AB) = \mathrm{tr}(BA)$.

The character of a matrix representation has this property, but we need not restrict the image of the function to be the scalars.

Any diagram T in H_n can be *closed* to give a diagram $\hat{T}$ in the annulus, as shown, with the property that $\widehat{AB} = \widehat{BA}$. This procedure respects the skein relations, and so determines a Λ-linear map $\widehat{\ }: H_n \to \mathcal{C}$ to the skein of the annulus. Now $\mathcal{C}$ is a commutative algebra, so the closure map is a trace function on H_n, and its composite with any linear function on $\mathcal{C}$ will determine further trace functions. Indeed it is possible to construct all irreducible characters of H_n by suitable linear functions on $\mathcal{C}$.

Write $\mathcal{C}_n \subset \mathcal{C}$ for the image of H_n, and define the *meridian map* $\varphi : \mathcal{C}_n \to \mathcal{C}_n$ diagrammatically by

$\varphi(X) =$  .

Thus if $X = \hat{A}$ then $\varphi(X) = \widehat{AT^{(n)}}$. If $AT^{(n)} = cA$ then $\hat{A}$ is an eigenvector of φ with eigenvalue c.

Theorem 3.1. *The meridian map φ has no repeated eigenvalues.*

Aiston and I[5] gave a direct proof of this by exhibiting suitable choices of A. The result can be interpreted as a different angle on Chakrabarti's observation about the action of the sum of the Murphy operators on H_n.

Indeed when the meridian map is extended over all diagrams in the annulus to give $\varphi : \mathcal{C} \to \mathcal{C}$ it still has no repeated eigenvalues[6]. In $\mathcal{C}_n$ the eigenvectors correspond to partitions of n, and the subspace of $\mathcal{C}$ spanned by the union of $\mathcal{C}_n$ for all n can be interpreted as the representation ring of $SL(N)$ for large N. In this context the eigenvectors match up well with the irreducible representations, and give well-adapted skein theoretic elements Q_λ for each $\lambda \vdash n$. These can be used to provide a 2-variable invariant of a knot for each partition λ that yields the irreducible 1-variable quantum $SL(N)$ invariants for each N by a simple substitution. Eigenvectors for the meridian map in the whole skein of the annulus correspond to pairs λ, μ of partitions, and again give natural 2-variable invariants which are well-adapted to quantum group interpretations[6].

References

1. R. Dipper and G.D. James, Blocks and idempotents of Hecke algebras of general linear groups. *Proc. London Math. Soc.* 54 (1987), 57-82.

2. A. Jucys, Factorization of Young's projection operators for symmetric groups. *Litovsk. Fiz. Sb.* 11 (1971), 1-10.
3. J. Katriel, B. Abdesselam and A. Chakrabarti, The fundamental invariant of the Hecke algebra $H_n(q)$ characterizes the representations of $H_n(q)$, S_n, $SU_q(N)$ and $SU(N)$. *J. Math. Phys.* 36 (1995), 5139-5158.
4. H.R.Morton, Skein theory and the Murphy operators. *J. Knot Theory Ramifications* 11 (2002), 475-492.
5. H.R.Morton and A.K.Aiston, Idempotents of Hecke algebras of type A. *J. Knot Theory Ramifications* 7 (1998), 463-487.
6. H.R.Morton and R.J.Hadji, A basis for the full Homfly skein of the annulus. ArXiv math.GT/0408078. To appear in *Math. Proc. Camb. Philos. Soc.*
7. H.R.Morton and P.Traczyk, H.R.Morton and P.Traczyk, Knots and algebras. In *Contribuciones Matematicas en homenaje al profesor D. Antonio Plans Sanz de Bremond*, ed. E. Martin-Peinador and A. Rodez Usan, University of Zaragoza, (1990), 201-220.
8. G.E. Murphy, A new construction of Young's seminormal representation of the symmetric groups. *J. Algebra* 69 (1981), 287-297.
9. A. Ram, Seminormal representations of Weyl groups and Iwahori-Hecke algebras. *Proc. London Math. Soc.* 75 (1997), 99-133.

Bethe Ansatz for the Open XXZ Chain from Functional Relations at Roots of Unity

Rafael I. Nepomechie*

Physics Department, P.O. Box 248046, University of Miami
Coral Gables, FL 33124, USA
E-mail: nepomechie@physics.miami.edu

We briefly review Bethe Ansatz solutions of the integrable open spin-$\frac{1}{2}$ XXZ quantum spin chain derived from functional relations obeyed by the transfer matrix at roots of unity.

1. Introduction

A long standing problem has been to solve the open spin-$\frac{1}{2}$ XXZ quantum spin chain with general integrable boundary terms, defined by the Hamiltonian [1,2]

$$\mathcal{H} = \frac{1}{2}\Big\{ \sum_{n=1}^{N-1} \left(\sigma_n^x \sigma_{n+1}^x + \sigma_n^y \sigma_{n+1}^y + \operatorname{ch}\eta\ \sigma_n^z \sigma_{n+1}^z \right) \tag{1.1}$$
$$+ \operatorname{sh}\eta \Big[\operatorname{cth}\alpha_- \operatorname{th}\beta_- \sigma_1^z + \operatorname{csch}\alpha_- \operatorname{sech}\beta_- \left(\operatorname{ch}\theta_- \sigma_1^x + i \operatorname{sh}\theta_- \sigma_1^y \right)$$
$$- \operatorname{cth}\alpha_+ \operatorname{th}\beta_+ \sigma_N^z + \operatorname{csch}\alpha_+ \operatorname{sech}\beta_+ \left(\operatorname{ch}\theta_+ \sigma_N^x + i \operatorname{sh}\theta_+ \sigma_N^y \right) \Big] \Big\},$$

where $\sigma^x, \sigma^y, \sigma^z$ are the standard Pauli matrices, η is the bulk anisotropy parameter, $\alpha_\pm, \beta_\pm, \theta_\pm$ are arbitrary boundary parameters, and N is the number of spins. Determining the energy eigenvalues in terms of solutions of a system of Bethe Ansatz equations is a fundamental problem, which has important applications in integrable quantum field theory as well as condensed matter physics and statistical mechanics [3], and perhaps also string theory. (For an introduction to Bethe Ansatz, see e.g. Refs. 4, 5, 6.)

The basic difficulty in solving (1.1) is that, in contrast to the special case of diagonal boundary terms (*i.e.*, $\alpha_\pm$ or $\beta_\pm \to \pm\infty$, in which case $\mathcal{H}$ has

*Work supported in part by the National Science Foundation under Grant PHY-0244261.

a $U(1)$ symmetry) which was solved long ago [7–9], a simple pseudovacuum state does *not* exist. For instance, the state with all spins up is not an eigenstate of the Hamiltonian. Hence, many of the techniques which have been developed to solve integrable models cannot be applied.

We observed some time ago [10,11] that, for bulk anisotropy parameter values

$$\eta = \frac{i\pi}{p+1}\,, \qquad p = 1\,,2\,,\ldots \tag{1.2}$$

(hence $q \equiv e^{\eta}$ is a root of unity, satisfying $q^{p+1} = -1$) and arbitrary values of the boundary parameters, the model's transfer matrix $t(u)$ (see Sec. 2) obeys a functional relation of order $p+1$. For example, for the case $p = 2$, the functional relation is

$$\begin{aligned} t(u)t(u+\eta)t(u+2\eta) - \delta(u-\eta)t(u+\eta) - \delta(u)t(u+2\eta) \\ - \delta(u+\eta)t(u) = f(u)\,, \end{aligned} \tag{1.3}$$

where $\delta(u)$ and $f(u)$ are known scalar functions which depend on the boundary parameters. (Expressions for these functions in terms of the boundary parameters in (1.1) are given in Ref. 18.) Similar results had been known earlier for closed spin chains.[12–14]

By exploiting these functional relations, we have obtained Bethe Ansatz solutions of the model for various special cases of the bulk and boundary parameters:

(i) [Refs. 15, 16, 17] The bulk anisotropy parameter has values (1.2); and the boundary parameters satisfy the constraint

$$\alpha_- + \beta_- + \alpha_+ + \beta_+ = \pm(\theta_- - \theta_+) + \eta k\,, \tag{1.4}$$

where $k \in [-(N+1)\,, N+1]$ is even (odd) if N is odd (even), respectively.

(ii) [Ref. 18] The bulk anisotropy parameter has values (1.2) with p *even*; and either
   (a) at most one of the boundary parameters is nonzero, or
   (b) any two of the boundary parameters $\{\alpha_-, \alpha_+, \beta_-, \beta_+\}$ are arbitrary, the remaining boundary parameters from this set are either η or $i\pi/2$, and $\theta_- = \theta_+$.

(iii) [Ref. 19] The bulk anisotropy parameter has values (1.2) with p *odd*; at most two of the boundary parameters $\{\alpha_-, \alpha_+, \beta_-, \beta_+\}$ are nonzero, and $\theta_- = \theta_+$.

All of these cases have the property that the quantity $\Delta(u)$, defined by

$$\Delta(u) = f(u)^2 - 4 \prod_{j=0}^{p} \delta(u + j\eta) \,, \tag{1.5}$$

is a perfect square.

Solutions for generic values of the bulk anisotropy parameter and for boundary parameters obeying a constraint similar to (1.4) have been discussed in Refs. 20, 21, 22.

Here we briefly review our results for the cases (i) - (iii).

2. Transfer matrix

The transfer matrix $t(u)$ of the open XXZ chain with general integrable boundary terms, which satisfies the fundamental commutativity property $[t(u)\,, t(v)] = 0$, is given by [9]

$$t(u) = \mathrm{tr}_0\, K_0^+(u)\; T_0(u)\; K_0^-(u)\; \hat{T}_0(u)\,, \tag{2.1}$$

where $T_0(u)$ and $\hat{T}_0(u)$ are the monodromy matrices

$$T_0(u) = R_{0N}(u) \cdots R_{01}(u)\,, \qquad \hat{T}_0(u) = R_{01}(u) \cdots R_{0N}(u)\,, \tag{2.2}$$

and tr_0 denotes trace over the "auxiliary space" 0. The R matrix is given by

$$R(u) = \begin{pmatrix} \mathrm{sh}(u+\eta) & 0 & 0 & 0 \\ 0 & \mathrm{sh}\, u & \mathrm{sh}\, \eta & 0 \\ 0 & \mathrm{sh}\, \eta & \mathrm{sh}\, u & 0 \\ 0 & 0 & 0 & \mathrm{sh}(u+\eta) \end{pmatrix} , \tag{2.3}$$

where η is the bulk anisotropy parameter; and $K^{\mp}(u)$ are 2×2 matrices whose components are given by [1,2]

$$\begin{aligned} K_{11}^-(u) &= 2\, (\mathrm{sh}\, \alpha_-\, \mathrm{ch}\, \beta_-\, \mathrm{ch}\, u + \mathrm{ch}\, \alpha_-\, \mathrm{sh}\, \beta_-\, \mathrm{sh}\, u) \\ K_{22}^-(u) &= 2\, (\mathrm{sh}\, \alpha_-\, \mathrm{ch}\, \beta_-\, \mathrm{ch}\, u - \mathrm{ch}\, \alpha_-\, \mathrm{sh}\, \beta_-\, \mathrm{sh}\, u) \\ K_{12}^-(u) &= e^{\theta_-}\, \mathrm{sh}\, 2u\,, \qquad K_{21}^-(u) = e^{-\theta_-}\, \mathrm{sh}\, 2u\,, \end{aligned} \tag{2.4}$$

and

$$K^+(u) = K^-(-u-\eta)\Big|_{\alpha_- \to -\alpha_+\,,\ \beta_- \to -\beta_+\,,\ \theta_- \to \theta_+} , \tag{2.5}$$

where $\alpha_\mp\,, \beta_\mp\,, \theta_\mp$ are the boundary parameters. The Hamiltonian (1.1) is proportional to $t'(0)$ plus a constant.

The transfer matrix also has $i\pi$ periodicity

$$t(u+i\pi) = t(u)\,, \tag{2.6}$$

crossing symmetry

$$t(-u-\eta) = t(u)\,, \tag{2.7}$$

and the asymptotic behavior

$$t(u) \sim -\operatorname{ch}(\theta_- - \theta_+)\frac{e^{u(2N+4)+\eta(N+2)}}{2^{2N+1}}\mathbb{I} + \ldots \qquad \text{for} \qquad u\rightarrow\infty\,. \tag{2.8}$$

3. Case (i)

Our main objective is to determine the eigenvalues $\Lambda(u)$ of the transfer matrix $t(u)$ (2.1), from which the energy eigenvalues can readily be computed. The functional relations for the transfer matrix (e.g., (1.3)) evidently imply corresponding relations for $\Lambda(u)$. Following Ref. 23, we observe that the latter relations can be written as

$$\det \mathcal{M}(u) = 0\,, \tag{3.1}$$

where $\mathcal{M}(u)$ is the $(p+1)\times(p+1)$ matrix

$$\mathcal{M}(u) = \begin{pmatrix} \Lambda(u) & -\frac{\delta(u)}{h(u+\eta)} & 0 & \ldots & 0 & -h(u) \\ -h(u+\eta) & \Lambda(u+\eta) & -\frac{\delta(u+\eta)}{h(u+2\eta)} & \ldots & 0 & 0 \\ \vdots & \vdots & \vdots & \ddots & \vdots & \vdots \\ -\frac{\delta(u-\eta)}{h(u)} & 0 & 0 & \ldots & -h(u+p\eta) & \Lambda(u+p\eta) \end{pmatrix} \tag{3.2}$$

if there exists an $i\pi$-periodic function $h(u)$ such that

$$f(u) = \prod_{j=0}^{p} h(u+j\eta) + \prod_{j=0}^{p} \frac{\delta(u+j\eta)}{h(u+j\eta)}\,. \tag{3.3}$$

To solve for $h(u)$, we set $z(u) \equiv \prod_{j=0}^{p} h(u+j\eta)$, and observe that (3.3) implies that $z(u)$ satisfies a quadratic equation

$$z(u)^2 - z(u)f(u) + \prod_{j=0}^{p}\delta(u+j\eta) = 0\,, \tag{3.4}$$

whose solution is evidently given by

$$z(u) = \frac{1}{2}\left(f(u) \pm \sqrt{\Delta(u)}\right)\,, \tag{3.5}$$

where $\Delta(u)$ is defined in (1.5). If the boundary parameters satisfy the constraint (1.4), then $\Delta(u)$ is a perfect square, and two solutions of (3.3) are

$$h^{(\pm)}(u) = -4\,\mathrm{sh}^{2N}(u+\eta)\frac{\mathrm{sh}(2u+2\eta)}{\mathrm{sh}(2u+\eta)} \times \mathrm{sh}(u\pm\alpha_-)\,\mathrm{ch}(u\pm\beta_-)\,\mathrm{sh}(u\pm\alpha_+)\,\mathrm{ch}(u\pm\beta_+)\,. \tag{3.6}$$

Let us now label the corresponding matrices (3.2) by $\mathcal{M}^{(\pm)}(u)$.

The condition (3.1) implies that $\mathcal{M}^{(\pm)}(u)$ has a null eigenvector $v^{(\pm)}(u)$,

$$\mathcal{M}^{(\pm)}(u)\; v^{(\pm)}(u) = 0\,, \tag{3.7}$$

Note that the matrix $\mathcal{M}^{(\pm)}(u)$ has the symmetry

$$S\mathcal{M}^{(\pm)}(u)S^{-1} = \mathcal{M}^{(\pm)}(u+\eta)\,, \tag{3.8}$$

where

$$S = \begin{pmatrix} 0\,1\,0 \ldots 0\,0 \\ 0\,0\,1 \ldots 0\,0 \\ \vdots\;\vdots\;\vdots\;\ddots\;\vdots\;\vdots \\ 0\,0\,0 \ldots 0\,1 \\ 1\,0\,0 \ldots 0\,0 \end{pmatrix}, \qquad S^{p+1} = \mathbb{I}\,. \tag{3.9}$$

It follows that the null eigenvector $v^{(\pm)}(u)$ satisfies $Sv^{(\pm)}(u) = v^{(\pm)}(u+\eta)$. Thus, its components can be expressed in terms of a function $Q^{(\pm)}(u)$,

$$v^{(\pm)}(u) = \left(Q^{(\pm)}(u)\,, Q^{(\pm)}(u+\eta)\,, \ldots\,, Q^{(\pm)}(u+p\eta)\right), \tag{3.10}$$

with $Q^{(\pm)}(u+i\pi) = Q^{(\pm)}(u)$. We make the Ansatz

$$Q^{(\pm)}(u) = \prod_{j=1}^{M^{(\pm)}} \mathrm{sh}(u-u_j^{(\pm)})\,\mathrm{sh}(u+u_j^{(\pm)}+\eta)\,, \tag{3.11}$$

which has the crossing symmetry $Q^{(\pm)}(u) = Q^{(\pm)}(-u-\eta)$. Substituting the expressions for $\mathcal{M}^{(\pm)}(u)$ (3.2) and $v^{(\pm)}(u)$ (3.10) into the null eigenvector equation (3.7) yields the result for the transfer matrix eigenvalues

$$\Lambda^{(\pm)}(u) = h^{(\pm)}(u)\frac{Q^{(\pm)}(u-\eta)}{Q^{(\pm)}(u)} + h^{(\pm)}(-u-\eta)\frac{Q^{(\pm)}(u+\eta)}{Q^{(\pm)}(u)}\,. \tag{3.12}$$

The asymptotic behavior (2.8) implies that $M^{(\pm)} = \frac{1}{2}(N-1\pm k)$, where k is the integer appearing in the constraint (1.4). Analyticity of the eigenvalues (3.12) implies the Bethe Ansatz equations

$$\frac{h^{(\pm)}(u_j^{(\pm)})}{h^{(\pm)}(-u_j^{(\pm)}-\eta)} = -\frac{Q^{(\pm)}(u_j^{(\pm)}+\eta)}{Q^{(\pm)}(u_j^{(\pm)}-\eta)}\,, \qquad j = 1\,, \ldots\,, M^{(\pm)}\,. \tag{3.13}$$

In short, for case (i), the eigenvalues of the transfer matrix (2.1) are given by (3.12), where $h^{(\pm)}(u)$ and $Q^{(\pm)}(u)$ are given by (3.6), (3.11) and (3.13).

In Ref. 17, we have verified numerically that this solution holds also for generic values of η, which is consistent with Refs. 20, 21, 22; and that this solution gives the complete set of 2^N eigenvalues. To illustrate how completeness is achieved, let us consider the case $N = 4$. The integer k in the constraint (1.4) must therefore be odd, with $-5 \le k \le 5$. The six possibilities are summarized in Table 1.

Table 1. Completeness for $N = 4$. For each k, there are 2^4 eigenvalues.

k	# eigenvalues given by $\Lambda^{(+)}(u)$	# eigenvalues given by $\Lambda^{(-)}(u)$
5	16	0
3	15	1
1	11	5
-1	5	11
-3	1	15
-5	0	16

4. Case(ii)

A key feature of case (i) is that the quantity $\Delta(u)$ (1.5) is a perfect square. We therefore look for additional such cases. For p even, we find that $\Delta(u)$ is also a perfect square if either (a) at most one of the boundary parameters is nonzero; or (b) any two of the boundary parameters $\{\alpha_-, \alpha_+, \beta_-, \beta_+\}$ are arbitrary, the remaining boundary parameters from this set are either η or $i\pi/2$, and $\theta_- = \theta_+$. For definiteness, we focus here on the subcase (b) with $\alpha_\pm$ arbitrary, $\beta_\pm = \eta$ and N even. Unfortunately, the resulting $z(u)$ (3.5) is not consistent. To surmount this difficulty, we use a matrix $\mathcal{M}(u)$ which is different from (3.2), namely [18]

$$\mathcal{M}(u) =$$
$$\begin{pmatrix} \Lambda(u) & -h(u) & 0 & \ldots & 0 & -h(-u+p\eta) \\ -h(-u) & \Lambda(u+p\eta) & -h(u+p\eta) & \ldots & 0 & 0 \\ \vdots & \vdots & \vdots & \ddots & \vdots & \vdots \\ -h(u+p^2\eta) & 0 & 0 & \ldots & -h(-u-p(p-1)\eta) & \Lambda(u+p^2\eta) \end{pmatrix} \tag{4.1}$$

where $h(u)$ is $2i\pi$-periodic. This matrix has the symmetry

$$S\mathcal{M}(u)S^{-1} = \mathcal{M}(u+p\eta)\,, \tag{4.2}$$

where S is given by (3.9). By arguments similar to those used in Sec. 3, we find that the transfer matrix eigenvalues are given by

$$\Lambda(u) = h(u)\frac{Q(u+p\eta)}{Q(u)} + h(-u+p\eta)\frac{Q(u-p\eta)}{Q(u)}\,, \tag{4.3}$$

where $h(u)$ is given by

$$h(u) = 4\,\mathrm{sh}^{2N}(u+\eta)\frac{\mathrm{sh}(2u+2\eta)}{\mathrm{sh}(2u+\eta)}\,\mathrm{ch}^2(u-\eta) \tag{4.4}$$
$$\times\,\mathrm{sh}(u-\alpha_-)\,\mathrm{sh}(u+\alpha_+)\frac{\mathrm{ch}\left(\frac{1}{2}(u+\alpha_-+\eta)\right)\,\mathrm{ch}\left(\frac{1}{2}(u-\alpha_++\eta)\right)}{\mathrm{ch}\left(\frac{1}{2}(u-\alpha_--\eta)\right)\,\mathrm{ch}\left(\frac{1}{2}(u+\alpha_+-\eta)\right)}\,,$$

and $Q(u)$ is given by

$$Q(u) = \prod_{j=1}^{M}\mathrm{sh}\left(\frac{1}{2}(u-u_j)\right)\mathrm{sh}\left(\frac{1}{2}(u+u_j-p\eta)\right)\,, \tag{4.5}$$

with $M = N+2p+1$; and the Bethe Ansatz equations are

$$\frac{h(u_j)}{h(-u_j+p\eta)} = -\frac{Q(u_j-p\eta)}{Q(u_j+p\eta)}\,, \qquad j = 1\,,\ldots\,,M\,. \tag{4.6}$$

We have verified numerically the completeness of this solution. The other subcases (a) and (b) are mostly similar. †

5. Case(iii)

For p odd, we find that the quantity $\Delta(u)$ (1.5) is also a perfect square if at most two of the boundary parameters $\{\alpha_-, \alpha_+, \beta_-, \beta_+\}$ are nonzero, and $\theta_- = \theta_+$. For definiteness, we focus here on the case with $\alpha_\pm$ arbitrary, $\beta_\pm = 0$ and N even. As in case (ii), the resulting $z(u)$ (3.5) is not consistent. To surmount this difficulty, we again use a matrix $\mathcal{M}(u)$ which is different from (3.2), namely [19]

$$\mathcal{M}(u) =$$
$$\begin{pmatrix} \Lambda(u) & -\frac{\delta(u)}{h^{(1)}(u)} & 0 & \ldots & 0 & -\frac{\delta(u-\eta)}{h^{(2)}(u-\eta)} \\ -h^{(1)}(u) & \Lambda(u+\eta) & -h^{(2)}(u+\eta) & \ldots & 0 & 0 \\ \vdots & \vdots & \vdots & \ddots & \vdots & \vdots \\ -h^{(2)}(u-\eta) & 0 & 0 & \ldots & -h^{(1)}(u+(p-1)\eta) & \Lambda(u+p\eta) \end{pmatrix} \tag{5.1}$$

† The exception is the subcase (a) with $\theta_\pm$ nonzero, for which $Q(u) = \prod_{j=1}^{2M}\mathrm{sh}(u-u_j)$, which is not crossing symmetric. See Sec. 3.3 in Ref. 18.

where $h^{(1)}(u)$ and $h^{(2)}(u)$ are $i\pi$-periodic. It has the reduced symmetry

$$T\mathcal{M}(u)T^{-1} = \mathcal{M}(u+2\eta)\,, \tag{5.2}$$

where $T = S^2$, and S is given by (3.9). (While (3.8) implies (5.2), the converse is not true.) The condition $\det \mathcal{M}(u) = 0$ implies that $\mathcal{M}(u)$ has a null eigenvector $v(u)$,

$$\mathcal{M}(u)\; v(u) = 0\,, \tag{5.3}$$

where $v(u)$ satisfies $Tv(u) = v(u+2\eta)$. Thus, its components are expressed in terms of *two* independent functions $Q_1(u)$, $Q_2(u)$:

$$v(u) = (Q_1(u)\,, Q_2(u)\,, \ldots\,, Q_1(u-2\eta)\,, Q_2(u-2\eta))\;. \tag{5.4}$$

We make the Ansätze

$$Q_1(u) = \prod_{j=1}^{M_1} \mathrm{sh}(u-u_j^{(1)})\,\mathrm{sh}(u+u_j^{(1)}+\eta)\,,$$
$$Q_2(u) = \prod_{j=1}^{M_2} \mathrm{sh}(u-u_j^{(2)})\,\mathrm{sh}(u+u_j^{(2)}+3\eta)\,. \tag{5.5}$$

Substituting the expressions for $\mathcal{M}(u)$ (5.1) and $v(u)$ (5.4) into the null eigenvector equation (5.3) yields *two* expressions for the transfer matrix eigenvalues,

$$\begin{aligned}\Lambda(u) &= \frac{\delta(u)}{h^{(1)}(u)}\frac{Q_2(u)}{Q_1(u)} + \frac{\delta(u-\eta)}{h^{(2)}(u-\eta)}\frac{Q_2(u-2\eta)}{Q_1(u)}\,,\\ &= h^{(1)}(u-\eta)\frac{Q_1(u-\eta)}{Q_2(u-\eta)} + h^{(2)}(u)\frac{Q_1(u+\eta)}{Q_2(u-\eta)}\,.\end{aligned} \tag{5.6}$$

Analyticity of these expressions leads to the Bethe Ansatz equations

$$\begin{aligned}\frac{\delta(u_j^{(1)})\; h^{(2)}(u_j^{(1)}-\eta)}{\delta(u_j^{(1)}-\eta)\; h^{(1)}(u_j^{(1)})} &= -\frac{Q_2(u_j^{(1)}-2\eta)}{Q_2(u_j^{(1)})}\,, \qquad j=1\,,2\,,\ldots\,,M_1\,,\\ \frac{h^{(1)}(u_j^{(2)})}{h^{(2)}(u_j^{(2)}+\eta)} &= -\frac{Q_1(u_j^{(2)}+2\eta)}{Q_1(u_j^{(2)})}\,, \qquad j=1\,,2\,,\ldots\,,M_2\,.\end{aligned} \tag{5.7}$$

We expect that there are sufficiently many equations to determine all the zeros $\{u_j^{(1)}\,, u_j^{(2)}\}$ of $Q_1(u)\,, Q_2(u)$, respectively. Functions $h^{(1)}(u)$ (with $h^{(2)}(u) = h^{(1)}(-u-2\eta)$) which ensure the condition $\det \mathcal{M}(u) = 0$ are given by

$$\begin{aligned}&h^{(1)}(u) = 4\,\mathrm{sh}^{2N}(u+2\eta)\,, \quad M_2 = \frac{1}{2}N + \frac{1}{2}(3p-1)\,, \quad M_1 = M_2 + 2\,,\\ &\quad p = 3\,,7\,,11\,,\ldots\end{aligned} \tag{5.8}$$

and

$$h^{(1)}(u) = \begin{cases} -2\,\mathrm{ch}(2u)\,\mathrm{sh}^2 u\,\mathrm{sh}^{2N}(u+2\eta)\,, \; M_1 = M_2 = \frac{1}{2}N + 2p - 1\,, \\ \qquad p = 9\,, 17\,, 25\,, \ldots \\ 2\,\mathrm{ch}(2u)\,\mathrm{sh}^2 u\,\mathrm{sh}^{2N}(u+2\eta)\,, \quad M_1 = M_2 = \frac{1}{2}N + \frac{3}{2}(p-1)\,, \\ \qquad p = 5\,, 13\,, 21\,, \ldots \\ 2\,\mathrm{ch}(2u)\,\mathrm{sh}^2 u\,\mathrm{sh}^{2N}(u+2\eta)\,, \quad M_1 = M_2 = \frac{1}{2}N + 2\,, \\ \qquad p = 1\,. \end{cases} \tag{5.9}$$

We have verified numerically the completeness of this solution. Similar results hold for the case with α_-, β_- arbitrary and $\alpha_+ = \beta_+ = 0$, *etc.*

We observe that this solution represents a generalization of the famous Baxter $T-Q$ relation [4], which schematically has the form

$$t(u)\; Q(u) = Q(u') + Q(u'')\,. \tag{5.10}$$

Indeed, our result (5.6) has the structure

$$\begin{aligned} t(u)\; Q_1(u) &= Q_2(u') + Q_2(u'')\,, \\ t(u)\; Q_2(u) &= Q_1(u') + Q_1(u'')\,. \end{aligned} \tag{5.11}$$

Such generalized $T-Q$ relations, involving two or more independent $Q(u)$'s, may also appear in other integrable models.

6. Conclusions

We have seen that Bethe Ansatz solutions of the open spin-$\frac{1}{2}$ XXZ quantum spin chain are available for the cases (i)-(iii), for which the quantity $\Delta(u)$ (1.5) is a perfect square. There may be further special cases for which $\Delta(u)$ is a perfect square, in which case it should not be difficult to find the corresponding Bethe Ansatz solution. Our solution for case (iii) involves more than one $Q(u)$. This is a novel structure, which should be further understood. The general case that $\Delta(u)$ is not a perfect square and/or that $\eta \neq i\pi/(p+1)$ also remains to be understood.

Acknowledgments

Some of the work described here was done in collaboration with R. Murgan and F. Ravanini. I am grateful to the conference organizers for their wonderful hospitality and for the opportunity to present this work. I am also grateful to many of the participants, in particular F.C. Alcaraz, L. Faddeev, J.-M. Maillard, G. Sierra, F.Y. Wu and W.-L. Yang, for their questions or comments.

References

1. H.J. de Vega and A. González-Ruiz, *J. Phys.* **A26**, L519 (1993) [hep-th/9211114]
2. S. Ghoshal and A. B. Zamolodchikov, *Int. J. Mod. Phys.* **A9**, 3841 (1994) [hep-th/9306002]
3. M.T. Batchelor, in *Proceedings of the 22nd International Colloquium on Group Theoretical Methods in Physics*, eds S.P. Corney et al., (International Press, 1999) 261 [cond-mat/9811165]
4. R.J. Baxter, *Exactly Solved Models in Statistical Mechanics* (Academic Press, 1982).
5. V.E. Korepin, N.M. Bogoliubov, and A.G. Izergin, *Quantum Inverse Scattering Method, Correlation Functions and Algebraic Bethe Ansatz* (Cambridge University Press, 1993).
6. L.D. Faddeev, in *Les Houches 1995, Relativistic gravitation and gravitational radiation*, eds J.A. Marck and J.P. Lasota (Cambridge University Press, 1997) 149 [hep-th/9605187]
7. M. Gaudin, *Phys. Rev.* **A4**, 386 (1971); *La fonction d'onde de Bethe* (Masson, 1983).
8. F.C. Alcaraz, M.N. Barber, M.T. Batchelor, R.J. Baxter and G.R.W. Quispel, *J. Phys.* **A20**, 6397 (1987).
9. E.K. Sklyanin, *J. Phys.* **A21**, 2375 (1988).
10. R.I. Nepomechie, *J. Phys.* **A34**, 9993 (2001) [hep-th/0110081]
11. R.I. Nepomechie, *Nucl. Phys.* **B622**, 615 (2002); Addendum, *Nucl. Phys.* **B631**, 519 (2002) [hep-th/0110116]
12. R.J. Baxter, *Adv. Stud. Pure Math.* **19**, 95 (1989).
13. V.V. Bazhanov, S.L. Lukyanov and A.B. Zamolodchikov, *Commun. Math. Phys.* **177**, 381 (1996) [hep-th/9412229]; *Commun. Math. Phys.* **200**, 297 (1999) [hep-th/9805008]
14. A. Kuniba, K. Sakai and J. Suzuki, *Nucl. Phys.* **B525**, 597 (1998).
15. R.I. Nepomechie, *J. Stat. Phys.* **111**, 1363 (2003) [hep-th/0211001]
16. R.I. Nepomechie, *J. Phys.* **A37**, 433 (2004) [hep-th/0304092]
17. R.I. Nepomechie and F. Ravanini, *J. Phys.* **A36**, 11391 (2003); Addendum, *J. Phys.* **A37**, 1945 (2004) [hep-th/0307095]
18. R. Murgan and R.I. Nepomechie, *J. Stat. Mech.* **P05007** (2005); Addendum, *J. Stat. Mech.*, in press [hep-th/0504124]
19. R. Murgan and R.I. Nepomechie, *J. Stat. Mech.* **P08002** (2005) [hep-th/0507139]
20. J. Cao, H.-Q. Lin, K.-J. Shi and Y. Wang, cond-mat/0212163; *Nucl. Phys.* **B663**, 487 (2003).
21. J. de Gier and P. Pyatov, *J. Stat. Mech.* **P03002** (2004) [hep-th/0312235]
22. W.-L. Yang, Y.-Z. Zhang and M. Gould, *Nucl. Phys.* **B698**, 503 (2004) [hep-th/0411048]
23. V.V. Bazhanov and N.Yu. Reshetikhin, *Int. J. Mod. Phys.* **A4**, 115 (1989).

Separation Between Spin And Charge in SU(2) Yang-Mills Theory

A. J. Niemi

Department of Theoretical Physics, Uppsala University
BOX 803, S-75 108 Uppsala, Sweden
and
Nankai Institute of Mathematics
Tianjin 300071, P.R. China
and
Laboratoire de Mathematiques et Physique Theorique
CNRS UMR 6083, Universite de Tours
Parc de Grandmont, F37200, Tours, France
E-mail: Antti.Niemi@teorfys.uu.se

A Yang-Mills gauge field with gauge group $SU(2)$ can be decomposed into a single charge neutral complex vector and two spinless charged scalar fields. Under normal circumstances these constituents are tightly confined into each other by a compact $U(1)$ interaction, and the Yang-Mills Lagrangian describes the dynamics of asymptotically free pointlike gauge particles. But in a low energy finite density environment the interaction between the constituents can become weak, and a spin-charge separation may occur. It could be that this separation between the spin and charge, in combination with a condensation of the charge carriers, occurs when the Yang-Mills theory enters its confinement phase.

1. Introduction

Color confinement by strong nuclear forces and superconductivity in high temperature cuprate superconductors are both among the outstanding physical quagmires. Curiously, it seems that these two apparently very distinct phenomena have much in common. In both cases there is a well-defined theoretical framework that is at least in principle capable of explaining all of their physical properties. In the case of strong nuclear forces we have the Yang-Mills theory, while the description of high-T_c superconductivity employs the $t - J$ (Hubbart) model. In both cases the fundamental theoretical problem is also very similar: The lack of a natural condensate. In the case of the Yang-Mills theory, we desire a gauge invariant operator that can describe the mass gap. In the $t - J$ model there is no apparent Cooper pair, that could yield an explanation of superconductivity by the BCS-mechanism.

In the case of the $t - J$ model, it has been proposed that a Cooper pair condensate can be constructed in a manner which, if correct, has far reaching consequences to our understanding of the fundamental structure of Matter. This proposal states [1–3] that in the strongly correlated environment of cuprate superconductors, an electron ceases to be a fundamental particle. Instead it is postulated that an electron is a bound state of two other particles, which are called spinon and holon. The spinon is a fermion that carries the spin degree of freedom of the electron. It does not directly couple to Maxwell's electrodynamics. The holon is a spinless, complex boson and it carries the electric charge of the electron.

To introduce the decomposition of the electron into its spinon and holon constituents, we consider a charge neutral, spin-1/2 fermionic operator $f_{i\sigma}$ where i is the site label and $\sigma = \uparrow, \downarrow$ is the spin index. This operator corresponds to the spinon, it it the carrier of the (statistical) spin degree of freedom of the electron.

The holon is described by a spinless bosonic operator $b_i^T = (b_{i1}, b_{i2})$ and it carries the electric charge of the electron.

In terms of the spinon and holon, the electron operator $c_{i\sigma}$ decomposes according to

$$c_{i\sigma} = \frac{1}{\sqrt{2}} b_i^\dagger \psi_{i\sigma} \quad , \tag{1.1}$$

where we have combined the spinon operators as

$$\psi_{i\sigma}^T = (f_{i\sigma}, \epsilon_{\sigma\tilde{\sigma}} f_{i\tilde{\sigma}}^\dagger) \quad . \tag{1.2}$$

But the decomposition (1.1) also introduces an internal $U(1)$ gauge symmetry, as it remains invariant under the simultaneous phase rotations

$$b_i \to e^{i\theta} b_i, \qquad \psi_{i\sigma} \to e^{i\theta} \psi_{i\sigma}. \tag{1.3}$$

As a result there is a compact $U(1)$ gauge interaction between the spinon and holon.

Under normal circumstances one expects that the strength of the internal $U(1)$ interaction increases with increasing energy. As a consequence at high energies the spinon and holon are confined into a (point-like) electron, consistent with experimental observations at high energies. But in a strongly correlated environment, such as in a cuprate superconductor, the spin and the charge of the electron can become independent excitations

[1–3]. This leads to a very complicated phase diagram, with several different regions [3]. This phase structure is usually inspected using a mean-field theory, which is obtained by integrating over the fermions $\psi_{i\sigma}$. In this way one finds that (d-wave) superconductivity occurs when the remaining bosonic holon field b_i condenses,

$$\langle b_i^\dagger b_i \rangle = \Delta_b \neq 0. \tag{1.4}$$

Of substantial interest is also the possibility that the system can enter a pseudogap phase[3], which is a precursor to the superconducting phase; It has the characteristic property that even though the underlying symmetry is broken the effective bosonic order parameter Δ_b vanishes due to quantum fluctuations.

2. Spin-Charge Separation

We now proceed to describe the spin-charge decomposition of the $SU(2)$ gauge field[4,5]. For this we represent the gauge field as a linear combination

$$A_\mu = A_{\mu i}\sigma^i = C_\mu \sigma^3 + X_{\mu +}\sigma^+ + X_{\mu -}\sigma^- \tag{2.1}$$

where

$$X_{\mu\pm} = A_{\mu 1} \mp i A_{\mu 2}$$

The spin-charge decomposition of A_μ entails a decomposition of $X_{\mu\pm}$ into its spin and charge constituents. Therefore, we introduce a complex vector field e_μ which we normalize according to

$$\begin{aligned} \vec{e}^{\,2} &= 0 \\ \vec{e}\cdot\vec{e}^{\,*} &= 1 \end{aligned} \tag{2.2}$$

With ψ_1 and ψ_2 two complex scalars we can then write $X_{\mu\pm}$ as [4]

$$X_{\mu +} = X_{\mu -}^* = i\psi_1 e_\mu - \; i\psi_2^* e_\mu^* \tag{2.3}$$

Indeed, *any* four component complex vector can always be represented as a linear combination of the form (2.3). For this, it suffices to observe that an arbitrary, unconstrained four component complex vector describes eight independent real field degrees of freedom. On the other hand, the two complex fields ψ_1 and ψ_2 describe four, and the complex vector $\vec{e}$ when subject to the conditions (2.2) describes five independent field degrees of freedom. But one of these corresponds to the internal $U(1)$ rotation

$$\begin{aligned} \vec{e} &\longrightarrow e^{-i\xi}\vec{e} \\ \psi_1 &\longrightarrow e^{i\xi}\psi_1 \\ \psi_2 &\longrightarrow e^{i\xi}\psi_2 \end{aligned} \tag{2.4}$$

which leaves the *r.h.s.* of (2.3) intact. As a consequence, in the general case the *r.h.s.* of (2.3) also describes eight independent field degrees of freedom.

In general, the present decomposition of the gauge field is not gauge invariant. But it turns out that in a proper gauge the decomposition can be given a gauge invariant meaning. In particular the combination

$$\rho^2 = \rho_1^2 + \rho_2^2 = \langle|\psi_1|^2\rangle + \langle|\psi_2|^2\rangle \tag{2.5}$$

of the holons of the gauge field becomes a gauge invariant quantity. For this we introduce [5,6]

$$\int \rho^2 = \int (\rho_1^2 + \rho_2^2) = \int \left[(A_\mu^1)^2 + (A_\mu^2)^2\right] = \int X_\mu X_\mu^*. \tag{2.6}$$

This is in general a gauge dependent quantity. But if we consider the extrema of (2.6) along the gauge orbits with respect to the full $SU(2)$ gauge transformations, these extrema are by construction gauge independent quantities. Moreover, the gauge orbit extrema of (2.6) correspond to field configurations X_μ which are subject to a background version of the maximal abelian gauge [5,6],

$$(\partial_\mu + igC_\mu)X_\mu = 0, \tag{2.7}$$

This gauge condition is widely used in lattice studies [7]. In the following we shall assume that the gauge fixing condition (2.7) has been introduced. The spin-charge decomposition of the gauge field then acquires a gauge invariant meaning. In particular, the condensate (2.5) is a gauge invariant quantity.

The internal $U(1)$ invariance determines a compact version of the $U(1)$ gauge structure. A compact $U(1)$ gauge theory is known to be confining when the coupling is sufficiently strong [8]. The confining phase is separated by a first order phase transition from the deconfined weak coupling phase. Furthermore, since the running of the β-function of the compact $U(1)$ leads to an increase of the coupling with increasing energy, we expect that at high energy the spin and charge of the gauge field become confined by an increasingly strong compact $U(1)$ interaction to the effect that the high energy Yang-Mills theory describes asymptotically free and pointlike gluons, as it should.

But at low energy and in a strongly correlated environment, maybe in the interior of hadronic particles, the internal $U(1)$ gauge interaction can become weak and the spin and the color-charge degrees of freedom of the gluon can separate from each other. If in analogy with (1.4) the spinless color-carriers then condense

$$\rho^2 = \rho_1^2 + \rho_2^2 = \langle \psi_1^\dagger \psi_1 \rangle + \langle \psi_2^\dagger \psi_2 \rangle = \Delta_\psi \neq 0 \quad ,$$

we have a mass gap and the theory is in a phase which is very similar to the holon condensation phase of cuprate superconductors.

We observe, that when we use the condition (2.7) and solve for ρ_1 and ρ_2, we introduce *no* restrictions on the complex vector $\vec{e}$. Nor do we introduce any restrictions on the phases of the complex fields ψ_1 and ψ_2. In particular, this means that the internal symmetry (2.4) remains intact when we evaluate the absolute values ρ_1 and ρ_2 at their gauge invariant extrema along the gauge orbit.

We note that in general there are Gribov ambiguities in the maximal abelian gauge condition. Consequently the extrema values of ρ_1 and ρ_2 on the orbit are not unique. In this article we will not analyze the consequences that Gribov ambiguities might have.

The diagonal $U(1) \subset SU(2)$ gauge transformation of the original gauge group acts on the complex fields $\psi_{1,2}$ as follows,

$$\begin{aligned} \psi_1 &\to e^{2i\omega}\psi_1 \\ \psi_2 &\to e^{-2i\omega}\psi_2 \end{aligned} \tag{2.8}$$

Here the phases differ from those in (2.4) by a relative sign. Since this $U(1)$ transformation leaves the vector $\vec{e}$ intact, only the complex fields ψ_1 and ψ_2 couple to the Cartan subgroup $U(1) \subset SU(2)$. On the other hand, the components e_μ transform as a vector under Lorenz transformations while the fields ψ_1 ad ψ_2 are scalars. This means that (2.3) entails a decomposition of $X_{\mu\pm}$ into two qualitatively very different sets of fields: The scalar fields ψ_1 and ψ_2 couple nontrivially to the abelian component of the $SU(2)$ gauge transformations *i.e.* carry a charge but have no spin. The complex vector $\vec{e}$ is neutral *w.r.t.* the abelian component of the gauge transformation but it carries the spin degrees of freedom of the $X_{\mu\pm}$.

Onviously, for consistency of the decomposition (2.3) we must assume that *both* condensates $\rho_{1,2}$ are nontrivial. This means, that in the quantum Yang-Mills theory we need the expectation values

$$< \rho_{1,2} >= \Delta_{1,2} \tag{2.9}$$

to be nonvanishing. This condition then specifies the physical environment where the separation between the spin and the charge of a gauge field can occur.

Numerical simulations [9] indicate, that in the confinement region of $SU(2)$ gauge theory *both* Δ_1 and Δ_2 are non-vanishing. It would be very interesting, if this indeed implies that the $\Delta_{1,2}$ can be viewed as the order parameters that characterize when the Yang-Mills theory displays confinement.

It is apparent that the present spin-charge decomposition of the gauge field is fully analogous to the spin-charge decomposition of the electron: In both cases, the decomposition entails a separation between the carriers of spin, and the carriers of charge. Furthermore, in both cases the separation can only occur in a finite density environment. In the case of electron we need the μ in (1.4) to be non-vanishing and in the case of gauge field we need the condensates (2.9) to be non-vanishing. Furthermore, in both cases the decomposition introduces an internal, compact $U(1)$ that can be employed to argue that asymptotically in the short distance limit both the gauge field and the electron must behave like structureless point particles, with the spinon and holon confined to each other by the strong internal force. The internal spin-charge structures can be visible only in the infrared region and in a finite density environment, when the internal $U(1)$ interaction becomes weak.

3. Conclusion

In conclusion, the spin-charge decomposition that has been widely employed in the theory of cuprate superconductors, can also be introduced in Yang-Mills theories. The decomposition of a gauge field turns out to be very similar to that of an electron in the context of high-T_c supercoductivity. Furthermore, if both fermions and gauge fields are dcomposed in a theory such as QCD or more generally the Standard Model, this could lead to a very rich gauge structure. Since the conditions for a decomposition to occur are quite similar to those expected in a confining environment, it is of interest to understand whether confinnement allows for a natural interpretation in terms of the decomposed structures. Indeed, the widely accepted intuitive picture of confinement as a dual Meissner effect relies heavily on the BCS approach to superconductivity. But the BCS theory is based on the existence of a natural condensate, the Cooper pair, which is absent in Yang-Mills theories. The failure of BCS theory due to the absence of a natural Cooper pair in theories of cuprate superconductors originally led to the introduction of a spin-charge decomposition in that context. Maybe a similar remedy turns out to be applicable also in the case of strong interaction physics?

Acknowledgments

This work is supported by a VR grant 2003-3466, and by the STINT Thunberg Stipend

References

1. P.W. Andersson, Science **235**, 1196 (1987); L.D Faddeev and L.A. Takhtajan, Phys. Lett. **A85**, 375 (1981).
2. G. Baskaran, Z. Zhou and P.W. Andersson, Solid State Comm. **63**, 973 (1987).
3. P.A. Lee, N. Nagaosa and X.-G. Wen, `cond-mat/0410445`.
4. L.D. Faddeev and A.J. Niemi, Phys. Lett. **B525** (2002) 195; L.D. Faddeev and A.J. Niemi, Phys. Lett. **B449** (1999); L.D. Faddeev and A.J. Niemi, Phys. Rev. Lett. **82** (1999) 1624
5. A.J. Niemi and N. Walet, `hep-ph/0504034`; A.J. Niemi, JHEP **0408** (2004) 035
6. F.V. Gubarev, L. Stodolsky and V.I. Zakharov, Phys. Rev. Lett. **86**, 2220 (2001); L Stodolsky, P. van Baal and V.I. Zakharov, Phys. Lett. **B552**, 214 (2003).
7. M.N. Chernodub and M.I. Polikarpov, in *Confinement, duality, and nonperturbative aspects of QCD*, P. van Baal, Ed. (Plenum Press, New York 1998); T. Suzuki, Prog. Theor. Phys. Suppl. **131**, 633 (1998); R.W. Haymaker, Phys. Rept. **315**, 153 (1999).
8. A.M. Polyakov, Phys. Lett. **59B**, 122 (1978); M.E. Peskin, Ann. Phys. **113**, 122 (1978).
9. M. Chernodub and M. Polikarpov, private communication

On Solutions of the One-dimensional Holstein Model *

Feng Pan

Department of Physics, Liaoning Normal University,
Dalian 116029, P. R. China
E-mail: daipan@dlut.edu.cn

Jerry P. Draayer

Department of Physics and Astronomy, Louisiana State University,
Baton Rouge, LA 70803-4001, USA
E-mail: draayer@sura.org

The one-dimensional Holstein model of spinless or hard-core fermions interacting with dispersionless phonons is proved to be exactly solvable. Excitation energies and the corresponding wavefunctions of the model are obtained by using a simple extended Bethe ansatz.

Models of interacting electrons with phonons have been attracting much attention as they are helpful in understanding superconductivity in many aspects, such as in fullerenes, bismuth oxides, and the high-T_c superconductors.[1] Unlike conventional metals these materials are not necessarily in the weak-coupling regime where perturbation theory can be used or the strong-coupling regime in which a polaronic treatment is possible. Neither are they necessarily in the adiabatic regime in which characteristic phonon energies are much less than characteristic electronic energies. This challenge has led to numerical studies of the Holstein (or molecular crystal) model of electrons interacting with dispersionless phonons in infinite dimensions, two dimensions, one dimension and on just two sites.[1,2] The one-dimensional case is important because of the wide range of quasi-one-dimensional materials which undergo a Peierls or charge-density-wave (CDW) instability due to the electron-phonon interaction. Most theoretical

*Support from the U. S. National Science Foundation (0140300), the Southeastern Universities Research Association, the Natural Science Foundation of China (10175031), the Education Department of Liaoning Province (202122024), and the LSU-LNNU joint research program (C164063)

treatments assume the adiabatic limit and treat the phonons in a mean-field approximation. However, it has been argued that in many CDW materials the quantum lattice fluctuations are important [3].

In this talk, we present a study of the one-dimensional Holstein model of spinless or hard-core fermions with an algebraic approach. This model is particularly interesting because at a finite fermion-phonon coupling there is a quantum phase transition from a Luttinger liquid (metallic) phase to an insulating phase with CDW long-range order [4,5]. This illustrates how quantum fluctuations can destroy the Peierls state.

The Hamiltonian is

$$H = \omega \sum_{i} b_i^\dagger b_i - t \sum_{\langle i,j \rangle} f_i^\dagger f_j + g \sum_{i} f_i^\dagger f_i (b_i^\dagger + b_i), \tag{1}$$

where f_i destroys a fermion on site i, b_i destroys a local phonon of frequency ω, t is the hopping integral, g is the fermion-phonon coupling and a periodic chain of N sites is assumed. The phase transition occurs at a critical coupling g_c separating metallic $(0 \leq g \leq g_c)$ and CDW insulating phases $(g > g_c)$ [4,5]. In the strong coupling limit $(g^2 \gg \omega_t)$ (1) can be mapped onto the anisotropic, antiferromagnetic Heisenberg (XXZ) model [4] which is exactly solvable. The transition occurs at the spin isotropy point, is of the Kosterlitz-Thouless (K-T) type, and the Luttinger liquid parameters can be found in the metallic phase [2].

In order to diagonalized the Hamiltonian (1), let use consider the simpler hard-core Fermi-Hubbard model[6] with

$$H = \sum_{i=1}^{p} h_i \hat{n}_i - t \sum_{i=1}^{p-1} (f_i^\dagger f_{i+1} + f_{i+1}^\dagger f_i) - t(f_1^\dagger f_p + f_p^\dagger f_1)(1 - \delta_{p2}), \tag{2}$$

where $\{h_i\}$ are a set of parameters independent of the number of fermions, and the last term keeps (2) satisfying the periodic condition. It has been known that (2) is simply exactly solvable.[6] For k-particle excitation, the eignstates are

$$|k; \eta\rangle = \sum_{i_1 < i_2 < \cdots < i_k} C^{(\eta)}_{i_1 i_2 \cdots i_k} f_{i_1}^\dagger f_{i_2}^\dagger \cdots f_{i_k}^\dagger |0\rangle, \tag{3}$$

where

$$C^{(\eta)}_{i_1 i_2 \cdots i_k} = \begin{vmatrix} g^{(\eta_1)}_{i_1} & g^{(\eta_1)}_{i_2} & \cdots & g^{(\eta_1)}_{i_k} \\ g^{(\eta_2)}_{i_1} & g^{(\eta_2)}_{i_2} & \cdots & g^{(\eta_2)}_{i_k} \\ \vdots & \cdots & \cdots & \vdots \\ g^{(\eta_k)}_{i_1} & g^{(\eta_k)}_{i_2} & \cdots & g^{(\eta_k)}_{i_k} \end{vmatrix}, \tag{4}$$

in which $\{g^{(\eta_\mu)}_i\}$ should satisfy the following eigen-equation for a $p \times p$ matrix T with

$$\sum_j T_{ij}(p) g^{(\eta_\mu)}_j = E^\eta g^{(\eta_\mu)}_i, \tag{5}$$

where $\{T_{ij}(p)\}$ are elements of the matrix

$$T(2) = \begin{pmatrix} h_1 & -t \\ -t & h_2 \end{pmatrix}, \quad T(p) = \begin{pmatrix} h_1 & -t & 0 & \cdots & -t \\ -t & h_2 & -t & 0 & \cdots \\ 0 & \ddots & \ddots & \ddots & \ddots \\ \vdots & \cdots & -t & h_{p-1} & -t \\ -t & 0 & \cdots & -t & h_p \end{pmatrix} \quad \text{for } p \geq 3, \tag{6}$$

which is tridiagonal except the elements $T_{1p} = T_{p1} = -t$ for $p \geq 3$ originating from the last term in (2) needed in order to satisfy the periodic condition.

Let us introduce the differential realization for the boson operators with

$$b^\dagger_i \Rightarrow y_i, \quad b_i \Rightarrow \frac{\partial}{\partial y_i} \tag{7}$$

for $i = 1, 2, \cdots, p$. Then, the Hamiltonian (1) is mapped into the following form:

$$H = \omega \sum_{i=1}^{p} y_i \frac{\partial}{\partial y_i} - t \sum_{\langle i,j \rangle} f^\dagger_i f_j + g \sum_{i=1}^{p} f^\dagger_i f_i (y_i + \frac{\partial}{\partial y_i}). \tag{8}$$

According to the digonalization procedure used to solve the eigenvalue problem (2), the one fermion excitation states can be assumed to be the following ansatz form:

$$|n_f = 1\rangle = \sum_{\mu=1}^{p} q_\mu(x_1, x_2, \cdots, x_p)\, e^{-\frac{g}{\omega} y_\mu} f_\mu^\dagger |0\rangle, \tag{9}$$

where $|0\rangle$ is the fermion vacuum state and

$$x_1 = y_1 - y_2, \ \cdots, \ x_{k-1} = y_{k-1} - y_k, \ x_k = y_k - y_{k+1}, \cdots,$$

$$x_{p-1} = y_{p-1} - y_p, \ x_p = y_p - y_1. \tag{10}$$

By using the expressions of (8) and (9), the energy eigen-equation becomes

$$\sum_\mu \left(\sum_i \omega y_i \frac{\partial q_\mu}{\partial y_i} e^{-\frac{g}{\omega} y_\mu} + g \frac{\partial q_\mu}{\partial y_\mu} e^{-\frac{g}{\omega} y_\mu} \right) f_\mu^\dagger |0\rangle$$

$$-t \sum_{\langle i,j \rangle} f_i^\dagger f_j \sum_\mu q_\mu e^{-\frac{g}{\omega} y_\mu} f_\mu^\dagger |0\rangle = (E + \frac{g^2}{\omega}) \sum_\mu q_\mu e^{-\frac{g}{\omega} y_\mu} f_\mu^\dagger |0\rangle, \tag{11}$$

which results in the following set of the extended Bethe ansatz equations:

$$\sum_{i=1}^{p} \omega y_i \frac{\partial q_\mu}{\partial y_i} + g \frac{\partial q_\mu}{\partial y_\mu} - \frac{t}{1+\delta_{2p}} q_{\mu-1} e^{-\frac{g}{\omega} x_{\mu-1}} - \frac{t}{1+\delta_{2p}} q_{\mu+1} e^{\frac{g}{\omega} x_\mu} = (E + \frac{g^2}{\omega}) q_\mu \tag{12}$$

for $\mu = 1, 2, \cdots, p$, which is a set of coupled rank-1 Partial Differential Equations (PDEs). Eq. (12) completely determines the eigenenergies and the corresponding coefficients $\{q_\mu \equiv q_\mu(x_1, x_2, \cdots, x_p)\}$. Once the above PDEs are solved for one-fermion excitation, according to the procedure used for solving the hard-core Fermi-Hubbard model, the k-fermion excitation wavefunction can be organized into the following form:

$$|n_f = k; \zeta\rangle = \sum_{i_1 < i_2 < \cdots < i_k} C_{i_1 i_2 \cdots i_k}^{(\eta)} e^{-\frac{g}{\omega} \sum_{\mu=1}^{k} y_{i_\mu}} f_{i_1}^\dagger f_{i_2}^\dagger \cdots f_{i_k}^\dagger |0\rangle \tag{13}$$

with

$$C_{i_1 i_2 \cdots i_k}^{(\eta)} = \begin{vmatrix} q_{i_1}^{(\eta_1)} & q_{i_2}^{(\eta_1)} & \cdots & q_{i_k}^{(\eta_1)} \\ q_{i_1}^{(\eta_2)} & q_{i_2}^{(\eta_2)} & \cdots & q_{i_k}^{(\eta_2)} \\ \vdots & \cdots & \cdots & \vdots \\ q_{i_1}^{(\eta_k)} & q_{i_2}^{(\eta_k)} & \cdots & q_{i_k}^{(\eta_k)} \end{vmatrix}. \tag{14}$$

The corresponding k-fermion excitation energy is given by

$$E_k^{(\eta)} = \sum_{\nu=1}^{k} E^{(\eta_\nu)} - kg^2/\omega, \tag{15}$$

in which $E^{(\eta_\nu)}$ is the ν-th eigenvalue of Eq. (12).

In summary, general solutions of the 1-dim Holstein model are derived based on an algebraic approach similar to that used in solving 1-dim hard-core Fermi-Hubbard model. A set of the extended Bethe ansatz equations are coupled rank-1 Partial Differential Equations (PDEs), which completely determine the eigenenergies and the corresponding wavefunctions of the model. Though we still do not know whether the PDEs are exactly solvable or not, at least, these PDEs should be quasi-exactly solvable.

References

1. A. S. Alexandrov and N. Mott, *Polarons and Bipolarons* (World Scientific, Singapore, 1995)
2. R. H. McKenzie, C.J. Hamer and D.W. Murray, *Phys. Rev.* **B53**, 9676 (1996)
3. R. H. McKenzie and J.W. Wilkins, *Phys. Rev. Lett.* **69**, 1085 (1992)
4. J. E. Hirsch and E. Fradkin, *Phys. Rev.* **B27**, 4302 (1983)
5. G. Benfatto, G. Gallavotti and J. L. Lebowitz, *Helv. Phys. Acta.* **68**, 312 (1995).
6. Feng Pan and J. P. Draayer, *J. Phys.* **A33**, 9095 (2000)

Recent Developments on Ising and Chiral Potts Model

Jacques H. H. PERK and Helen AU-YANG

Department of Physics,
Oklahoma State University,
Stillwater, OK 74078-3072, USA
E-mail: perk@okstate.edu

After briefly reviewing selected Ising and chiral Potts model results, we discuss a number of properties of cyclic hypergeometric functions which appear naturally in the description of the integrable chiral Potts model and its three-dimensional generalizations.

1. Ising Model and Integrable Chiral Potts Model

1.1. *Z-Invariant Ising Model*

Baxter's Z-invariant Ising model is the prototype integrable lattice model in statistical mechanics. It is "exactly solvable" for two reasons, namely because of a complete parametrization in terms of Yang–Baxter rapidities but also because of reformulations in terms of free fermions. This does not mean that the calculation of its pair-correlation or its susceptibility is a straightforward exercise. A more detailed description of the singularity structure of the zero-field susceptibility of the square-lattice Ising model has been obtained only recently.[1]

Both integrability features were exploited in our recent studies of the pair-correlation function and the wavevector-dependent susceptibility of Ising models with quasiperiodic coupling constants[2,3] and of the pentagrid Ising model[4] of Korepin.

1.2. *Integrable Chiral Potts Model*

An N-state generalization of the Ising model with fermions replaced by cyclic parafermions is given by the integrable chiral Potts model.[5–7] One version of this model is given in terms of a square lattice of horizontal and vertical rapidity lines with rapidities q and p, respectively pointing left and up. After black-and-white checkerboard coloring of the faces, Potts spins are placed on the black faces. Boltzmann weights $W(a-b)$ and $\overline{W}(a-b)$ are assigned to each nearest-neighbor pair of spins in states ω^a and ω^b,

$$\omega \equiv \mathrm{e}^{2\pi \mathrm{i}/N}, \tag{1.1}$$

$(a, b = 1, \ldots, N)$, as in Fig. 1.1. Here the difference $a-b$ is to be taken

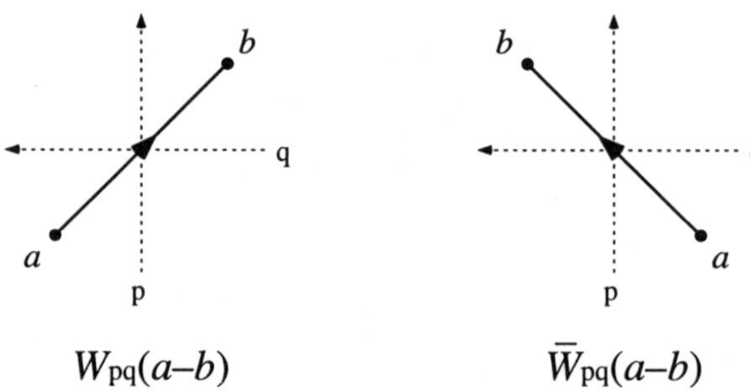

Fig. 1.1. Chiral Potts Model Boltzmann Weights.

mod N. The Boltzmann weights W and $\overline{W}$ can be parametrized as

$$\frac{W_{\mathrm{pq}}(n)}{W_{\mathrm{pq}}(0)} = \Big(\frac{\mu_{\mathrm{p}}}{\mu_{\mathrm{q}}}\Big)^n \prod_{j=1}^{n} \frac{y_{\mathrm{q}} - x_{\mathrm{p}}\omega^j}{y_{\mathrm{p}} - x_{\mathrm{q}}\omega^j}, \quad \frac{\overline{W}_{\mathrm{pq}}(n)}{\overline{W}_{\mathrm{pq}}(0)} = (\mu_{\mathrm{p}}\mu_{\mathrm{q}})^n \prod_{j=1}^{n} \frac{\omega x_{\mathrm{p}} - x_{\mathrm{q}}\omega^j}{y_{\mathrm{q}} - y_{\mathrm{p}}\omega^j}. \tag{1.2}$$

The rapidities p and q lie on a higher genus curve with moduli k, k′, with $\mathrm{k}^2 + \mathrm{k}'^2 = 1$. The p-curve is parametrized by $(x_{\mathrm{p}}, y_{\mathrm{p}}, \mu_{\mathrm{p}})$ satisfying the algebraic equations

$$y_{\mathrm{p}}^N = (1 - \mathrm{k}'\lambda_{\mathrm{p}})/\mathrm{k}, \quad x_{\mathrm{p}}^N = (1 - \mathrm{k}'/\lambda_{\mathrm{p}})/\mathrm{k}, \quad \mu_{\mathrm{p}}^N = \lambda_{\mathrm{p}}, \tag{1.3}$$

$$\lambda_{\mathrm{p}} + \lambda_{\mathrm{p}}^{-1} = (1 + \mathrm{k}'^2 - \mathrm{k}^2 {t_p}^N)/\mathrm{k}', \quad t_{\mathrm{p}} = x_{\mathrm{p}} y_{\mathrm{p}}, \tag{1.4}$$

which follow from the two mod N conditions $W_{\mathrm{pq}}(N) = W_{\mathrm{pq}}(0)$ and $\overline{W}_{\mathrm{pq}}(N) = \overline{W}_{\mathrm{pq}}(0)$. Given a value of t_{p} one can choose $|\lambda_{\mathrm{p}}| > 1$ or $|\lambda_{\mathrm{p}}| < 1$. Then x_{p}, y_{p}, μ_{p} are given by (1.3) up to powers of ω.

1.3. *Chiral Potts Free Energy and Order Parameters*

Baxter has derived several exact results for the free energy of the chiral Potts model. Most of his work is based on a set of functional equations for the transfer matrices.[8] An account with results for all four regimes, with each of $|\lambda_{\rm p}|$ and $|\lambda_{\rm q}| > 1$ or < 1, can be found in Ref. 9. Baxter also obtained results for the interfacial tension, which can be much simplified in the symmetric case.[10]

For the order parameters of the integrable chiral Potts model we have

$$\langle\sigma_0^n\rangle = (1-k'^2)^{\beta_n}, \quad \beta_n = \frac{n(N-n)}{2N^2}, \quad (1 \le n \le N-1, \quad \sigma_0^N = 1), \tag{1.5}$$

which was conjectured[11] early in 1988 and proved only very recently by Baxter.[12,13]

2. Cyclic Hypergeometric Functions

2.1. *Basic Hypergeometric Series at Root of Unity*

The basic hypergeometric hypergeometric series is defined as

$${}_{p+1}\Phi_p\left[\begin{matrix}\alpha_1,\cdots,\alpha_{p+1}\\ \beta_1,\cdots,\beta_p\end{matrix};z\right] = \sum_{l=0}^{\infty}\frac{(\alpha_1;q)_l\cdots(\alpha_{p+1};q)_l}{(\beta_1;q)_l\cdots(\beta_p;q)_l(q;q)_l}\,z^l, \tag{2.1}$$

where

$$(x;q)_l \equiv \prod_{j=1}^{l}(1-xq^{j-1}), \quad l \ge 0. \tag{2.2}$$

Setting first $\alpha_{p+1} = q^{1-N}$ and then $q \to \omega \equiv \mathrm{e}^{2\pi\mathrm{i}/N}$, we get

$${}_{p+1}\Phi_p\left[\begin{matrix}\omega,\alpha_1,\cdots,\alpha_p\\ \beta_1,\cdots,\beta_p\end{matrix};z\right] = \sum_{l=0}^{N-1}\frac{(\alpha_1;\omega)_l\cdots(\alpha_p;\omega)_l}{(\beta_1;\omega)_l\cdots(\beta_p;\omega)_l}\,z^l. \tag{2.3}$$

We note

$$(x;\omega)_{l+N} = (1-x^N)(x;\omega)_l \quad \text{and} \quad (\omega;\omega)_l = 0, \quad l \ge N. \tag{2.4}$$

Requiring

$$z^N = \prod_{j=1}^{p}\gamma_j^N, \quad \gamma_j{}^N = \frac{1-\beta_j^N}{1-\alpha_j^N}, \tag{2.5}$$

we obtain from (2.3) the "cyclic hypergeometric function" with summand periodic mod N. Of special importance is the Saalschütz case, defined by

$$z = q = \frac{\beta_1\cdots\beta_p}{\alpha_1\cdots\alpha_{p+1}} \qquad \text{or} \quad \omega^2\alpha_1\alpha_2\cdots\alpha_p = \beta_1\beta_2\cdots\beta_p, \quad z=\omega. \tag{2.6}$$

The theory of cyclic hypergeometric series is intimately related with the theory of the integrable chiral Potts model and its generalizations in three dimensions. We note that our notations differ from those of Bazhanov and Baxter[14,15] and of others,[16–19] who have an upside-down version of the q-Pochhammer symbol $(x;q)_l$.

2.2. *Integrable Chiral Potts Model Weights*

The weights W and $\overline{W}$ of the integrable chiral Potts model can be written in product form[10]

$$\frac{W(n)}{W(0)} = \gamma^n \frac{(\alpha;\omega)_n}{(\beta;\omega)_n}, \qquad \gamma^N = \frac{1-\beta^N}{1-\alpha^N}. \tag{2.7}$$

This is periodic with period N.

The dual weights are given by Fourier transform, *i.e.*[20]

$$\hat{W}(k) = \sum_{n=0}^{N-1} \omega^{nk}\, W(n) = {}_2\Phi_1\left[\begin{matrix}\omega,\alpha\\ \beta\end{matrix};\gamma\,\omega^k\right] W(0). \tag{2.8}$$

They have the same structure as the original weights[7,20]

$$\frac{\hat{W}(n)}{\hat{W}(0)} = \hat{\gamma}^n \frac{(\hat{\alpha};\omega)_n}{(\hat{\beta};\omega)_n}, \quad \text{with} \quad \hat{\alpha} = \gamma, \quad \hat{\beta} = \frac{\omega\alpha\gamma}{\beta}, \quad \hat{\gamma} = \frac{\omega}{\beta}. \tag{2.9}$$

2.3. *Summation Formula for* ${}_2\Phi_1$

The ${}_2\Phi_1$ is exactly summable as a product.[20] More precisely, we introduce the functions

$$\Delta(z) \equiv (1-z^N)^{1/N}, \quad p(z) \equiv \prod_{j=1}^{N-1}(1-\omega^j z)^{j/N}, \tag{2.10}$$

$$p_0(z) \equiv \frac{p(z)}{\Delta(z)^{(N-1)/2}} = \prod_{j=1}^{N-1}\left[\frac{(1-\omega^j z)}{\Delta(z)}\right]^{j/N}, \tag{2.11}$$

with all have cuts for $z^N \geq 1$ real, with the exception that $p(z)$ is regular on the positive real z-axis, where $p(1) = \sqrt{N}\Phi_0, \quad \Phi_0 \equiv \omega^{(N-1)(N-2)/24}$.

With these definitions,

$${}_2\Phi_1\left[\begin{matrix}\omega,\alpha\\ \beta\end{matrix};\gamma\right] = F_*\,\omega^{-\frac{1}{2}k(k+1)-mk}\,\frac{N}{\gamma^{\frac{1}{2}(N-1)}}\,\frac{p(\beta)p(\gamma)p(\varepsilon)}{p(\alpha)p(1)p(\delta)}, \tag{2.12}$$

where

$$m \equiv \left\lfloor \frac{N}{2\pi}\arg\alpha \right\rfloor, \quad n \equiv \left\lfloor \frac{N}{2\pi}\arg\beta \right\rfloor, \tag{2.13}$$

with $\lfloor x \rfloor$ the floor of x and

$$\gamma \equiv \omega^k \frac{\Delta(\beta)}{\Delta(\alpha)}, \quad \delta \equiv \frac{\beta}{\alpha}, \quad \varepsilon \equiv \frac{\beta}{\alpha\gamma}. \tag{2.14}$$

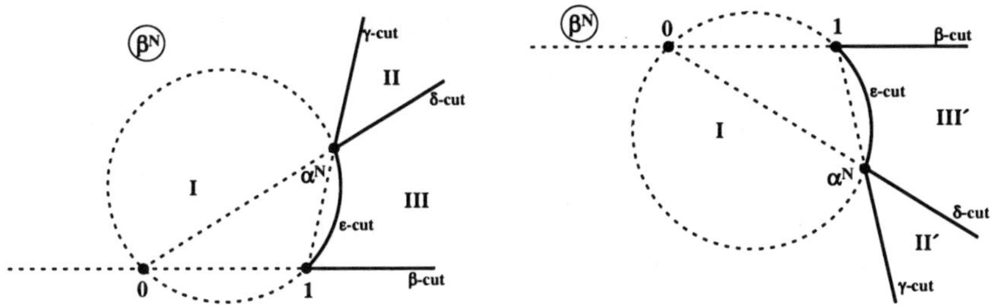

Fig. 2.1. Cut structure in β^N-plane for $\operatorname{Im}\alpha^N > 0$, respectively $\operatorname{Im}\alpha^N < 0$.

The phase factor F_* can take several values. If we keep α fixed and move β in the complex plane, we encounter the cuts in Fig. 2.1. From a detailed analysis at each cut we find

$$\begin{aligned} &F_{\mathrm{I}} = 1, \quad F_{\mathrm{II}} = \omega^k, \quad F_{\mathrm{III}} = \omega^{m-n+k}, \qquad \text{if } \operatorname{Im}\alpha^N > 0, \\ &F_{\mathrm{I}} = 1, \quad F_{\mathrm{II'}} = \omega^{-k}, \quad F_{\mathrm{III'}} = \omega^{n-m-k}, \qquad \text{if } \operatorname{Im}\alpha^N < 0. \end{aligned} \tag{2.15}$$

Noting

$$(z;\omega)_n \equiv \prod_{j=1}^{n} (1-\omega^{j-1}z), \quad z = 0,\ldots,N-1. \tag{2.16}$$

we see that

$$\frac{p(\omega^n z)}{p(z)} = \frac{p_0(\omega^n z)}{p_0(z)} = \frac{(z;\omega)_n}{\Delta(z)^n} \equiv (\!(z;\omega)\!)_n, \tag{2.17}$$

which is a "cyclic Pochhammer symbol" $(\!(z;\omega)\!)_{n+N} = (\!(z;\omega)\!)_n$. On the principal sector $0 < \arg z < 2\pi/N$, we find

$$p_0(z)p_0(\omega/z) = \omega^{(N^2-1)/12} = \Phi_0^2\omega^{(N-1)/4}, \qquad \frac{\Delta(\omega/z)}{\Delta(z)} = \frac{\omega^{n+\frac{1}{2}}}{z}. \tag{2.18}$$

2.4. $\mathbb{Z}_4$ *Symmetry of* ${}_2\Phi_1$

The Fourier transform (2.9) defines a transformation μ,

$$\mu : \begin{cases} \alpha \to \quad \gamma \quad \to \dfrac{\omega}{\beta} \to \dfrac{\beta}{\alpha\gamma} \to \alpha, \\ \\ \beta \to \dfrac{\omega\alpha\gamma}{\beta} \to \dfrac{\omega}{\alpha} \to \dfrac{\omega}{\gamma} \to \beta. \end{cases} \tag{2.19}$$

From (2.8) we may infer

$$\frac{\hat{W}(0)}{W(0)} = {}_2\Phi_1\left[\begin{matrix}\omega,\alpha\\ \beta\end{matrix};\gamma\right]. \tag{2.20}$$

Using this and applying Fourier transform μ four times, we find

$$
{}_2\Phi_1\left[\begin{matrix}\omega,\alpha\\ \beta\end{matrix};\gamma\right] = \frac{N}{{}_2\Phi_1\left[\begin{matrix}\omega,\gamma\\ \omega\alpha\gamma/\beta\end{matrix};\omega/\beta\right]} = {}_2\Phi_1\left[\begin{matrix}\omega,\omega/\beta\\ \omega/\alpha\end{matrix};\beta/\alpha\gamma\right]
$$

$$
= \frac{N}{{}_2\Phi_1\left[\begin{matrix}\omega,\beta/\alpha\gamma\\ \omega/\gamma\end{matrix};\alpha\right]} = {}_2\Phi_1\left[\begin{matrix}\omega,\alpha\\ \beta\end{matrix};\gamma\right], \tag{2.21}
$$

which is a $\mathbb{Z}_4$ symmetry.

2.5. *The ${}_3\Phi_2$ identities*

Using the convolution theorem, we find

$$
{}_3\Phi_2\left[\begin{matrix}\omega,\alpha_1,\alpha_2\\ \beta_1,\beta_2\end{matrix};\gamma_1\gamma_2\right]
= N^{-1}\sum_{k=0}^{N-1} {}_2\Phi_1\left[\begin{matrix}\omega,\alpha_1\\ \beta_1\end{matrix};\omega^{-k}\gamma_1\right] {}_2\Phi_1\left[\begin{matrix}\omega,\alpha_2\\ \beta_2\end{matrix};\omega^{k}\gamma_2\right], \tag{2.22}
$$

where $\gamma_i = \Delta(\beta_i)/\Delta(\alpha_i)$, $i = 1, 2$. We can use the recurrence relation[7,20]

$$
\frac{\hat{W}(n)}{\hat{W}(0)} = \frac{{}_2\Phi_1\left[\begin{matrix}\omega,\alpha\\ \beta\end{matrix};\gamma\omega^n\right]}{{}_2\Phi_1\left[\begin{matrix}\omega,\alpha\\ \beta\end{matrix};\gamma\right]} = \hat{\gamma}^n\,\frac{(\hat{\alpha};\omega)_n}{(\hat{\beta};\omega)_n}. \tag{2.23}
$$

to find

$$
{}_3\Phi_2\left[\begin{matrix}\omega,\alpha_1,\alpha_2\\ \beta_1,\beta_2\end{matrix};\gamma_1\gamma_2\right] = A\ {}_3\Phi_2\left[\begin{matrix}\omega,\ \beta_1/\alpha_1\gamma_1,\ \gamma_2\\ \omega/\gamma_1,\ \omega\alpha_2\gamma_2/\beta_2\end{matrix};\omega\alpha_1\beta_2\right], \tag{2.24}
$$

with

$$
A \equiv N^{-1}\,{}_2\Phi_1\left[\begin{matrix}\omega,\alpha_1\\ \beta_1\end{matrix};\gamma_1\right] {}_2\Phi_1\left[\begin{matrix}\omega,\alpha_2\\ \beta_2\end{matrix};\gamma_2\right]. \tag{2.25}
$$

More generally, one can generate the symmetry relations of the cube in the Baxter–Bazhanov model under the 48 elements of the symmetry group of the cube, see also the work of Sergeev *et al.*[18]

The group is generated by two generators. The first one is ι_α: $\alpha_1 \leftrightarrow \alpha_2$ resulting in

$$
{}_3\Phi_2\left[\begin{matrix}\omega,\alpha_1,\alpha_2\\ \beta_1,\beta_2\end{matrix};\gamma_1\gamma_2\right] = {}_3\Phi_2\left[\begin{matrix}\omega,\alpha_2,\alpha_1\\ \beta_1,\beta_2\end{matrix};\gamma_3\gamma_4\right], \tag{2.26}
$$

with $\gamma_3 = \Delta(\beta_1)/\Delta(\alpha_2)$ and $\gamma_4 = \Delta(\beta_2)/\Delta(\alpha_1)$, so that $\gamma_3\gamma_4 = \gamma_1\gamma_2$. The second generator is $M = \mu^{-1} \otimes \mu$, which results in

$$
{}_3\Phi_2\left[\begin{matrix}\omega,\alpha_1,\alpha_2\\ \beta_1,\beta_2\end{matrix};\gamma_1\gamma_2\right] = \frac{N\;{}_3\Phi_2\left[\begin{matrix}\omega,\tilde\alpha_2,\tilde\alpha_1\\ \tilde\beta_1,\tilde\beta_2\end{matrix};\tilde\gamma_1\tilde\gamma_2\right]}{{}_2\Phi_1\left[\begin{matrix}\omega,\tilde\alpha_1\\ \tilde\beta_1\end{matrix};\tilde\gamma_1\right]{}_2\Phi_1\left[\begin{matrix}\omega,\tilde\alpha_2\\ \tilde\beta_2\end{matrix};\tilde\gamma_2\right]}, \qquad (2.27)
$$

with

$$
\begin{aligned}
\tilde\alpha_1 &= \frac{\beta_1}{\alpha_1\gamma_1}, \quad \tilde\beta_1 = \frac{\omega}{\gamma_1}, \quad \tilde\gamma_1 = \alpha_1, \\
\tilde\alpha_2 &= \gamma_2, \quad \tilde\beta_2 = \frac{\omega\alpha_2\gamma_2}{\beta_2}, \quad \tilde\gamma_2 = \frac{\omega}{\beta_2}.
\end{aligned} \qquad (2.28)
$$

This is the inverse of (2.24). We can use (2.12) to evaluate the ${}_2\Phi_1$'s, but this will lead to a phase factor depending on the positions of the α's and β's with respect to the cuts defined by Fig 2.1. Eqs. (2.24), (2.26) and (2.28) are valid in general, independent of choices of Riemann sheets or branch cuts.

2.6. *Connection with Sergeev, Mangazeev and Stroganov*

In several of the Russian works[16–19] one uses points, p, p', *etc.*, from the Fermat curve Γ in homogeneous notation, *i.e.*,

$$
p \in \Gamma \quad \leftrightarrow \quad p = (x,y,z) \quad \text{with} \quad x^N + y^N = z^N. \qquad (2.29)
$$

In our affine notation, $p \leftrightarrow \alpha$, $p' \leftrightarrow \beta$, *etc.*, we would identify

$$
\alpha \equiv \frac{\omega x}{z}, \quad \Delta(\alpha) \equiv \frac{y}{z} = \left(1-\alpha^N\right)^{1/N}. \qquad (2.30)
$$

The assignment of Riemann sheets and branch cuts is more subtle in their homogeneous notation. They deal with that by breaking up the curve Γ in parts Γ_l^m,

$$
\begin{gathered}
p \in \Gamma_{l;m} \equiv \Gamma_l^m \quad \leftrightarrow \quad \alpha \equiv \frac{\omega^{m+1}x}{z}, \quad \Delta(\alpha) \equiv \frac{\omega^{-l}y}{z}, \\
-\frac{\pi}{N} + l\,\frac{2\pi}{N} < \arg\frac{y}{z} < +\frac{\pi}{N} + l\,\frac{2\pi}{N}, \quad 0 < \arg\frac{x}{z} < \frac{2\pi}{N},
\end{gathered} \qquad (2.31)
$$

and by using the notation (p,m) for points in Γ_0^m. How their notations translate into ours is also indicated in (2.31).

The ω-Pochhammer symbol $(x;\omega)_l$ is defined upside-down and is not even unique in the various Russian papers. It is to be translated as

$$w(x,y,z|l) \equiv \prod_{s=1}^{l} \frac{y}{z-x\omega^s} = \left(\frac{y}{z}\right)^l \frac{1}{(\omega x/z;\omega)_l} \tag{2.32}$$

in Ref. 16. However, for the work of Sergeev *et al.*[18] one must identify

$$w(p'|m'+\sigma) \equiv \frac{1}{p_0(\omega^\sigma\alpha)}, \quad w(p|m+\sigma) \equiv \frac{1}{p_0(\omega^\sigma\beta)}, \tag{2.33}$$

$$\alpha = \omega^{m'+1}\,\frac{x'}{z'}, \quad \Delta(\alpha) = \frac{y'}{z'}, \quad \beta = \omega^{m+1}\,\frac{x}{z}, \quad \Delta(\beta) = \frac{y}{z}, \tag{2.34}$$

with $p_0(z)$ defined in (2.11), as they normalize $\prod_l w(x|l) = 1$, not $w(p|0) = 1$. Therefore, for the appendix of Ref. 18 we have to make the translation

$${}_r\Psi_r\left(\begin{matrix}(p_1,m_1),\ldots,(p_r,m_r)\\(p'_1,m'_1),\ldots,(p'_r,m'_r)\end{matrix}\middle|\, n\right) = C\ {}_{r+1}\Phi_r\left[\begin{matrix}\omega,\alpha_1,\ldots,\alpha_r\\\beta_1,\ldots,\beta_r\end{matrix};z\right], \tag{2.35}$$

$$C \equiv \frac{1}{\sqrt{N}}\,\frac{p_0(\alpha_1)\cdots p_0(\alpha_r)}{p_0(\beta_1)\cdots p_0(\beta_r)}, \qquad z \equiv \omega^n\,\frac{\Delta(\beta_1)\cdots\Delta(\beta_r)}{\Delta(\alpha_1)\cdots\Delta(\alpha_r)}. \tag{2.36}$$

2.7. *Other Identities for Cyclic Hypergeometric Functions*

One can derive many other identities for the cyclic hypergeometric function (2.3), (2.5). Without giving explicit expressions, we list some of the types of identities in Table 1.

Table 1. Cyclic hypergeometric identities.

Conditions	${}_{p+1}\Phi_p = \prod/\prod$	${}_{p+1}\Phi_p \propto {}_{p+1}\Phi_p$
None	${}_2\Phi_1$	${}_3\Phi_2$
$z=\omega$	${}_3\Phi_2$	${}_4\Phi_3$
Saalschütz	${}_4\Phi_3$	${}_5\Phi_4$

One type of identity is the evaluation of ${}_{p+1}\Phi_p$ in terms of a ratio of products. This is shown in the middle column of Table 1. Another type of identity is the proportionality of two ${}_{p+1}\Phi_p$'s where the proportionality factor can be expressed in terms of ${}_2\Phi_1$'s or, equivalently, products. This is shown in the last column of Table 1. The conditions under which such identities can be found are listed in the first column.

The two cases where there are no further conditions have been discussed in previous subsections. Other cases requiring the conditions $z = \omega$ and the more restrictive Saalschütz condition (2.6) have also been discussed in Ref. 21. The star–triangle equation of the integrable chiral Potts model is a special case of the Saalschützian ${}_4\Phi_3$ identities.[20,21]

It must be noted that identities of all six types in Table 1 have been derived by Sergeev, Mangazeev and Stroganov in the appendix of Ref. 18. However, one needs the translation (2.35) to see the connections with more standard basic hypergeometric notations and with the Saalschütz condition.

Many other identities can be derived. For example, Watson's analogue of Whipple's theorem for ${}_8\Phi_7$ reduces to ${}_7\Phi_6 \propto {}_4\Phi_3$. Moreover, new identities can be found in the $N \to \infty$ limit.[22]

3. Final Remarks

We have presented several results on the deep connection of the integrable chiral Potts model with the theory of cyclic hypergeometric functions. Eq. (2.12) with F_* as specified in Sec. 2.3 is new and is easier to use than a formulation with multiple Riemann sheets, especially when doing numerical computations with it. Finally, translation (2.35) is also new and may make the results of Sergeev *et al.*[18] more accessible to a wider audience familiar with basic hypergeometric series.

Acknowledgments

It is a pleasure to thank Dr. Molin Ge, Dr. Chengming Bai and the Nankai Institute of Mathematics for their hospitality and support.

References

1. W. P. Orrick, B. Nickel, A. J. Guttmann and J. H. H. Perk, *J. Stat. Phys.* **102**, 795–841 (2001).
2. H. Au-Yang, B.-Q. Jin and J. H. H. Perk, *J. Stat. Phys.* **102**, 501–543 (2001).
3. H. Au-Yang and J. H. H. Perk, in *MathPhys Odyssey 2001: Integrable Models and Beyond*, M. Kashiwara and T. Miwa, eds., (Birkhäuser, Boston, 2002), pp. 1–21 and 23–48.
4. H. Au-Yang and J. H. H. Perk, *J. Stat. Phys.* [cond-mat/0409577].
5. H. Au-Yang, B. M. McCoy, J. H. H. Perk, S. Tang and M.-L. Yan, *Phys. Lett.* **A123**, 219–223 (1987).
6. R. J. Baxter, J. H. H. Perk and H. Au-Yang, *Phys. Lett.* **A128**, 138–142 (1988).
7. H. Au-Yang and J. H. H. Perk, in *Advanced Studies in Pure Mathematics*, Vol. 19, (Kinokuniya-Academic, Tokyo, 1989), pp. 57–94.

8. R. J. Baxter, V. V. Bazhanov and J. H. H. Perk, *Int. J. Mod. Phys.* **B4**, 803–870 (1990).
9. H. Au-Yang, B.-Q. Jin and J. H. H. Perk, *J. Stat. Phys.* **102**, 471–499 (2001).
10. H. Au-Yang and J. H. H. Perk, *J. Stat. Phys.* **78**, 17–78 (1995).
11. G. Albertini, B. M. McCoy, J. H. H. Perk and S. Tang, *Nucl. Phys.* **B314**, 741–763 (1989).
12. R. J. Baxter, *Phys. Rev. Lett.* **94**, 130602 (2005).
13. R. J. Baxter, *J. Stat. Phys.* **120**, 1–36 (2005).
14. V. V. Bazhanov and R. J. Baxter, *J. Stat. Phys.* **69**, 453–485 (1992).
15. V. V. Bazhanov and R. J. Baxter, *J. Stat. Phys.* **71**, 839–864 (1993).
16. R. M. Kashaev, V. V. Mangazeev and Yu. G. Stroganov, *Int. J. Mod. Phys.* **A8**, 1399–1409 (1993).
17. V. V. Mangazeev, S. M. Sergeev and Yu. G. Stroganov, *Mod. Phys. Lett.* **A10**, 279–287 (1995).
18. S. M. Sergeev, V. V. Mangazeev and Yu. G. Stroganov, *J. Stat. Phys.* **82**, 31–49 (1996).
19. S. M. Sergeev, G. E. Boos, V. V. Mangazeev and Yu. G. Stroganov, *Mod. Phys. Lett.* **A11**, 491–498 (1996).
20. H. Au-Yang and J. H. H. Perk, *Int. J. Mod. Phys.* **B11**, 11–26 (1997).
21. H. Au-Yang and J. H. H. Perk, *Int. J. Mod. Phys.* **B16**, 1853–1865 (2002).
22. H. Au-Yang and J. H. H. Perk, *Physica* **A268**, 175–206 (1999).

Bethe Ansatz and Symmetry in Superintegrable Chiral Potts Model and Root-of-unity Six-vertex Model

Shi-shyr Roan*

Institute of Mathematics
Academia Sinica, Taipei , Taiwan
E-mail: maroan@gate.sinica.edu.tw

We examine the Onsager algebra symmetry of $\tau^{(j)}$-matrices in the superintegrable chiral Potts model. The comparison of Onsager algebra symmetry of the chiral Potts model with the sl_2-loop algebra symmetry of six-vertex model at roots of unity is made from the aspect of functional relations using the Q-operator and fusion matrices. The discussion of Bethe ansatz for both models is conducted in a uniform manner through the evaluation parameters of their symmetry algebras.

1. Introduction

The symmetry of quantum spin chains and the related lattice models has recently attracted certain attention due to their close connection with diverse areas of physics as well as mathematics. However, up to the present stage, only limited knowledge is available about the symmetry of lattice vertex models, and few exact results are obtained in this area. Even the sl_2-loop algebra symmetry of the six-vertex model at roots of unity, found in [11], has not been fully understood till now, given that much accomplishment has been made on the study of evaluation parameters related to the symmetry algebra representation. Some conjectures supported by numerical evidence remain to be answered, (see [12] [14] and references therein). Though the understanding of the symmetry of eight-vertex model in [15] [16] is still rudimentary in the present stage, the conjectural functional-relation-analogy discovered in the study on the eight-vertex model and the N-state chiral Potts model (CMP) in [7] did lead to exact results about the Onsager algebra symmetry of $\tau^{(j)}$-models in the superintegrable CPM [20] [21]. In the study of CPM,

*This work is supported in part by National Science Council of Taiwan under Grant No NSC 94-2115-M-001-013.

the lack difference-property for the rapidities is considered as the characteristic nature which distinguishes CPM from other known solvable lattice models. Nevertheless, progress made on the transfer matrix of CPM for the past fifteen years, culminating in the recent Baxter's proof of the order parameter [6], has provided the sufficient knowledge for the understanding of the detailed nature about the symmetry of superintegrable CPM. By this, the study of CPM could suggest a promising method to help the symmetry study about the six-vertex model at roots of unity as the limit case of eight-vertex model from the approach of functional relations, a scheme proposed in [16]. In this paper, we examine the similarity of the symmetry structure of two lattice models: the superintegrable CPM and the six-vertex model at roots of unity. The symmetry of superintegrable CPM is described by the Onsager algebra, obtained in [20] [21], with a short explanation in Sec. 2. The six-vertex model at roots of unity possesses the sl_2-loop algebra symmetry by the works in [11] [12] [14]. We present a discussion of symmetry of six-vertex model, parallel to the theory of CPM, through Bethe roots and evaluation parameters in Sec. 3, and give some concluding remarks in Sec. 4.

Notation. N is an integer ≥ 2, $\omega = e^{\frac{2\pi i}{N}}$, $\mathrm{i} = \sqrt{-1}$, and X, Z the Weyl operators of $\mathbf{C}^N$:$X|n\rangle = |n+1\rangle$, $Z|n\rangle = \omega^n|n\rangle$ for $n \in \mathbf{Z}_N = \mathbf{Z}/N\mathbf{Z}$.

2. The N-state Chiral Potts Model

2.1. *Rapidity and functional Relation of chiral Potts model*

In the study of CPM as a descendent of the six-vertex model, Bazhanov and Stroganov obtained the following 3-parameter family of Yang-Baxter solutions for the inhomogeneous R-matrix of six-vertex model,

$$R(t) = \begin{pmatrix} t\omega - 1 & 0 & 0 & 0 \\ 0 & t-1 & \omega - 1 & 0 \\ 0 & t(\omega-1) & \omega(t-1) & 0 \\ 0 & 0 & 0 & t\omega - 1 \end{pmatrix},$$

with the $\mathbf{C}^N$-operator entries parametrized by a four-vector ratio $p = [a, b, c, d]$ [8] [20]:

$$b^2 G_p(t) = \begin{pmatrix} b^2 - td^2X & (bc - \omega adX)Z \\ -t(bc - adX)Z^{-1} & -tc^2 + \omega a^2 X \end{pmatrix}, \quad t \in \mathbf{C}, \tag{2.1}$$

which satisfy the Yang-Baxter equation: $R(t/t')G_{p,1}(t)G_{p,2}(t') = G_{p,2}(t')G_{p,1}(t)R(t/t')$. Hence the same relation holds for the monodromy matrix of size L, $\bigotimes_{\ell=1}^L G_{p,\ell}(t)$, with $G_{p,\ell}(t) = G_p(t)$ at the site ℓ. Therefore, the $\tau_p^{(2)}$-matrix,

$$\tau_p^{(2)}(t) := \mathrm{tr}_{aux}(\bigotimes_{\ell=1}^{L} G_{p,\ell}(\omega t)) \quad \text{for } t \in \mathbf{C}, \tag{2.2}$$

form a commuting family of $(\overset{L}{\otimes} \mathbf{C}^N)$-operators. The $\mathbf{Z}_N$-operators X, Z of $\mathbf{C}^N$ at the site j give rise to the Weyl operators $X_j Z_j$ of $\overset{L}{\otimes} \mathbf{C}^N$, The spin-shift operator of $\overset{L}{\otimes} \mathbf{C}^N$, $X := \prod_{j=1}^{L} X_j$, defines the $\mathbf{Z}_N$-charge Q, and commutes with $\tau_p^{(2)}(t)$. The rapidities of N-state CMP are elements in the genus $(N^3 - 2N^2 + 1)$ curve W in the projective 3-space $\mathbf{P}^3$, defined by the equivalent sets of equations:

$$\mathsf{W}: \begin{cases} ka^N + k'c^N = d^N \\ kb^N + k'd^N = c^N \end{cases} \leftrightarrow \begin{cases} a^N + k'b^N = kd^N \\ k'a^N + b^N = kc^N \end{cases} \leftrightarrow \begin{cases} kx^N = 1 - k'\mu^{-N} \\ ky^N = 1 - k'\mu^N \end{cases} \tag{2.3}$$

where $[a,b,c,d] \in \mathbf{P}^3$, $(x,y,\mu) = (\frac{a}{d}, \frac{b}{c}, \frac{d}{c}) \in \mathbf{C}^3$, k' is a parameter with $k^2 = 1 - k'^2 \neq 0, 1$. By eliminating the variable μ^N in the last set of above equations , and using the variables $t := xy, \lambda := \mu^N$, one arrives the hyperelliptic curve of genus $N-1$, $t^N = \frac{(1-k'\lambda)(1-k'\lambda^{-1})}{k^2}$, as a N^2-unramified quotient of (2.3). The rapidities possess a large symmetry group, in which the following two will be needed in our later discussion,

$$U : (x, y, \mu) \mapsto (\omega x, y, \mu), \; C : (x, y, \mu) \mapsto (y, x, \mu^{-1}). \tag{2.4}$$

The Boltzmann weights $W_{p,q}, \overline{W}_{p,q}$ of the CPM, depending on two rapidities $p, q \in \mathsf{W}$, are two N-cyclic vectors, defined by $\frac{W_{p,q}(n)}{W_{p,q}(0)} = \prod_{j=1}^{n} \frac{d_p b_q - a_p c_q \omega^j}{b_p d_q - c_p a_q \omega^j}$, $\frac{\overline{W}_{p,q}(n)}{\overline{W}_{p,q}(0)} = \prod_{j=1}^{n} \frac{\omega a_p d_q - d_p a_q \omega^j}{c_p b_q - b_p c_q \omega^j}$. The CPM transfer matrix of size L with periodic boundary condition, $L+1 = 1$, is the $(\overset{L}{\otimes} \mathbf{C}^N)$-operator defined by

$$T_{\rm cp}(p;q)_{\sigma_1,\ldots,\sigma_L}^{\sigma'_1,\ldots,\sigma'_L} = \prod_{l=1}^{L} \overline{W}_{p,q}(\sigma_l - \sigma'_l) W_{p,q}(\sigma_l - \sigma'_{l+1}) \;, \quad \sigma_l, \sigma'_l \in \mathbf{Z}_N. \tag{2.5}$$

For a fixed $p \in \mathsf{W}$, $\{T_{\rm cp}(p;q)\}_{q \in \mathsf{W}}$ form a commuting family of operators by the well-known star-triangle relation of Boltzmann weights $W_{pq}, \overline{W}_{pq}$. Then $T_{\rm cp}(p;q)$ commutes with X and the spatial translation operator S_R (which defines the total momentum $P \in \mathbf{Z}_L$). Denote $\widehat{T}_{\rm cp}(p;q) := T_{\rm cp}(p;q) S_R$. The transfer matrix $T_{\rm cp}(p;q)$ can be derived from $\tau_p^{(2)}(t_q)$ with $p \in \mathsf{W}$ as the auxiliary "Q"-operator, as discussed in [2] on the TQ-relation of the eight-vertex model. One arrives $\tau^{(2)}T_{\rm cp}$-relation ((4.20) in [7]) using the automorphism U in (2.4):

$$\tau_p^{(2)}(t_q)T_{\rm cp}(p;Uq) = \varphi_p(q)T_{\rm cp}(p(;q) + \overline{\varphi}_p(Uq)XT_{\rm cp}(p;U^2q) \quad (2.6)$$
$$\Longleftrightarrow \tau_p^{(2)}(t_q) = (\varphi_p(q)T_{\rm cp}(p;q) + \overline{\varphi}_p(Uq)XT_{\rm cp}(p;U^2q))T_{\rm cp}(p;Uq)^{-1} \quad (2.7)$$

where $\varphi_p(q) := (\frac{(y_p-\omega x_q)(t_p-t_q)}{y_p^2(x_p-x_q)})^L$, $\overline{\varphi}_p(q) := (\frac{\omega\mu_p^2(x_p-x_q)(t_p-t_q)}{y_p^2(y_p-\omega x_q)})^L$. By (2.7) and the commutativity of $T_{\rm cp}(p;*)$, $[\tau_p^{(2)}(t_q), T_{\rm cp}(p;q')] = 0$ for $p,q,q' \in \mathsf{W}$. The fusion operators $\tau^{(j)}(t)$ for $0 \leq j \leq N$ are determined by the following $T_{\rm cp}\widehat{T}_{\rm cp}$-relation for $0 \leq j \leq N$, ((3.46) for $(l,k) = (j,0)$ in [7]) with $\tau_p^{(0)} := 0$, $\tau_p^{(1)} := I$,

$$T_{\rm cp}(p;q)\widehat{T}_{\rm cp}(p;CU^jq) = rh_j(\tau_p^{(j)}(t_q) + \frac{z(t_q)z(\omega t_q)\cdots z(\omega^{j-1}t_q)X^j\tau_p^{(N-j)}(\omega^j t_q)}{\alpha_p(\lambda_q)}) \quad (2.8)$$

where $z(t) = (\frac{\omega\mu_p^2(x_py_p-t)^2}{y_p^4})^L$, $\alpha_p(\lambda_q) = (\frac{k'(1-\lambda_p\lambda_q)^2}{\lambda_q(1-k'\lambda_p)^2})^L$ and $r = (\frac{N(x_p-x_q)(y_p-y_q)(t_p^N-t_q^N)}{(x_p^N-x_q^N)(y_p^N-y_q^N)(t_p-t_q)})^L$, $h_j = (\prod_{m=1}^{j-1}\frac{y_p^2(x_p-\omega^m x_q)}{(y_p-\omega^m x_q)(t_p-\omega^m t_q)})^L$. By (2.8), one can derive the fusion relations of $\tau^{(j)}$s ((4.27) of [7]) for $1 \leq j \leq N$:

$$\begin{aligned}\tau_p^{(j)}(t)\tau_p^{(2)}(\omega^{j-1}t) &= z(\omega^{j-1}t)X\tau_p^{(j-1)}(t) + \tau_p^{(j+1)}(t),\\ \tau_p^{(N+1)}(t) &:= z(t)X\tau_p^{(N-1)}(\omega t) + u(t)I,\end{aligned} \quad (2.9)$$

where $u(t) := \alpha_p(\lambda)+\alpha_p(\lambda^{-1})$. Note that with $\tau_p^{(2)}(t)$ in (2.2) for $p \in \mathbf{P}^3$, the validity of fusion relation (2.9) provides a characterization of the rapidity constraint (2.3) for $p \in \mathsf{W}$, (Thm 1 in [20]). Using (2.7) and (2.9), one can express $\tau_p^{(j)}$ in terms of $T_{\rm cp}(p;q)$, hence the $\tau^{(j)}T_{\rm cp}$-relations ((4.34) in [7]):

$$\tau_p^{(j)}(q) = \sum_{m=0}^{j-1}(\prod_{i=0}^{m-1}\varphi_p(U^iq)\prod_{k=m+1}^{j-1}\overline{\varphi}_p(U^kq))\frac{T_{\rm cp}(p;q)T_{\rm cp}(p;U^jq)}{T_{\rm cp}(p;U^mq)T_{\rm cp}(p;U^{m+1}q)}. \quad (2.10)$$

Substituting (2.10) in (2.8), the functional equation of $T_{\rm cp}$ follows ((4.40) of [7]):

$$\begin{aligned}\widehat{T}_{\rm cp}(p;q) &= \sum_{m=0}^{N-1}C_{m;p}(q)T_{\rm cp}(p;q)T_{\rm cp}(p;U^mq)^{-1}T_{\rm cp}(p;U^{m+1}q)^{-1}X^{-m-1},\\ C_{m;p}(q) &:= (\prod_{i=0}^{m-1}\varphi_p(U^iq)\prod_{k=m+1}^{N-1}\overline{\varphi}_p(U^kq))(\frac{Ny_p^{2N-2}(y_p-y_q)(y_p-x_q)}{(y_p^N-y_q^N)(y_p^N-x_q^N)})^L.\end{aligned} \quad (2.11)$$

2.2. *Bethe Ansatz and Onsager algebra symmetry in superintegrable chiral Potts model*

For CPM in the superintegrable case, i.e., the rapidity p given by $\mu_p = 1$, $x_p = y_p = \eta^{\frac{1}{2}}$, where $\eta := (\frac{1-k'}{1+k'})^{\frac{1}{N}}$, simplification occurs for the functional relations. We shall omit the label p appeared in all operators for the superintegrable case, simply write $\tau^{(2)}(t)$, $T_{\rm cp}(q)$, *etc.* As q tends to p, to the first order of small ε, one has ((1.11) in [1]): $T_{\rm cp}(q) = \mathbf{1}[1+\varepsilon(N-1)L]+\varepsilon H(k')$, where $H(k') = H_0 + k'H_1$ is the $\mathbf{Z}_N$-quantum chain in [17 18], with the expression: $H_0 = -2\sum_{\ell=1}^{L}\sum_{n=1}^{N-1}\frac{X_\ell^n}{1-\omega^{-n}}$, $H_1 = -2\sum_{\ell=1}^{L}\sum_{n=1}^{N-1}\frac{Z_\ell^n Z_{\ell+1}^{-n}}{1-\omega^{-n}}$, which satisfy the Dolan-Grady relation [13], hence generate the Onsager algebra representation where only the spin-$\frac{1}{2}$ subrepresentation occurs as irreducible factors [1 3 5 10 19]. Through the gauge transform by $M = {\rm dia.}[1, \eta^{\frac{1}{2}}]$, the monodromy matrix $G(t)$ in (2.1) becomes the solution (2.1) for $a = b = c = d = 1$:

$$\widetilde{G}(\widetilde{t}) = \begin{pmatrix} 1-\widetilde{t}X & (1-\omega X)Z \\ -\widetilde{t}(1-X)Z^{-1} & -\widetilde{t}+\omega X \end{pmatrix}, \quad \widetilde{t} = \eta^{-1}t.$$

Hence $\tau^{(2)}(t) = \widetilde{\tau}^{(2)}(\widetilde{t})$, where $\widetilde{\tau}^{(2)}(\widetilde{t})$ is the trace of the L-size monodromy matrix associated to $G(\omega\widetilde{t})$. Write $\tau^{(j)}(t) = \widetilde{\tau}^{(j)}(\widetilde{t})$, the fusion relation (2.9) has the form:

$$\begin{aligned} \widetilde{\tau}^{(j)}(\widetilde{t})\widetilde{\tau}^{(2)}(\omega^{j-1}\widetilde{t}) &= (1-\omega^{j-1}\widetilde{t})^{2L}\ \widetilde{\tau}^{(j-1)}(\widetilde{t})\ \omega^L X\ + \widetilde{\tau}^{(j+1)}(\widetilde{t}),\ 1 \leq j \leq N\ , \\ \widetilde{\tau}^{(N+1)}(\widetilde{t}) &= (1-\widetilde{t})^{2L}\ \widetilde{\tau}^{(N-1)}(\omega\widetilde{t})\omega^L X\ + 2(1-\widetilde{t}^N)^L. \end{aligned} \tag{2.12}$$

By examining commutators of H_k with the entries of the monodromy matrix constructed from $G(\widetilde{t})$, one can show $[H_k, \tau^{(2)}(t)] = 0$ for $k = 0, 1$. It follows the Onsager algebra symmetry of $\tau^{(j)}$-model (Thm 1 of [21]). However, the understanding of the detailed nature of Onsager algebra symmetry in the superintegrable CPM still requires the full knowledge about eigenvalues of CPM transfer matrix, which was solved by the Bethe-ansatz method in [1 4 5] as follows. For parameters $v_1, \ldots, v_{m_p}$ with $(-v_i)^N \neq 0, 1$ and $v_i v_j^{-1} \neq 1, \omega$ for $i \neq j$, consider the rational function

$$\mathsf{P}(\widetilde{t}) = \omega^{-P_b}\sum_{j=0}^{N-1}\frac{(1-\widetilde{t}^N)^L(\omega^j\widetilde{t})^{-P_a-P_b}}{(1-\omega^j\widetilde{t})^L F(\omega^j\widetilde{t})F(\omega^{j+1}\widetilde{t})}, \quad F(\widetilde{t}) := \prod_{i=1}^{m_p}(1+\omega v_i\widetilde{t}) \tag{2.13}$$

where P_a, P_b are integers satisfying $0 \leq r(:= P_a + P_b) \leq N-1$, $P_b - P_a \equiv Q + L \pmod N$. $\mathsf{P}(\widetilde{t})$ is invariant under $\widetilde{t} \mapsto \omega\widetilde{t}$, hence depending only on $\widetilde{t}^N$. The criterion of $\mathsf{P}(\widetilde{t})$ as a $\widetilde{t}$-polynomial is the following constraint for v_js, ((4.4) in [1], (6.22) in [4]):

$$\left(\frac{v_i + \omega^{-1}}{v_i + \omega^{-2}}\right)^L = -\omega^{-r}\prod_{l=1}^{m_p}\frac{v_i - \omega^{-1}v_l}{v_i - \omega v_l}, \quad i = 1, \ldots, m_p. \tag{2.14}$$

Here the non-negative integer m_p satisfies the relation $LP_b \equiv m_p(Q - 2P_b - m_p) \pmod N$. The total momentum P is given by $e^{{\rm i}P} = \omega^{-P_b}\prod_{i=1}^{m_p}\frac{1+\omega v_i}{1+\omega^2 v_i}$.

The above relation is indeed the Bethe equation of $\tau^{(2)}$-model (Thm 3 of [21]). Then $\mathsf{P}(\widetilde{t})$ is a simple $\widetilde{t}^N$-polynomial of degree $m_E = [\frac{(N-1)L-r-2m_p}{N}]$ with negative real roots, (Thm 2 of [21]). Let $s_1, \ldots s_{m_E}$ be the $\widetilde{t}^N$-zeros of $\mathsf{P}(\widetilde{t})$, and define $\mathsf{G}(\lambda) = \prod_{j=1}^{m_E} \frac{\lambda+1\pm(\lambda-1)w_j}{2\lambda}$ where $w_j := (\frac{s_j-\eta^{-2N}}{s_j-1})^{\frac{1}{2}}$. Then $\frac{\mathsf{P}(\widetilde{t})}{\mathsf{P}(1)} = \mathsf{G}(\lambda)\mathsf{G}(\lambda^{-1})$. One has the following expression of $T_{\rm cp}(q)(= e^{-{\rm i}P}\widehat{T}_{\rm cp}(q))$-eigenvalues ((1.11) in [1] and (21) in [5]):

$$T_{\rm cp}(q) = N^L \frac{(\eta^{\frac{-1}{2}}x_q-1)^L}{(\eta^{\frac{-N}{2}}x_q^N-1)^L}(\eta^{\frac{-1}{2}}x_q)^{P_a}(\eta^{\frac{-1}{2}}y_q)^{P_b}\mu_q^{-P_\mu}\frac{F(\widetilde{t}_q)}{F(1)}\mathsf{G}(\lambda_q), \qquad (2.15)$$

which gives the energy value of $H(k')$ ((2.23) of [1]) of the form: $\alpha + \beta k' + N(1-k')\sum_{j=1}^{m_E} \pm w_j$ for $\alpha, \beta \in \mathbf{R}$. Therefore the $\tau^{(2)}$-degenerate states associated to the Bethe roots v_is form an irreducible Onsager-algebra-representation space of dimension 2^{m_E}, which we associate the following normalized CPM transfer matrix: $Q(q)(= e^{-{\rm i}P}\widehat{Q}(q)) = \frac{T_{\rm cp}(q)(1-\eta^{\frac{-N}{2}}x_q^N)^L}{N^L(1-\eta^{\frac{-1}{2}}x_q)^L(\eta^{\frac{-1}{2}}x_q)^{P_a}(\eta^{\frac{-1}{2}}y_q)^{P_b}\mu_q^{-P_\mu}}$, related to Q_{cp} in [21] by $(\eta^{\frac{-1}{2}}x_q)^{P_a}(\eta^{\frac{-1}{2}}y_q)^{P_b}\mu_q^{-P_\mu}Q(q) = Q_{cp}(q)$. Then the Q-eigenvalues and the functional equation (2.11) become

$$Q(q) = \frac{F(\widetilde{t}_q)}{F(1)}\mathsf{G}(\lambda_q), \quad \widehat{Q}(Cq) = e^{{\rm i}P}\frac{F(\widetilde{t}_q)}{F(1)}\mathsf{G}(\lambda_q^{-1}), \qquad (2.16)$$

$$\widehat{Q}(Cq) = \frac{\omega^{-P_b}}{N^L}\sum_{m=0}^{N-1}\frac{(1-\widetilde{t}_q^N)^L\omega^{-mr}}{(1-\omega^m\widetilde{t}_q)^L}Q(q)Q(U^mq)^{-1}Q(U^{m+1}q)^{-1}. (2.17)$$

The relations, (2.6) (2.10) (2.8), now become the following ones for $0 \le j \le N$:

$$\widetilde{\tau}^{(2)}(\widetilde{t}_q)Q(Uq) = \omega^{-P_a}(1-\widetilde{t}_q)^LQ(q) + \omega^{P_b}(1-\omega\widetilde{t}_q)^LQ(U^2q), \qquad (2.18)$$

$$\widetilde{\tau}^{(j)}(\widetilde{t}_q) = \omega^{(j-1)P_b-r}\sum_{m=0}^{j-1}\left(\frac{\prod_{k=0}^{j-1}(1-\omega^k\widetilde{t}_q)^L}{(1-\omega^m\widetilde{t}_q)^L}\frac{Q(q)Q(U^jq)}{Q(U^mq)Q(U^{m+1}q)}\right), \qquad (2.19)$$

$$\widetilde{t}_q^rQ(q)\widehat{Q}(CU^jq) = \frac{\omega^{-jP_b}(1-\widetilde{t}_q^N)^L\widetilde{\tau}^{(j)}(\widetilde{t}_q)}{N^L\prod_{m=0}^{j-1}(1-\omega^m\widetilde{t}_q)^L} + \frac{\omega^{-jP_a}(1-\widetilde{t}_q^N)^L\widetilde{\tau}^{(N-j)}(\omega^j\widetilde{t}_q)}{N^L\prod_{m=j}^{N-1}(1-\omega^m\widetilde{t}_q)^L}. \qquad (2.20)$$

By (2.16), the relation (2.18) yield ((6.18) in [4]):

$$\widetilde{\tau}^{(2)}(\widetilde{t}_q)F(\omega\widetilde{t}_q) = \omega^{-P_a}(1-\widetilde{t}_q)^LF(\widetilde{t}_q) + \omega^{P_b}(1-\omega\widetilde{t}_q)^LF(\omega^2\widetilde{t}_q). \qquad (2.21)$$

Using (2.19) and (2.20), follows the $\widetilde{\tau}^{(j)}$-polynomial (Thm 3 of [21]):

$$\widetilde{\tau}^{(j)}(\widetilde{t}) = \omega^{(j-1)P_b}\prod_{k=0}^{j-1}(1-\omega^k\widetilde{t})^L\sum_{m=0}^{j-1}\frac{F(\widetilde{t})F(\omega^j\widetilde{t})\omega^{-m(P_a+P_b)}}{(1-\omega^m\widetilde{t})^LF(\omega^m\widetilde{t})F(\omega^{m+1}\widetilde{t})},$$

$$\widetilde{t}^rF(\widetilde{t})F(\omega^j\widetilde{t})\mathsf{P}(\widetilde{t}) = \frac{\omega^{-jP_b}(1-\widetilde{t}^N)^L}{\prod_{m=0}^{j-1}(1-\omega^m\widetilde{t})^L}\widetilde{\tau}^{(j)}(\widetilde{t}) + \frac{\omega^{-jP_a}(1-\widetilde{t}^N)^L}{\prod_{m=j}^{N-1}(1-\omega^m\widetilde{t})^L}\widetilde{\tau}^{(N-j)}(\omega^j\widetilde{t}). \qquad (2.22)$$

3. Six-vertex Model at Roots of Unity

3.1. *Bethe equation of the six-vertex model*

The transfer matrix of the six-vertex model of an *even* size L is the $(\overset{L}{\otimes} \mathbf{C}^2)$-operator constructed from the Yang-Baxter solution,

$$L = \begin{pmatrix} z^{\frac{1}{2}}\mathsf{q}^{\frac{-\sigma^Z}{2}} - z^{\frac{-1}{2}}\mathsf{q}^{\frac{\sigma^Z}{2}}, & (\mathsf{q}-\mathsf{q}^{-1})\sigma_- \\ (\mathsf{q}-\mathsf{q}^{-1})\sigma_+, & z^{\frac{1}{2}}\mathsf{q}^{\frac{\sigma^Z}{2}} - z^{\frac{-1}{2}}\mathsf{q}^{\frac{-\sigma^Z}{2}} \end{pmatrix}, \quad \sigma^Z, \sigma_\pm : \text{Pauli matrix},$$

for the R-matrix

$$R(z) = \begin{pmatrix} z^{\frac{-1}{2}}\mathsf{q} - z^{\frac{1}{2}}\mathsf{q}^{-1} & 0 & 0 & 0 \\ 0 & z^{\frac{-1}{2}} - z^{\frac{1}{2}} & \mathsf{q}-\mathsf{q}^{-1} & 0 \\ 0 & \mathsf{q}-\mathsf{q}^{-1} & z^{\frac{-1}{2}} - z^{\frac{1}{2}} & 0 \\ 0 & 0 & 0 & z^{\frac{-1}{2}}\mathsf{q} - z^{\frac{1}{2}}\mathsf{q}^{-1} \end{pmatrix},$$

as the trace of monodromy matrix: $T(z) = \mathrm{tr}_{aux}(\bigotimes_{\ell=1}^{L} L_\ell(z))$ for $z \in \mathbf{C}$. The logarithmic $z\frac{d}{dz}$-derivative of $T(z)$ at $z = \mathsf{q}^{-1}$ gives rise to the XXZ chain with the periodic boundary condition: $H_{XXZ} = -\frac{1}{2}\sum_{\ell=1}^{L}(\sigma_\ell^1\sigma_{\ell+1}^1 + \sigma_\ell^2\sigma_{\ell+1}^2 + \Delta\sigma_\ell^3\sigma_{\ell+1}^3)$ with $\Delta = \frac{1}{2}(\mathsf{q}+\mathsf{q}^{-1})$, a well-studied Hamiltonian initiated by Bethe [9]. The ground state energy for the value $S^Z (= \frac{1}{2}\sum_\ell \sigma_\ell^Z)$ is determined by an appropriate solution of the following Bethe equation for $v := -z^{-1}$:

$$\left(\frac{v_i + \mathsf{q}^{-1}}{v_i + \mathsf{q}}\right)^L = -\mathsf{q}^{-L+2m}\prod_{l=1}^{m}\frac{v_i - \mathsf{q}^{-2}v_l}{v_i - \mathsf{q}^2 v_l}, \qquad m = \frac{L}{2} - |S^Z|. \tag{3.1}$$

The Bethe-equation technique was further extended to the method of Baxter's TQ-relation in eight-vertex model; when applying to the six-vertex model, there exists a non-degenerated commuting family of Q-operators with the following relation (see, Chapter 9 of [2]):

$$T(z)Q(s) = \mathsf{q}^{-2|S^Z|}(1-\mathsf{q}z)^L Q(U^{-1}s) + (1-\mathsf{q}^{-1}z)^L Q(Us). \tag{3.2}$$

Here s is a suitable multi-valued coordinate of z, and U is a s-automorphism inducing the transformation sending z to $\mathsf{q}^2 z$. Note that there are many such Q-operators, however all give the same Bethe equation (3.1) through Eq.(3.2).

3.2. *Evaluation polynomial and fusion relation of six-vertex model at roots of unity*

For the root of unity case with $\mathrm{q}^{2N} = 1$, i.e. $\mathrm{q}^2 = \omega$, the six-vertex model possesses the sl_2-loop algebra symmetry [11]. The Bethe state corresponding to the Bethe roots is the "highest weight" vector of an irreducible representation of the sl_2-loop algebra, with the evaluation parameters characterized by the Drinfeld polynomial [12 14]. By studying the creation sl_2-loop current operator in the ABCD-algebra, the Drinfeld polynomial for a Bethe root $\{v_i\}_{i=1}^m$ of (3.1) is given by Eq. (3.9) in [14]. Denote $\widetilde{t} = \mathsf{q}z$, and define the integer r by $r \equiv \frac{L}{2} - m \pmod N$, $0 \leq r \leq N-1$. The Drinfeld polynomial is indeed the $\widetilde{t}^N$-polynomial associated the following polynomial $P(\widetilde{t})$,

$$P(\widetilde{t}) = \sum_{j=0}^{N-1} \frac{(1-\omega^j\widetilde{t})^L(\omega^j\widetilde{t})^{-r}}{F(\omega^j\widetilde{t})F(\omega^{j+1}\widetilde{t})}, \quad F(\widetilde{t}) := \prod_{i=1}^m (1+\mathsf{q}^{-1}v_i\widetilde{t}), \tag{3.3}$$

which has a similar form as (2.13). Indeed with $F(\widetilde{t})$ in (2.13) or (3.3), let $H(\widetilde{t}) = \frac{1-\widetilde{t}^N}{1-\widetilde{t}}, 1-\widetilde{t}$ in CPM, six-vertex model resp. The function $P(\widetilde{t}) := \sum_{j=0}^{N-1} \frac{H(\omega^j\widetilde{t})^L(\omega^j\widetilde{t})^{-r}}{F(\omega^j\widetilde{t})F(\omega^{j+1}\widetilde{t})}$ is invariant under $\widetilde{t} \mapsto \omega\widetilde{t}$. The condition on roots of $F(\widetilde{t})$ so that $P(\widetilde{t})$ is a polynomial is provided by Bethe equation (2.14), (3.1) resp. Define the $T^{(2)}$-operator by $T^{(2)}(\widetilde{t}) = z^{\frac{L}{2}}T(z)$ in the six-vertex model, and $\frac{\omega^{-P_b}(1-\widetilde{t}^N)^L \tau^{(2)}(\omega^{-1}\widetilde{t})}{(1-\omega^{-1}\widetilde{t})^L(1-\widetilde{t})^L}$ in CPM. Then Eqs. (2.18), (3.2) are combined into one $T^{(2)}Q$-relation:

$$T^{(2)}(\widetilde{t})Q(q) = \omega^{-r}H(\widetilde{t})^L Q(U^{-1}q) + H(\omega^{-1}\widetilde{t})^L Q(Uq), \quad U^N = 1. \tag{3.4}$$

The $T^{(j)}$-operators are defined recursively through the following fusion relation for $j \geq 1$ by setting $T^{(0)} = 0$, $T^{(1)} = H(\omega^{-1}\widetilde{t})^L$,

$$T^{(j)}(\widetilde{t})T^{(2)}(\omega^{j-1}\widetilde{t}) = \omega^{-r}H(\omega^{j-1}\widetilde{t})^L T^{(j-1)}(\widetilde{t}) + H(\omega^{j-2}\widetilde{t})^L T^{(j+1)}(\widetilde{t}) \tag{3.5}$$

By (3.4), the induction-argument yields the $T^{(j)}Q$-relation for $j \geq 0$:

$$T^{(j)}(\widetilde{t}) = \sum_{k=0}^{j-1} \omega^{-kr} H(\omega^{k-1}\widetilde{t})^L \frac{Q(U^{-1}q)Q(U^{j-1}q)}{Q(U^{k-1}q)Q(U^k q)}. \tag{3.6}$$

By this, one obtains the boundary condition of the fusion relation:

$$T^{(N+1)}(\widetilde{t}) = \omega^{-r}T^{(N-1)}(\omega\widetilde{t}) + 2H(\omega^{-1}\widetilde{t})^L. \tag{3.7}$$

In CPM case, with the identification $T^{(j)}(\widetilde{t}) = \frac{\omega^{-(j-1)P_b}(1-\widetilde{t}^N)^L \tau^{(j)}(\omega^{-1}\widetilde{t})}{\prod_{k=-1}^{j-2}(1-\omega^k\widetilde{t})^L}$, Eqs.(3.5)-(3.7) are the same as Eqs.(2.12), (2.19). While in six-vertex model, the fusion relation and $T^{(j)}Q$-relation hold for any Q-operator satisfying $T^{(2)}Q$-relation (3.4). For a polynomial $F(\widetilde{t})$ with Bethe roots v_is, by (3.6) the corresponding $T^{(2)}$-eigenvalue is determined by the relation

$$T^{(2)}(\widetilde{t})F(\widetilde{t}) = \omega^{-r}H(\widetilde{t})^L F(\omega^{-1}\widetilde{t}) + H(\omega^{-1}\widetilde{t})^L F(\omega\widetilde{t}). \tag{3.8}$$

Using Eq.(3.5), one obtains the form of $T^{(j)}$-eigenvalues from Eq.(3.8):

$$T^{(j)}(\widetilde{t}) = F(\omega^{-1}\widetilde{t})F(\omega^{j-1}\widetilde{t})\sum_{k=0}^{j-1}\frac{H(\omega^{k-1}\widetilde{t})^L\omega^{-kr}}{F(\omega^{k-1}\widetilde{t})F(\omega^k\widetilde{t})}, \quad j \geq 1, \qquad (3.9)$$

which implies

$$\widetilde{t}^r F(\omega^{-1}\widetilde{t})F(\omega^{j-1}\widetilde{t})P(\widetilde{t}) = T^{(j)}(\widetilde{t}) + \omega^{-jr}T^{(N-j)}(\omega^j\widetilde{t}), \quad 0 \leq j \leq N. \quad (3.10)$$

Eqs.(3.8)-(3.10) in CPM case are the same as Eqs.(2.21),(2.22). Note that Eq.(3.10) is a consequence of the QQ-relation (2.20) in CPM case, which encodes the detailed nature of Onsager algebra symmetry in the derivation of Eq.(2.16). However in the case of roots-of-unity six-vertex model, the QQ-relation has yet been found, even though the sl_2-loop algebra symmetry together with evaluation parameters has already been known [11 12 14]. Based on the understanding in the CPM case, we now describe a similar, but speculated, structure about the QQ-relation in six-vertex case as follows. Consider the curve $W : w^2 = \widetilde{t}^N$, and its symmetries, $U(w, \widetilde{t}) = (w, \omega\widetilde{t})$,$C(w, \widetilde{t}) = (-w, \widetilde{t})$. For odd N, the curve is parametrized by $s = \widetilde{t}^{\frac{-1}{2}}$, and the automorphism $\varphi(s)(:= \mathrm{q}s)$ gives rise to the above symmetries by $U = \varphi^{-2[\frac{N}{2}]}$, $C = \varphi^N$. The polynomial $P(\widetilde{t})$ in (3.3) are expected, (true for $r = 0$ by [12]), to have the simple $\widetilde{t}^N$-roots $\{s_k\}_{k=1}^M$ with $P(0) \neq 0$. Define $G(w) = \prod_{j=1}^M(\sqrt{s_j} - w)$, then $\frac{G(w)G(-w)}{G(0)^2} = \frac{P(\widetilde{t})}{P(0)}$. In the eigen-space of $T(z)$ corresponding to $F(\widetilde{t})$ determined by a Bethe root, The Q-operator has the Q-eigenvalues:$Q(q) = F(\widetilde{t})\frac{G(w)}{G(0)}$, $Q(Cq) = F(\widetilde{t})\frac{G(-w)}{G(0)}$ for $q = (w, \widetilde{t})$. The above conditions reveal the sl_2-loop symmetry of six-vertex model, as the role of Eq.(2.16) for the Onsager algebra symmetry in CPM. Hence such a Q-matrix, if exits, must possess certain constraints in order to incorporate the symmetry of six-vertex model as discussed in [2] Secs. 9.1-9.5.

4. Discussion

In this paper, we have examined the symmetry structure of the superintegrable CPM and the six-vertex model at roots of unity by the method of functional relations. For the superintegrable CPM, exact results about the Onsager algebra symmetry of the $\tau^{(j)}$-models are obtained using the explicit form of eigenvalues of the CPM-transfer matrix. Based on common features related to evaluation parameters of the symmetry algebra representation, we discussed the Bethe ansatz of both theories in a unified manner. By this, in the six-vertex model at roots of unity, we obtained the fusion relation of $T^{(j)}$-matrices, $T^{(j)}Q$-relation from the TQ-relation, and further indicate the special nature of Q-operator in accord with the required sl_2-loop algebra symmetry of the six-vertex model.

Acknowledgements

The author thanks the conference organizers for the invitation, and the local committee at Nankai University for the hospitality and hard work. I also wish to thank Professor F. Hirzebruch for the warm hospitality in November, 2005 during the author's one-month stay at Max-Planck-Institut für Mathematik in Bonn, where this report was completed.

References

1. G. Albertini, B. M. McCoy, and J. H. H. Perk, in *Adv. Stud. Pure Math.* **19** 1 (1989) (Tokyo: Kinokuniya Academic press).
2. R. J. Baxter, *Exactly solved models in statistical mechanics,* 1982 (New York: Academic Press).
3. R. J. Baxter, *Phys. Letts. A* **133** 185 (1988).
4. R. J. Baxter, *J. Stat. Phys.* **73** 461 (1993).
5. R. J. Baxter, *J. Phys. A: Math. Gen.* **27** 1837 (1994).
6. R. J. Baxter, *Phys. Rev. Lett.* **94** 130602 (2005), cond-mat/0501227.
7. R. J. Baxter, V.V. Bazhanov and J.H.H. Perk, *Int. J. Mod. Phys. B* **4** 803 (1990).
8. V.V. Bazhanov and Yu.G. Stroganov, *J. Stat. Phys.* **59** 799 (1990).
9. H. A. Bethe, Z. Physik **71** 205 (1931).
10. B. Davies, *J. Phys. A: Math. Gen.* **23** 2245 (1990) ; *J. Math. Phys.* **32** 2945 (1991).
11. T. Deguchi, K. Fabricius and B. M. McCoy, *J. Stat. Phys.* **102** 701 (2001), cond-mat/9912141.
12. T. Deguchi: cond-mat/0503564.
13. L. Dolan and M. Grady, *Phys. Rev. D* **25** 1587 (1982).
14. K. Fabricius and B. M. McCoy, in *Progress in Mathematical Physics* **23** 119 (2002) (Boston: Birkhauser), cond-mat/0108057.
15. K. Fabricius and B. M. McCoy, *J. Stat. Phys.* **111** 323 (2003), cond-mat/0207177.
16. K. Fabricius and B. M. McCoy, *Publ. RIMS* **40** 905 (2004), cond-mat/0311122.
17. G. von Gehlen and R. Rittenberg, *Nucl. Phys. B* **257** 351 (1985).
18. S. Howes, L.P. Kadanoff and M. den Nijs, *Nucl. Phys. B* **215** 169 (1983).
19. S. S. Roan, 1991 Preprint MPI 91-70 (Max-Planck-Inst. für Math., Bonn); E. Date and S. S. Roan, *J. Phys. A: Math. Gen.* **33** 3275 (2000), math.QA/9911018.

20. S. S. Roan, *J. Phys. A: Math. Gen.* **38**, 7483 (2005), cond-mat/0410011.
21. S. S. Roan, *J. Stat. Mech.* P09007 (2005), cond-mat/0505698.

The Cyclic Renormalization Group

German Sierra*

Instituto de Física Teórica, CSIC-UAM, Madrid, Spain
email: german.sierra@uam.es

We present a brief introduction to the cyclic Renormalization Group concept and we illustrate it with quantum mechanical and many body examples.

1. Brief History

(1) In 1971 Wilson suggested the possible existence of limit cycles and chaotic behaviour in RG flows involving two or more coupling constants [1].

(2) In 1998 Bedaque, Hammer and van Kolck studied a Hamiltonian in Nuclear Physics with two and three body delta function potentials exhibiting limit cycle behaviour [2].

(3) In 2001 Bernard and LeClair found cyclic Kosterlitz-Thouless flows in the anisotropic WZW model [3].

(4) In 2002 Glazek and Wilson defined a discrete QM Hamiltonian with two couplings whose RG has limit cycles and chaotic behaviour [4].

(5) In 2002 LeClair, Román, Sierra proposed a BCS model of superconductivity with RG cycles (Russian Doll model) [5]. This model was shown to be integrable by Dunning and Links [6] and the exact solution employed to study the elementary excitations in ref. [7].

(6) In 2003-4 LeClair, Román, Sierra proposed a sine-Gordon model with RG cycles (S matrix) and computed the finite size effects [8,9].

2. Limit cycles and the Russian Doll property

Let $H(g_1, g_2, \Lambda)$ be a Hamiltonian with coupling constants g_1, g_2 and cutoff Λ. Integrating the high energy modes yields the renormalized Hamiltonian

*Work supported by the CICYT of Spain under contract BFM2003-05316-C02-01, the EC Grant HPRN-CT-2002-00325 and the ESF Science Programme INSTANS 2005-2010.

$$H(g_1, g_2, \Lambda) \to H(g_1(s), g_2(s), e^{-s}\Lambda) \tag{2.1}$$

where s is the RG scaling factor. The usual situations correspond to fixed points (attractive, repulsive or unstable):

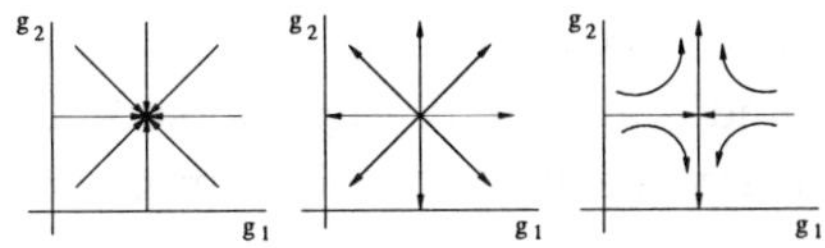

Fig. 2.1. Three different types of RG flows with a fixed point in the spectrum

However there is also the possibility of RG flows with *limit cycles* or *centers*:

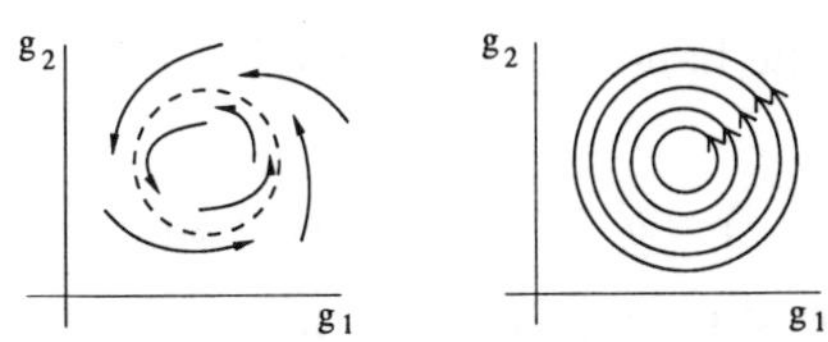

Fig. 2.2. RG flows with a limit cycle (left) or centers (right)

For RG cycles, the relevant quantities are not critical exponents but the periods:

$$g_a(s) = g_a(s + \lambda), \quad a = 1, 2 \tag{2.2}$$

where λ depends on g_1, g_2. What is the physical consequence of this fact? The RG preserves the low energy spectrum of the Hamiltonian. Hence after a complete RG cycle the spectrum is self-similar, e.g.

$$E_n(g_1, g_2, e^{-\lambda}\Lambda) = E_{n+1}(g_1, g_2, \Lambda) \tag{2.3}$$

Suppose that $H(g_1, g_2, \Lambda)$ contains low energy states with energies

$$E_n(g_1, g_2, \Lambda) = \Lambda f_n(g_1, g_2) \tag{2.4}$$

then from the selfsimilarity of the spectrum one obtains,

$$e^{-\lambda}\Lambda f_n(g_1, g_2) = \Lambda f_{n+1}(g_1, g_2) \rightarrow E_n(g_1, g_2, \Lambda) \sim \Lambda e^{-n\lambda} \tag{2.5}$$

If the Hamiltonian has bound states one gets

$$E_0 < E_1 < \cdots < E_\infty = 0 \tag{2.6}$$

This is a generic feature of the models constructed so far in Nuclear Physics, Quantum Mechanics and Many Body Physics (superconductivity).
The Russian doll property is expected to arise in models with infrared limit cycles, but there exist also some Field Theoretical models, like the cyclic sine-Gordon model, where the limit cycles appear in the ultraviolet [8,9]. In the latter model the periodicity appears in some physical quantities like the effective central charge [9].

3. The Glazek-Wilson model

The Hamiltonian is the half-infinite matrix [4]

$$H_{n,m}(g_N, h_N) = b^{n+m}(\delta_{n,m} - g_N - ih_N \, \text{sign}(n-m)) \tag{3.1}$$

where $b > 1$ and $-\infty < n, m \leq N$. N is the cutoff. The Gauss elimination of the component ϕ_N of the wave function defines a new Hamiltonian with,

$$g_{N-1} = g_N + \frac{g_N^2 + h_N^2}{1 - g_N}, \quad h_{N-1} = h_N \equiv h \tag{3.2}$$

After p-iterations one gets

$$g_{N-p} = h \, \tan\left(\tan^{-1}(\frac{g_N}{h}) + p\beta\right), \quad \beta = \tan^{-1} h \tag{3.3}$$

If $\pi/\beta = p$ there is a cycle with period p, i.e. $g_{N-p} = g_N$. If π/β is irrational the flow of g_N is chaotic. The model has an infinite number of bound states from 0 to $-\infty$ with Russian doll scaling.

4. Russian Doll superconductors

The BCS Hamiltonian for s-wave pairing is,

$$H_{BCS} = \sum_{j=1}^{N} \varepsilon_j b_j^\dagger b_j - G \sum_{j,j'=1}^{N} b_j^\dagger b_{j'} \tag{4.1}$$

where $b_j = c_{j,-}c_{j,+}$ are Cooper pair operators and ε_j are equally spaced energy levels $-\omega < \varepsilon_j < \omega$ with level spacing 2δ. The ground state (condensate) is characterized by the energy gap

$$\Delta_0 \sim 2\omega\, e^{-1/g}, \quad g = G/\delta << 1 \tag{4.2}$$

The RD Hamiltonian is a generalization of the BCS one:

$$H = \sum_{j=1}^{N} \varepsilon_j b_j^\dagger b_j - \sum_{j,j'=1}^{N} (G + i\eta\, \mathrm{sign}(j-j'))\, b_j^\dagger b_{j'} \tag{4.3}$$

H is hermitean but breaks the time reversal symmetry. The gap function is complex and its modulus satisfies the Russian dolls scaling,

$$\Delta_n \sim \Delta_0\, e^{-n\pi/h} \to \Delta_{n+1} = e^{-\lambda}\Delta_n \tag{4.4}$$

with $\lambda = \frac{\pi}{h}$. We shall call the solutions with $n = 1, 2, , \ldots$ dolls. The corresponding Cooper pairs being larger and larger (see figure 4.1).

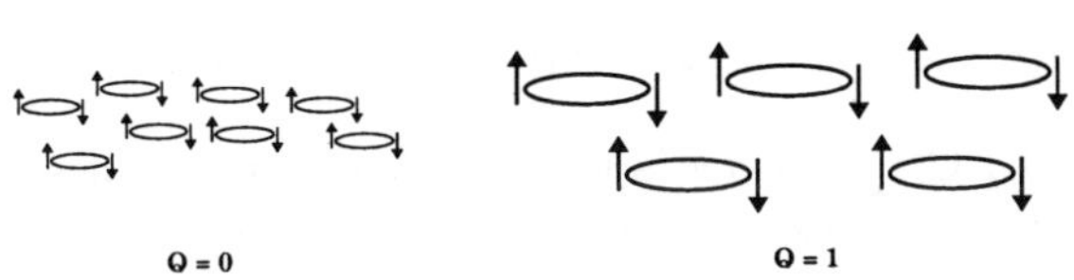

Fig. 4.1. Pictorial representation of the ground state (left) and the excited state (right) corresponding to the $Q = 1$ solution of the gap equation of the RD model. In the latter case the Cooper pairs are bigger than those forming the ground state.

4.1. *Renormalization Group of the RD model*

The Gauss elimination of the highest component leads:

$$H(G_N, \eta_N) \to H(G_{N-1}, \eta_{N-1}) \tag{4.5}$$

$$G_{N-1} + i\eta_{N-1} = G_N + i\eta_N + \frac{1}{N\delta}(G_N + i\eta_N)(G_N - i\eta_N) \tag{4.6}$$

Hence $\eta_N = \eta_{N-1}$ is an RG invariant. In the large N limit one can define a variable $s = \log N_0/N$, where N_0 is the initial size of the system.

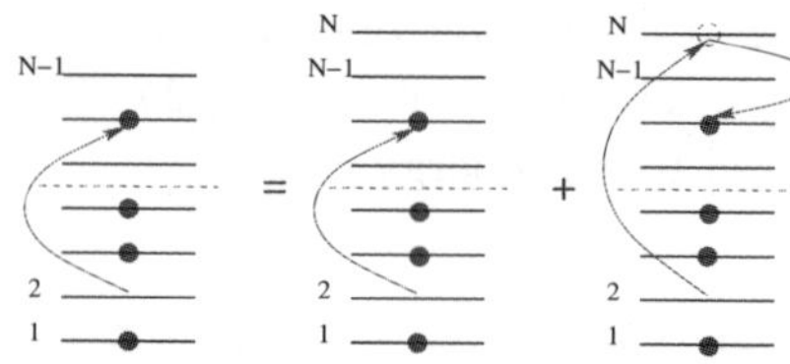

Fig. 4.2. Graphical representation of the RG equation 4.6.

$$\frac{dg}{ds} = (g^2 + h^2), \qquad s \equiv \log \frac{N_0}{N}. \tag{4.7}$$

The effect of the coupling h is to *accelerate* the running of g with the scale. The solution to the above equation is

$$g(s) = h \tan\left[hs + \tan^{-1}\left(\frac{g_0}{h}\right)\right], \quad g_0 = g(N_0). \tag{4.8}$$

which exhibits a periodic behaviour

$$g(s+\lambda) = g(s) \iff g(e^{-\lambda}N) = g(N), \quad \lambda \equiv \frac{\pi}{h} \tag{4.9}$$

The number of RG cycles n_C is equal to the number of solutions of the BCS gap equation. In every cycle the size of the system is reduced by the scaling factor $e^{-\lambda}$. Hence n_C satisfies

$$e^{-n_C\lambda}N \sim 1 \to n_C = \frac{h}{\pi}\log N + \text{corrections} \tag{4.10}$$

In a practical example for h of order 1 and N of order 10^{23}, the number of cycles, i.e. dolls, will also be of order 1.

4.2. *Numerical Work- One Cooper Pair problem*

The RD Hamiltonian for one Cooper pair becomes the N-dimensional matrix:

$$H_{j,k} = \varepsilon_j\,\delta_{j,k} - (G + i\,\eta\,\text{sign}(j-k)) \tag{4.11}$$

In the large N limit there are many bound states with wave function and energies,

$$\psi(\varepsilon) = \frac{1}{(\varepsilon - E_n)^{1-ih}}, \qquad E_n \sim E_0 \, e^{-2\pi n/h}, \quad n = 0, 1, \ldots \tag{4.12}$$

The RG period is now $\lambda_1 = 2\pi/h$. Numerical diagonalization of the Hamiltonian gives the exact eigenstates of one-pair Hamiltonian for N levels ($N_0 = 500$ down to 30). The vertical lines are at the values $N_n = e^{-n\lambda_1} N_0$.

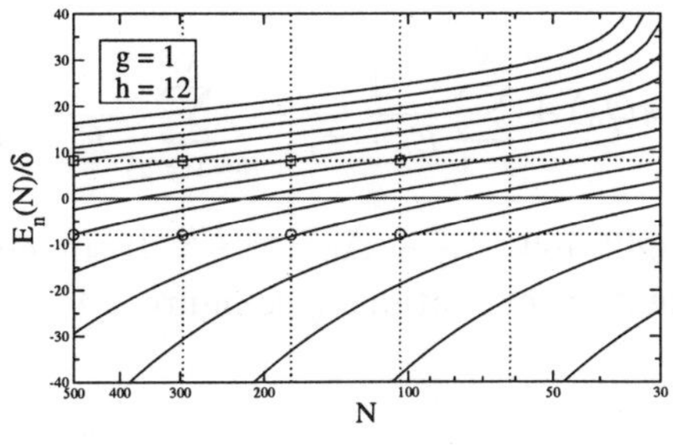

Fig. 4.3. Self-similarity of the spectrum.

Self-similarity of the spectrum:

$$E_{n+1}(N) = E_n(e^{-\lambda_1} N), \qquad \lambda_1 = \frac{2\pi}{h} \tag{4.13}$$

Russian Doll property

$$E_{n+1}(N) = e^{-\lambda_1} E_n(N) \tag{4.14}$$

The two properties are related: $E_n(N) \sim N \, e^{-\lambda_1 n}$.

4.3. *Integrability of the Russian doll model*

The RD model is exactly solved à la Bethe using the inhomegenous XXX vertex model with a boundary matrix [6]

$$K_0 = \exp(-i\alpha \, \sigma^z), \quad \alpha = \tan^{-1}\left(\frac{\eta}{G}\right) \tag{4.15}$$

The Hamiltonian appears in an expansion of the transfer matrix in the inverse of the spectral parameter. The BAE's are given by

$$e^{2i\alpha} \prod_{j=1}^{N} \frac{E_a - \varepsilon_j + i\eta}{E_a - \varepsilon_j - i\eta} = \prod_{b=1(\neq a)}^{M} \frac{E_a - E_b + 2i\eta}{E_a - E_b - 2i\eta} \tag{4.16}$$

and the total energy is $E = \sum_a E_a$. In the semiclassical limit $\eta \to 0$, the RD model becomes the usual BCS model. In the large N limit where $\eta = h\delta \to 0$ we get

$$\frac{1}{\eta}(\alpha + \pi Q_a) + \sum_{j=1}^{N} \frac{1}{E_a - \varepsilon_j} - \sum_{b=1(\neq a)}^{M} \frac{2}{E_a - E_b} = 0 \tag{4.17}$$

If $Q_a = Q, \ \forall a \Rightarrow Q^{\text{th}}$-doll, $\Delta_Q \sim \Delta_0 \, e^{-\pi Q/h}$. If some $Q'_a s \neq 0$ we expect a new type of elementary excitations (see figure 4.4).

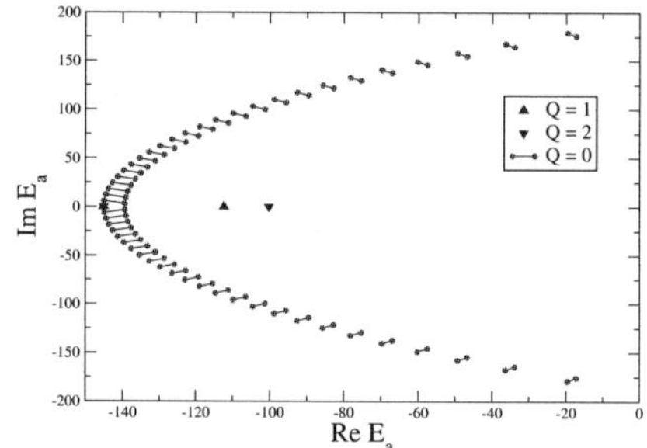

Fig. 4.4. Numerical solutions E_a corresponding to three choices: i) $\{Q_a = 0\}_{a=1}^{M}$, ii) $Q_1 = 1$ and $\{Q_a = 0\}_{a=2}^{M}$ and iii) $Q_1 = 2$ and $\{Q_a = 0\}_{a=2}^{M}$.

Questions concerning the RD model

(1) Nature of the new excitations. The Cooper pairs can be excited without breaking them. Q_a appears as a principal quantum number. What is their dispersion relation? What is their statistics? It seems bosonic in contrast with the fermionic character of the standard BCS quasiparticles.

(2) Find the phase diagram at finite temperature. Are there new phases?

(3) Find a macroscopic derivation of the RD Hamiltonian. The problem is that H_{RD} breaks time reversal symmetry (T) while the usual phonon or other interaction processes do not break T? Some possibilities are external magnetic fields or spontaneous T breaking.

5. The Riemann hypothesis and the cyclic RG

Riemann Hypothesis: all the non trivial zeros of the zeta function $\zeta(s)$ lie in the critical axis $s = \frac{1}{2} + i\tau$.
Polya-Hilbert conjecture: there exist a Hamiltonian H_R whose spectrum E_a are all the non trivial zeros:

$$H_R\, \psi_a = E_a\, \psi_a \Longleftrightarrow \zeta(\frac{1}{2} + iE_a) = 0 \tag{5.1}$$

This is supported by the fact that the E_a are randomly distributed according to the GUE, which suggest that H_R breaks time reversal.
Quantum Chaos conjecture: H_R is the quantization of a classical Hamiltonian which has stable periodic orbits labelled by the prime numbers. Berry and Keating proposed $H_R = px$ which reproduces semiclassically the counting formula for the non trivial zeros.

$$N(T) \sim \frac{T}{2\pi} \log \frac{T}{2\pi} - \frac{T}{2\pi} \tag{5.2}$$

In Connes's work on adeles the zeros appear as absortion lines in the spectrum of a Hamiltonian.

Hints on the relation RH-RD model

Generalize the RD Hamiltonian as follows:

$$H_{n,m} = \varepsilon_n\, \delta_{n,m} - f_n\, (g + ih\, \text{sign}(n-m))\, f_m, \tag{5.3}$$

The previous choice is ($\kappa = 1$):

$$\varepsilon_n = n,\ f_n = 1 \Longrightarrow E_n \sim \begin{cases} -Ne^{-\frac{2\pi n}{h}}, & n = 0, 1, \dots \\ N(1 - e^{\frac{2\pi n}{h}}), & n = -1, -2, \dots \end{cases} \tag{5.4}$$

Consider the new choice ($\kappa = 0$):

$$\varepsilon_n = 0,\ f_n = \frac{1}{\sqrt{n}} \Longrightarrow E_n \sim -h\, \frac{\log N}{n}, n = 0, \pm 1, \dots \tag{5.5}$$

For $N \to \infty$ the spectrum becomes continuous converging towards $E = 0$ algebraically (gapless RD scaling) and not exponentially (gaped RD scaling). The eigenstates of the new model are given by

$$\psi_n \sim \frac{1}{n^{\frac{1}{2} - i\tau_E}},\ \ n >> 1 \tag{5.6}$$

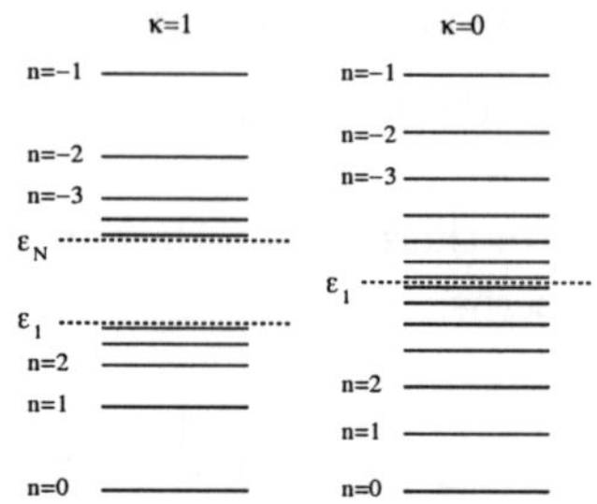

Fig. 5.1. Comparison of spectrums with gapped (right) or gapless (right) RD scaling.

where τ_E depends on the energy E as

$$\tau_E = -\frac{h}{E} \tag{5.7}$$

The key point is that the number of RG cycles

$$n_C \sim \frac{\mathcal{T}}{2\pi} \log N - \frac{\mathcal{T}}{2\pi} \log \frac{\mathcal{T}}{2} + \frac{\mathcal{T}}{2\pi} \tag{5.8}$$

is almost the semiclassical Riemann's formula

$$N(T) \sim \frac{T}{2\pi} \log \frac{T}{2\pi} - \frac{T}{2\pi} \tag{5.9}$$

which suggest that

$$\text{Riemann zeros} \Longleftrightarrow \text{Missing RG cycles} \tag{5.10}$$

These ideas are explained in detail in reference [10] where a consistent quantization of the hamiltonian $H = xp$ is proposed and the connection with the RD model is established.

Acknowledgements I would like to thank Professors Weiping Zhang and Mo-Lin Ge for the kind invitation to participate in the XXIII International Conference of DGMTP and in the opening of the new building of the Nankai Institute of Mathematics.

Fig. 5.2. Picture of a collection of Russian dolls or Matrioskas. They are choosen to symbolize the scaling behaviour typical of a model with RG limit cycles.

References

1. K. G. Wilson, "Renormalization Group and Strong Interactions", Phys. Rev. **D3** (1971) 1818.
2. P. F. Bedaque, H.-W. Hammer, and U. van Kolck, "Renormalization of the Three-Body System with Short-Range Interactions", Phys. Rev. Lett. **82** (1999) 463, nucl-th/9809025.
3. D. Bernard and A. LeClair, "Strong-weak coupling duality in anisotropic current interactions", Phys.Lett. **B512** (2001) 78; hep-th/0103096.
4. S. D. Glazek and K. G. Wilson, "Limit cycles in quantum theories", Phys. Rev. Lett. **89** (2002) 230401, hep-th/0203088;
5. A. LeClair, J.M. Román and G. Sierra, "Russian doll Renormalization Group and Superconductivity", Phys. Rev. **B69** (2004) 20505; cond-mat/0211338.
6. C. Dunning and J. Links, "Integrability of the Russian doll BCS model", Nucl. Phys. **B702** (2004) 481, cond mat/0406234.
7. "The elementary excitations of the exactly solvable Russian doll BCS model of superconductivity", A. Anfossi, A. LeClair, G. Sierra, J. Stat. Mech. (2005) P05011; cond-mat/0503014.
8. A. LeClair, J.M. Román and G. Sierra, "Russian doll Renormalization Group and Kosterlitz-Thouless Flows", Nucl. Phys. **B675** (2003) 584; hep-th/0301042.
9. A. LeClair, J.M. Román and G. Sierra, "Log-periodic behaviour of finite size effects in field theory models with cyclic renormalization group", Nucl. Phys. **B700** [FS] (2004) 407; hep-th/0312141.
10. G. Sierra, "The Riemann zeros and the cyclic Renormalization Group" math.NT/0510572, to appear in JSTAT.

Bohm-Aharonov Type Effects in Dissipative Atomic Systems*

Allan I. Solomon

Department of Physics and Astronomy
The Open University, Milton Keynes MK7 6AA, UK
E-mail: a.i.solomon@open.ac.uk,

Sonia G. Schirmer

DAMTP, Cambridge University , UK
E-mail: sgs29@cam.ac.uk

A state in quantum mechanics is defined as a positive operator of norm 1. For finite systems, this may be thought of as a positive matrix of trace 1. This constraint of positivity imposes severe restrictions on the allowed evolution of such a state. From the mathematical viewpoint, we describe the two forms of standard dynamical equations - global (Kraus) and local (Lindblad) - and show how each of these gives rise to a semi-group description of the evolution. We then look at specific examples from atomic systems, involving 3-level systems for simplicity, and show how these mathematical constraints give rise to non-intuitive physical phenomena, reminiscent of Bohm-Aharonov effects. In particular, we show that for a multi-level atomic system it is generally impossible to isolate the levels, and this leads to observable effects on the population relaxation and decoherence.

1. Introduction

The standard description of a quantum state suitable for an open system is by means of a density matrix ρ, a positive matrix of trace 1. For a *hamiltonian* (non-dissipative) system one obtains a unitary evolution of the state. For a non-dissipative system the time evolution of the density matrix $\rho(t)$ with $\rho(t_0) = \rho_0$ is governed by

$$\rho(t) = U(t)\rho_0 U(t)^\dagger, \tag{1.1}$$

*Presented at DGMTP XXIII, Nankai Institute, Tianjin: 25 August 2005 where $U(t)$ is the time-evolution operator satisfying the Schrodinger equation

$$i\hbar\frac{d}{dt}U(t) = HU(t), \qquad U(0) = I, \tag{1.2}$$

where I is the identity operator. The state $\rho(t)$ equivalently satisfies the quantum Liouville equation

$$i\hbar\frac{d}{dt}\rho(t) = [H, \rho(t)] \equiv H\rho(t) - \rho(t)H. \tag{1.3}$$

H is the total Hamiltonian of the system. (In the context of *Quantum Control* theory, we may assume that $H \equiv H(\vec{f})$ depends on a set of control fields f_m:

$$H(\vec{f}) = H_0 + \sum_{m=1}^{M} f_m(t)H_m, \tag{1.4}$$

where H_0 is the internal Hamiltonian and H_m is the interaction Hamiltonian for the field f_m for $1 \leq m \leq M$.) The advantage of the Liouville equation (1.3) over the unitary evolution equation (1.1) is that it can easily be adapted for dissipative systems by adding a dissipation (super-)operator $L_D[\rho(t)]$:

$$i\hbar\dot{\rho}(t) = [H, \rho(t)] + i\hbar L_D[\rho(t)]. \tag{1.5}$$

In general, uncontrollable interactions of the system with its environment lead to two types of dissipation: phase decoherence (dephasing) and population relaxation (decay). The former occurs when the interaction with the enviroment destroys the phase correlations between states, which leads to a decay of the off-diagonal elements of the density matrix:

$$\dot{\rho}_{kn}(t) = -\frac{i}{\hbar}([H, \rho(t)])_{kn} - \Gamma_{kn}\rho_{kn}(t) \tag{1.6}$$

where Γ_{kn} (for $k \neq n$) is the dephasing rate between $|k\rangle$ and $|n\rangle$. The latter happens, for instance, when a quantum particle in state $|n\rangle$ spontaneously emits a photon and decays to another quantum state $|k\rangle$, which changes the populations according to

$$\dot{\rho}_{nn}(t) = -\frac{i}{\hbar}([H, \rho(t)])_{nn} + \sum_{k\neq n} [\gamma_{nk}\rho_{kk}(t) - \gamma_{kn}\rho_{nn}(t)] \tag{1.7}$$

where $\gamma_{kn}\rho_{nn}$ is the population loss for level $|n\rangle$ due to transitions $|n\rangle \to |k\rangle$, and $\gamma_{nk}\rho_{kk}$ is the population gain caused by transitions $|k\rangle \to |n\rangle$. The population relaxation rate γ_{kn} is determined by the lifetime of the state $|n\rangle$, and for multiple decay pathways, the relative probability for the transition $|n\rangle \rightarrow |k\rangle$. Phase decoherence and population relaxation lead to a dissipation superoperator (represented by an $N^2 \times N^2$ matrix) whose non-zero elements are

$$\begin{aligned} (L_D)_{kn,kn} &= -\Gamma_{kn} & k \neq n \\ (L_D)_{nn,kk} &= +\gamma_{nk} & k \neq n \\ (L_D)_{nn,nn} &= -\textstyle\sum_{n\neq k} \gamma_{kn} & \end{aligned} \tag{1.8}$$

where Γ_{kn} and γ_{kn} are positive numbers, with Γ_{kn} symmetric in its indices. The $N^2 \times N^2$ matrix superoperator L_D may be thought of as acting on the N^2-vector V obtained from ρ by

$$V_{[(i-1)N+j]} \equiv \rho_{ij}. \tag{1.9}$$

The resulting vector equation is

$$\dot{V} = LV = (L_H + L_D)V \tag{1.10}$$

where L_H is the anti-hermitian matrix derived from the hamiltonian H.

The values of the relaxation and dephasing parameters may be determined by experiment, or simply chosen to supply a model for the dissipation phenomenon. But they may not be chosen arbitrarily; the condition of positivity for the state ρ imposes constraints on their values, as does their deduction from rigorous theory. We illustrate this by demonstrating the constraint for a two-level system.

2. Two-level systems

2.1. *Unitary evolution*

The general hamiltonian for a two-level system is given, up to an additive constant, by

$$H = w \begin{bmatrix} 1 & 0 \\ 0 & -1 \end{bmatrix} + f_x \begin{bmatrix} 0 & 1 \\ 1 & 0 \end{bmatrix} + f_y \begin{bmatrix} 0 & -i \\ i & 0 \end{bmatrix} \tag{2.1}$$

which we assume here to be time-independent.

This corresponds to the (superoperator form) L_H, where

$$L_H = \begin{bmatrix} 0 & i\,(f_x + if_y) & i\,(-f_x + if_y) & 0 \\ -i\,(-f_x + if_y) & -2\,iw & 0 & i\,(-f_x + if_y) \\ -i\,(f_x + if_y) & 0 & 2\,iw & i\,(f_x + if_y) \\ 0 & -i\,(f_x + if_y) & -i\,(-f_x + if_y) & 0 \end{bmatrix}. \tag{2.2}$$

Note the useful rule for obtaining the equivalent $N^2 \times N^2$ matrix action

$$A\rho B \Longleftrightarrow A \otimes B^T\ V.$$

The corresponding evolution equation for the 4-vector V corresponding to the state ρ is

$$\dot{V} = L_H V. \tag{2.3}$$

This is equivalent to Eq.(1.1), which clearly preserves the trace of ρ, and also its positivity, using the definition of a positive matrix as one of the form $MM^\dagger$. (Of course this result is true in general.)

2.2. *Pure dissipation*

The dissipation (super-)operator is

$$L_D = \begin{bmatrix} -\gamma_{21} & 0 & 0 & \gamma_{12} \\ 0 & -\Gamma & 0 & 0 \\ 0 & 0 & -\Gamma & 0 \\ \gamma_{21} & 0 & 0 & -\gamma_{12} \end{bmatrix}. \tag{2.4}$$

The corresponding evolution equation

$$\dot{V} = L_D V. \tag{2.5}$$

has solution

$$V(t) = \exp(L_D t)V(0) \tag{2.6}$$

which corresponds to a value of the state $\rho(t)$

$$\begin{bmatrix} \frac{\rho_{11}(\gamma_{12}+\gamma_{21}E)+\gamma_{12}\rho_{22}(1-E)}{\gamma_{21}+\gamma_{12}} & e^{-t\Gamma}\rho_{12} \\ e^{-t\Gamma}\rho_{21} & \frac{\gamma_{21}\rho_{11}(1-E)+\rho_{22}(\gamma_{21}+\gamma_{12}E)}{\gamma_{21}+\gamma_{12}} \end{bmatrix} \tag{2.7}$$

where $E = e^{-t(\gamma_{21}+\gamma_{12})}$, for which it may readily be checked that $\mathrm{Tr}\rho(t) = \rho_{11} + \rho_{22} = 1$. Additionally, $\det \rho(t)$ is given by

$$\rho_{11}\rho_{22}e^{-t(\gamma_{21}+\gamma_{12})} - \left(e^{-2t\Gamma}\right)\rho_{12}\rho_{21} + 2\,\frac{\rho_{11}\gamma_{12}\rho_{22}\gamma_{21}\left(1 - e^{-t(\gamma_{21}+\gamma_{12})}\right)^2}{\left(\gamma_{21}+\gamma_{12}\right)^2} \tag{2.8}$$

which is clearly positive for all t when

$$2\Gamma \geq \gamma_{12} + \gamma_{21} \tag{2.9}$$

since $\det \rho(t) \geq e^{-t(\gamma_{21}+\gamma_{12})} \det \rho(0) \geq 0$. Conversely, when the condition Eq.(2.9) is violated, it is easy to display examples for which the evolution does not produce a state. For example, for a pure state, which satisfies $\rho_{11}\rho_{22} - \rho_{12}\rho_{21} = 0$, choosing $\gamma_{12} > 2\Gamma, \gamma_{21} = 0$, Eq.(2.8) is clearly negative.

2.3. *General dissipation*

When the hamiltonian matrix L_H and the dissipation matrix L_D commute, the conclusions of the previous two subsections produce the same constraint for the solution of Eq.(1.10). In the general case these matrices do not commute; they do however generate a local semi-direct group. More accurately the Lie algebra is locally a semi-direct sum†, which then generates a *semi-group.* In this case also, general theory, which we discuss in the next section, shows that the trace and determinant conditions of Eq.(2.9) remain unchanged.

3. Rigorous formulations

3.1. *Kraus formalism and semi-groups*

The global form of the evolution equation Eq.(1.1) in the presence of dissipation is due to Kraus[2]. The evolution of the state ρ is given by

$$\rho(t) = \sum_i W_i(t) \rho_0 W_i(t)^\dagger, \tag{3.1}$$

with

$$\sum_i W_i(t)^\dagger W_i(t) = I. \tag{3.2}$$

Equation(3.1) and the condition Eq.(3.2) clearly guarantee both positivity and unit trace.

Further, though less obviously, this system implies the existence of a *semi-group* description of the evolution. For if we consider the set G whose elements are the *sets* $\{w_i\}$ satisfying Eq.(3.2), then if $g = \{w_i\}$ and $g' = \{w_i'\}$ are two elements of G, then so too is gg', where the product is taken in the sense of set multiplication. Although closed under composition, the only elements of G which possess inverses are the singleton sets $\{U\}$, where U is unitary.

---

†In the present two-level case, the local Lie algebra is the 12-element $gl(3, R) \oplus R^3$, and in general $gl(N^2 - 1, R) \oplus R^{N^2-1}$, as discussed in[1].

3.2. *Lindblad formalism*

In so far as the Kraus formalism provides an analogue of the unitary evolution equation Eq.(1.1), the Lindblad[3] formalism gives an analogue of the Schroedinger equation Eq.(1.3):

$$\begin{aligned}\dot{\rho}(t) &= L[\rho(t)]\rho(t) \\ &= -i[H,\rho(t)] + \frac{1}{2}\sum_k \left([V_k\rho(t), V_k^\dagger] + [V_k, \rho(t)V_k^\dagger]\right)\end{aligned} \tag{3.3}$$

where the V_k are $N \times N$ matrices, but otherwise arbitrary‡. It may be proved that the dissipation superoperator L_D arising from Eq.(3.3) has negative eigenvalues. Since the evolution dynamics arises from exponentiation of $L_D t$ it follows that operators $\exp(L_D t)$ in the theory will become unbounded for arbitrary negative t. This means that not all operators will have inverses and implies a semi-group character to the evolution, as in the Kraus formalism.

3.3. 2×2 *Lindblad example*

Choosing four independent complex V-matrices

$$V_1 = \begin{bmatrix} a_1 & 0 \\ 0 & 0 \end{bmatrix} \quad V_2 = \begin{bmatrix} 0 & a_2 \\ 0 & 0 \end{bmatrix} \quad V_3 = \begin{bmatrix} 0 & 0 \\ a_3 & 0 \end{bmatrix} \quad V_4 = \begin{bmatrix} 0 & 0 \\ 0 & a_4 \end{bmatrix}$$

we obtain for the dissipation superoperator L_D

$$\begin{bmatrix} -|a_3|^2 & 0 & 0 & |a_2|^2 \\ 0 & -1/2\,A & 0 & 0 \\ 0 & 0 & -1/2\,A & 0 \\ |a_3|^2 & 0 & 0 & -|a_2|^2 \end{bmatrix}.$$

where $A = |a_1|^2 + |a_2|^2 + |a_3|^2 + |a_4|^2$, which on comparison with Eq.(2.4) gives, defining $\tilde{\Gamma} \equiv \frac{1}{2}(|a_1|^2 + |a_4|^2)$

$$\gamma_{21} = |a_3|^2, \quad \gamma_{12} = |a_2|^2, \quad \Gamma = \tilde{\Gamma} + \frac{1}{2}(\gamma_{12} + \gamma_{21})$$

whence the constraint Eq.(2.9). Note that $(\gamma_{12} + \gamma_{21})/2$ is the phase decoherence forced by population relaxation and $\tilde{\Gamma}$ is the contribution of pure dephasing.

‡We may also choose an arbitrary number of matrices V_k.

3.4. *General $N \times N$ Lindblad case*

A convenient choice for the V_k matrices may be made by defining

$$V_{[i,j]} = a_{[i,j]} E_{ij}$$

where E_{ij} is the standard basis for $N \times N$ matrices, with $(E_{ij})_{\alpha\beta} = \delta_{ia}\delta_{j\beta}$ and we use the index notation $[i,j] \equiv (i-1)N + j$. The relaxation and decoherence parameters are defined by

$$\begin{aligned}
\gamma_{ij} &= \left|a_{[i,j]}\right|^2 \quad (i \neq j) \\
\tilde{\Gamma}_{ij} &= \frac{1}{2}(\left|a_{[i,i]}\right|^2 + \left|a_{[j,j]}\right|^2) \quad (i \neq j) \\
\Gamma_{ij} &= \frac{1}{2}\sum_k(\left|a_{[k,i]}\right|^2 + \left|a_{[k,j]}\right|^2) \quad (i \neq j)
\end{aligned} \tag{3.4}$$

4. Bohm-Aharonov type effects

What we mean by *Bohm-Aharonov type effects* in the title of this note, and of this section, is the impossibility of isolation of quantum subsystems. We illustrate this type of effect by considering the use of a two-level atomic system as, say, a qubit, when this is a subsystem of a multi-level system.

We consider the case of pure dissipation as discussed in subsection 2.2. Choosing values $\gamma_{21} = 0, \quad \gamma_{12} = \gamma, \quad \Gamma = \frac{1}{2}\gamma$, which satisfy the constraint Eq.(2.9), the state evolution is given by

$$\rho(t) = \begin{bmatrix} \rho_{11} + \rho_{22}\left(1 - e^{-t\gamma}\right) & e^{-1/2\,t\gamma}\rho_{12} \\ e^{-1/2\,t\gamma}\rho_{21} & \rho_{22}e^{-t\gamma} \end{bmatrix} \tag{4.1}$$

where the initial state is

$$\rho(0) = \begin{bmatrix} \rho_{11} & \rho_{12} \\ \rho_{21} & \rho_{22} \end{bmatrix}$$

We now assume that our two-level system is embedded in a three level system, so that the state's evolution is given by

$$\rho(t) = \begin{bmatrix} \rho_{11} + \rho_{22}\left(1 - e^{-t\gamma}\right) & e^{-1/2\,t\gamma}\rho_{12} & \rho_{13} \\ e^{-1/2\,t\gamma}\rho_{21} & \rho_{22}e^{-t\gamma} & \rho_{23} \\ \rho_{31} & \rho_{32} & \rho_{33} \end{bmatrix}. \tag{4.2}$$

Now consider three examples for the state evolution. In all cases we start off with a pure state, in the first case with the third level not being populated.

4.1. *Unpopulated third level*

Assume an initial pure state represented by the 3-vector $v = [1/\sqrt{2}, 1/\sqrt{2}, 0]$ corresponding to the density matrix

$$\frac{1}{2}\begin{bmatrix} 1 & 1 & 0 \\ 1 & 1 & 0 \\ 0 & 0 & 0 \end{bmatrix}.$$

Assuming that the third level is unaffected, the state evolution is given by (measuring t in units of $1/\gamma$)

$$\rho(t) = \begin{bmatrix} 1 - 1/2\, e^{-t} & 1/2\, e^{-1/2\, t} & 0 \\ 1/2\, e^{-1/2\, t} & 1/2\, e^{-t} & 0 \\ 0 & 0 & 0 \end{bmatrix}. \tag{4.3}$$

In this case the naive picture of the evolution is justified, as the third level remains unpopulated, the eigenvalues remain positive (≥ 0), and the extra levels are not affected by the dissipative dynamics. The third level plays no role in the evolution. However, in general an upper level will not be totally unpopulated; and in this case the constraints play a role.

4.2. *Equally populated third level*

We take the initial pure state vector to be

$$v = [1/\sqrt{3}, 1/\sqrt{3}, 1/\sqrt{3}] \tag{4.4}$$

giving the evolution

$$\rho(t) = \frac{1}{3}\begin{bmatrix} 2 - e^{-t} & e^{-1/2\, t} & 1 \\ e^{-1/2\, t} & e^{-t} & 1 \\ 1 & 1 & 1 \end{bmatrix}. \tag{4.5}$$

As in subsection 4.1 we have assumed that the third levels are *not* affected by the dissipative dynamics. However, a numerical calculation shows that the eigenvalues of $\rho(t)$ are not all positive; therefore the *assumed* evolution does not give a state, and so the naive assumption that the other levels remain unaffected is *false*.

4.3. *Pure dephasing*

Population relaxation is not the only source of constraints on the decoherence rates for $N > 2$. Even if there is no population relaxation at all, i.e.,

$\gamma_{kn} = 0$ for all k, n, and the system experiences only pure dephasing, we cannot choose the decoherence rates Γ_{kn} arbitrarily. For example, setting $\Gamma_{12} \neq 0$ and $\Gamma_{23} = \Gamma_{13} = 0$ for our three-level system gives

$$\rho(t) = \begin{bmatrix} \rho_{11} & e^{-\Gamma_{12}t}\rho_{12} & \rho_{13} \\ e^{-\Gamma_{12}t}\rho_{21} & \rho_{22} & \rho_{23} \\ \rho_{31} & \rho_{32} & \rho_{33} \end{bmatrix}. \tag{4.6}$$

Choosing $\rho(0)$ as in Eq. (4.4) we again obtain a density operator $\rho(t)$ with negative eigenvalues, as a simple calculation will reveal. This shows that there must be additional constraints on the decoherence rates to ensure that the state of the system remains physical.

5. Conclusions

We have shown that it is impossible to isolate a two-level system from a multi-level system in the sense of assuming that the other levels will not be affected by relaxation and decoherence effects in the "isolated" system. A more general treatment of the effects noted here may be found elsewhere[4]; in that paper the constraints are explicitly described for some multilevel systems, and the effects of these constraints are discussed.

References

1. A.I. Solomon and S.G. Schirmer, *Dissipative 'Groups' and the Bloch Ball*, Institute of Physics Conference Series **173** (Ed: J Gazeau et al), 485 (arXiv:quant-ph/0211027) (2002).
2. K. Kraus, Ann. Phys. **64**, 311 (1971).
3. G. Lindblad, Comm. Math. Phys. **48**, 119 (1976).
   G. Lindblad, Comm. Math. Phys. **40**, 147 (1975).
   V. Gorini, A. Kossakowski, and E. C. G. Sudarshan, J. Math. Phys. **17**, 821 (1976).
4. S.G. Schirmer and A.I. Solomon, *Phys. Rev.* **A 70**, 022107 (2004).

Noncommutative Procedures in Spontaneous Symmetry Breaking and Quantum Differentiation

Masuo Suzuki

Department of Applied Physics, Tokyo University of Science,
1-3, Kagurazaka, Shinjuku-ku, Tokyo 162-8601 Japan
E-mail: msuzuki@rs.kagu.tus.ac.jp

The present paper mainly reviews the following two topics: Part I; topological perturbation to study spontaneous symmetry breaking, and Part II; quantum analysis, q-derivative and exponential splitting. In both parts, the concept of noncomutativity of procedures or operation is essentially important. This effect appears in taking some limits in Part I, and also in quantum differentiation in Part II.

1. Part I : Topological Perturbation to Study Spontaneous Symmetry Breaking

1.1. *Topological Interaction Method*

As is well known, a continuous (namely second-order) phase transition occurs only in the thermodynamic limit namely in the limit $N \to \infty$ for the system size N. The order parameter to characterize this symmetry breaking appears below the critical point and it is obtained by taking the limit $N \to \infty$ first and then taking the limit $H \to +0$ for the external symmetry breaking field H, as was explicitly performed by C. N. Yang[1] in the two-dimensional Ising Model. Thus, he obtained exactly the uniform spontanious magnetization m_s in the same model. Another exact solution was given by McCoy and Wu[2] for the surface spontaneous magnetization in a similar limiting procedure.

Recently we have studied the surface spontaneous magnetization by calculating the boundary-boundary correlation function and by taking the limit[3]

$$\lim_{|i-j|\to\infty} < S_i S_j >_{\text{boundary}} = m_b^2, \tag{1.1}$$

where S_i and S_j denote the spins in opposite boundaries and m_b denotes the surface spontaneous magnetization. The boundary-boundary correlation functions can be evaluated using the topological interaction method proposed by the present author[4] For example, the topological perturbation or interaction J' is introduced[4-6] in the two-dimentional Ising model as shown in Fig.1.

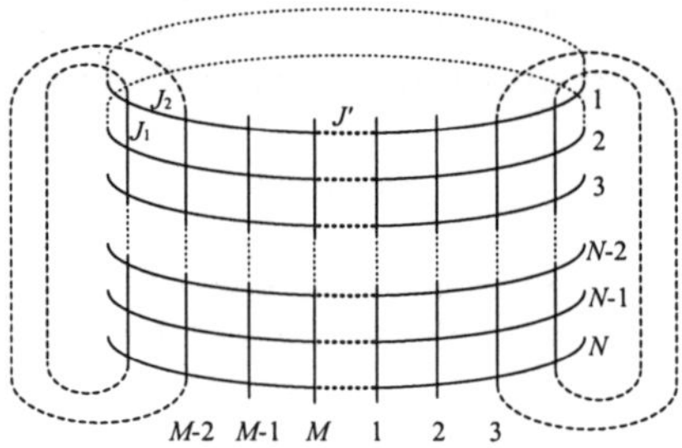

Fig. 1. Topological interaction J' to connect the two opposite boundary spins $S_{1,k}$ and $S_{M,k}$.

Our Hamiltonian including the topological perturbation is given by

$$\mathcal{H} = \mathcal{H}_0 - J' \sum_{k=1}^{N} S_{1,k} S_{M,k};\ , \tag{1.2}$$

where

$$\mathcal{H}_0 = -J \sum_{k=1}^{N} \left(\sum_{j=1}^{M} S_{j,k} S_{j,k+1} + \sum_{j=1}^{M-1} S_{j,k} S_{j+1,k} \right). \tag{1.3}$$

with $S_{j,k} = \pm 1$ and $S_{j,N+1} = S_{j,1}$. Clearly, J' changes the topology of the system. If $J' \neq 0$, the quantity $\langle S_{1,k} S_{M,k} \rangle$ denotes short-range correlation. If $J' = 0$, it denotes long-range correlation for $M \to \infty$ and consequently it yields the square of the boundary spontaneous magnetization as shown in (1). Thus, the topology of this system changes drastically according to the situation whether $J' \neq 0$ or $J' = 0$. The bounadary-boundary correlation function $C_M(0)$ (where $C_M(J') \equiv \langle S_{1,k} S_{M,k} \rangle_{J'}$) is given by

$$C_M(0) = \lim_{J' \to +0} \lim_{N \to \infty} \frac{1}{\beta N} \frac{\partial \log Z(J')}{\partial J'}, \tag{1.4}$$

where $Z(J')$ is the partition function of the relevant system, N denotes the length of the relevant system in the vertical direction in Figure 1 and β is the inverse temperature (i.e., $\beta = 1/k_{\mathrm{B}}T$).

After lengthy calculation[5,6], we obtain

$$C_M(0) = \frac{1}{2\pi z_2}\int_0^{2\pi}\frac{1}{f_M(0)}d\theta \tag{1.5}$$

where

$$f_M(\theta) = f(\alpha(\theta)\alpha(\theta)^M + f(\frac{1}{\alpha(\theta)})\alpha(\theta)^{-M}, \tag{1.6}$$

$$f(\alpha) = \frac{(\alpha - z_2/x_1)(\alpha - z_2 x_1)}{(1-z_2^2)(\alpha^2-1)}\,;\, x_1 \equiv \frac{1-z_1}{1+z_1}, \tag{1.7}$$

and $\alpha(\theta)$ is the larger solution of the following equation

$$(1+z_1^2)(1+z_2^2) - 2z_1(1-z_2^2\cos\theta) - z_2(1-z_1^2)(\alpha+\alpha^{-1}) = 0, \tag{1.8}$$

where $z_j = \tanh K_j (j = 1, 2)$, and $K_j = \beta J_j$. The correct branch of the solution differes above and below the critical point T_c. The functions $f(\alpha(\theta))$ and $f(1/\alpha(\theta))$ are shown[5] to have the following properties:

$$a)\ \ \mathit{For}\ \ T > T_c, f(\alpha(0)) = 1, f(1/\alpha(\theta)) = O(\theta^2), \tag{1.9}$$

and

$$b)\ \ \text{for}\ \ T < T_c, f(1/\alpha(0)) = 1, f(\alpha(\theta)) = O(\theta^2). \tag{1.10}$$

Using the above properties, we can evaluate the asymptotic form of the correlation function $C_M(0)$ for large M. For $T > T_c$, the correlation function $C_M(0)$ is given in the form[5]

$$C_M(0) \simeq \frac{A_+(T)}{\sqrt{M}}\exp(-\frac{M}{\xi}) \tag{1.11}$$

where the correlation length ξ is given by

$$\xi = \left[\log(\frac{1-z_1}{z_2(1+z_1)})\right]^{-1} \sim \frac{1}{T-T_c} \to \infty \tag{1.12}$$

as $T \to T_c$. For $T < T_c, C_M(0)$ is given in the form[5]

$$C_M(0) \simeq m_b^2 + \left(-A_-(T)\sqrt{M} + B(T) + \frac{C(T)}{\sqrt{M}}\right)\exp\left(-\frac{M}{\xi}\right), \tag{1.13}$$

using the renormalized evaluation method of singular integrals[7], where $A_-(T) > 0$. The existence of the negative sign in front of $A_-(T)$ in (1.13) shows the non-monotonic behaviour of $C_M(0)$ with respect to M even in the ferromagnetic case ($J_1 > 0$ and $J_2 > 0$). This is an unexpeted result.

This happens only for the boundary-boundary correlation functions because of the cut-off effect of long-range correlation contribution from the region wider than the system size M compared to the infinite system. At $T = T_c$, we have

$$C_M(0) \sim \frac{A(T_c)}{M}; A(T_c) = \left(\frac{(1+z_1)^2}{4z_1} \right)_{T=T_c}. \qquad (1.14)$$

The order of the two limits $N \to \infty$ (first) and $M \to \infty$ is vital in order to obtain the boundary spontaneous magnetization m_b in (1.13).

1.2. *Applications to Quantum Spin Systems*

The above topological interaction method can be also applied to quantum spin chains such as the XY-model and transverse Ising model[8]. Our Hamiltonian is given by

$$\mathcal{H} = -\sum_{i=1}^{N-1} \left(J_x \sigma_i^x \sigma_{i+1}^x + J_y \sigma_i^y \sigma_{i+1}^y \right) - \mu_B H \sum_{i=1}^{N} \sigma_i^z. \qquad (1.15)$$

We introduce the dimensionless parameters γ and η through the relations $J_{x,y} = J(1 \pm \gamma)$ and $\eta = \mu_B H/J$. The correlation function $C_N^{xx} \equiv \langle \sigma_1^x \sigma_N^x \rangle_{T=0}$ is shown to approach the following limit

$$m_{b,x}^2 = \frac{4\gamma}{(1+\gamma)^2}(1-\eta^2) \qquad (1.16)$$

for $\eta \leq 1$ at $T = 0$. When $\gamma = 1$, we have $m_{b,x}^2 = 1 - \eta^2$. This agrees with the result by Barouch and McCoy[9], Pfeuty[10], Suzuki[11] and Peschel[12].

1.3. *Symmetry Breaking by Local Fields*

It is also interesting to study the following situation in which an external field is applied to a finite (local) region Ω, and to ask what happens, namely to evaluate the total magnetization[13]. We have derived the result that the total magnetization $M(T, H)$ is given in the form

$$M(T,H) = N\, m_s^2\, \mathcal{F}_\Omega(T,H) \qquad (1.17)$$

with the uniform spontaneous magnetization m_s for the large total number of spins, N, where

$$\mathcal{F}_\Omega(T,H) = \frac{< \sinh(h\, M_\Omega) >_+}{< \cosh(h\, M_\Omega) >_+} \times \frac{1}{m_s}; M_\Omega = \sum_{j \epsilon \Omega} S_j \qquad (1.18)$$

for $S_j = \pm 1$ and $h = \beta\mu_B H$ with the uniform spontaneous magnetization m_s per spin. Here, $< \cdots >_+$ denotes the average over the state Ψ_+ which is a symmetry-broken state obtained in the limit $H \to +0$ after taking the thermodynamic limit $N \to \infty$. The above factor $\mathcal{F}_\Omega(T, H)$ in (1.17) is shown[5] to be finite and non-vanishing even at T_c for $H \neq 0$, because the average $< \sinh(hM_\Omega) >_+$ is proportional to m_s near the critical point for a finite domain Ω. In particular, when Ω is given by a single site j, we have [13,14] $< \sinh(hM_\Omega) >_+ = (\sinh h)m_s$ and $< \cosh(hM_\Omega) >_+ = \cosh h$. Therefore, we arrive at[13] $\mathcal{F}_j(T, H) = \tanh(\beta\mu_B H)$. This yeilds the formula

$$M(T, H) = Nm_s^2 \tanh(\beta\mu_B H). \tag{1.19}$$

For $H \to \infty$, we have $M(T, \infty) = Nm_s^2$, which agrees with C. N. Yang's result[1]. The above general formula (1.17) is also useful in evaluating the critical exponent of the uniform spontaneous magnetization numerically (for example, using Monte Carlo simulations) in general dimensions[15].

2. Part II : Quantum Analysis, q-Derivative and Exponential Splitting

2.1. *Quantum Analysis and Exponential Splitting*

We discuss here the quantum derivative[16–21] of an operator function $f(A)$ with respect to the operator A itself. Our quantum analysis is based on the differential $df(A)$, which depends on its definition. A typical one is given by the following Gâteau differential

$$df(A) = \lim_{h \to 0} \frac{f(A + hdA) - f(A)}{h} \tag{2.1}$$

Another one is given by the commutator

$$df(A) = [H, f(A)] \tag{2.2}$$

for a certain fixed operator H. These differentials both satisfy the Leibniz rule,

$$d(f(A)g(A)) = (df(A))g(A) + f(A)dg(A). \tag{2.3}$$

Then, it is easily shown[21] that using the inner derivation δ_A defined by $\delta_A Q = [A, Q] = AQ - QA$, we have

$$\delta_A df(A) = \delta_{f(A)} dA \tag{2.4}$$

because $d(Af(A)) = d(f(A)A)$, namely

$$(dA)f(A) + Adf(A) = (df(A))A + f(A)dA. \tag{2.5}$$

The above relation (2.4) can be written formally as

$$df(A) = \frac{\delta_{f(A)}}{\delta_A} dA. \tag{2.6}$$

In fact, the ratio of the two hyperoperators $\delta_{f(A)}$ and δ_A always exists in the space of hyperoperators $L_A(\equiv A\,\times)$ and δ_A. Thus, we defined[16–21] the quantum derivative by

$$\frac{df(A)}{dA} = \frac{\delta_{f(A)}}{\delta_A} = \frac{f(A) - f(A - \delta_A)}{\delta_A}, \tag{2.7}$$

which is a function of A and δ_A in our quantum analysis. These hyperoperators A and δ_A commute with each other and consequently this function of A and δ_A can be easily treated in analytical calculations. Quantum corrections can be also easily obtained in our formulation, as shown below. This is one of the merits of our quantum analysis[16–21] compared to the other formulation[22] based on Feyman's indices[23]. In fact, it is easily shown that

$$\frac{df(A)}{dA} = \int_0^1 f^{(1)}(A - t\delta_A)dt, \tag{2.8}$$

where $f^{(n)}(x)$ denotes the n-th derivative of $f(x)$.

The n-th order quantum derivative $d^n f(A)/dA^n$ is similarly expressed by

$$\frac{d^n f(A)}{dA^n} = n! \int_0^1 dt_1 \int_0^{t_1} dt_2 \cdots \int_0^{t_{n-1}} dt_n f^{(n)}(A - \sum_{j=1}^{n} t_j \delta_j), \tag{2.9}$$

where δ_j is a hyperoperator defined using δ_A as

$$\delta_j \,:\, B^n = \delta_j \,:\, B \cdots\cdots B = B^{j-1}(\delta_A B)B^{n-j}. \tag{2.10}$$

The following general operator Taylor expansion formula holds[16–21]

$$f(A + xB) = \sum_{n=0}^{\infty} \frac{x^n}{n!} \frac{d^n f(A)}{dA^n} \,:\, B^n = f(A)$$

$$+ \sum_{n=1}^{\infty} x^n \int_0^1 dt_1 \int_0^{t_1} dt_2 \cdots \int_0^{t_{n-1}} dt_n f^{(n)}(A - \sum_{j=1}^{n} t_j \delta_j) \,:\, B^n. \tag{2.11}$$

From this general formula, we can easily obtain[16] the well known Feynman expansion formula on $e^{t(A+xB)}$ as

$$e^{t(A+xB)} = \sum_{n=0}^{\infty} \frac{x^n}{n!} \frac{d^n e^{tA}}{dA^n} : B^n$$

$$= e^{tA} \sum_{n=0}^{\infty} x^n \int_0^t dt_1 \int_0^{t_1} dt_2 \cdots \int_0^{t_{n-1}} dt_n B(t_1)B(t_2)\cdots B(t_n), \quad (2.12)$$

where $B(t)$ is defined by $B(t) = e^{-t\delta_A} B = e^{-tA} B e^{tA}$.

There are many other applications of the present formulation (namely quantum analysis) in physics[16–21], and also in the derivation of higher-order exponential splitting formulas[24–40] such as

$$e^{x(A+B)} = e^{t_1 xA} e^{t_2 xB} e^{t_3 xA} e^{t_4 xB} \cdots e^{t_M xA} + O(x^{m+1}). \quad (2.13)$$

The splitting parameters $\{t_j\}$ are obtained using the above quantum analysis[16–21], or using the recursive scheme proposed by the present author[16].

2.2. *An Integral Representation of q-Derivative*

The q-derivative D_q is defined by[41]

$$D_q f(x) = \frac{f(qx) - f(x)}{(q-1)x} \equiv \frac{d_q f(x)}{d_q x} \quad (2.14)$$

for an ordinary function $f(x)$. This derivative is related to Euler's identities, the Jacobic identity, and the Ramanujan formula[41]. Clearly we have $D_{q \to 1} f(x) = f^{(1)}(x)$, when $f(x)$ is analytic. It is easy to show that

$$D_q^2 f(x) = \frac{f(q^2 x) - (q+1) f(qx) + q f(x)}{q(q-1)^2 x^2} \quad (2.15)$$

The formal similarity between (2.7) and (2.14) yields, in general, the following integral representation of $D_q^n f(x)$:

$$D_q^n f(x) = [n]_q! \int_0^1 dt_1 \int_0^{t_1} dt_2 \cdots \int_0^{t_{n-1}} dt_n f^{(n)}\Big(\{1 + (q-1) \sum_{j=1}^{n} t_j \, q^{j-1}\} x \Big), \quad (2.16)$$

where $[n]_q = 1 + q + q^2 + \cdots + q^{n-1}$, and $[n]_q! = [1]_q \times [2]_q \times \cdots \times [n]_q$. For example, we have

$$D_q f(x) = \int_0^1 f^{(1)}\Big((1 + (q-1)t)x \Big) dt. \quad (2.17)$$

It is interesting to remark that the quantum derivative and q-derivative have similar integral representations, while the former is useful for non-commutative operator functions and the latter is defined for ordinary functions. The quantum analysis is also useful in evaluating the commutator $[f(A), g(B)]$, which includes the commutators $[e^{xA}, e^{yB}]$. In fact, we have

$$[f(A),\, g(B)] = \frac{df(A)}{dA}\frac{dg(B)}{dB}\,[A, B\,]. \tag{2.18}$$

This is useful in Kubo's linear response theory[42] in the form

$$\begin{aligned}[e^{-\beta\mathcal{H}}, A] &= e^{-\beta\mathcal{H}}\int_0^\beta e^{\lambda\mathcal{H}}[A,\mathcal{H}]e^{-\lambda\mathcal{H}}d\lambda \\ &\equiv \frac{de^{-\beta\mathcal{H}}}{d\mathcal{H}}[\mathcal{H}, A] = -i\hbar\frac{de^{-\beta\mathcal{H}}}{d\mathcal{H}}\dot{A}. \end{aligned} \tag{2.20}$$

For some applications, modifications and extensions of the quantum analysis proposed by the present author, see References 43-48.

References

1. C. N. Yang, *Phys. Rev.* **85**, 808 (1952).
2. B. M. McCoy and T. T. Wu, *Phys. Rev.* **162**, 436 (1967).
3. C. N. Yang, *Rev. Mod. Phys.* **34**, 694 (1962).
4. M. Suzuki, *Prog. Theor. Phys.* **113**, 1391 (2005).
5. M. Suzuki, H. Suzuki, and S-C.Chang, *J. Math. Phys.* **46** 033301 (2005). For the duality of boundary-boundary correlations, see F. Y. Wu, *Phys. Lett.* **A223**, 43 (1997) and W. T. Lu and F. Y. Wu, *J. Phys. A: Math. Gen.* **31**, 2823 (1998).
6. H. Suzuki and M. Suzuki, *Physica* **A353**, 309 (2005).
7. M. Suzuki, in preparation.
8. A. Sugiyama, H. Suzuki and M. Suzuki, *Physica* **A353** 271 (2005).
9. E. Barouch and B. M. McCoy, *Phys. Rev.* **A3** 786 (1971).
10. P. Pfeuty, *Ann. Phys.* **57**, 79 (1970).
11. M. Suzuki, *Prog. Theor. Phys.* **46**, 1337 (1971).
12. I. Peschel, *Phys. Rev.* **B30**, 6783 (1984). See also H. W. Blöte, H. J. Hilhorst, *Phys. Rev. Lett.* **51**, 2015 (1983), and references cited therein.
13. M. Suzuki and H. Suzuki, *J. Phys. Soc. Jpn.* **73**, 3299 (2004).
14. M. Suzuki, *J. Phys. Soc. Jpn.* **69**, Suppl. A, p.156 (2000) .
15. M. Suzuki and M. Harabayashi, in preparation.
16. M. Suzuki, *Commun. Math. Phys.* **183**, 339 (1997).
17. M. Suzuki, *Int. J. Mod. Phys. B* **10**, 1637 (1996).
18. M. Suzuki, *J. Math. Phys.* **38**, 1183 (1997).
19. M. Suzuki, *Phys. Lett. A* **224**, 337 (1997).
20. M. Suzuki, *Prog. Theor. Phys.* **100**, 475 (1998).

21. M. Suzuki, *Rev. Math. Phys.* **11** 243(1999).
22. V. E. Nazaikinskii, V. E. Shatalov and B. Yu. Sternin, *Methods of Noncommutative Analysis,* Walter de Gruter (1996).
23. R. P. Feynman, *Phys. Rev.* **84**, No.2 108 (1951).
24. M. Suzuki, *Phys. Lett. A* **146**, 319 (1990).
25. M. Suzuki, *J. Math. Phys.* **32**, 400 (1991).
26. M. Suzuki, *Phys. Lett. A* **165**, 387 (1992).
27. M. Suzuki, *J. Phys. Soc. Jpn.* **61**, 3015 (1992).
28. M. Suzuki and K. Umeno, in *Computer Simulation Studies in Condensed-Matter Physics VI,* ed. D. P. Landau, K. K. Mon, H. B. Schüttler (Berlin:Springer) 74 (1993).
29. H. Kobayashi, N. Hatano and M. Suzuki, *Physica A* **211**, 234 (1994).
30. M Suzuki, *Proc. Japan Acad.* Ser. B **69**, 161 (1993).
31. Z. Tsuboi and M. Suzuki, *Int. J. Mod. Phys B* **9**, 3241 (1995).
32. M. Suzuki, *Commun. Math. Phys.* **163**, 491 (1994).
33. K. Aomoto, *J. Math. Soc. Jpn.* **48**, 493 (1996).
34. M. Suzuki, *Physica* **A 321**, 334 (2003).
35. E. Hairer, C. Lubich and G. Wanner, *Geometric Numerical Integration* (Berlin: Springer) 39 (2002).
36. N. Hatano and M. Suzuki, *Prog. Theor. Phys.* **85**, 481 (1991).
37. N. Hatano and M. Suzuki, in *Quantum Annealing and related Optimization Methods,* Lecture Notes Phys. **679** edited by A. Das and B. K. Chakrabarti, Springer (2005), and references cited therein.
38. R. I. McLachlan, *BIT* **35**, 258 (1995).
39. R. I. McLachlan and G. R. W. Quispel, *Acta Numerica* **11**,241 (2002).
40. S. A. Chin and C. R. Chin, *J. Chem. Phys.* **117**, 1409 (2002) and references cited therein.
41. V. Kac and P. Cheung, *Quantum Calculus*, Springer (2000).
42. R. Kubo, *J. Phys. Soc. Jpn.* **12**, 570 (1957).
43. R. Bhatia, D. Singh and K. B. Sinha, *Commun. Math. Phys.* **191**, 603 (1998).
44. R. Bhatia and K. B. Sinha, *Lin. Algeb. and its Appl.* **303**, 231 (1999).
45. R. Bhatia and J. A. T. da Silva, *Lin. Algeb. and its Appl.* **341**, 391 (2002).
46. H. Hasegawa, *Infinite Dimensional Analysis, Quantum Probability and Related Topics* **6 (3)**, 413 (2003).
47. N. Hatano, *J. Phys. Soc. Jpn.* **74**, 3093 (2005).
48. W. A. Majewski and M. Marciniak, *On Quantum Lyapunov Exponent*, quant-phys/0510224.

Löwner Equations and Dispersionless Hierarchies

Kanehisa Takasaki*

Graduate School of Human and Environmental Sciences,
Kyoto University,
Kyoto 606-8502, Japan
E-mail: takasaki@math.h.kyoto-u.ac.jp

Takashi Takebe†

Department of Mathematics, Ochanomizu University
Otsuka 2-1-1, Bunkyo-ku
Tokyo, 112-8610, Japan
E-mail: takebe@math.ocha.ac.jp

Reduction of a dispersionless type integrable system (dcmKP hierarchy) to the radial Löwner equation is presented.

1. Introduction

Recently reductions and hodograph solutions of dispersionless/Whitham type integrable systems are intensively studied[1–4]. In this article we report another example; reduction of the dispersionless coupled modified KP (dcmKP) hierarchy to the (radial) Löwner equation.

The dcmKP hierarchy introduced by Teo[5] is an extension of the dispersionless mKP hierarchy[6] with an additional degree of freedom, or in other words, a "half" of the dispersionless Toda lattice hierarchy[7,8].

The Löwner equation was introduced by K. Löwner[9] in an attempt to solve the Bieberbach conjecture. It is an evolution equation of the conformal mapping from (a chain of) subdomains of the unit disk onto the unit disk. We can also define the same kind of equation with different normalization

*Work partially supported by Grant-in-Aid for Scientific Research by Japan Society for the Promotion of Science, No. 16340040.
†Work partially supported by Grant-in-Aid for Scientific Research by Japan Society for the Promotion of Science, No. 15540014.

which is called the "*chordal* Löwner equation". See Lawler, Schramm and Werner[10] §2.3 for details. The original Löwner equation is, therefore, often called the "*radial* Löwner equation".

The reduction of the dispersionless KP hierarchy[11,12,8] to the chordal Löwner equation (and its generalization) has been studied by Gibbons and Tsarev[1], Yu and Gibbons[2], Mañas, Martínez Alonso and Medina[3] and others. Our question is: how about the radial Löwner equation? The answer is that there appears another degree of freedom and the resulting system turns out to be the dcmKP hierarchy.

In the following two sections we review the two ingredients, the Löwner equation and the dcmKP hierarchy. The main result is presented in the last section. Details including proofs will be published in the forthcoming paper.

2. Radial Löwner equation

In this section we review the (radial) Löwner equation and introduce related notions. Since we are interested in algebro-analytic nature of the system, we omit reality/positivity conditions which are essential in the context of the complex analysis.

The Löwner equation is a system of differential equations for a function

$$w = g(\boldsymbol{\lambda}, z) = e^{-\phi(\boldsymbol{\lambda})} z + b_0(\boldsymbol{\lambda}) + b_1(\boldsymbol{\lambda}) z^{-1} + b_2(\boldsymbol{\lambda}) z^{-2} + \cdots \tag{2.1}$$

where $\boldsymbol{\lambda} = (\lambda_1, \ldots, \lambda_N)$ and z are independent variables. In the complex analysis the variable z moves in a subdomain of the compliment of the unit disk and the variables λ_i parametrize the subdomain. In our context $g(\boldsymbol{\lambda}, z)$ is considered as a generating function of the unknown functions $\phi(\boldsymbol{\lambda})$ and $b_n(\boldsymbol{\lambda})$. We assume that for each $i = 1, \ldots, N$ a *driving function* $\kappa_i(\boldsymbol{\lambda})$ is given. The *Löwner equation* is the following system:

$$\frac{\partial g}{\partial \lambda_i}(\boldsymbol{\lambda}; z) = g(\boldsymbol{\lambda}; z) \frac{\kappa_i(\boldsymbol{\lambda}) + g(\boldsymbol{\lambda}; z)}{\kappa_i(\boldsymbol{\lambda}) - g(\boldsymbol{\lambda}; z)} \frac{\partial \phi(\boldsymbol{\lambda})}{\partial \lambda_i}, \qquad i = 1, \ldots, N. \tag{2.2}$$

(The original Löwner equation[9] is the case $N = 1$.)

Later the inverse function of $g(\boldsymbol{\lambda}, z)$ with respect to the z-variable will be more important than g itself. We denote it by $f(\boldsymbol{\lambda}, w)$:

$$z = f(\boldsymbol{\lambda}, w) = e^{\phi(\boldsymbol{\lambda})} w + c_0(\boldsymbol{\lambda}) + c_1(\boldsymbol{\lambda}) w^{-1} + c_2(\boldsymbol{\lambda}) w^{-2} + \cdots. \tag{2.3}$$

It satisfies $g(\boldsymbol{\lambda}, f(\boldsymbol{\lambda}, w)) = w$ and $f(\boldsymbol{\lambda}, g(\boldsymbol{\lambda}, z)) = z$, from which we can determine the coefficients $c_n(\boldsymbol{\lambda})$'s in terms of $\phi(\boldsymbol{\lambda})$ and $b_n(\boldsymbol{\lambda})$'s. The Löwner equation (2.2) is rewritten as the equation for $f(\boldsymbol{\lambda}, w)$ as follows:

$$\frac{\partial f}{\partial \lambda_i}(\boldsymbol{\lambda}; w) = w \frac{w + \kappa_i(\boldsymbol{\lambda})}{w - \kappa_i(\boldsymbol{\lambda})} \frac{\partial \phi(\boldsymbol{\lambda})}{\partial \lambda_i} \frac{\partial f}{\partial w}(\boldsymbol{\lambda}; w). \tag{2.4}$$

The compatibility condition for the system (2.2) or (2.4) is:

$$\frac{\partial \kappa_j}{\partial \lambda_i} = -\kappa_j \frac{\kappa_j + \kappa_i}{\kappa_j - \kappa_i} \frac{\partial \phi}{\partial \lambda_i}, \tag{2.5}$$

$$\frac{\partial^2 \phi}{\partial \lambda_i \partial \lambda_j} = \frac{4\kappa_i \kappa_j}{(\kappa_i - \kappa_j)^2} \frac{\partial \phi}{\partial \lambda_i} \frac{\partial \phi}{\partial \lambda_j}, \tag{2.6}$$

for any i, j $(i \neq j)$.

The *Faber polynomials* are defined as follows[13]:

$$\Phi_n(\boldsymbol{\lambda}, w) := (f(\boldsymbol{\lambda}, w)^n)_{\geq 0}. \tag{2.7}$$

Here $(\cdot)_{\geq 0}$ is the truncation of the Laurent series in w to its polynomial part.

3. dcmKP hierarchy

We give a formulation of the dcmKP hierarchy different from Teo[5]. The equivalence (up to a gauge factor) will be explained in a forthcoming paper.

The independent variables of the system is (s, x, t) where $t = (t_1, t_2, \dots)$ is a series of infinitely many variables. The variables x and t_1 appear in the equations only as the combination $x + t_1$, so we often omit x. Namely, "t_1" should be understood as the abbreviation of $x + t_1$. The unknown functions $\phi(s, t)$ and $u_n(s, t)$ $(n = 0, 1, 2, \dots)$ are encapsulated in the series

$$\mathcal{L}(s, t; w) = e^{\phi(s,t)} w + u_0(s, t) + u_1(s, t) w^{-1} + u_2(s, t) w^{-2} + \cdots, \tag{3.1}$$

where w is a formal variable. The *dispersionless coupled modified KP hierarchy* (dcmKP hierarchy) is the following system of differential equations:

$$\frac{\partial \mathcal{L}}{\partial t_n} = \{\mathcal{B}_n, \mathcal{L}\}, \qquad n = 1, 2, \dots. \tag{3.2}$$

Here the Poisson bracket $\{,\}$ is defined by

$$\{f(s, x), g(s, x)\} := w \frac{\partial f}{\partial w} \frac{\partial g}{\partial s} - w \frac{\partial f}{\partial s} \frac{\partial g}{\partial w}, \tag{3.3}$$

and $\mathcal{B}_n$ is the polynomial in w defined by

$$\mathcal{B}_n := (\mathcal{L}^n)_{>0} + \frac{1}{2} (\mathcal{L}^n)_0, \tag{3.4}$$

where $(\cdot)_{>0}$ is the positive power part in w and $(\cdot)_0$ is the constant term with respect to w.

It is easy to construct a theory for this system similar to those for the dispersionless KP hierarchy or the dispersionless Toda hierarchy[8].

4. Main results

In this section we show that a specialization of the variables $\boldsymbol{\lambda}$ in $f(\boldsymbol{\lambda}, w)$ gives a solution of the dcmKP hierarchy.

Suppose $\boldsymbol{\lambda}(s,t) = (\lambda_1(s,t), \ldots, \lambda_N(s,t))$ satisfies the equations

$$\frac{\partial \lambda_i}{\partial t_n} = v_i^n(\boldsymbol{\lambda}(s,t)) \frac{\partial \lambda_i}{\partial s}, \tag{4.1}$$

where $v_j^n(\boldsymbol{\lambda})$ are defined by

$$v_j^n(\boldsymbol{\lambda}) := \kappa_j(\boldsymbol{\lambda}) \frac{\partial \Phi_n}{\partial w}(\boldsymbol{\lambda}, \kappa_j(\boldsymbol{\lambda})) = \left. \frac{\partial \Phi_n}{\partial \log w}(\boldsymbol{\lambda}, w) \right|_{w=\kappa_j(\boldsymbol{\lambda})}. \tag{4.2}$$

They satisfy the equations

$$\frac{\partial v_j^n}{\partial \lambda_i} = V_{ij}(v_i^n - v_j^n), \tag{4.3}$$

where

$$V_{ij} := \frac{2\kappa_i \kappa_j}{(\kappa_i - \kappa_j)^2} \frac{\partial \phi}{\partial \lambda_i}. \tag{4.4}$$

The hydrodynamic type equations (4.1) can be solved by the generalized hodograph method of Tsarev[14]: Let $F_i(\boldsymbol{\lambda})$ be functions satisfying

$$\frac{\partial F_j}{\partial \lambda_i} = V_{ij}(F_i - F_j). \tag{4.5}$$

Then the *hodograph relation*

$$F_i(\boldsymbol{\lambda}(s,t)) = s + \sum_{n=1}^{\infty} v_i^n(\boldsymbol{\lambda}(s,t))\, t_n \tag{4.6}$$

determines the solution of (4.1), $\boldsymbol{\lambda}(s,t)$, as the implicit function.

Our main result is as follows: let $f(\boldsymbol{\lambda}, w)$ be a solution of the radial Löwner equation (2.4) of the form (2.3) and $\boldsymbol{\lambda}(s,t)$ be a solution of (4.1). Then the function $\mathcal{L} = \mathcal{L}(s,t;w)$ defined by

$$\begin{aligned} \mathcal{L}(s,t;;w) &:= f(\boldsymbol{\lambda}(s,t), w) \\ &= e^{\phi(\boldsymbol{\lambda}(s,t))} w + c_0(\boldsymbol{\lambda}(s,t)) + c_1(\boldsymbol{\lambda}(s,t)) w^{-1} + c_2(\boldsymbol{\lambda}(s,t)) w^{-2} + \cdots \end{aligned} \tag{4.7}$$

is a solution of the dcmKP hierarchy (3.2).

In the proof we construct the S-function[7,8], following the method by Mañas, Martínez Alonso and Medina[3].

If we start from the chordal Löwner equation instead of the radial Löwner equation, we obtain a solution of the dispersionless KP hierarchy. This is due to Gibbons and Tsarev[1], Yu and Gibbons[2], Mañas, Martínez Alonso and Medina[3]. The generalization to the Whitham hierarchies is considered by Guil, Mañas and Martínez Alonso[4]. Note that their generalization does not contain the radial Löwner case, because of the normalization at the infinity.

References

1. J. Gibbons and S. P. Tsarev, *Phys. Let.* **258A**, 263–271 (1999).
2. L. Yu and J. Gibbons, *Inverse Problems* **16**, 605–618 (2000).
3. M. Mañas, L. Martínez Alonso and E. Medina, *J. Phys. A: Math. Gen.* **35**, 401–417 (2002).
4. F. Guil, M. Mañas and L. Martínez Alonso, *J. Phys. A: Math. Gen.* **36**, 4047–4062 (2003).
5. L.-P. Teo, On dispersionless coupled modified KP hierarchy, preprint `nlin.SI/0304007`.
6. T. Takebe, *Lett. Math. Phys.* **p59**, 157-172 (2002).
7. K. Takasaki and T. Takebe, *Lett. Math. Phys.* **23**, 205–214 (1991).
8. K. Takasaki and T. Takebe, *Rev. Math. Phys.* **7**, 743-803 (1995).
9. K. Löwner, *Math. Ann.* **89**, 103–121 (1923).
10. G. F. Lawler, O. Schramm and W. Werner, *Acta Math.* **187**, 237–273 (2001).
11. D. Lebedev and Yu. Manin, *Phys.Lett.* **74A**, 154–156 (1979); V. E. Zakharov, *Physica* **3D**, 193–202 (1981); Y. Kodama, *Phys. Lett.* **129A**, 223–226 (1988); Y. Kodama and J. Gibbons, *Phys. Lett.* **135A**, 167–170 (1989); Y. Kodama, *Phys. Lett.* **147A**, 477–482 (1990); I. M. Krichever, *Commun. Math. Phys.* **143**, 415–426 (1991).
12. K. Takasaki and T. Takebe, in: Proceedings of RIMS Research Project 1991 “Infinite Analysis,” *Int. J. Mod. Phys.* **A7, Suppl. 1B**, 889–922 (1992).
13. L.-P. Teo, *Lett. Math. Phys.* **64**, 75–92 (2003).
14. S. P. Tsarëv, (Russian) *Izv. Akad. Nauk SSSR Ser. Mat.* **54**, 1048–1068 (1990); translation in *Math. USSR-Izv.* **37**, 397–419 (1991).

Multiparameter Quantum Deformations of Jordanian Type for Lie Superalgebras *

V.N. Tolstoy

Institute of Nuclear Physics, Moscow State University,
119992 Moscow, RUSSIA
E-mail:tolstoy@nucl-th.sinp.msu.ru

We discuss quantum deformations of Jordanian type for Lie superalgebras. These deformations are described by twisting functions with support from Borel subalgebras and they are multiparameter in the general case. The total twists are presented in explicit form for the Lie superalgebras $\mathfrak{sl}(m|n)$ and $\mathfrak{osp}(1|2n)$. We show also that the classical r-matrix for a light-cone deformation of $D = 4$ super-Poincare algebra is of Jordanian type and a corresponding twist is given in explicit form.

1. Introduction

The Drinfeld's quantum group theory roughly includes two classes of Hopf algebras: quasitriangular and triangular. The (standard) q-deformation of simple Lie algebras belongs to the first class. The simplest example of the triangular (non-standard) deformation is the Jordanian deformation of $\mathfrak{sl}(2)$. In the case of simple Lie algebras of rank ≥ 2 some non-standard deformations were constructed by Kulish, Lyakhovsky et al.[1–4]. These deformations are described by twisting functions (which are extensions of the Jordanian twist) with support from Borel subalgebras, and they are multiparameter in the general case. We call their as the deformations of Jordanian type. Total twists of Jordanian type were constructed for all complex Lie algebras of the classical series A_n, B_n, C_n and D_n.

In this paper we discuss quantum deformations of Jordanian type for Lie superalgebras. The total twists are presented in explicit form for the Lie superalgebras $\mathfrak{sl}(m|n)$ and $\mathfrak{osp}(1|2n)$. We show also that the classical

*This work is supported by the grants RFBR-05-01-01086 and INTAS-OPEN-03-51-3350.

r-matrix for a light-cone deformation of $D=4$ super-Poincare algebra is of Jordanian type and a corresponding twist is given in explicit form.

2. Classical r-matrices of Jordanian type

Let $\mathfrak{g}$ be any finite-dimensional complex simple Lie superalgebra then $\mathfrak{g} = \mathfrak{n}_- \oplus \mathfrak{h} \oplus \mathfrak{n}_+$, where $\mathfrak{n}_\pm$ are maximal nilpotent subalgebras and $\mathfrak{h}$ is a Cartan subalgebra. The subalgebra $\mathfrak{n}_+$ ($\mathfrak{n}_-$) is generated by the positive (negative) root vectors e_β ($e_{-\beta}$) for all $\beta \in \Delta_+(\mathfrak{g})$. The symbol $\mathfrak{b}_+$ will denote the Borel subalgebra of $\mathfrak{g}$, $\mathfrak{b}_+ := \mathfrak{h} \oplus \mathfrak{n}_+$. Let θ be a maximal root of $\mathfrak{g}$, and let a Cartan element $h_\theta \in \mathfrak{h}$ and a root vector $e_\theta \in \mathfrak{n}_+$ satisfies the relation

$$[h_\theta,\, e_\theta] = e_\theta. \tag{2.1}$$

The elements h_θ and e_θ are homogeneous, i.e.

$$\deg(h_\theta) = 0, \quad \deg(e_\theta) = 0, \text{ or } 1\ . \tag{2.2}$$

Moreover, let homogeneous elements $e_{\pm i}$ indexed by the symbols i and $-i$, $(i = 1, 2, \ldots, N)$, satisfy the relations

$$\begin{aligned} &[h_\theta,\, e_{-i}] = \ t_i\, e_{-i}, \qquad [h_\theta,\, e_i] = (1-t_i)\, e_i \quad (t_i \in \mathbb{C}), \\ &[e_i,\, e_{-j}] = \delta_{ij}\, e_\theta, \quad [e_{\pm i},\, e_{\pm j}] = 0\ , \quad [e_{\pm i},\, e_\theta] = 0\ , \end{aligned} \tag{2.3}$$

provided that

$$\deg(e_\theta) = \deg(e_i) + \deg(e_{-i}) \pmod 2. \tag{2.4}$$

For the Lie superalgebra $\mathfrak{g}$ the brackets $[\cdot,\,\cdot]$ always denote the super-commutator:

$$[x, y] := xy - (-1)^{\deg(x)\deg(y)} yx \tag{2.5}$$

for any homogeneous elements x and y.

Consider the even skew-symmetric two-tensor

$$r_{\theta,N}(\xi) = \xi \Big(h_\theta \wedge e_\theta + \sum_{i=1}^{N} (-1)^{\deg(e_i)\,\deg(e_{-i})} e_i \wedge e_{-i} \Big) \tag{2.6}$$

where

$$\deg(\xi) = \deg(e_\theta) = \deg(e_i) + \deg(e_{-i}) \pmod 2, \tag{2.7}$$

and we assume that the operation " $\wedge$ " in (2.6) is graded:

$$e_i \wedge e_{-i} := e_i \otimes e_{-i} - (-1)^{\deg(e_i)\,\deg(e_{-i})} e_{-i} \otimes e_i. \tag{2.8}$$

It is not hard to check that the element (2.6) satisfies the classical Yang-Baxter equation (CYBE),

$$[r_{\theta,N}^{12}(\xi), r_{\theta,N}^{13}(\xi) + r_{\theta,N}^{23}(\xi)] + [r_{\theta,N}^{13}(\xi), r_{\theta,N}^{23}(\xi)] = 0 \ , \tag{2.9}$$

and it is called the extended Jordanian r-matrix of N-order. Let N be maximal order, i.e. we assume that another elements $e_{\pm j} \in \mathfrak{n}_+$, $j > N$, which satisfy the relations (2.3), do not exist. Such element (2.6) is called the extended Jordanian r-matrix of maximal order[5].

Consider a maximal subalgebra $\mathfrak{b}'_+ \in \mathfrak{b}_+$ which co-commutes with the maximal extended Jordanian r-matrix (2.6), $\mathfrak{b}'_+ := \operatorname{Ker}\delta \in \mathfrak{b}_+$:

$$\xi\delta(x) = [x \otimes 1 + 1 \otimes x,\, r_{\theta,N}(\xi)] = [\Delta(x),\, r_{\theta,N}(\xi)] = 0 \tag{2.10}$$

for $\forall x \in \mathfrak{b}'_+$. Let $r_{\theta_1,N_1}(\xi_1) \in \mathfrak{b}'_+ \otimes \mathfrak{b}'_+$ is also a extended Jordanian r-matrix of the form (2.5) with a maximal root $\theta_1 \in \mathfrak{h}'$ and maximal order N_1. Then the sum

$$r_{\theta,N;\,\theta_1,N_1}(\xi, \xi_1) := r_{\theta,N}(\xi) + r_{\theta_1,N_1}(\xi_1) \tag{2.11}$$

is also a classical r-matrix.

Further, we consider a maximal subalgebra $\mathfrak{b}''_+ \in \mathfrak{b}'_+$ which co-commutes with the maximal extended Jordanian r-matrix $r_{\theta_1,N_1}(\xi_1)$ and we construct a extended Jordanian r-matrix of maximal order, $r_{\theta_2,N_2}(\xi_2)$. Continuing this process as result we obtain a canonical chain of subalgebras

$$\mathfrak{b}_+ \supset \mathfrak{b}'_+ \supset \mathfrak{b}''_+ \cdots \supset \mathfrak{b}_+^{(k)} \tag{2.12}$$

and the resulting r-matrix

$$r_{\theta,N;\ldots;\theta_k,N_k}(\xi, \xi_1, \cdots, \xi_k) = r_{\theta,N}(\xi) + r_{\theta_1,N_1}(\xi_1) + \cdots + r_{\theta_k,N_k}(\xi_k). \tag{2.13}$$

If the chain (2.12) is maximal, i.e. it is constructed in corresponding with the maximal orders $N, N_1, \ldots N_k$, then the r-matrix (2.13) is called the maximal classical r-matrix of Jordanian type for the Lie superalgebra $\mathfrak{g}$.

3. Multiparameter twists of Jordanian type

The twisting two-tensor $F_{\theta,N}(\xi)$ corresponding to the r-matrix (2.6) has the form

$$F_{\theta,N}(\xi) = \mathcal{F}_N(\xi)F_J(\sigma_\theta), \tag{3.1}$$

where the two-tensor F_J is the Jordanian twist and $\mathcal{F}_N$ is extension of the Jordanian twist (see[5]). These two-tensors are given by the formulas

$$F_J(\sigma_\theta) = \exp(2h_\theta \otimes \sigma_\theta) \ , \tag{3.2}$$

$$\begin{aligned}\mathcal{F}_N(\xi) &= \Big(\prod_{i=1}^{N'} \exp\big(\xi(-1)^{\deg(e_i)\deg(e_{-i})} e_i \otimes e_{-i}\, e^{-2t_i\sigma_\theta}\big)\Big)\mathcal{F}_s(\sigma_\theta) \\ &= \exp\Big(\xi \sum_{i=1}^{N'} (-1)^{\deg(e_i)\deg(e_{-i})} e_i \otimes e_{-i}\, e^{-2t_i\sigma_\theta}\Big)\,\mathcal{F}_s(\sigma_\theta),\end{aligned} \tag{3.3}$$

where

$$\mathcal{F}_s(\sigma_\theta) = \Big(1 - \xi \frac{e_{\theta/2}}{e^{\sigma_\theta}+1} \otimes \frac{e_{\theta/2}}{e^{\sigma_\theta}+1}\Big)\sqrt{\frac{(e^{\sigma_\theta}+1)\otimes(e^{\sigma_\theta}+1)}{2(e^{\sigma_\theta}\otimes e^{\sigma_\theta}+1)}}\ , \tag{3.4}$$

if $\theta/2$ is a root, $e_{\theta/2}^2 = e_\theta$, $N' = N-1$, and

$$\mathcal{F}_s(\sigma_\theta) = 1\ , \tag{3.5}$$

if $\theta/2$ is not any root, $N' = N$. Moreover

$$\deg(\xi) = \deg(e_\theta) = \deg(e_i) + \deg(e_{-i}) \pmod 2\ , \tag{3.6}$$

$$\sigma_\theta := \frac{1}{2}\ln(1+\xi e_\theta). \tag{3.7}$$

It should be noted that if the root vector e_θ is odd then $\sigma_\theta = \frac{1}{2}\xi e_\theta$.

We can check that the twisting two-tensor (3.1) defined by the formulas (3.2)–(3.7) satisfies the cocycle equation

$$F^{12}(\Delta\otimes \mathrm{id})(F) = F^{23}(\mathrm{id}\otimes\Delta)(F) \tag{3.8}$$

and the "unital" normalization condition

$$(\epsilon\otimes\mathrm{id})(F) = (\mathrm{id}\otimes\epsilon)(F) = 1\ . \tag{3.9}$$

The twisted coproduct $\Delta_\xi(\,\cdot\,) := F_{\theta,N}(\xi)\Delta(\,\cdot\,)F^{-1}_{\theta,N}(\xi)$ and the corresponding antipode S_ξ for elements in (2.3) are given by the formulas

$$\Delta_\xi(e^{\pm\sigma_\theta}) = e^{\pm\sigma_\theta}\otimes e^{\pm\sigma_\theta},\quad \Delta_\xi(e_{\theta/2}) = e_{\theta/2}\otimes 1 + e^{\sigma_\theta}\otimes e_{\theta/2}\ , \tag{3.10}$$

$$\begin{aligned}\Delta_\xi(h_\theta) &= h_\theta\otimes e^{-2\sigma_\theta} + 1\otimes h_\theta + \frac{\xi}{4} e_{\theta/2}e^{-\sigma_\theta}\otimes e_{\theta/2}e^{-2\sigma_\theta} \\ &\quad -\xi\sum_{i=1}^{N'}(-1)^{\deg e_i \deg e_{-i}} e_i\otimes e_{-i}e^{-2(t_{\gamma_i}+1)\sigma_\theta},\end{aligned} \tag{3.11}$$

$$\Delta_\xi(e_i) = e_i\otimes e^{-2t_i\sigma_\theta} + 1\otimes e_i, \tag{3.12}$$

$$\Delta_\xi(e_{-i}) = e_{-i}\otimes e^{2t_i\sigma_\theta} + e^{2\sigma_\theta}\otimes e_{-i}, \tag{3.13}$$

$$S_\xi(e^{\pm\sigma_\theta}) = e^{\mp\sigma_\theta}, \qquad S_\xi(e_{\theta/2}) = -e_{\theta/2}e^{-\sigma_\theta}, \tag{3.14}$$

$$S_\xi(h_\theta) = -h_\theta e^{2\sigma_\theta} + \frac{1}{4}(e^{2\sigma_\theta} - 1) - \xi\sum_{i=1}^{N'}(-1)^{\deg(e_i)\deg(e_{-i})}e_i\, e_{-i}, \tag{3.15}$$

$$S_\xi(e_i) = -e_i e^{2t_i\sigma_\theta}, \quad S_\xi(e_{-i}) \;=\; -e_{-i}e^{-2(t_i+1)\sigma_\theta}. \tag{3.16}$$

If $\theta/2$ is not any root, the third term in (3.11) and the second term in (3.15) should be removed.

The twisted deformation of $U(\mathfrak{g})$ with the new coproduct $\Delta_\xi(\,\cdot\,)$ and the antipode S_ξ is denoted by $U_\xi(\mathfrak{g})$.

In order to construct the twist corresponding to the r-matrix (2.9) we can not apply the second twist $F_{\theta_1,N_1}(\xi_1)$ directly in the form (3.1)–(3.4) to the twisted superalgebra $U_\xi(\mathfrak{g})$ because the deformed coproduct for the elements of subalgebra $\mathfrak{b}'_+$ can be not trivial, i.e.

$$\Delta_\xi(x) = x\otimes 1 + 1\otimes x + \text{something}\,, \quad x\in\mathfrak{b}'_+. \tag{3.17}$$

However, there exists a similarity automorphism w_ξ which trivializes (makes trivial) the twisted coproduct $\Delta_\xi(\,\cdot\,)$ for elements of the subalgebra $\mathfrak{b}'_+$, i.e.

$$\Delta_\xi(w_\xi x w_\xi^{-1}) := w_\xi x w_\xi^{-1}\otimes 1 + 1\otimes w_\xi x w_\xi^{-1}, \quad x\in\mathfrak{b}'_+. \tag{3.18}$$

The automorphism w_ξ is connected with the Hopf "folding" of the two-tensor (3.3) and it is given by the following formula (see[5]):

$$w_\xi = \exp\Bigl(\frac{-\xi\sigma_\theta}{e^{2\sigma_\theta}-1}\sum_{i=1}^{N'}(-1)^{\deg(e_i)\deg(e_i)}e_i e_{-i}\Bigr)w_s, \tag{3.19}$$

where $w_s = \exp(\frac{1}{4}\sigma_\theta)$ if $\theta/2$ is a root, and $w_s = 1$ if $\theta/2$ is not any root.

With the help of the automorphism w_ξ the total twist chain corresponding to the r-matrix (2.11) can be presented as follows

$$F_{\theta,N;\theta_1,N_1}(\xi,\xi_1) = F_{\theta_1,N_1}(\xi;\xi_1)F_{\theta,N}(\xi)\ , \tag{3.20}$$

where

$$F_{\theta_1,N_1}(\xi;\xi_1) := (w_\xi\otimes w_\xi)F_{\theta_1,N_1}(\xi_1)(w_\xi^{-1}\otimes w_\xi^{-1}). \tag{3.21}$$

Here the two-tensors $F_{\theta,N}(\xi)$ and $F_{\theta_1,N_1}(\xi_1)$ are given by the formulas of type (3.1)–(3.5).

Iterating the formula (3.21) we obtain the total twist corresponding to the r-matrix (2.13):

$$\begin{aligned} F_{\theta,N;\theta_1,N_1;\ldots;\theta_k,N_k}(\xi,\xi_1,\ldots,\xi_k) &= F_{\theta_k,N_k}(\xi,\xi_1,\ldots,\xi_{k-1};\xi_k)\cdots \\ &\quad\times F_{\theta_2,N_2}(\xi,\xi_1;\xi_2)F_{\theta_1,N_1}(\xi;\xi_1)F_{\theta,N}(\xi)\ , \end{aligned} \tag{3.22}$$

where $(i = 1, \ldots, k)$

$$F_{\theta_i,N_i}(\xi,\xi_1,\ldots,\xi_{i-1};\xi_i) := (w_{\xi_{i-1}}\otimes w_{\xi_{i-1}})\cdots(w_{\xi_1}\otimes w_{\xi_1})(w_{\xi}\otimes w_{\xi})$$
$$\times F_{\theta_i,N_i}(\xi_i)(w_{\xi}^{-1}\otimes w_{\xi}^{-1})(w_{\xi_1}^{-1}\otimes w_{\xi_1}^{-1})\cdots(w_{\xi_{i-1}}^{-1}\otimes w_{\xi_{i-1}}^{-1}). \tag{3.23}$$

Now we consider specifically the multiparameter twists for the classical superalgebras $\mathfrak{gl}(m|n)$ and $\mathfrak{osp}(1|2n)$.

4. Quantum deformation of Jordanian type for $\mathfrak{gl}(m|n)$

Let e_{ij} $(i,j = 1,2,\ldots,m+n)$ be standard $(n+m)\times(n+m)$-matrices, where $(e_{ij})_{kl} = \delta_{ik}\delta_{jl}$. For such matrices we define a supercommutator as follows

$$[e_{ij},\, e_{kl}] := e_{ij}e_{kl} - (-1)^{\deg(e_{ij})\deg(e_{kl})}e_{kl}e_{ij}, \tag{4.1}$$

where $\deg(e_{ij}) = 0$ for $i,j \le n$ or $i,j > n$, and $\deg(e_{ij}) = 0$ in another cases. It is easy to check that

$$[e_{ij},\, e_{kl}] = \delta_{jk}e_{il} - (-1)^{\deg(e_{ij})\deg(e_{kl})}\delta_{il}e_{kj}. \tag{4.2}$$

The elements e_{ij} $(i,j = 1,2,\ldots,N := m+n)$ with the relations (4.2) are generated the Lie superalgebra $\mathfrak{gl}(m|n)$.

The maximal r-matrix of Jordanian type for the Lie superalgebra $\mathfrak{gl}(m|n)$ has the form [5]

$$r_{1,\ldots,[N/2]}(\xi_1,\ldots,\xi_{[N/2]}) = r_1(\xi_1) + \cdots + r_{[N/2]}(\xi_{[N/2]}), \tag{4.3}$$

where $(i = 1,2,\ldots,[N/2])$

$$r_i(\xi_i) = \xi_i\Big(\frac{1}{2}(e_{ii} - e_{N+1-i,N+1-i})\wedge e_{i,N+1-i} + \sum_{k=i+1}^{N-i}(-1)^{\deg(e_{ik})\deg(e_{kN+1-i})}e_{ik}\wedge e_{kN+1-i}\Big). \tag{4.4}$$

Consider the first twist corresponding to the r-matrix $r_1(\xi_1)$

$$F_{1,N-2}(\xi_1) \;=\; \mathcal{F}_{N-2}(\xi_1)F_J(\sigma_1), \tag{4.5}$$

where

$$F_J(\sigma_1) = e^{(e_{11}-e_{NN})\otimes\sigma_1}, \tag{4.6}$$

$$\mathcal{F}_{N-2}(\xi_1) = \exp\Big(\xi_1\sum_{k=2}^{N-1}(-1)^{\deg(e_{1k})\deg(e_{kN})}e_{1k}\otimes e_{kN}\, e^{-2\sigma_1}\Big), \tag{4.7}$$

$$\sigma_1 := \frac{1}{2}\ln(1+\xi_1 e_{1N}). \tag{4.8}$$

The corresponding automorphism w_ξ is connected with the Hopf "folding" of the two-tensor (4.7) and is given as follows

$$w_{\xi_1} = \exp\Bigl(\frac{-\xi_1\sigma_1}{e^{2\sigma_1}-1}\sum_{k=2}^{N-1}(-1)^{\deg(e_{1k})\deg(e_{kN})}e_{1k}e_{kN}\Bigr). \tag{4.9}$$

It is easy to see that

$$w_{\xi_1}e_{ij}w_{\xi_1}^{-1} = e_{ij} \tag{4.10}$$

for all i, j satisfying the condition $2 \le i,\, j \le N-2$, therefore (see the formula (3.18)) deformed coproducts $\Delta_{\xi_1}(\,\cdot\,) := F_{1,N-2}(\xi_1)\Delta(\,\cdot\,)F_{1,N-2}^{-1}(\xi_1)$ for these elements are trivial:

$$\Delta_{\xi_1}(e_{ij}) = e_{ij}\otimes 1 + 1\otimes e_{ij}, \quad 2\le i,\, j\le N-1. \tag{4.11}$$

This means that the automorphism w_{ξ_1} in the formula (3.21) for the case $\mathfrak{gl}(m|n)$ acts trivially and therefore the total twist corresponding to the r-matrix (4.3) is given as follows

$$\begin{aligned} F_{1,N-2;2,N-4;\ldots;k,N-2k}(\xi_1,\xi_2,\ldots,\xi_k) &= F_{k,N-2k}(\xi_k)\cdots \\ &\quad\times F_{2,N-4}(\xi_2)F_{1,N-2}(\xi_1)\,, \end{aligned} \tag{4.12}$$

where $(i = 1,\ldots,[N/2])$

$$\begin{aligned} F_{i,N-2i}(\xi_i) &= \exp\Bigl(\xi_i\sum_{k=i+1}^{N-i}(-1)^{\deg(e_{ik})\deg(e_{kN-2i})}e_{ik}\otimes e_{kN-2i}\,e^{-2\sigma_i}\Bigr) \\ &\quad\times\exp\bigl((e_{ii}-e_{N-2i,N-2i})\otimes\sigma_i\bigr)\,. \end{aligned} \tag{4.13}$$

5. Quantum deformation of Jordanian type for $\mathfrak{osp}(1|2n)$

In order to obtain compact formulas describing the commutation relations for generators of the orthosymplectic superalgebra $C(n)\simeq\mathfrak{osp}(1|2n)$ we use embedding of this superalgebra in the general linear superalgebra $\mathfrak{gl}(1|2n)$. Let a_{ij} $(i,j = 0,\pm1,\pm2,\ldots,\pm n)$ be a standard basis of the superalgebra $\mathfrak{gl}(1|2n)$ (see the previous Section 4) with the standard supercommutation relations

$$[a_{ij},\, a_{kl}] = \delta_{jk}a_{il} - (-1)^{\deg(e_{ij})\deg(e_{kl})}\delta_{il}a_{kj}, \tag{5.1}$$

where $\deg(e_{ij}) = 1$ when one index i or j is equal to 0 and another takes any value $\pm1,\ldots,\pm n$; $\deg(e_{ij}) = 0$ in the remaining cases. The superalgebra

$\mathfrak{osp}(1|2n)$ is embedded in $\mathfrak{gl}(1|2n)$ as a linear envelope of the following generators:
(i) the even (boson) generators spanning the symplectic algebra $\mathfrak{sp}(2n)$:

$$e_{ij} := a_{i-j} + \mathrm{sign}(ij)\, a_{j-i} = \mathrm{sign}(ij)\, e_{ji} \quad (i,j = \pm 1, \pm 2, \ldots, \pm n)\ ; \quad (5.2)$$

(ii) the odd (fermion) generators extending $\mathfrak{sp}(2n)$ to $\mathfrak{osp}(1|2n)$:

$$e_{0i} := a_{0-i} + \mathrm{sign}(i)\, a_{i0} = \mathrm{sign}(i)\, e_{i0} \quad (i = \pm 1, \pm 2, \ldots, \pm n)\ . \qquad (5.3)$$

We also set $e_{00} = 0$ and introduce the sign function: $\mathrm{sign}\, x = 1$ if a real number $x \geq 0$ and $\mathrm{sign}\, x = -1$ if $x < 0$. One can check that the elements (5.2) and (5.3) satisfy the following relations:

$$[e_{ij}, e_{kl}] = \delta_{j-k} e_{il} + \delta_{j-l}\, \mathrm{sign}(kl)\, e_{ik} - \delta_{i-l} e_{kj} - \delta_{i-k}\, \mathrm{sign}(kl)\, e_{lj}, \quad (5.4)$$

$$[e_{ij}, e_{0k}] = \delta_{j-k}\, \mathrm{sign}(k)\, e_{i0} - \delta_{i-k} e_{0j}, \qquad (5.5)$$

$$\{e_{0i}, e_{0k}\} = \mathrm{sign}(i)\, e_{ik} \qquad (5.6)$$

for all $i, j, k, l = \pm 1, \pm 2, \ldots, \pm n$, where the bracket $\{\cdot, \cdot\}$ means anti-commutator.

The elements e_{ij} $(i, j = 0, \pm 1, \pm 2)$ are not linearly independent (we have for example, $e_{1-2} = -e_{-21}$) and we can choose from them the Cartan-Weyl basis as follows

$$\text{rising generators}: e_{i\pm j}, e_{kk}, e_{0k} \ (1 \leq i < j \leq n,\ 1 \leq k \leq n); \qquad (5.7)$$

$$\text{lowering generators}: e_{\pm j-i}, e_{-k-k}, e_{-k0} \ (1 \leq i < j \leq n,\ 1 \leq k \leq n); \quad (5.8)$$

$$\text{Cartan generators}: h_i := e_{k-k} \ \ (1 \leq k \leq n). \qquad (5.9)$$

Maximal classical r-matrix of Jordanian type for the Lie superalgebra $\mathfrak{osp}(1|2n)$ has the form[5]

$$r_{1,2,\ldots,n}\,(\xi_1, \xi_2, \cdots, \xi_n) = r_1(\xi_1) + r_2(\xi_2) + \cdots + r_n(\xi_n)\ . \qquad (5.10)$$

where

$$r_i\,(\xi_i) := \xi_i\Big(\frac{1}{2} e_{i-i} \wedge e_{ii} - 2 e_{0i} \otimes e_{0i} + \sum_{k=i+1}^{n} e_{i-k} \wedge e_{ik}\Big)\ , \qquad (5.11)$$

The total twist corresponding to the r-matrix (5.10) is given as follows

$$\begin{aligned} F_{1,n;2,n-1;\ldots;n,1}(\xi, \xi_1, \ldots, \xi_n) = {} & F_{n,1}(\xi_1, \xi_2, \ldots, \xi_{n-1}; \xi_n) \cdots \\ & \times F_{2,n-1}(\xi_1; \xi_2) F_{1,n}(\xi_1)\ . \end{aligned} \qquad (5.12)$$

Here $(i = 1, \ldots, k)$

$$\begin{aligned} F_{i,n+1-i}(\xi_1, \ldots, \xi_{i-1}; \xi_i) &:= (w_{\xi_{i-1}} \otimes w_{\xi_{i-1}}) \cdots w_{\xi_2})(w_{\xi_1} \otimes w_{\xi_1}) \\ &\times F_{i,n+1-i}(\xi_i)(w_{\xi_1}^{-1} \otimes w_{\xi_1}^{-1}) \cdots (w_{\xi_{i-1}}^{-1} \otimes w_{\xi_{i-1}}^{-1}) . \end{aligned} \tag{5.13}$$

$$F_{i,n+1-i}(\xi_i) = \exp\Big(\xi_i \sum_{k=i+1}^{n+1-i} e_{i-k} \otimes e_{k,n+1-i}\, e^{-2\sigma_i}\Big) \mathcal{F}_s(\sigma_i)\, e^{e_{i-i} \otimes \sigma_i}, \tag{5.14}$$

where $\mathcal{F}_s(\sigma_i)$ is defined by the formula (3.4), and

$$w_{\xi_i} = \exp\Big(\frac{-\xi_i \sigma_i}{e^{2\sigma_i} - 1} \sum_{k=i+1}^{n+1-i} e_{i-k} e_{k,n+1-i}\Big), \quad \sigma_i := \frac{1}{2} \ln(1 + \xi_1 e_{i-i}). \tag{5.15}$$

6. Light-cone κ-deformation of the super-Poincaré algebra $\mathcal{P}(3,1|1)$

The Poincaré algebra $\mathcal{P}(3,1)$ of the 4-dimensional space-time is generated by 10 elements, M_j, N_j, P_j, P_0 $(j = 1, 2, 3)$ with the standard commutation relations:

$$\begin{aligned} &[M_j, M_k] = i\epsilon_{jkl} M_l, \quad [M_j, N_k] = i\epsilon_{jkl} N_l, \quad [N_j, N_k] = -i\epsilon_{jkl} M_l, \\ &[M_j, P_k] = i\epsilon_{jkl} P_l, \quad [M_j, P_0] = 0, \\ &[N_j, P_k] = -i\delta_{jk} P_0, \quad [N_j, P_0] = -iP_j, \quad [P_\mu, P_\nu] = 0 . \end{aligned} \tag{6.1}$$

The super-Poincaré algebra $\mathcal{P}(3,1|1)$ is generated by the algebra $\mathcal{P}(3,1)$ and four real supercharges Q_α $(\alpha = \pm 1, \pm 2))$ with the commutation relations

$$\begin{aligned} &[M_j, Q_\alpha^{(\pm)}] = -\tfrac{i}{2}(\sigma_j)_{\alpha\beta} Q_\beta^{(\pm)}, \\ &[N_j, Q_\alpha^{(\pm)}] = \mp\tfrac{i}{2}(\sigma_j)_{\alpha\beta} Q_\beta^{(\pm)}, \quad [P_\mu, Q_\alpha^{(\pm)}] = 0, \end{aligned} \tag{6.2}$$

and moreover

$$\{Q_\alpha^{(\pm)}, Q_\beta^{(\pm)}\} = 0, \quad \{Q_\alpha^{(+)}, Q_\beta^{(-)}] = 2\big(\delta_{\alpha\beta} P_0 - (\sigma_j)_{\alpha\beta} P_j\big), \tag{6.3}$$

where we use the denotations $Q_1^{(\pm)} := Q_1 \pm iQ_2$, $Q_2^{(\pm)} := Q_{-1} \pm iQ_{-2}$, and σ_j $(j = 1, 2, 3)$ are 2×2 σ-matrices. It should be noted that the spinor $\mathbf{Q}^{(+)} := (Q_1^{(+)}, Q_2^{(+)})$ transformes as the left-regular representation and the spinor $\mathbf{Q}^{(-)} := (Q_1^{(-)}, Q_2^{(-)})$ provides the right-regular one with respect to $\mathcal{P}(3,1)$.

Using the commutation relations (6.1) and (6.2), (6.3) it is easy to check that the elements iN_3, $P_+ := P_0 + P_3$, P_1, $i(N_1 + M_2)$, P_2, $i(N_2 - M_1)$, and Q_α ($\alpha = 1,2$) satisfy the relations (2.1)–(2.4), namely, $\{h_{\gamma_0}, e_{\gamma_0}\} \to \{iN_3, P_+\}$, $\{e_1, e_{-1}\} \to \{P_1, i(N_1+M_2)\}$, $\{e_2, e_{-2}\} \to \{P_2, i(N_2-M_1)\}$, $e_{\pm3} \to Q_1$, $e_{\pm4} \to Q_2$. Therefore the two-tensor

$$r = \tfrac{1}{\kappa}\Big(P_1 \wedge (N_1+M_2) + P_2 \wedge (N_2 - M_1) + P_+ \wedge N_3 + \\ +2(Q_1 \wedge Q_1 + Q_2 \wedge Q_2)\Big)\ , \tag{6.4}$$

is a classical r-matrix of Jordanian type. It is called the classical r-matrix for light-cone κ-deformation of $D = 4$ super-Poincaré. Specializing the general formula (3.3) to our case $\mathcal{P}(3,1|1)$ we immediately obtain the twisting two-tensor corresponding to this r-matrix

$$F_\kappa(\mathcal{P}(3,1|1) := \mathfrak{F}_\kappa(Q_2)\mathfrak{F}_\kappa(Q_1)F_\kappa(\mathcal{P}(3,1))\ , \tag{6.5}$$

where $F_\kappa(\mathcal{P}(3,1))$ is the twisting two-tensor of the light-cone κ-deformation of the Poincaré algebra $\mathcal{P}(3,1)$

$$F_\kappa(\mathcal{P}(3,1)) := e^{\frac{i}{\kappa}P_1 \otimes (N_1+M_2)e^{-2\sigma_+}}\, e^{\frac{i}{\kappa}P_2 \otimes (N_2-M_1)e^{-2\sigma_+}}\, e^{2iN_3 \otimes \sigma_+} \tag{6.6}$$

and the super-factors $\mathfrak{F}_\kappa(Q_\alpha)$ ($\alpha = 1,2$) are given by the formula

$$\mathfrak{F}_\kappa(Q_\alpha) = \sqrt{\frac{(1+e^{\sigma_+}) \otimes (1+e^{\sigma_+})}{2(1+e^{\sigma_+} \otimes e^{\sigma_+})}}\left(1 + \frac{2}{\kappa}\,\frac{Q_\alpha}{1+e^{\sigma_+}} \otimes \frac{Q_\alpha}{1+e^{\sigma_+}}\right), \tag{6.7}$$

$$\sigma_+ := \frac{1}{2}\ln\Big(1 + \frac{1}{\kappa}P_+\Big)\ . \tag{6.8}$$

The formulas (6.5)–(6.8) were obtained by a suitable contraction of the quantum deformation of Jordanian type[6].

References

1. P.P. Kulish, V.D. Lyakhovsky and A.I. Mudrov, *Journ. Math. Phys.* **40**, 4569 (1999).
2. P.P. Kulish, V.D. Lyakhovsky and M.A. del Olmo, *Journ. Phys. A: Math. Gen.* **32**, 8671 (1999).
3. V.D. Lyakhovsky, S. Stolin and P.P. Kulish, *J. Math. Phys. Gen.* **42**, 5006 (2000).
4. D.N. Ananikyan, P.P. Kulish and V.D. Lyakhovsky, *St.Petersburg Math. J.* **14**, 385 (2003).
5. V.N. Tolstoy, in: Proc. of Internat. Workshop "Supersymmetries and Quantum Symmetries (SQS'03)", Eds: E. Ivanov and A. Pashnev, publ. JINR, Dubna, (2004), p. 242; `arXiv:math.QA/0402433`.
6. A. Borowiec, J. Lukierski, V.N. Tolstoy, Mod. Phys. Lett. **18A**, (2003), 1157; `hep-th/0301033`; *Eur. Phys. J.* **C44** (2005), 139; `arXiv:hep-th/0412131`;

A Correlation-Function Bell Inequality with Improved Visibility for 3 Qubits

Chunfeng Wu

Department of Physics, National University of Singapore, 2 Science Drive 3, Singapore 117542

Jing-Ling Chen

Theoretical Physics Division, Nankai Institute of Mathematics, Nankai University, Tianjin 300071, P. R. China

L. C. Kwek

Department of Physics, National University of Singapore, 2 Science Drive 3, Singapore 117542

Nanyang Technological University, National Institute of Education, 1, Nanyang Walk, Singapore 637616

C. H. Oh*

Department of Physics, National University of Singapore, 2 Science Drive 3, Singapore 117542

We construct a Bell inequality in terms of correlation functions for three qubits. The inequality is violated by quantum mechanics for all pure entangled states of 3 qubits. The strength of the violation is stronger than the result given in published literature, ref. [13].

1. Introduction

By now it is well-known that no local and realistic theory can be compatible with all predictions of quantum mechanics [2] by the Bell inequalities. Local realism implies experimentally variable constraints on the statistical measurement on two or more physically separated systems. These constraints, the Bell inequalities, can be violated by the predictions of quantum mechan-

*Electronic address: phyohch@nus.edu.sg

ics. Thus, Bell inequalities made it possible for the first time to eliminate local realistic description of quantum mechanics. Since then, violation of Bell inequalities has also become an effective method to detect entanglement.

If Bell inequalities are violated by all pure entangled states, these Bell inequalities can be used to characterize entanglement. Characterizing entanglement based on Bell inequality is an important issue in quantum information theory. There are several important recent developments in characterizing entanglement based on Bell inequalities. In 1991, Gisin [4] demonstrated that every pure entangled state of two qubits violates the CHSH inequality. This is Gisin's theorem. One year later, Bell inequalities for N qubits were first developed by Mermin-Ardehali-Belinskii-Klyshko (MABK) [5–7]. However, soon after, Scarani and Gisin [8] noticed that there exist pure states of N qubits that do not violate any of the inequalities. These states are the generalized Greenberger-Horne-Zeilinger (GHZ) [9] states given by

$$|\psi\rangle_{\rm GHZ} = \cos\xi|0\cdots 0\rangle + \sin\xi|1\cdots 1\rangle, \tag{1.1}$$

with $0 \leq \xi \leq \pi/4$. The GHZ states [9] are for $\xi = \pi/4$. In Ref. [8], Scarani and Gisin noticed that for $\sin 2\xi \leq 1/\sqrt{2^{N-1}}$ the states (1.1) do not violate the MABK inequalities. This observation prompted Scarani and Gisin to write that "this analysis suggests that MK (in Ref. [10], MABK) inequalities, and more generally the family of Bell's inequalities with two observables per qubit, may not be the 'natural' generalizations of the CHSH inequality to more than two qubits" [8]. Recently, Żukowski [11] and Werner [12] independently found the more general correlation-Bell inequalities (the ŻB inequalities) for N qubits. Using the ŻB inequalities, Żukowski et al in Ref. [10] showed that (a) For N =even, although the generalized GHZ states do not violate MABK inequalities, they violate the ŻB inequalities and (b) For N =odd and $\sin 2\xi \leq 1/\sqrt{2^{N-1}}$, the generalized GHZ states satisfy all known Bell inequalities for correlation functions. Thus it seems that Gisin's theorem is invalid for N (odd numbers) qubits.

In Ref. [13], we developed Bell inequalities in terms of both probabilities and correlation functions for three qubits. These inequalities are violated by all pure entangled states and hence the return of Gisin's theorem for 3-qubit systems. Indeed Bell inequalities are sensitive to the presence of noise and above a certain amount of noise the Bell inequalities will cease to be violated by a quantum system [14]. When noise is present, the considered state is described by $\rho = V|\psi\rangle\langle\psi| + (1-V)\rho_{\rm noise}$, where $\rho_{\rm noise} = \frac{1}{8}$ for three qubits. V is the visibility which is bounded by 0 and 1. For $V = 0$, no violation of local realism occurs and for $V = 1$, local realism description does not exist. Thus, there exists a quantity V_{thr}, called the threshold visibility, above which the state cannot be described by local realism. It seems that the inequalities in Ref. [13] are not good enough to the resistance of noise. For the three-qubit GHZ state, the threshold visibility is $V_{thr}^{GHZ} = 4\sqrt{3}/9 = 0.7698$ and for W state, the threshold visibility is $V_{thr}^{W} = 0.7312$. Our recent work shows that there is one new Bell inequality for 3 qubits that can be derived in terms of correlation functions. We demonstrate that the inequality is violated by quantum mechanics for any pure state of three qubits. The violation strength of the GHZ state is stronger than that predicted in Ref. [13]. Hence the inequality is more resistant to noise than those given in Ref. [13] are.

2. A new Bell Inequality involving Correlation Functions for 3 Qubits

Consider 3 observers, Alice, Bob and Charlie. Suppose they are each allowed to choose between two dichotomic observables, parameterized by $\vec{n}_1$ and $\vec{n}_2$. Each observer can choose independently two arbitrary directions. The outcomes of observer X's measurement on the observable defined by $\vec{n}_1$ and $\vec{n}_2$ are represented by $X(\hat{n}_1)$ and $X(\hat{n}_2)$ (with $X = A, B, C$). Each outcome can take values +1 or -1 under the assumption of local realism. In a specific run of the experiment the correlations between all 3 observers can be represented by the product $A(\hat{n}_i)B(\hat{n}_j)C(\hat{n}_k)$, where $i, j, k = 1, 2$. For convenience, we write $A(\hat{n}_i)B(\hat{n}_j)C(\hat{n}_k)$ as $A_iB_jC_k$. In a local realistic theory, the three-particle correlation function of the measurements performed by the three observers is the average over many runs of the experiment

$$Q(A_iB_jC_k) = \langle A(\hat{n}_i)B(\hat{n}_j)C(\hat{n}_k)\rangle = \langle A_iB_jC_k\rangle. \tag{2.1}$$

Similarly, two-particle correlation functions are given as

$$\begin{aligned} Q(A_iB_j) &= \langle A(\hat{n}_i)B(\hat{n}_j)\rangle = \langle A_iB_j\rangle, \\ Q(A_iC_k) &= \langle A(\hat{n}_i)C(\hat{n}_k)\rangle = \langle A_iC_k\rangle, \\ Q(B_jC_k) &= \langle B(\hat{n}_j)C(\hat{n}_k)\rangle = \langle B_jC_k\rangle, \end{aligned} \tag{2.2}$$

and one-particle correlation functions are given as

$$\begin{aligned} Q(A_i) &= \langle A(\hat{n}_i)\rangle = \langle A_i\rangle, \\ Q(B_j) &= \langle B(\hat{n}_j)\rangle = \langle B_j\rangle, \\ Q(C_k) &= \langle C(\hat{n}_k)\rangle = \langle C_k\rangle. \end{aligned} \tag{2.3}$$

The following inequality holds for the predetermined results:

$$-Q(A_1B_1C_1) + Q(A_2B_2C_1) + Q(A_1B_2C_2) + Q(A_2B_1C_2) - Q(A_2B_2C_2)$$
$$-Q(A_1B_1) - Q(A_2B_1) - Q(A_1B_2) - Q(A_1C_1) - Q(A_2C_1) - Q(A_1C_2)$$
$$-Q(B_1C_1) - Q(B_2C_1) - Q(B_1C_2) + Q(A_2) + Q(B_2) + Q(C_2) \leq 3. \quad (2.4)$$

The above inequality (2.4) is symmetric under the permutations of A_j, B_j and C_j. The proof consists of enumerating all the possible values of $A_i, B_j, C_k (i, j, k = 1, 2)$. This is easily done by fixing values of A_2, B_2, C_2 first. By fixing the values of A_2, B_2 and C_2, the inequality (2.4) is shown to be always satisfied under a local realistic description in the following.

1. For the case that A_2, B_2 and C_2 are all plus one, the inequality (2.4) becomes

$$-A_1B_1C_1 - A_1B_1 - A_1C_1 - B_1C_1 - A_1 - B_1 - C_1 - 1 \leq 0. \quad (2.5)$$

If $C_1 = 1$, we have $-2(A_1 + 1)(B_1 + 1) \leq 0$ from inequality (2.5). Because A_1 and B_1 can be either plus one or minus one, $-2(A_1 + 1)(B_1 + 1)$ will be -8 or 0. These two values are no larger than 0. If $C_1 = -1$, from inequality (2.5) we have $0 \leq 0$, which is obviously satisfied.

2. For the case that $A_2 = B_2 = 1$ and $C_2 = -1$, the inequality (2.4) becomes

$$-A_1B_1C_1 - A_1B_1 - A_1C_1 - B_1C_1 - A_1 - B_1 - C_1 - 1 \leq 0. \quad (2.6)$$

The inequality is the same as inequality (2.5). Seen from the first case, no matter which values A_1, B_1 and C_1 take, the inequality (2.6) is always correct. Similar conclusions can be drawn for the cases that $A_2 = C_2 = 1$ and $B_2 = -1$, and $B_2 = C_2 = 1$ and $A_2 = -1$ because the inequality (2.4) is symmetric under the permutations of A, B and C.

3. For the case that $A_2 = B_2 = -1$ and $C_2 = 1$, the inequality (2.4) becomes

$$-A_1B_1C_1 - A_1B_1 - A_1C_1 - B_1C_1 - A_1 - B_1 + 3C_1 - 5 \leq 0. \quad (2.7)$$

If $C_1 = 1$, we have $-2(A_1+1)(B_1+1) \leq 0$ from inequality (2.7). Because A_1 and B_1 can be either plus one or minus one, $-2(A_1+1)(B_1+1)$ will be -8 or 0. These two values are no larger than 0. If $C_1 = -1$, from inequality (2.7) we have $-8 \leq 0$, which is obviously correct whichever values A_1, B_1 and C_1 take. Similar conclusions can be drawn for the cases that $A_2 = C_2 = -1$ and $B_2 = 1$, and $B_2 = C_2 = -1$ and $A_2 = 1$ because the inequality (2.4) is symmetric under the permutations of A, B and C.

4. For the case that A_2, B_2 and C_2 are all minus one, the inequality (2.4) becomes

$$-A_1B_1C_1 - A_1B_1 - A_1C_1 - B_1C_1 + 3A_1 + 3B_1 + 3C_1 - 5 \le 0. \quad (2.8)$$

If $C_1 = 1$, we have $-2(A_1 - 1)(B_1 - 1) \le 0$ from inequality (2.8). Because A_1 and B_1 can be either plus one or minus one, $-2(A_1 - 1)(B_1 - 1)$ will be -8 or 0. These two values are no larger than 0. If $C_1 = -1$, from inequality (2.8) we have $4(A_1 + B_1 - 2) \le 0$, which is satisfied because A_1, B_1 can be either plus one or minus one and hence $4(A_1 + B_1 - 2)$ will be -16, -8 or 0.

Thus, the inequality (2.4) is always satisfied under a local realistic description whichever values A_i, B_j and C_k take. When setting $C_1 = 1, C_2 = -1$, the inequality (2.4) reduces directly to an equivalent form of the CHSH inequality for two qubits

$$Q(A_2B_2) - Q(A_2B_1) - Q(A_1B_2) - Q(A_1B_1) \le 2.$$

3. Quantum Violation of the Bell Inequality for 3 Qubits

Quantum mechanically, the above inequality is violated by all pure entangled states of three qubits. To test the quantum violation of any Bell inequalities, observables and quantum states are first specified. We consider the Bell type experiment in which three spatially separated observers Alice, Bob, and Charlie each measure two noncommuting observables $A_i = \hat{n}_{a_i} \cdot \vec{\sigma}(i = 1, 2)$ for Alice, $B_j = \hat{n}_{b_j} \cdot \vec{\sigma}(j = 1, 2)$ for Bob, and $C_k = \hat{n}_{c_k} \cdot \vec{\sigma}(k = 1, 2)$ for Charlie on a quantum entangled state $|\psi\rangle$ of three qubits. For each set of observables A_i, B_j, and C_k,

$$\begin{aligned}
A_i &= \hat{n}_{a_i} \cdot \vec{\sigma} = \begin{pmatrix} \cos\theta_{a_i} & \sin\theta_{a_i} e^{-i\phi_{a_i}} \\ \sin\theta_{a_i} e^{i\phi_{a_i}} & -\cos\theta_{a_i} \end{pmatrix}, \\
B_j &= \hat{n}_{b_j} \cdot \vec{\sigma} = \begin{pmatrix} \cos\theta_{b_j} & \sin\theta_{b_j} e^{-i\phi_{b_j}} \\ \sin\theta_{b_j} e^{i\phi_{b_j}} & -\cos\theta_{b_j} \end{pmatrix}, \\
C_k &= \hat{n}_{c_k} \cdot \vec{\sigma} = \begin{pmatrix} \cos\theta_{c_k} & \sin\theta_{c_k} e^{-i\phi_{c_k}} \\ \sin\theta_{c_k} e^{i\phi_{c_k}} & -\cos\theta_{c_k} \end{pmatrix},
\end{aligned} \quad (3.1)$$

where $i, j, k = 1, 2$, the following correlation functions are resulted,

$$\begin{aligned}
Q(A_iB_jC_k) &= \langle\psi|A_i \otimes B_j \otimes C_k|\psi\rangle, \\
Q(A_iB_j) &= \langle\psi|A_i \otimes B_j \otimes 1|\psi\rangle, \\
Q(B_jC_k) &= \langle\psi|1 \otimes B_j \otimes C_k|\psi\rangle, \\
Q(A_iC_k) &= \langle\psi|A_i \otimes 1 \otimes C_k|\psi\rangle, \\
Q(A_i) &= \langle\psi|A_i \otimes 1 \otimes 1|\psi\rangle, \\
Q(B_j) &= \langle\psi|1 \otimes B_j \otimes 1|\psi\rangle, \\
Q(C_k) &= \langle\psi|1 \otimes 1 \otimes C_k|\psi\rangle.
\end{aligned} \tag{3.2}$$

Pure states of three qubits constitute a five-parameter family, with equivalence up to local unitary transformations. This family has the representation [15]

$$\begin{aligned}
|\psi\rangle = &\sqrt{\mu_0}|000\rangle + \sqrt{\mu_1}e^{i\phi}|100\rangle + \sqrt{\mu_2}|101\rangle \\
&+\sqrt{\mu_3}|110\rangle + \sqrt{\mu_4}|111\rangle,
\end{aligned} \tag{3.3}$$

with $\mu_i \geq 0$, $\sum_i \mu_i = 1$ and $0 \leq \phi \leq \pi$. Numerical results show that this Bell inequality (2.4) is violated by all pure entangled states of three-qubit systems. However, no analytical proof of the conclusion can be given. In the following, some special cases will be given to show the inequality (2.4) is violated by all pure entangled states. The first example considered is the family of generalized GHZ states $|\psi\rangle_{\text{GHZ}} = \cos\xi|000\rangle + \sin\xi|111\rangle$. The inequality (2.4) is violated by the generalized GHZ states for the whole region except $\xi = 0, \pi/2$. The variation of the violation with ξ is shown in Figure 3.1. For the GHZ state with $\xi = \pi/4$, the quantum violation reaches its maximum value 4.40367. Another set of states considered are generalized W states $|\psi\rangle_W = \sin\beta\cos\xi|100\rangle + \sin\beta\sin\xi|010\rangle + \cos\beta|001\rangle$. By fixing the value of β, quantum violation of the inequality (2.4) varies with ξ (see Figure 3.2). The inequality (2.4) is violated by generalized W states except the cases with $\beta = \frac{\pi}{2}$, $\xi = 0$ and $\xi = \frac{\pi}{2}$. The states in these cases are product states which do not violate any Bell inequality. For the standard W state, quantum violation of the inequality (2.4) approaches 4.54086.

Hence inequality (2.4) is also one candidate to generalize the theorem of Gisin to three-qubit systems. One of the interests of the new inequality for three qubits is that it is highly resistant to noise. The inequality (2.4) is violated by the GHZ state, the threshold visibility is $V_{thr}^{GHZ} = 0.68125$. The inequality (2.4) is also violated by the W state, the threshold visibility is $V_{thr}^{W} = 0.660668$. We plot the variation of quantum violation for the generalized GHZ states with angle ξ for inequality given in Ref. [13] and inequal-

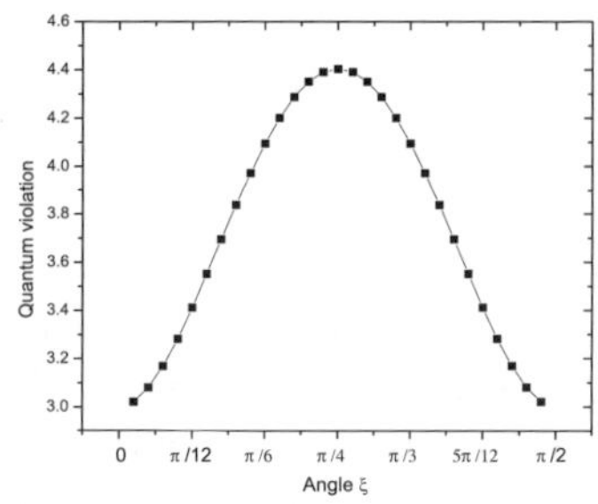

Fig. 3.1. Numerical results for the generalized GHZ states $|\psi\rangle_{\rm GHZ} = \cos\xi|000\rangle + \sin\xi|111\rangle$, which violate the inequality (2.4) except $0, \pi/2$.

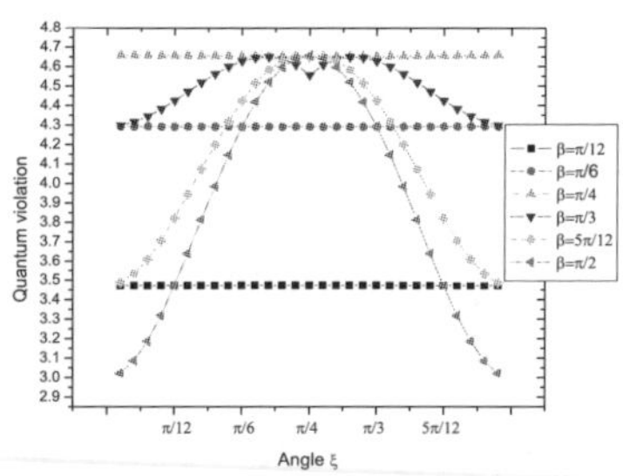

Fig. 3.2. Numerical results for the generalized W states $|\psi\rangle_W = \sin\beta\cos\xi|100\rangle + \sin\beta\sin\xi|010\rangle + \cos\beta|001\rangle$ which violate the inequality (2.4) for different ξ and β. Here the cases $\beta = \pi/12, \pi/6, \pi/4, \pi/3, 5\pi/12$ and $\pi/2$ are considered.

ity (2.4), see Figure 3.3. In plotting the figure, we rewrite the expressions of these two inequalities as

$$\begin{aligned}
\frac{1}{4}\,[\,& Q(A_1B_1C_1) - Q(A_1B_2C_2) - Q(A_2B_1C_2) - Q(A_2B_2C_1) + 2Q(A_2B_2C_2) \\
& -Q(A_1B_1) - Q(A_1B_2) - Q(A_2B_1) - Q(A_2B_2) + Q(A_1C_1) + Q(A_1C_2) \\
& +Q(A_2C_1) + Q(A_2C_2) + Q(B_1C_1) + Q(B_1C_2) + Q(B_2C_1) + Q(B_2C_2)] \\
& \leq 1, \qquad (3.4)
\end{aligned}$$

$$\frac{1}{3}\,[\,-Q(A_1B_1C_1)+Q(A_1B_1C_2)+Q(A_1B_2C_1)+Q(A_2B_1C_1)$$
$$-Q(A_2B_2C_2)-Q(A_1B_2)-Q(A_2B_1)-Q(A_2B_2)-Q(A_1C_2)$$
$$-Q(A_2C_1)-Q(A_2C_2)-Q(B_1C_2)-Q(B_2C_1)-Q(B_2C_2)+Q(A_1)$$
$$+Q(B_1)+Q(C_1)]\leq 1, \qquad (3.5)$$

respectively. In these forms, the violation degrees of the two inequalities can be compared directly. Comparing the results of the inequality given in Ref. [13], the new inequality (2.4) is really more resistant to noise. Although inequality (2.4) is more resistant to noise than the ones given in [13], the visibility of the GHZ state is still not optimal. The visibility of the inequality for three qubits given by Żukouski-Brukner for the GHZ state is 0.5. The improvement of this paper is that a Bell inequality involving corraltion functions, which is more resistant to noise than the previous ones, is constructed. However, there is no inequality which is not only maximally violated by the GHZ state, but also violated by all pure entangled states of three qubits. To develop such a new Bell inequality for three qubits is still an open problem.

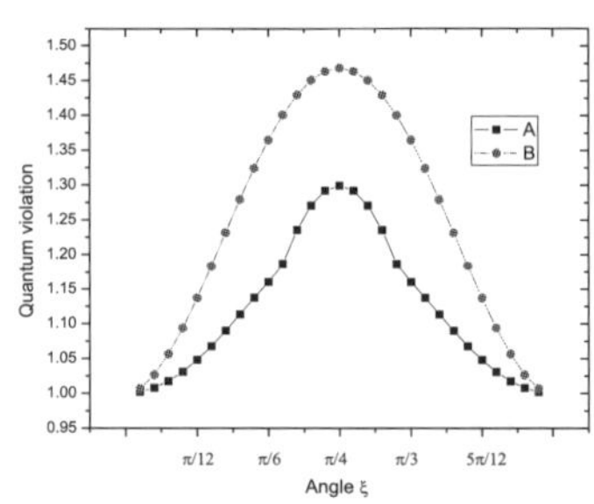

Fig. 3.3. Violation of two Bell inequalities for three qubits with different value of ξ, where curve A is for inequality given in Ref.[13] and curve B is for the new inequality (2.4).

4. Conclusion

We have presented a Bell inequality involving correlation functions for three qubits. The inequality is violated by all pure entangled states of 3 qubits, although it is not maximally violated by the GHZ state. The visibility of the inequality for the GHZ state is $V_{thr}^{GHZ} = 0.68125$, which is the less than that ($\sqrt{3}/9$) of the inequalities given by us in Ref. [13]. Thus the inequality (2.4) is more resistant to noise than the inequalities given before in [13].

This work is supported by NUS academic research Grant No. WBS: R-144-000-089-112.

References

1. A. Einstein, B. Podolsky and N. Rosen, Phys. Rev. **47**, 777 (1935).
2. J. S. Bell, Physics 1, 195 (1964).
3. D. Bohm, *Quantum Physics* (Prentice Hall, 1951).
4. N. Gisin, Phys. Lett. A **154**, 201 (1991); N. Gisin and A. Peres, Phys. Lett. A **162**, 15 (1992).
5. N. D. Mermin, Phys. Rev. Lett. **65**, 1838 (1990).
6. M. Ardehali, Phys. Rev. A **46**, 5375 (1992).
7. A. V. Belinskii and D. N. Klyshko, Phys. Usp. **36**, 653 (1993).
8. V. Scarani and N. Gisin, J. Phys. A **34**, 6043 (2001).
9. D. M. Greenberger, M. Horne, A. Shimony, and A. Zeilinger, Am. J. Phys. **58**, 1131 (1990).
10. M. Żukowski, Č. Brukner, W. Laskowski, and M. Wiesniak, Phys. Rev. Lett. **88**, 210402 (2002).
11. M. Żukowski and Č. Brukner, Phys. Rev. Lett. **88**, 210401 (2002).
12. R. F. Werner and M. M. Wolf, Phys. Rev. A **64**, 032112 (2001).
13. J. L. Chen, C. F. Wu, L. C. Kwek and C. H. Oh, Phys. Rev. Lett. **93**, 140407 (2004); C. F. Wu, J. L. Chen, L. C. Kwek and C. H. Oh, Proceedings of the 1st Asia-Pacific Conference on Quantum Information Science in Taiwan.
14. D. Kaszlikowski, P. Ganciński, M. Żukowski, W. Mislaszewski and A. Zeilinger, Phys. Rev. Lett. **85**, 4418 (2000).
15. A. Acín, A. Andrianov, L. Costa, E. Jané, J. I. Latorre, and R. Tarrach, Phys. Rev. Lett. **85**, 1560 (2000).

Topological Aspects of the Spin Hall Effect

Yong-Shi Wu

Department of Physics, University of Utah
Salt Lake City, UT 84112, USA

I review some recent developments in understanding topological aspects of the spin Hall effect, particularly a joint work with X.L. Qi and S.C. Zhang on topological quantization of the spin Hall conductivity, as a first Chern number, in certain planar insulating systems. I devote this talk to the memory of Prof. S.S. Chern for the great, profound and ever-lasting impacts of his mathematics on physics.

1. Introduction to the Spin Hall Effect

It is well-known that the electron, as an elementary particle, has both charge and spin. So the motion of electrons may lead to transport of both charge and spin. The study of charge transport has had a long history in classical electromagnetism since Volta's invention of the first battery to generate electric currents. But the study of spin transport becomes focus of attention only recently because of the surge of the interests in spintronics, which requires better understanding how spin can be manipulated in and transported across solid-state devices.

Compared to charge transport, spin transport has very different symmetry properties. This is because under rotations charge is a scalar, while spin is an axial vector. So charge current density J_j (as a vector) is odd under time reversal T, while spin current density J^i_j, as a tensor, is even under T. Thus, in the absence of magnetic field, the charge current driven by an electric field is longitudinal, obeying the Ohm's law

$$J_j = \sigma_c E_j; \tag{1.1}$$

so it breaks T-invariance and is dissipative. On the other hand, the spin current induced by an electric field is transverse:

$$J^i_j = \sigma_s \epsilon_{ijk} E_k, \tag{1.2}$$

and it respects T-invariance and is dissipationless. This is the so-called spin Hall effect which, if exploited for the purposes of semiconductor spintronics, is believed to have the following advantages: 1) It provides efficient generation of spins inside the devices and avoids the usual problem with inefficient spin injection from outside. 2) It allows information processing with *no* heat dissipation, thus overcoming the increasing heat dissipation problem for ordinary semiconductor devices upon miniaturization. 3) It allows using electric fields, instead of magnetic fields, to access and manipulate individual spins at nanometer scales. 4) Perhaps it may allow *quantum* information processing. For these reasons the spin Hall effect becomes a focus of current theoretical and experimental efforts in spintronics.

Theoretically two different mechanisms have been proposed for the spin Hall effect. The *extrinsic* mechanism involves disorder, such as impurities and imperfections. It is based on spin-dependent scattering of electrons by impurities. The *intrinsic* mechanism is independent of disorder, giving rise to dissipationless spin current in a perfect crystal. Below I will present a brief review of the topological aspects of these mechanisms, particularly of a joint work of mine with Qi and Zhang[1] on quantization of the spin Hall conductivity, as the first Chern number, in 2D insulating systems.

A historic remark seems appropriate here, as the present conference is devoted to the memory of Prof. S.S. Chern for the great impacts of his mathematics in physics. The recent resurgence of interests in the spin Hall effect[2,3] was partly due to the theoretical discovery[2] of the intrinsic dissipationless spin Hall current (1.2) as the dimensional reduction of a non-abelian Hall current in 4D disc when restricted to the 3D edge. The study of the 4D quantum Hall effect had originated from considerations of the second Hopf bundle that carries a non-trivial second Chern number[4]. This relationship was enough to motivate me to study the topological aspects of the spin Hall effect, including quantization of the spin Hall conductivity.

2. Spin-Orbit Coupling and Berry Curvature in k-Space

The spin Hall effect happens due to spin-dependent response of electrons to an external electric field. However, an electric field couples directly to charge, not to spin. How can electrons with different spin orientations respond differently? The answer is through the spin-orbit (SO) coupling, which makes the motion of the electron depend on its spin orientation.

In quantum mechanics the SO coupling is known due to a relativistic effect: In the rest frame of the electron, its spin magnetic moment couples to a (momentum dependent) magnetic field, which originates from the electric field in the lab frame. In a real sample, either atomic, crystalline, impurity or even gate electric fields can give rise to SO coupling. The magnitude of the SO coupling for free electrons in a solid state sample, compared with that in vacuum, can be greatly enhanced. The study of the SO coupling is at the heart of the study of electric-field driven spin transport, and especially of the spin Hall effect in solid state devices. Depending on the symmetries, two typical SO Hamiltonians for electrons in 3D crystals are the Rashba term and the Dresselhaus term. The effects of SO coupling on spin transport critically depend on the form of the SO coupling. For a recent brief review on SO coupling and spin transport, see ref. 5.

In addition to giving rise to spin-dependent scattering of an electron against impurities, SO coupling also modifies the Berry curvature in **k**-space. For simplicity, we consider a non-interacting electron system on a 2D lattice. In the absence of SO coupling, the energy eigenstates of an electron are labeled by quasi-momentum **k** and the band index n. We denote by $u_{n\mathbf{k}}(\mathbf{x})$ the periodic part of the Bloch wave function. Then the Berry connection (or vector potential) in **k**-space is defined to be

$$A_{ni}(\mathbf{k}) = (-i) < n\mathbf{k}|\frac{\partial}{\partial k_i}|n\mathbf{k} >= (-i) \int d^2 x u^*_{n\mathbf{k}} \frac{\partial u_{n\mathbf{k}}}{\partial k_i}. \tag{2.1}$$

The corresponding curvature (or field strength) is $\mathbf{B}_n(\mathbf{k}) = \nabla_{\mathbf{k}} \times \mathbf{A}_n(\mathbf{k})$. In the presence of SO coupling, $\mathbf{B}_n(\mathbf{k})$ becomes spin-dependent.

It is this Berry curvature in **k**-space that plays a fundamental role in studying topological aspects of planar electron systems. In particular, the first Chern number of the Berry curvature normally gives a topological number that characterizes a completely filled band[6,7]. In addition, $\mathbf{B}_n(\mathbf{k})$ also plays a role in dynamics by modifying the semi-classical trajectory of the electron as a wave packet. The semi-classical equations of motion in phase space for a wave packet of the Bloch electron in the presence of an external electric field are given by[8]

$$\hbar\dot{\mathbf{k}} = e\mathbf{E}, \qquad \dot{\mathbf{x}} = \frac{\hbar\mathbf{k}}{m} - \dot{\mathbf{k}} \times \mathbf{B}(\mathbf{k}). \tag{2.2}$$

Here **x** and **k** are the central position of the wave packet in coordinate and momentum space, respectively. The last term in the right-hand side of the second equation is nothing but the anomalous velocity first found by Karplus and Luttinger[9] half century ago, now identified as being induced by the Berry curvature in **k**-space. With SO coupling, the spin-dependent anomalous velocity, which for narrow-band semiconductors induces a drift that is both perpendicular to spin vector and to the electric field (the "side-jump" effect[10]).

3. Extrinsic Spin Hall Effect and Anomalous Hall Effect

The SO coupling arising from *impurity* electric fields leads to the so-called *extrinsic* spin Hall effect. When an electron is scattered by an impurity, SO coupling induces a spin-dependent amplitude, which is known as the skew Mott scattering[11]. This amplitude, giving rise to different scattering angles for spin-up and spin-down electrons, will make a contribution to the spin Hall current[10]. Moreover, as we saw in last section, the band structure of Bloch electrons also induces a Berry curvature in **k**-space, which together with the impurity electric field gives rise to a spin-dependent side-jump effect for the wave-packet trajectory in **x**-space. The latter will contribute to the spin Hall conductivity too[10].

The estimation of the magnitude of the extrinsic spin Hall effect, which is material and sample dependent, is crucial to interpreting experimental observations of spin accumulation in real samples. It is a very delicate job, which unfortunately I do not have time to address any more in this talk.

A close cousin of the spin Hall effect is the anomalous (charge) Hall effect. Usually the (charge) Hall effect occurs in an applied magnetic field. However, in magnetic systems (with broken T-symmetry) or in systems with SO coupling, the *anomalous* (charge) Hall effect may happen in the *absence* of an external magnetic field. Theoretically, depending on whether the underlying mechanism involves disorders or not, the anomalous Hall effect is said to be extrinsic or intrinsic. In the cases with SO coupling, one may consider two spin species of electrons with spin fully aligned along x_i-direction; then the spin current, J^i_j, is related to the anomalous Hall current of each species, $J^{\uparrow,\downarrow}_j$, by

$$J^i_j = e^{-1}\left(J^{\uparrow}_j - J^{\downarrow}_j\right). \tag{3.1}$$

This relation applies both to extrinsic and intrinsic contributions.

In the following I will report on a recent joint work[1] with Qi and Zhang on the *impurity independent* or *intrinsic* anomalous Hall and spin Hall effects in *planar insulating* systems, especially on our suggestion of topological quantization of the anomalous Hall conductivity and spin Hall conductivity.

4. Quantized Anomalous Hall Effect (QAHE)

In ref. 1 we proposed a model for a wide class of 2D magnetic semiconductors, whose ground state is a bulk magnetic insulator (with the Fermi level lying in the band gap) with gapless edge states responsible for the transport with quantized Hall conductivity. The most general two-band Hamiltonian describing a 2D system with SO coupling and magnetic moment is of the form: ($\alpha = 1, 2, 3$)

$$H(\mathbf{k}) = \epsilon(\mathbf{k}) + V d_\alpha(\mathbf{k})\sigma^\alpha \,. \tag{4.1}$$

where σ^α are Pauli matrices and $\mathbf{k} = (k_x, k_y)$ Bloch wavevector. Parity (or time reversal) is broken if one of the $d_\alpha(\mathbf{k})$'s is odd (or even) in $\mathbf{k}$. The two-band spectrum is given by $E_\pm(\mathbf{k}) = \epsilon(\mathbf{k}) \pm V d(\mathbf{k})$, where $d(\mathbf{k})$ is the norm of the 3-vector d_α. When the coupling constant V is large enough, the two bands with energies $E_\pm(\mathbf{k})$ will be separated by a full gap. In this case, when the chemical potential lies in the gap, the system is a bulk insulator with the lower band completely filled and the upper band empty. Using Kubo's formula the charge Hall conductivity is shown to be

$$\sigma_{ij} = -\frac{1}{8\pi^2} \int\int_{\text{FBZ}} d^2k\, \hat{\mathbf{d}} \cdot \partial_\mathbf{i} \times \partial_\mathbf{j} \hat{\mathbf{d}}, \tag{4.2}$$

where $\hat{\mathbf{d}}$ is the unit vector in the direction of d_α. This is known to be the winding number of the map from the first Brillioun zone (FBZ) to the 2-sphere, $\hat{d}_\alpha(\mathbf{k}) : FBZ \to S^2$. So it is a topological invariant, independent of the details of the band structure parameters provided the band gap does not close; and the value of σ_{ij} is always quantized: $\sigma_{xy} = -n/2\pi$ when the map $\hat{d}_\alpha(\mathbf{k})$ covers S^2 n times. Although the single-electron states in this system are very different from those in the Landau levels in the usual integer quantum Hall effect (IQHE), the quantization of Hall conductivity in the two systems shares the same topological origin, as the first Chern number of Berry curvature in $\mathbf{k}$-space. Our formula (4.2) for the QAHE generalizes the TKNN formula[6] for the IQHE to the cases without a magnetic field.

For an explicit discussion on the QAHE and on the characters of associated edge states, as an example we choose: $d_x = \sin k_y$, $d_y = -\sin k_x$ and $d_z = c\,(2 - \cos k_x - \cos k_y - e_s)$, and consider the tight-binding model on a square lattice, which describes a magnetic semiconductor with SO coupling and uniform magnetization, with the $\mathbf{k}$-space Hamiltonian of the form (4.1):

$$\begin{aligned} H = &-t \sum_{\langle ij \rangle} \left(c_i^\dagger c_j + h.c. \right) + \frac{iV}{2} \sum_i \left(c_i^\dagger \sigma_y c_{i+\hat{x}} - c_i^\dagger \sigma_x c_{i+\hat{y}} - h.c. \right) \\ &- \frac{cV}{2} \sum_{\langle ij \rangle} \left(c_i^\dagger \sigma^z c_j + h.c. \right) + (2 - e_s) V c \sum_i c_i^\dagger \sigma^z c_i . \end{aligned} \tag{4.3}$$

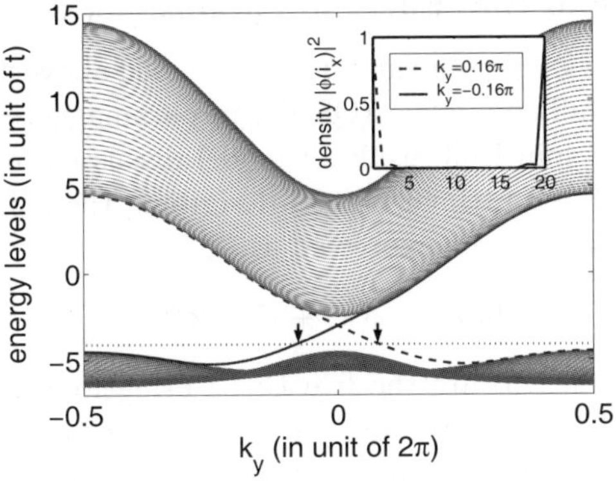

Fig. 4.1. Energy spectrum of the system (4.3) on a cylinder, with parameters $c = 1, t/V = 1/3, e_s = 0.5$. The solid and dashed lines between two bands are the edge states on the right and left edge, respectively. Inset: the density distribution of the two edge states at the Fermi surface, calculated for a 50×50 lattice.

When V/t is large enough and $c > 0$, the Hall conductivity is shown to be

$$\sigma_{xy} = \begin{cases} 1/2\pi, & 0 < e_s < 2 \\ -1/2\pi, & 2 < e_s < 4 \\ 0, & e_s > 4 \text{ or } e_s < 0\,. \end{cases} \tag{4.4}$$

Thus, the bulk topological number for the first two cases is $n = \pm 1$.

To show the behavior of edge states, put this system on a strip with periodic boundary condition in y-direction and open boundary condition in x-direction (with the wave function vanishing at $x = 0, L+1$). In this case k_y is a good quantum number and the single-particle energy spectrum can be obtained as $E_m(k_y), m = 1, .., 2L$. The energy spectrum is shown in Fig. 4.1. For a given k_y, there are $2L$ states, two of them being localized and the rest extended. When the Fermi level, represented by the horizontal dotted line, lies in the bulk energy gap the only gapless excitations are edge states (marked by the arrows). Similar to the usual IQHE case, the edge states have definite chirality. In the present case, all the left-edge states move with velocity $v_y < 0$ and all the right ones with $v_y > 0$, as seen from their dispersion relation. Generally, when $\sigma_{xy} = n/2\pi$, there are $|n|$ edge states on each edge, where the charge current is right-handed for $n > 0$ and left-handed for $n < 0$. We leave the discussion of the relation between edge states and bulk transport to next section.

5. Quantized Spin Hall Effect in Paramagnetic Insulators

The QAHE of a magnetic semiconductor discussed above can be generalized to the quantum Hall spin effect (QSHE) in paramagnetic semiconductors. We start with the spin Hall insulator model discussed in ref. 12, and specialize it to 2D. The Luttinger model describing the spin $S = 3/2$ heavy and light hole bands can be expressed as

$$H(\mathbf{k}) = \epsilon(\mathbf{k}) + V d_a(\mathbf{k})\Gamma^a \,, \tag{5.1}$$

where Γ^a $(a = 1, 2, .., 5)$ are the five Dirac Γ-matrices forming an $SO(5)$ Clifford algebra. In the continuum limit, $d_a(\mathbf{k})$ are the five d-wave combinations of $\mathbf{k}$. If there is a *mirror symmetry*: $z \leftrightarrow -z$, we can consistently set $\langle k_z \rangle \equiv 0$ and $\langle k_z^2 \rangle \equiv e_s$. In this quasi-2D case $d_1 = d_2 = 0$, and Γ_a $(a = 3, 4, 5)$ form a representation of an $SO(3)$ Clifford sub-algebra. The Hamiltonian (5.1) preserves T-invariance. Its energy spectrum is exactly the same as that of (4.1), but with each energy level now doubly degenerate due to the Kramers theorem.

Suppose that V is large enough so that a full gap is open between the two energy bands. With the Fermi level in the gap the system is in an insulating phase; by using Kubo's formula, the spin Hall conductivity σ^z_{xy} for the *conserved* spin current defined in ref. 13 can be written as

$$\sigma^{z(c)}_{xy} = \frac{1}{8\pi^2} \int\!\!\int_{\mathrm{FBZ}} d^2k\, \hat{\mathbf{d}} \cdot \partial_{\mathbf{x}} \hat{\mathbf{d}} \times \partial_{\mathbf{y}} \hat{\mathbf{d}}. \tag{5.2}$$

So the conserved spin Hall conductivity in the quasi-2D systems (5.1) is always quantized in units of $1/2\pi$, and its value, as a topological invariant, gives a *characterization* of the *topological order* in the insulating phase. The topological quantization of the spin Hall conductivity in this model is easy to understand: The present spin-3/2 Hamiltonian (5.1) is essentially two decoupled copies of the previous QAHE Hamiltonian (4.1), with $d_a(\mathbf{k})$ to be opposite in the two copies. Each copy is labeled by the eigenvalue ± 1 of Γ_{12}, which commutes with Γ_a and the Hamiltonian and thus serves as a "conserved spin quantum number" even in the presence of the SO coupling. The spin Hall conductivity of the system is the difference between the anomalous Hall conductivity for each copy.

For a numerical example, we choose $d_a(\mathbf{k})$ to be

$$\begin{aligned} d_3(k) &= -\sqrt{3}\sin k_x \sin k_y \\ d_4(k) &= \sqrt{3}\,(\cos k_x - \cos k_y) \\ d_5(k) &= 2 - e_s - \cos k_x - \cos k_y, \end{aligned} \tag{5.3}$$

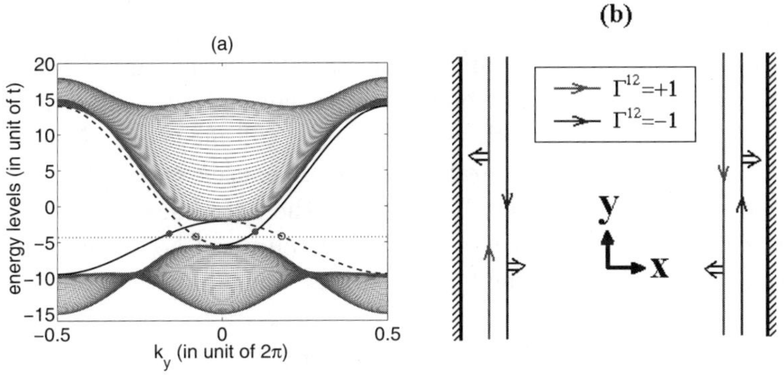

Fig. 5.1. (a) The energy spectrum for the Hamiltonian (5.3) with $t/V = 4$ and $e_s = 0.5$. The mid-gap solid (dashed) lines stand for the doubly-degenerate edge states, and the dotted line a typical in-gap Fermi level with $\mu = -4.2t$. Each crossing of the edge spectral curve with the Fermi level defines two edge states on the left and right boundary with opposite value of Γ^{12}. The solid (hollow) circles mark the particle (hole) edge excitations induced by adiabatic flux turning-on. (b) Schematic picture of edge states. Each red (blue) line stands for two edge states with $\Gamma^{12} = +1(-1)$. The double arrow shows the direction of current, carried by the edge state, induced by an electric field in y-direction.

which reduces to the continuum Luttinger model when $k_x, k_y \to 0$. Direct calculations show that

$$\sigma_{xy}^{z(c)} = \begin{cases} 1/\pi, & 0 < e_s < 4 \\ 0, & e_s > 4 \text{ or } e_s < 0 \,. \end{cases} \tag{5.4}$$

Here the non-zero topological charge is 2, as twice bigger as in the previous case (4.4).

To see how spin is transported, again we put the system on a cylinder (periodic in y-direction). Then we follow a Laughlin-type gauge argument[14], adiabatically turning on a magnetic flux threading the cylinder and keeping track of the evolution of the edge states near the open boundaries in the infinitesimal electric field.

At zero flux there are four edge states on each open boundary. For the states with $\Gamma^{12} = +1(-1)$, the $v_y > 0$ state is localized on the left (right) edge, while the $v_y < 0$ state is localized on the right (left) edge. The energy spectrum and the chirality of the edge states are shown in Fig. 5.1 (b).

When the Fermi level lies in the bulk energy gap, the insulating ground state (Fermi sea) consists of filled bulk and edge states $|mk_y\rangle$ below the Fermi energy. Now let the flux $\Phi(t)$ threading the cylinder changes adiabatically from $\Phi(0) = 0$ to $\Phi(T) = 2\pi$. The effect of flux threading is to replace $k_y \to k_y - A_y$ in the Hamiltonian (or to impose twisted boundary conditions[7]), which transforms each single-particle eigenstate $u_{mk_y}(x)e^{ik_y y}$ into $u_{m,k_y-A_y}(x)e^{i(k_y-A_y)y}$. Namely, the states in the Fermi sea get translated in momentum space. The bulk states remain in bulk, while each edge state on the Fermi surface with velocity $v_y > 0$ will move out of the Fermi sea and becomes a particle excitation, since $\delta E \simeq v_y \delta k = 2\pi v_y/L > 0$, and each edge state with $v_y < 0$ will move into the Fermi sea and leads to a hole excitation, as shown in Fig. 5.1 (a) by solid and hollow circles near the Fermi surface. So when the flux reaches 2π, the adiabatic evolution will result in

$$|m, k_y\rangle \to \left|m, k_y + \frac{2\pi}{L}\right\rangle . \tag{5.5}$$

And the the net effect is to transfer the edge states with $\Gamma^{12} = 1$ from the right edge to the left edge and to transfer the edge states with $\Gamma^{12} = -1$ in the opposite way. (The above analysis is a generalization of that in ref. 15 from the usual IQHE to QSHE.)

This leads to an accumulation of the Γ^{12}-spin on the boundaries, which in turn leads to a non-vanishing spin S^z density on the boundary, since Γ^{12} is related to S^z by $S^z = -\frac{1}{2}\Gamma^{12} - \Gamma^{34}$. On the other hand, such an accumulation can also be considered as a consequence of the spin Hall current j_x^z induced by the electric field E_y. So the *physically observed* spin Hall conductivity is proportional to the amplitude of spin accumulation after 2π-flux threading. Though spin Hall conductivity consists of conserved and non-conserved parts, only the conserved part $\sigma_{(c)}$ corresponds to a transport of Γ^{12}-spin carried by the motion of edge states, while the non-conserved part $\sigma_{(nc)}$ is just a precession effect due to the non-conserved nature of spin as represented by $\langle\Gamma^{34}\rangle$ for each edge state. Consequently, it is only $\sigma_{(c)}$ that counts genuine transport of quantum states in the system and is protected by the bulk topological order. These considerations constitute a physical justification of the conserved spin current operator defined in ref. 13.

6. Summary

Both the anomalous (charge) Hall effect and the spin Hall effect rely on the interplay between SO coupling and the Berry curvature in **k**-space. The latter accommodates intriguing topological aspects of quantum many-body theory, such as characterization of *topological orders in insulators* in terms of quantized charge and spin Hall conductivity, as well as *holographic* relation between bulk transport and edge states.

The QSHE models discussed in this paper can be experimentally realized in two classes of 2D semiconductors. One class is the (distorted) zero-gap semiconductors such as HgTe, HgSe, β-HgS and α-Sn. The other class is the narrow-gap semiconductors such as PbTe, PbSe and PbS. The quantized spin Hall effect is expected[1] to be observable in a wide temperature range, say $T << 100K$. Also it is predicted[1] that when the Fermi level changes in a ballistic regime, the quantum spin Hall insulator (5.3) should exhibit a plateau of residue longitudinal charge conductance $G = 8e^2/h$, due to the existence of 8 edge states as demanded by topological arguments.

In this talk, I did not address the effects of randomness and interactions in spin transport, critical for explaining experimental data. The study of random systems with SO coupling is still in its infancy. How to incorporate interactions is not clear either. Much work remains to be done.

7. Acknowledgment

I thank Prof. Wei-Ping Zhang, Prof. Mo-Lin Ge and the Nankai Institute of Mathematics for invitation and warm hospitality. Also I thank X.L. Qi and S.C. Zhang for pleasant collaboration. This work was supported in part by the U.S. NSF under Grant PHY-0457018.

References

1. X.L. Qi, Y.S. Wu and S.C. Zhang, ArXiv:cond-mat/0505308.
2. S. Murakami, N. Nagaosa and S.C. Zhang, *Science* **301**, 1348 (2003).
3. J. Sinova, D. Culcer, Q. Niu, T. Jungwirth and A. MacDonald, *Phys. Rev. Lett.* **92** 126603 (2004).
4. S.C. Zhang and J.P Hu, *Science* **294**, 823 (2001).
5. E.I. Rashba, ArXiv:cond-mat/0507007.
6. D. Thouless, M. Kohmoto, M.P. Nightingale, and M. den Nijs, *Phys. Rev. Lett.*, **49**, 405 (1982); M. Kohmoto, *Ann. of Phys.*, **160**, 355 (1985).
7. Q. Niu, D. Thouless and Y.S. Wu, *Phys. Rev.* **B 31**, 3372 (1985).
8. M.C. Chang and Q. Niu, *Phys. Rev.* **B 53**, 7010 (1996).
9. R. Kaplus and J.M. Luttinger, *Phys. Rev.*, **95**, 1154 (1954)
10. For the relevance to the latest experiments, see H.A. Engel, B.I. Halperin and E.I. Rashba, ArXiv:cond-mat/0505535.
11. N.F. Mott and H.S.W. Massey, *The Theory of Atomic Collisions* (Oxford Univ. Press, 1965).
12. S. Murakami, N. Nagaosa and S.C. Zhang, *Phys. Rev.* **B69**, 235206 (2004).
13. S. Murakami, N. Nagaosa and S.C. Zhang, *Phys. Rev. Lett.*, **93**, 156804 (2004).

14. R.B. Laughlin, *Phys. Rev.* **B 23**, 5632 (1981).
15. Y. Hatsugai, *Phys. Rev. Lett.* **71**, 3697 (1993).

Positive Mass Theorems and Calabi-Yau Compactification

Naqing Xie

Institute of Mathematics
School of Mathematical Sciences
Fudan University
Shanghai 200433, China
E-mail: nqxie@fudan.edu.cn

In this talk, we review the positive mass theorems in general relativity as well as discuss recent progress to their generalization for spaces with asymptotic Calabi-Yau compactification in string theory.

1. Introduction

In general relativity, our universe is modelled by a 4-dimensional Lorentzian manifold $(N^{1,3}, \widetilde{g})$ together with an energy-momentum tensor T which satisfies the Einstein field equations

$$\widetilde{Ric}(\widetilde{g}) - \frac{\widetilde{R}(\widetilde{g})}{2}\widetilde{g} = T. \tag{1.1}$$

Usually, a triple (M^3, g_{ij}, h_{ij}) is served as a Cauchy surface on the initial problem of the Einstein equations. Here M^3 is a 3-dimensional spacelike hypersurface with induced Riemannian metric g_{ij} and h_{ij} is a symmetric 2-tensor (e.g. the second fundamental form of M in N).

It is difficult to globally define the total energy, total linear momentum, and total angular momentum in general relativity. However, these basic quantities are well-studied for asymptotically flat initial data sets. Physicists believe, with some justification, that the total mass for a nontrivial isolated gravitational system must be positive. This was the famous positive mass conjecture which was first proved by Schoen and Yau in a series of papers[11–13] using minimal surface techniques and then by Witten[14,10] using spinors.

2. Positive Mass Theorems in General Relativity

Recall that an initial data set (M^3, g_{ij}, h_{ij}) is said to be asymptotically flat if there is a compact set K in M such that the end $M - K$ is diffeomorphic to $\mathbb{R}^3 - B_R(0)$ where $B_R(0)$ is the ball of radius R with center at the origin. Under this diffeomorphism, the metric on the end $M - K$ is of the form

$$g_{ij} = \delta_{ij} + O(r^{-\tau}),\ \partial_k g_{ij} = O(r^{-\tau-1}),\ \partial_k\partial_l g_{ij} = O(r^{-\tau-2}). \tag{2.1}$$

Furthermore, the second fundamental form h_{ij} satisfies

$$h_{ij} = O(r^{-\tau-1}),\ \partial_k h_{ij} = O(r^{-\tau-2}) \tag{2.2}$$

for the asymptotic order $\tau > \frac{1}{2}$.

For this space, the total mass and the total linear momentum are defined as follows[1]:

$$E = \lim_{R\to\infty} \frac{1}{16\pi} \int_{S_R} (\partial_j g_{ij} - \partial_i g_{jj}) * dx_i, \tag{2.3}$$

$$P_k = \lim_{R\to\infty} \frac{1}{8\pi} \int_{S_R} (h_{ki} - \delta_{ki} h_{jj}) * dx_i, \tag{2.4}$$

where S_R denotes the sphere of radius R.

When the asymptotic order $\tau > \frac{1}{2}$, these quantities are finite and independent of the asymptotic coordinates[2,5].

The positive mass theorem states

Theorem 2.1. (Schoen-Yau[11–13]; Witten[14,10]) *With the assumptions above and assuming that M satisfies the following dominant energy condition*

$$T^{00} \geq |T^{\alpha\beta}|,\ T^{00} \geq (-T_{0i}T^{0i})^{\frac{1}{2}}, \tag{2.5}$$

then one has

$$E - |P| \geq 0. \tag{2.6}$$

Moreover, if $E = 0$, then N is flat along M.

In Ref. 17, Yau asked what a good definition of total angular momentum and what the relationship would be with the total mass. Zhang answered this question in Ref. 18. The main idea is as follows. First, we define the local angular momentum density $\widetilde{h}^z_{ij}$ with respect to a point $z \in M$ by

$$\widetilde{h}^z_{ij} = \frac{1}{2}\epsilon_i^{uv}(\nabla_u \rho_z^2)(h_{vj} - g_{vj} tr_g(h)) \tag{2.7}$$

where ρ is the distance function of M with respect to z and ϵ_{ijk} are the components of the volume element. Note that this 2-tensor $\widetilde{h}^z_{ij}$ is not symmetric in general. Then, the total angular momentum with respect to the point z is defined[18] by

$$J_k(z) = \lim_{R\to\infty} \frac{1}{8\pi} \int_{S_R} \widetilde{h}^z_{ki} * dx_i. \tag{2.8}$$

(Here we need some 'regular' conditions in Ref. 18.) Zhang also proved a new positive mass theorem associated with nonsymmetric initial data p_{ij}. Denote

$$\mu = \frac{1}{2}(R + (\sum_i p_{ii})^2 - \sum_{i,j} p_{ij}^2), \tag{2.9}$$

$$\omega_j = \sum_i (\nabla_i p_{ji} - \nabla_j p_{ii}), \tag{2.10}$$

$$\chi_j = 2\sum_i \nabla_i (p_{ij} - p_{ji}), \tag{2.11}$$

where R is the scalar curvature of M. The total momentum is defined as the same as in (2.4) except to replace h_{ij} by p_{ij}.

Theorem 2.2. (Zhang[18]) *Let (M, g_{ij}, p_{ij}) be a 3-dimensional almost asymptotically flat initial data set. If M satisfies the dominant energy condition*

$$\mu \geq max\{\sqrt{\sum_j \omega_j^2}, \sqrt{\sum_j (\omega_j + \chi_j)^2}\}, \tag{2.12}$$

then one has

$$E - |P| \geq 0. \tag{2.13}$$

If $E = 0$ and g_{ij} is C^2, p_{ij} is C^1, then the following equations hold on M:

$$R_{ijkl} + p_{ik}p_{jl} - p_{il}p_{jk} = 0,\ \nabla_i p_{jk} - \nabla_j p_{ik} = 0,\ \sum_i \nabla_i (p_{ij} - p_{ji}) = 0. \tag{2.14}$$

Finally, by taking $p_{ij} = \widetilde{h}^z_{ij}$ and $p_{ij} = h_{ij} \pm \widetilde{h}^z_{ij}$ to a regular point z in Theorem 2.2, we obtain the positive mass theorems involving the total angular momentum. This theorem is also extended to higher dimensional spin asymptotically flat initial data sets[18].

We refer to Ref. 19 for an extensive and detailed survey of positive mass theorems in general relativity.

3. Calabi-Yau Compactification and New Positive Mass Theorems

According to string theory[4], our universe is really 10-dimensional, modelled on $\mathbb{R}^{1,3} \times X$ where X is a Calabi-Yau 3-fold. This is the so-called Calabi-Yau compactification. The spatial slices of such spacetime then asymptotically approach the product of the flat Euclidean space with a compact Calabi-Yau manifold. Hertog-Horowitz-Maeda constructed classical configuration which has regions of negative energy density as seen from four dimensional perspective[9]. Physically, the negative energy density leads to the possible violation of Cosmic Censorship and new thermal instability. This guides us to revisit the concept of the mass in string theory.

We consider the complete Riemannian manifold (M, g) such that $M = M_0 \cup M_\infty$ with M_0 compact and $M_\infty \simeq (\mathbb{R}^k - B_R(0)) \times X$ for some $R > 0$ and X a compact simply connected Calabi-Yau manifold. We will call (M, g) a space with asymptotic Calabi-Yau compactification if the metric on the end M_∞ satisfies the following asymptotic conditions

$$g = \mathring{g} + h, \ \mathring{g} = g_{\mathbb{R}^k} + g_X, \tag{3.1}$$

$$h = O(r^{-\tau}), \ \mathring{\nabla} h = O(r^{-\tau-1}), \ \mathring{\nabla}\mathring{\nabla} h = O(r^{-\tau-2}). \tag{3.2}$$

Here $\mathring{\nabla}$ is the Levi-Civita connection with respect to $\mathring{g}$, $\tau > \frac{k-2}{2}$ $(k \geq 3)$ is the asymptotic order, and r is the Euclidean distance to a base point.

For such a space (M, g), the total mass is defined[6] as

$$E = \lim_{R \to \infty} \frac{1}{4\omega_k vol(X)} \int_{S_R \times X} (\partial_i g_{ij} - \partial_j g_{aa}) * dx_j dvol(X), \tag{3.3}$$

where the $*$ operator is the one on the Euclidean factor, the index i, j run over the Euclidean factor while the index a runs over the full index of the manifold.

A positive mass theorem for such a space was established by Dai recently.

Theorem 3.1. **(Dai[6])** *Let (M, g) be a complete spin manifold as above. If M has nonnegative scalar curvature, then $E \geq 0$ and $E = 0$ if and only if $M = \mathbb{R}^k \times X$.*

The above positive mass theorem is applied to the study of stability of Ricci flat manifolds[8]. And the Lorentzian version of this theorem was discussed in Ref. 7.

Motivated by the study of total angular momentum in general relativity[18], we generalize Dai's positive mass theorem to an initial data set with nonsymmetric p_{ab}. Let (M,g) be as above and assume further that on the end M_∞ the nonsymmetric 2-tensor p satisfies

$$p_{\beta\alpha} = p_{\beta i} = p_{i\beta} = 0 \tag{3.4}$$

and

$$p = O(r^{-\tau-1}),\ \overset{\circ}{\nabla} p = O(r^{-\tau-2}). \tag{3.5}$$

Here the index α, β run over the compact factor while the index i runs over Euclidean part.

We also define the total momentum as

$$P_k = \lim_{R\to\infty} \frac{1}{4\omega_k vol(X)} \int_{S_R\times X} 2(p_{kj} - \delta_{kj} p_{ii}) * dx_j dvol(X). \tag{3.6}$$

Again, the $*$ operator is the one on the Euclidean factor, and the index i, j, k run over the Euclidean factor.

We say that (M,g,p) satisfies the dominant energy condition if

$$\mu \geq max\Big\{\sqrt{\sum_a (\omega_a)^2}, \sqrt{\sum_a (\omega_a + \chi_a)^2}\Big\} + \sqrt{\sum_{1\leq a\leq n-3} \kappa_a^2}. \tag{3.7}$$

Here, local energy density is defined as

$$\mu = \frac{1}{2}(R + (\sum_a p_{aa})^2 - \sum_{a,b} p_{ab}^2) \tag{3.8}$$

where R is the scalar curvature, and local momentum densities are defined as

$$\omega_a = \sum_b (\nabla_b p_{ab} - \nabla_a p_{bb}), \tag{3.9}$$

$$\chi_a = 2\sum_b \nabla_b \widetilde{p}_{ba}, \tag{3.10}$$

$$\kappa_a^2 = \sum_{b,c,d;c>d>b>a} (\widetilde{p}_{ab}\widetilde{p}_{cd} + \widetilde{p}_{ac}\widetilde{p}_{db} + \widetilde{p}_{ad}\widetilde{p}_{bc})^2, \tag{3.11}$$

where $\widetilde{p}_{ab} = p_{ab} - p_{ba}$.

The generalized positive mass theorem associated with nonsymmetric initial data p_{ab} for spaces with asymptotic Calabi-Yau compactification is

Theorem 3.2. **(Xie[16])** *Let (M, g, p) be a complete spin manifold as above. If (M, g, p) satisfies the dominant energy condition (3.7), then one has*

$$E - |P| \geq 0. \tag{3.12}$$

Our argument is also of Witten-type adapting the methods in Ref. 18 and Ref. 6, 7. The main idea is to use the modified Dirac-Witten operators in Ref. 18 and then our positive mass theorem is a consequence of two nice generalized Weitzenböck formulae.

4. Some Remarks

We remark that both Theorem 3.1 and Theorem 3.2 can be extended to spaces asymptotically approach the product of a flat Euclidean space with a compact simply connected manifold which admits a nonzero parallel spinors.

However, Witten observed that the positive mass theorems do not extend immediately to Kaluza-Klein theory[15]. The analytically continued Reissner-Nordström metric, explicitly constructed in Ref. 3, is a negative energy solution. The reason here might be that the end $\mathbb{R}^3 \times S^1$, and especially S^1, has the wrong spin structure. The reader is referred to Sec. 5 in Ref. 6 for additional discussions regarding the spin structures.

Finally, we should mention that the present short article is only a very restricted and compressed version of the talk and its references (in particular to physical discussions) are far from complete. We apologize to the authors of relevant papers which we have not cited and the reader can find out more elaborate treatments in the original articles.

Acknowledgements

The author would like to express his gratitude to Professors Xiao Zhang and Xianzhe Dai for sharing their ideas and for many useful discussions on the topic of positive mass theorems. He also particularly wishes to thank Professors Chaohao Gu and Hesheng Hu for encouragement.

References

1. S. Arnowitt, S. Deser, and C. Misner, *Coordinate invariance and energy expressions in general relativity*, Phys. Rev. **122**, 997-1006 (1961).
2. R. Bartnik, *The mass of an asymptotically flat manifold*, Comm. Pure. Appl. Math. **36**, 661-693 (1986).

3. D. Brill and G. Horowitz, *Negative energy in string theory*, Phys. Lett. **B262**, 437-443 (1991).
4. P. Candelas, G. Horowitz, A. Strominger, and E. Witten, *Vacuum configurations for superstrings*, Nucl. Phys. **B258**, 46-74 (1985).
5. P. Chruściel, *Boundary conditions at spatial infinity from a Hamiltonian point of view*, in Topological properties and global structure of space-time (Erice 1985), eds. P. Bergmann/V. de Sabbata, NATO, Adv. Sci. Inst. Ser. B, Plenum Press NY, 49-59 (1986).
6. X. Dai, *A Positive Mass Theorem for Spaces with Asymptotic SUSY Compactification*, Comm. Math. Phys. **244**, 335-345 (2004).
7. X. Dai, *A Note on Positive Energy Theorem for Spaces with Asymptotic SUSY Compactification*, J. Math. Phys. **46**, 042505 (2005).
8. X. Dai, X. Wang, and G. Wei, *On the stability of Riemannian manifold with parallel spinors*, Invent. Math. **161**, 151-176 (2005).
9. T. Hertog, G. Horowitz, and K. Maeda, *Negative energy density in Calabi-Yau compactifications*, JHEP **0305**, 060 (2003).
10. T. Parker and C. Taubes, *On Witten's proof of the positive energy theorem*, Comm. Math. Phys. **84**, 223-238 (1982).
11. R. Schoen and S.T. Yau, *On the proof of the positive mass conjecture in general relativity*, Comm. Math. Phys. **65**, 45-76 (1979).
12. R. Schoen and S.T. Yau, *The energy and the linear momentum of spacetimes in general relativity*, Comm. Math. Phys. **79**, 47-51 (1981).
13. R. Schoen and S.T. Yau, *Proof of the positive mass theorem. II*, Comm. Math. Phys. **79**, 231-260 (1981).
14. E. Witten, *A new proof of the positive energy theorem*, Comm. Math. Phys. **80**, 381-402 (1981).
15. E. Witten, *Instability of the Kaluza-Klein vacuum*, Nucl. Phys. **B195**, 481-492 (1982).
16. N. Xie, *A generalized positive energy theorem for spaces with asymptotic SUSY compactification*, J. Geom. Phys. **56**, 271-281 (2006).
17. S.T. Yau, *Problem section*, in Seminar on differential geometry, ed. S.T. Yau, Annals of Math. Stud. **102**, Princeton NJ, 699-706 (1982).
18. X. Zhang, *Angular momentum and positive mass theorem*, Comm. Math. Phys. **206**, 137-155 (1999).
19. X. Zhang, *The positive mass theorem in general relativity*, in Geometry and nonlinear partial differential equations (Hangzhou 2001), eds. S. Chen/S.T. Yau, AMS/IP Studies in Advanced Mathematics **29**, Providerce RI, 227-233 (2002).

Analytic Torsion and an Invariant of Calabi–Yau Threefold

Ken-Ichi Yoshikawa*

Graduate School of Mathematical Sciences,
University of Tokyo, Tokyo 153-8914, JAPAN
E-mail: yosikawa@ms.u-tokyo.ac.jp

1. Introduction

In mirror symmetry, it is expected that the analytic torsion of a Calabi–Yau threefold X provides an invariant $F_1(X)$. Following [2] and [9], we give a mathematical definition of $F_1(X)$, which we obtain using analytic torsion and a Bott–Chern secondary class. (See Sec. 2.) We write $\tau_{\rm BCOV}(X)$ for $F_1(X)$ and we call it the BCOV invariant of X. Then $\tau_{\rm BCOV}$ gives rise to a function on the moduli space of Calabi–Yau threefolds.

In [1], [2], Bershadsky–Cecotti–Ooguri–Vafa used mirror symmetry to study the function $\tau_{\rm BCOV}$ on the moduli space of quintic mirror threefolds. They gave a conjectural expression of $\tau_{\rm BCOV}$ as a generating function of the genus-one Gromov–Witten invariants of a general quintic hypersurface of $\mathbf{P}^4$. (See Sec. 3.) In [9], we gave an explicit formula for $\tau_{\rm BCOV}$ as a function on the moduli space of quintic mirror threefolds, which reduces the BCOV conjecture to a problem of symplectic geometry. (See Sec. 4.)

For a class of Calabi–Yau threefolds introduced by Borcea [6] and Voisin [17], Harvey–Moore studied the function $\tau_{\rm BCOV}$ on their moduli space. In [11], Harvey–Moore conjectured that $\tau_{\rm BCOV}$ is expressed as the norm of the denominator function of some generalized Kac–Moody superalgebra in these cases. In [22], we shall prove the Harvey–Moore conjecture for certain Calabi–Yau threefolds of Borcea–Voisin type. (See Sec. 5.)

In this article, we report a recent progress in the BCOV conjecture and the Harvey–Moore conjecture obtained in [9] and [22]. The results stated in Secs. 2 and 4 are based on the joint work with Hao Fang and Zhiqin Lu.

*partially supported by the Grants-in-Aid for Scientific Research for Young Scientists (B) 16740030, JSPS

2. Calabi–Yau threefold and the BCOV invariant

Let $\overline{X} = (X, g)$ be a compact Kähler manifold with Kähler form γ. Let $\square_{p,q}$ be the Laplacian of $\overline{X}$ acting on (p,q)-forms on X, and let $\zeta_{p,q}(s)$ be the spectral zeta function of $\square_{p,q}$. After Ray–Singer [16], Bismut–Gillet–Soulé [4], and Bershadsky–Cecotti–Ooguri–Vafa [2], we make the following

Definition 2.1. The *BCOV torsion* of $\overline{X}$ is defined as

$$\mathcal{T}_{\mathrm{BCOV}}(\overline{X}) := \exp[-\sum_{p,q\geq 0}(-1)^{p+q}pq\,\zeta'_{p,q}(0)].$$

Recall that a smooth, irreducible, compact Kähler n-fold X with canonical line bundle K_X is *Calabi–Yau* if the following hold:

$$(1)\quad K_X \cong \mathcal{O}_X, \qquad (2)\quad H^q(X, \mathcal{O}_X) = 0 \quad (0 < q < n).$$

Assume that X is a Calabi–Yau n-fold. Let $\mathrm{Vol}(\overline{X})$ be the volume of $\overline{X}$. Let $c_i(\overline{X})$ be the i-th Chern form of (TX, g). Then $\chi(X) = \int_X c_n(\overline{X})$ is the Euler number of X. Let η be a nowhere vanishing holomorphic n-form on X, whose L^2-norm is denoted by $\|\eta\|^2_{L^2}$. Define

$$\mathcal{A}(\overline{X}) := \mathrm{Vol}(\overline{X})^{\frac{\chi(X)}{12}} \exp\left[-\int_X \log\left(\frac{(\sqrt{-1})^{n^2}\eta\wedge\bar{\eta}}{\gamma^n/n!}\cdot\frac{\mathrm{Vol}(\overline{X})}{\|\eta\|^2_{L^2}}\right)\frac{c_n(\overline{X})}{12}\right].$$

By Hodge theory, $H^2(X, \mathbf{R})$ is equipped with the L^2-metric with respect to the Kähler class $[\gamma]$. Set $H^2(X, \mathbf{Z})_{\mathrm{fr}} := H^2(X, \mathbf{Z})/\mathrm{Torsion}$. Define $\mathrm{Vol}_{L^2}(\overline{H^2(X, \mathbf{Z})})$ as the volume of the real torus $H^2(X, \mathbf{R})/H^2(X, \mathbf{Z})_{\mathrm{fr}}$.

Definition 2.2. When $\dim X = 3$, define the *BCOV invariant* of X as

$$\tau_{\mathrm{BCOV}}(X) := \frac{\mathcal{A}(\overline{X})\,\mathcal{T}_{\mathrm{BCOV}}(\overline{X})}{\mathrm{Vol}(\overline{X})^3\,\mathrm{Vol}_{L^2}(\overline{H^2(X, \mathbf{Z})})}.$$

By the curvature formula for Quillen metrics [4], we have (cf. [9])

Theorem 2.1. *When* $\dim X = 3$, $\tau_{\mathrm{BCOV}}(X)$ *is independent of the choice of a Kähler metric on* X. *In particular,* $\tau_{\mathrm{BCOV}}(X)$ *is an invariant of* X.

When Calabi–Yau threefolds X and X' are birationally equivalent, their Hodge numbers coincide, i.e., $h^{p,q}(X) = h^{p,q}(X')$ for $p, q \geq 0$. As an analogue, we make the following

Conjecture 2.1. *If Calabi–Yau threefolds* X *and* X' *are birationally equivalent, then*

$$\tau_{\mathrm{BCOV}}(X) = \tau_{\mathrm{BCOV}}(X').$$

3. Mirror Symmetry and the BCOV conjecture

Let $p\colon \mathcal{X} = \{([z],\psi) \in \mathbf{P}^4 \times \mathbf{P}^1;\ F_\psi(z) = 0\} \ni ([z],\psi) \to \psi \in \mathbf{P}^1$ be the pencil of quintic hypersurfaces of $\mathbf{P}^4$ defined by the equation

$$F_\psi(z) := z_0^5 + z_1^5 + z_2^5 + z_3^5 + z_4^5 - 5\psi\, z_0 z_1 z_2 z_3 z_4,$$

where ψ is the inhomogeneous coordinate of $\mathbf{P}^1$. Set $X_\psi := p^{-1}(\psi)$ for $\psi \in \mathbf{P}^1$. Then X_ψ is a Calabi–Yau threefold when $\psi^5 \neq 1, \infty$. Let Ω_ψ be the holomorphic 3-form on X_ψ defined as

$$\Omega_\psi := \left(\frac{2\pi i}{5}\right)^{-3} 5\psi\, \frac{dz_0 \wedge dz_1 \wedge dz_2}{\partial F_\psi(z)/\partial z_3}.$$

Set $y_0(\psi) := \sum_{n=1}^{\infty} \frac{(5n)!}{(n!)^5 (5\psi)^{5n}}$ for $|\psi| \gg 1$. Define the Yukawa coupling by

$$\kappa_B\left(\frac{d}{d\psi}, \frac{d}{d\psi}, \frac{d}{d\psi}\right) := \int_{X_\psi} \frac{\Omega_\psi}{y_0(\psi)} \wedge \frac{\partial^3}{\partial\psi^3}\left(\frac{\Omega_\psi}{y_0(\psi)}\right) = \left(\frac{2\pi i}{5}\right)^3 \frac{5\psi^2}{1-\psi^5} \cdot \frac{1}{y_0(\psi)^2}.$$

Let $\mathfrak{H}$ be the complex upper half-plane. For $t \in \mathfrak{H}$, let $q := e^{2\pi i t}$ be the parameter of the unit disc of $\mathbf{C}$. Let $N_g(d)$ be the degree-d, genus-g Gromov–Witten invariant of a general quintic threefold of $\mathbf{P}^4$. Define the quantum cup-product by

$$\kappa_A\left(\frac{d}{dt}, \frac{d}{dt}, \frac{d}{dt}\right) := 5 + \sum_{d=1}^{\infty} N_0(d)\, \frac{d^3 q^d}{1-q^d}.$$

The mirror map is the identification of ψ^5 and q defined as

$$q := (5\psi)^{-5} \exp\left(\frac{5}{y_0(\psi)} \sum_{n=1}^{\infty} \frac{(5n)!}{(n!)^5} \left\{\sum_{j=n+1}^{5n} \frac{1}{j}\right\} \frac{1}{(5\psi)^{5n}}\right). \tag{3.1}$$

The following identity was conjectured by Candelas–de la Ossa–Green–Parkes [8], and it was proved by Givental [10], Lian–Liu–Yau [14].

Theorem 3.1. *Under the identification* (3.1), *the following identity holds:*

$$\kappa_A\left(\frac{d}{dt}, \frac{d}{dt}, \frac{d}{dt}\right) = \left(2\pi i\, q \frac{d\psi}{dq}\right)^3 \kappa_B\left(\frac{d}{d\psi}, \frac{d}{d\psi}, \frac{d}{d\psi}\right).$$

Bershadsky–Cecotti–Ooguri–Vafa extended the mirror symmetry conjecture to the genus-one Gromov–Witten invariants $\{N_1(d)\}_{d\geq 1}$ as follows.

Let $\mathbf{Z}_5 = \{\zeta \in \mathbf{C};\ \zeta^5 = 1\}$, which is a cyclic group of order 5. Set $G := \{[\mathrm{diag}(a_0, a_1, a_2, a_3, a_4)] \in PSL(\mathbf{C}^5);\ a_i \in \mathbf{Z}_5\}$, which acts fiberwise on $\mathcal{X}$. We have the induced family $p\colon \mathcal{X}/G \to \mathbf{P}^1$, whose general fiber is a Calabi–Yau orbifold X_ψ/G. Set $\mathcal{D}^* = \mathbf{Z}_5 \subset \mathbf{C}$ and $\mathcal{D} = \mathcal{D}^* \cup \{\infty\}$.

Definition 3.1. Let $f: \mathcal{W} \to \mathcal{X}/G$ be a resolution of the singularities of $\mathcal{X}/G$, and set $\pi := p \circ f$. Then $\pi: \mathcal{W} \to \mathbf{P}^1$ is called a *family of quintic mirror threefolds* if the following hold:
(1) For all $\psi \in \mathbf{P}^1 \setminus \mathcal{D}$, the map of fibers $f_\psi: W_\psi = \pi^{-1}(\psi) \to X_\psi/G$ induced by f is a resolution such that $K_{W_\psi} \cong \mathcal{O}_{W_\psi}$;
(2) $\operatorname{Sing} W_\psi$ consists of a unique ordinary double point if $\psi \in \mathcal{D}^*$.

By [15], there exists a family of quintic mirror threefolds. By the G-invariance of Ω_ψ, we identify Ω_ψ with the corresponding 3-form on X_ψ/G. Let Ξ_ψ be the holomorphic 3-form on W_ψ defined as $\Xi_\psi := f_\psi^* \Omega_\psi$. Let $K_{\mathcal{W}/\mathbf{P}^1}$ be the relative canonical bundle of the family $\pi: \mathcal{W} \to \mathbf{P}^1$. After Theorem 3.1, the line bundle $\pi_* K_{\mathcal{W}/\mathbf{P}^1}$ (resp. $T\mathbf{P}^1$) is trivialized by the section $\Xi_\psi / y_0(\psi)$ (resp. $d/dt = 2\pi i\, q\,(d/dq) = 2\pi i\, q\,(d\psi/dq)\,(d/d\psi)$) near $\psi = \infty$ (equivalently $q = 0$). Set $\widetilde{\eta}(q) := \prod_{n=1}^{\infty}(1-q^n)$.

In [1], [2], Bershadsky–Cecotti–Ooguri–Vafa made the following

Conjecture 3.1. *Up to a constant, the following identity of functions near $\psi = \infty$ holds under the identification* (3.1)*:*

$$\tau_{\mathrm{BCOV}}(W_\psi) = \left\| \left\{ q^{\frac{25}{12}} \prod_{d=1}^{\infty} \widetilde{\eta}(q^d)^{N_1(d)} (1-q^d)^{\frac{N_0(d)}{12}} \right\}^2 \left(\frac{\Xi_\psi}{y_0(\psi)} \right)^{\frac{62}{3}} \otimes q \frac{d}{dq} \right\|^2,$$

where $\pi_* K_{\mathcal{W}/\mathbf{P}^1}$ is equipped with the L^2-metric and $T\mathbf{P}^1$ is equipped with the Weil–Petersson metric.

This conjecture can be separated into the following two conjectures. Under the identification (3.1), define two functions $F_{1,B}^{\mathrm{top}}(\psi)$ and $F_{1,A}^{\mathrm{top}}(q)$ by

$$F_{1,B}^{\mathrm{top}}(\psi) := \left(\frac{\psi}{y_0(\psi)} \right)^{\frac{62}{3}} (\psi^5 - 1)^{-\frac{1}{6}}\, q \frac{d\psi}{dq}, \qquad F_{1,A}^{\mathrm{top}}(q) := F_{1,B}^{\mathrm{top}}(\psi(q)).$$

Conjecture 3.2. *(A) The following identity holds:*

$$q \frac{d}{dq} \log F_{1,A}^{\mathrm{top}}(q) = \frac{50}{12} - \sum_{n,d=1}^{\infty} N_1(d) \frac{2nd\, q^{nd}}{1 - q^{nd}} - \sum_{d=1}^{\infty} N_0(d) \frac{2d\, q^d}{12(1-q^d)}.$$

(B) Up to a constant, the following identity of functions near $\psi = \infty$ holds:

$$\tau_{\mathrm{BCOV}}(W_\psi) = \left\| \frac{1}{F_{1,B}^{\mathrm{top}}(\psi)} \left(\frac{\Xi_\psi}{y_0(\psi)} \right)^{\frac{62}{3}} \otimes q \frac{d}{dq} \right\|^2.$$

For Conjecture (A), we refer to [13].

4. An explicit formula for the BCOV invariant

Let $\mathcal{Y}$ be a (possibly singular) projective fourfold, and let $\pi\colon \mathcal{Y} \to \mathbf{P}^1$ be a surjective flat holomorphic map. Set $\mathcal{D} := \{\psi \in \mathbf{P}^1;\ \operatorname{Sing} Y_\psi \neq \emptyset\}$ and $\mathcal{D}^* := \{\psi \in \mathcal{D};\ \operatorname{Sing} Y_\psi$ consists of a unique ordinary double point$\}$.

In Sec. 4, we assume the following:

(i) $\mathcal{D}^*$ is a non-empty, finite subset of $\mathbf{P}^1$ such that $\mathcal{D} \setminus \mathcal{D}^* = \{\infty\}$;

(ii) if $\psi \in \mathbf{P}^1 \setminus \mathcal{D}$ (resp. $\psi \in \mathcal{D}^*$), then $Y_\psi := \pi^{-1}(\psi)$ is a (resp. singular) Calabi–Yau threefold with $h^2(\Omega^1_{Y_\psi}) = 1$.

For $\psi \in \mathbf{P}^1 \setminus \{\infty\}$, let $(\operatorname{Def}(Y_\psi), [Y_\psi])$ be the Kuranishi space of Y_ψ. By the universal property of the Kuranishi space, there exists a unique map of germs $\mu_\psi\colon (\mathbf{P}^1, \psi) \to (\operatorname{Def}(Y_\psi), [Y_\psi])$ such that the deformation germ $\pi\colon (\mathcal{Y}, Y_\psi) \to (\mathbf{P}^1, \psi)$ is induced from the universal family over $(\operatorname{Def}(Y_\psi), [Y_\psi])$ by μ_ψ. By (i), μ_ψ is not a constant map. By (ii), we have $(\operatorname{Def}(Y_\psi), [Y_\psi]) \cong (\mathbf{C}, 0)$. Let $r(\psi) \in \mathbf{Z}_{\geq 1}$ be the ramification index of μ_ψ at ψ. Write $\mathcal{D}^* = \{D_k\}_{k \in K}$ and $\{\psi \in \mathbf{P}^1 \setminus \{\infty\};\ r(\psi) > 1\} = \{R_j\}_{j \in J}$.

Let Ξ be a meromorphic section of $\pi_* K_{\mathcal{Y}/\mathbf{P}^1}$ defined on $\mathbf{P}^1$, and write $\operatorname{div}(\Xi) = \sum_{i \in I} m_i P_i + m_\infty P_\infty$, where $P_i \neq P_\infty$ for $i \in I$. Identify P_i, R_j, D_k with their coordinates $\psi(P_i), \psi(R_j), \psi(D_k)$, respectively. Let χ be the Euler number of a general fiber Y_ψ.

Theorem 4.1. *Up to a constant, the following identity of functions on* $\mathbf{P}^1$ *holds:*

$$\tau_{\mathrm{BCOV}}(Y_\psi) = \left\| \prod_{i \in I, j \in J, k \in K} \frac{(\psi - D_k)^{\frac{r(D_k)}{6}}}{(\psi - P_i)^{\frac{(48+\chi)m_i}{12}} (\psi - R_j)^{r(R_j)-1}} \, \Xi_\psi^{\frac{48+\chi}{12}} \otimes \frac{d}{d\psi} \right\|^2 .$$

See [9], [20] for the proof, in which the theory of Quillen metrics [3], [4], [5] plays the central role. Theorem 4.1, applied to the family of quintic mirror threefolds $\pi\colon \mathcal{W} \to \mathbf{P}^1$, yields the following result [9].

Theorem 4.2. *Conjecture (B) holds, i.e., the following identity of functions on* $\mathbf{P}^1$ *holds:*

$$\tau_{\mathrm{BCOV}}(W_\psi) = \mathrm{Const.} \left\| \psi^{-\frac{62}{3}} (\psi^5 - 1)^{\frac{1}{6}} (\Xi_\psi)^{\frac{62}{3}} \otimes \frac{d}{d\psi} \right\|^2 .$$

For other examples of one-parameter families of Calabi–Yau threefolds satisfying assumptions (i), (ii), we refer to [12].

5. BCOV invariant and the Borcherds product

Let S be a $K3$ surface, i.e., a two-dimensional Calabi–Yau manifold. Fix an even unimodular lattice $\mathbb{L}_{K3}$ with signature $(3,19)$. Then $H^2(S,\mathbf{Z})$ endowed with the cup-product is isometric to $\mathbb{L}_{K3}$. In Sec. 5, we assume the existence of a holomorphic involution $\theta\colon S \to S$ such that $\theta^* = -1$ on $H^0(S,K_S)$. Then S is an algebraic $K3$ surface. Let T be an elliptic curve. Let $-1_T\colon T \to T$ be the holomorphic involution that assigns $x \in T$ the inverse $-x \in T$. The involution $(\theta,-1_T)$ on $S\times T$ acts trivially on $H^0(S\times T, K_{S\times T})$. Set $\mathbf{Z}_2 := \mathbf{Z}/2\mathbf{Z}$. By identifying the generator of $\mathbf{Z}_2$ with θ, -1_T, $(\theta,-1_T)$, the group $\mathbf{Z}_2$ acts on S, T, $S\times T$, respectively. After Borcea [6] and Voisin [17], we make the following

Definition 5.1. For a $K3$ surface with involution (S,θ) and an elliptic curve T, let $X_{(S,\theta,T)}$ be the Calabi-Yau threefold defined as the blow-up of $S\times T/\mathbf{Z}_2$ along $\mathrm{Sing}\,(S\times T/\mathbf{Z}_2)$. Let $\pi_1\colon X_{(S,\theta,T)} \to S/\mathbf{Z}_2$ (resp. $\pi_2\colon X_{(S,\theta,T)} \to T/\mathbf{Z}_2$) be the projection induced from the projection $\mathrm{pr}_1\colon S\times T \to S$ (resp. $\mathrm{pr}_2\colon S\times T\to T$). The triple $(X_{(S,\theta,T)},\pi_1,\pi_2)$ is called the *Borcea–Voisin threefold associated with* (S,θ,T). Two Borcea–Voisin threefolds $(X_{(S,\theta,T)},\pi_1,\pi_2)$ and $(X_{(S',\theta',T')},\pi_1',\pi_2')$ are isomorphic if there exist isomorphisms of complex manifolds

$$f\colon X_{(S,\theta,T)} \to X_{(S',\theta',T')}, \qquad g\colon S/\mathbf{Z}_2 \to S'/\mathbf{Z}_2, \qquad h\colon T/\mathbf{Z}_2 \to T'/\mathbf{Z}_2$$

such that $\pi_1'\circ f = g\circ\pi_1$ and $\pi_2'\circ f = h\circ \pi_2$.

Definition 5.2. Let $\Lambda \subset \mathbb{L}_{K3}$ be a primitive 2-elementary sublattice of rank $r(\Lambda)$ with signature $(2, r(\Lambda)-2)$. A Borcea–Voisin threefold $(X_{(S,\theta,T)},\pi_1,\pi_2)$ is of type Λ if $H^2_-(S,\mathbf{Z}) := \{l \in H^2(S,\mathbf{Z});\ \theta^* l = -l\} \cong \Lambda$.

When $X_{(S,\theta,T)}$ is a Borcea–Voisin threefold of type Λ, (S,θ) is a 2-elementary $K3$ surface of type $\Lambda^\perp$ in the sense of [19].

Let $\Omega_\Lambda := \{[\eta] \in \mathbf{P}(\Lambda\otimes\mathbf{C});\ \langle\eta,\eta\rangle_\Lambda = 0,\ \langle\eta,\bar\eta\rangle_\Lambda > 0\}$ be the period domain for 2-elementary $K3$ surfaces of type $\Lambda^\perp$, which consists of two connected components Ω_Λ^+, Ω_Λ^-. Let $D_{IV,n}$ be a symmetric bounded domain of type IV of dimension n. Then $D_{IV,n} \cong \Omega_\Lambda^\pm$. Let $O(\Lambda)$ be the group of isometries of Λ, which acts projectively on Ω_Λ. There exists a subgroup $O^+(\Lambda) \subset O(\Lambda)$ of index 2 preserving $\Omega_\Lambda^\pm$. By [19], we have the following

Theorem 5.1. *The coarse moduli space of Borcea–Voisin threefolds of type Λ is isomorphic to a dense Zariski open subset of the locally symmetric variety $(O^+(\Lambda)\backslash D_{IV,r(\Lambda)-2}) \times (SL_2(\mathbf{Z})\backslash\mathfrak{H})$.*

Let 1_m be the identity $m \times m$-matrix. Let $\mathbb{I}_{1,m}(2)$ be the symmetric matrix of rank $m+1$ with signature $(1,m)$ defined as $\mathbb{I}_{1,m}(2) := 2\begin{pmatrix} 1 & 0 \\ 0 & -1_m \end{pmatrix}$. Identify $\mathbb{I}_{1,m}(2)$ with the corresponding Lorentzian lattice. For $1 \leq m \leq 9$, set $\mathbb{T}_m := \begin{pmatrix} 0 & 2 \\ 2 & 0 \end{pmatrix} \oplus \mathbb{I}_{1,m-1}(2)$ and identify $\mathbb{T}_m$ with the corresponding lattice of rank $m+2$ with signature $(2,m)$. There exists an isomorphism $\Omega_{\mathbb{T}_m} \cong \mathbb{I}_{1,m-1}(2) \otimes \mathbf{R} + i\, C_{\mathbb{I}_{1,m-1}(2)}$, where $C_{\mathbb{I}_{1,m-1}(2)}$ is the light cone of the Lorentzian vector space $\mathbb{I}_{1,m-1}(2) \otimes \mathbf{R}$. Let W_m be the Weyl chamber of $C_{\mathbb{I}_{1,m-1}(2)}$ containing the Weyl vector $\rho_m := \frac{1}{2}(3, -1, \cdots, -1) \in \mathbb{I}_{1,m-1}(2)^{\vee}$, where $\mathbb{I}_{1,m-1}(2)^{\vee} \subset \mathbb{I}_{1,m-1}(2) \otimes \mathbf{Q}$ is the dual lattice of $\mathbb{I}_{1,m-1}(2)$.

For an automorphic form Ψ on $D_{IV,m}$ or on $\mathfrak{H}$, let $\|\Psi\|$ be the Petersson norm of Ψ. We shall prove the following in [22]:

Theorem 5.2. *For $3 \leq m \leq 9$, there exists an automorphic form Φ_m on $D_{IV,m}$ for $O^+(\mathbb{T}_m)$ of weight $14 - m$ satisfying the following:*
(1) For every Borcea–Voisin threefold $(X_{(S,T)}, \pi_1, \pi_2)$ of type $\mathbb{T}_m$,

$$\tau_{\mathrm{BCOV}}(X_{(S,\theta,T)}) = \|\Phi_m(\varpi(S,\theta))\|^2 \, \|\Delta(\varpi(T))\|^2. \tag{5.1}$$

Here $\varpi(S,\theta) \in O^+(\mathbb{T}_m)\backslash D_{IV,m}$ (resp. $\varpi(T) \in SL_2(\mathbf{Z})\backslash\mathfrak{H}$) denotes the period of (S,θ) (resp. T), and $\Delta(\tau)$ is the Jacobi Δ-function.
(2) There exists a generalized Kac–Moody superalgebra $\mathfrak{g}_m$ such that Φ_m is the denominator function for $\mathfrak{g}_m$ up to a constant.
(3) For $z \in \mathbb{I}_{1,m-1}(2) \otimes \mathbf{R} + i\, W_m$ with $(\mathrm{Im}\, z) \cdot (\mathrm{Im}\, z) \gg 0$, the following identity holds up to a constant:

$$\Phi_m(z) = e^{2\pi i\, \rho_m \cdot z} \prod_{\delta \in \{0,1\}} \prod_{\mathbf{r} \in (\delta\rho_m + \mathbb{I}_{1,m-1}(2)) \cap W_m^{\vee}} \left(1 - e^{2\pi i\, \mathbf{r}\cdot z}\right)^{c_m^{(\delta)}(\mathbf{r}\cdot\mathbf{r}/2)},$$

where $W_m^{\vee} \subset \mathbb{I}_{1,m-1}(2) \otimes \mathbf{R}$ is the dual cone of W_m, and the series $\{c_m^{(\delta)}(l)\}_{l \in \mathbb{Z}+\delta/4}$, $\delta = 0, 1$, are defined by the generating functions

$$\sum_{l \in \mathbb{Z}+\delta/4} c_m^{(\delta)}(l)\, q^l := \begin{cases} \eta(\tau)^{-8}\eta(2\tau)^{8}\eta(4\tau)^{-8}\, \theta_{\mathbb{A}_1}(\tau)^{10-m} & (\delta = 0), \\ -8\,\eta(4\tau)^{8}\eta(2\tau)^{-16}\, \theta_{\mathbb{A}_1+1/2}(\tau)^{10-m} & (\delta = 1). \end{cases}$$

Here $\eta(\tau)$ is the Dedekind η-function and $\theta_{\mathbb{A}_1+\delta/2}(\tau) := \sum_{m \in \mathbf{Z}+\delta/2} q^{m^2}$.

Remark 5.1. Theorem 5.2 provides evidence for the Harvey–Moore conjecture [11] Sec. 7. When $\Lambda = \begin{pmatrix} 0 & 1 \\ 1 & 0 \end{pmatrix} \oplus \mathbb{I}_{1,9}(2)$, we proved a formula like (5.1) for the function τ_{BCOV} on the moduli space of Borcea–Voisin threefolds of type Λ, in which Φ_m should be replaced by the Borcherds Φ-function [7]. See [9], [11], [19] for more details.

In [19], we introduced an invariant $\tau_{\mathbb{T}_m^\perp}(S,\theta)$ of a 2-elementary $K3$ surface (S,θ) of type $\mathbb{T}_m^\perp$. By [19], [22], there exists a constant $C_{\mathbb{T}_m}$ with

$$\tau_{\mathbb{T}_m^\perp}(S,\theta) = C_{\mathbb{T}_m}\,\|\Phi_m(\varpi(S,\theta))\|^{-1/2}. \tag{5.2}$$

For every Borcea–Voisin threefold $(X_{(S,\theta,T)},\pi_1,\pi_2)$ of type $\mathbb{T}_m$, we get by (5.1), (5.2)

$$\tau_{\mathrm{BCOV}}(X_{(S,\theta,T)}) = C_{\mathbb{T}_m}^4\,\tau_{\mathbb{T}_m^\perp}(S,\theta)^{-4}\|\Delta(\varpi(T))\|^2. \tag{5.3}$$

As a generalization of (5.3), we make the following

Conjecture 5.1. *There exists a constant C_Λ depending only on the lattice Λ such that for every Borcea–Voisin threefold $(X_{(S,\theta,T)},\pi_1,\pi_2)$ of type Λ,*

$$\tau_{\mathrm{BCOV}}(X_{(S,\theta,T)}) = C_\Lambda\,\tau_{\Lambda^\perp}(S,\theta)^{-4}\|\Delta(\varpi(T))\|^2. \tag{5.4}$$

Remark 5.2. Equation (5.4) may be regarded as a blow-down formula for the BCOV invariant for the blow-down $X_{(S,\theta,T)} \to S\times T/\mathbf{Z}_2$. For the corresponding blow-down formula for Quillen metrics, we refer to [3]. Notice that one can not apply at once the result of Bismut [3] Sec. 8 to the blow-down $X_{(S,\theta,T)} \to S\times T/\mathbf{Z}_2$, because $S\times T/\mathbf{Z}_2$ is not smooth.

In the rest of Sec. 5, we study a class of Borcea–Voisin threefolds parametrized by some configuration spaces. Let $M(n,2n)$ be the set of all complex $n\times 2n$-matrices, and define $M^o(n,2n)\subset M(n,2n)$ as the open subset $\{(\mathbf{a}_1,\ldots,\mathbf{a}_{2n})\in M(n,2n);\ \det(\mathbf{a}_{i_1},\cdots,\mathbf{a}_{i_n})\neq 0,\ \forall i_1<\cdots<i_n\}$. Let $X^o(n,2n) := GL_n(\mathbf{C})\backslash M^o(n,2n)/(\mathbf{C}^*)^{2n}$ be the configuration space of ordered $2n$ hyperplanes of $\mathbf{P}^{n-1}$ in general position. Here an element of $(\mathbf{C}^*)^{2n}$ is regarded as a diagonal $2n\times 2n$-matrix. For $A\in M^o(n,2n)$, set

$$\Delta_n(A) := \prod_{\{i_1<\cdots<i_n\}\cup\{j_1<\cdots<j_n\}=\{1,\cdots,2n\}} \det(\mathbf{a}_{i_1},\ldots,\mathbf{a}_{i_n})\det(\mathbf{a}_{j_1},\ldots,\mathbf{a}_{j_n}),$$

whose norm gives rise to a function on $M^o(n,2n)$:

$$\|\Delta_n(A)\|^2 := \left\{\int_{\mathbf{P}^{n-1}} \frac{\left(\frac{i}{2\pi}\right)^{n-1} dx\wedge d\bar{x}}{\left|\prod_{i=1}^{2n}(a_{1i}x_1+\cdots+a_{ni}x_n)\right|}\right\}^{\binom{2n}{n}} |\Delta_n(A)|^2.$$

Here $dx = \sum_{i=1}^n (-1)^i x_i\, dx_1\wedge\cdots\wedge dx_{i-1}\wedge dx_{i+1}\wedge\cdots\wedge dx_n$. Then $\|\Delta_n\|$ descends to a function on $X^o(n,2n)$. We consider the cases $n=2,3$.

For $A = (a_{ij}) \in M^o(2,4)$ and $B = (b_{ij}) \in M^o(3,6)$, set

$$E_A := \{((x_1 : x_2), y) \in \mathcal{O}_{\mathbf{P}^1}(2);\ y^2 = \prod_{i=1}^{4}(a_{1i}\,x_1 + a_{2i}\,x_2)\},$$

$$S_B := \{((z_1 : z_2 : z_3), w) \in \mathcal{O}_{\mathbf{P}^2}(3);\ w^2 = \prod_{i=1}^{6}(b_{1i}\,z_1 + b_{2i}\,z_2 + b_{3i}\,z_3)\}.$$

Then E_A is an elliptic curve, and S_B is a singular $K3$ surface with 15 ordinary double points. Obviously, $E_A \cong E_{A'}$ (resp. $S_B \cong S_{B'}$) if $A = A'$ in $X(2,4)$ (resp. $B = B'$ in $X(3,6)$). Let $\widetilde{S}_B$ be the minimal resolution of S_B. The involution on S_B defined as $\theta_B(z,w) := (z,-w)$ for $(z,w) \in S_B$ induces an involution ι_B on $\widetilde{S}_B$. Then $\iota_B^* = -1$ on $H^0(\widetilde{S}_B, K_{\widetilde{S}_B})$. Similarly, the involution on E_A defined as $(x,y) \mapsto (x,-y)$ coincides with -1_{E_A}.

For $A \in M^o(2,4)$ and $B \in M^o(3,6)$, define

$$Z_{(A,B)} := X_{(\widetilde{S}_B, \iota_B, E_A)},$$

which is equipped with the projections $\pi_1 \colon Z_{(A,B)} \to \widetilde{S}_B/\iota_B$ and $\pi_2 \colon Z_{(A,B)} \to E_A/-1_{E_A}$. By e.g. [18], $(Z_{(A,B)}, \pi_1, \pi_2)$ is a Borcea–Voisin threefold of type $\mathbb{T}_4$. By (5.1), (5.2) and [21], we have the following

Theorem 5.3. *The following identity of functions on $X^o(2,4) \times X^o(3,6)$ holds:*

$$\tau_{\mathrm{BCOV}}(Z_{(A,B)}) = \mathrm{Const.}\, \|\Delta_2(A)\|^4\, \|\Delta_3(B)\|.$$

This algebraic expression of the BCOV invariant of $Z_{(A,B)}$ is an analogue of the Kronecker limit formula for elliptic curves. It seems to be an interesting problem to determine the constant in Theorem 5.3.

References

1. Bershadsky, M., Cecotti, S., Ooguri, H., Vafa, C. *Holomorphic anomalies in topological field theories*, Nuclear Phys. B **405** (1993), 279–304
2. Bershadsky, M., Cecotti, S., Ooguri, H., Vafa, C. *Kodaira–Spencer theory of gravity and exact results for quantum string amplitudes*, Commun. Math. Phys. **165** (1994), 311–427
3. Bismut, J.-M. *Quillen metrics and singular fibers in arbitrary relative dimension*, Jour. Algebr. Geom. **6** (1997), 19–149
4. Bismut, J.-M., Gillet, H., Soulé, C. *Analytic torsion and holomorphic determinant bundles I,II,III*, Commun. Math. Phys. **115** (1988), 49–78, 79–126, 301–351
5. Bismut, J.-M., Lebeau, G. *Complex immersions and Quillen metrics*, Publ. Math. IHES **74** (1991), 1–297

6. Borcea, C. *K3 surfaces with involution and mirror pairs of Calabi–Yau manifolds*, Mirror Symmetry II (B. Green and S.-T. Yau eds) AMS/International Press (1997), 717–743
7. Borcherds, R.E. *The moduli space of Enriques surfaces and the fake monster Lie superalgebra*, Topology **35** (1996), 699–710
8. Candelas, P., de la Ossa, X., Green, P., Parkes, L. *A pair of Calabi–Yau manifolds as an exactly solvable superconformal field theory*, Nuclear Phys. B **407** (1993), 115–154
9. Fang, H., Lu, Z., Yoshikawa, K.-I. *Analytic torsion for Calabi–Yau threefolds*, preprint, UTMS 2005-30, available at http://www.ms.u-tokyo.ac.jp/top/publication-preprint (2005)
10. Givental, A. *Equivariant Gromov–Witten invariants*, IMRN **13** (1996), 613–663
11. Harvey, J., Moore, G. *Exact gravitational threshold correction in the Ferrara–Harvey–Strominger–Vafa model*, Phys. Rev. D **57** (1998), 2329–2336
12. Klemm, A., Theisen, S. *Considerations of one-modulus Calabi–Yau compactifications: Picard–Fuchs equations, Kähler potentials and mirror maps*, Nuclear Phys. B **389** (1993), 153–180
13. Li, J., Zinger, A. *On the genus-one Gromov–Witten invariants of complete intersection threefolds*, math.AG/0406105 (2004)
14. Lian, B., Liu, K., Yau, S.-T. *Mirror principle I*, Asian Jour. Math. **1** (1997), 729–763
15. Morrison, D. *Mirror symmetry and rational curves on quintic threefolds: A quick guide for mathematicians*, Jour. Amer. Math. Soc. **6** (1993), 223-247
16. Ray, D.B., Singer, I.M. *Analytic torsion for complex manifolds*, Ann. of Math. **98** (1973), 154–177
17. Voisin, C. *Miroirs et involutions sur les surfaces K3*, Journée de Géométrie Algébrique d'Orsay (Orsay, 1992), Astérisque **218** (1993), 273–323
18. Yoshida, M. *Hypergeometric Functions, My Love*, Aspects of Math. Vieweg, Braunschweig (1997)
19. Yoshikawa, K.-I. *K3 surfaces with involution, equivariant analytic torsion, and automorphic forms on the moduli space*, Invent. Math. **156** (2004), 53–117
20. Yoshikawa, K.-I. *On the singularity of Quillen metrics*, preprint, UTMS 2005-31, available at http://www.ms.u-tokyo.ac.jp/top/publication-preprint (2005)
21. Yoshikawa, K.-I. *Discriminant of certain K3 surfaces*, preprint, UTMS 2005-37, available at http://www.ms.u-tokyo.ac.jp/top/publication-preprint (2005)
22. Yoshikawa, K.-I. *Borcea–Voisin threefolds, analytic torsion, and automorphic forms*, in preparation

Differential Galois Groups of High Order Fuchsian ODE's[1].

N. Zenine[§], S. Boukraa[†], S. Hassani[§], J.-M. Maillard[‡]

[§] C.R.N.A., Bld Frantz Fanon, BP 399, 16000 Alger, Algeria

[†] Université de Blida, Institut d'Aéronautique, Blida, Algeria

[‡] LPTMC, Univ. Paris VI, Tour 24, case 121,

4 Place Jussieu, 75252 Paris, France[2]

We present a simple, but efficient, way to calculate connection matrices between sets of independent local solutions, defined at two neighboring singular points, of Fuchsian differential equations of quite large orders, such as those found for the third and fourth contribution ($\chi^{(3)}$ and $\chi^{(4)}$) to the magnetic susceptibility of square lattice Ising model. We use the previous connection matrices to get the exact explicit expressions of all the monodromy matrices of the Fuchsian differential equation for $\chi^{(3)}$ (and $\chi^{(4)}$) expressed in the same basis of solutions. These monodromy matrices are the generators of the differential Galois group of the Fuchsian differential equations for $\chi^{(3)}$ (and $\chi^{(4)}$), whose analysis is just sketched here.

PACS: 05.50.+q, 05.10.-a, 02.30.Hq, 02.30.Gp, 02.40.Xx

AMS Classification scheme numbers: 34M55, 47E05, 81Qxx, 32G34, 34Lxx, 34Mxx, 14Kxx

Key-words: Susceptibility of the Ising model, Fuchsian differential equations, apparent singularities, rigid local systems, differential Galois group, monodromy group, Clausen function, polylogarithms, Riemann zeta function.

1. Introduction

Since the work of T.T. Wu, B. M. McCoy, C.A. Tracy and E. Barouch [1], it is known that the expansion in n-particle contributions to the zero field

[1]2005 Nankai conference on differential geometry in the honor of Professor Shiing Shen Chern.

[2]maillard@lptmc.jussieu.fr, sboukraa@wissal.dz, njzenine@yahoo.com

susceptibility of the square lattice Ising model at temperature T can be written as an infinite sum:

$$\chi(T) = \sum_{n=1}^{\infty} \chi^{(n)}(T) \tag{1.1}$$

of $(n-1)$-dimensional integrals [2–7], the sum being restricted to odd (respectively even) n for the high (respectively low) temperature case.

As far as regular singular points are concerned (physical or non-physical singularities in the complex plane), and besides the known $s = \pm 1$ and $s = \pm i$ singularities, B. Nickel showed [6] that $\chi^{(2n+1)}$ is singular for the following finite values of $s = sh(2\,J/kT)$ lying on the $|s = 1|$ *unit circle* ($m = k = 0$ excluded):

$$\begin{aligned} 2 \cdot \left(s + \frac{1}{s}\right) &= u^k + \frac{1}{u^k} + u^m + \frac{1}{u^m} \\ u^{2n+1} &= 1, \qquad -n \le m,\; k \le n \end{aligned} \tag{1.2}$$

In the following we will call these singularities: "Nickel singularities". When n increases, the singularities of the higher-particle components of $\chi(s)$ *accumulate on the unit circle* $|s| = 1$. The existence of such a *natural boundary* for the total $\chi(s)$, shows that $\chi(s)$ is *not D-finite* (non holonomic[3] as a function of s).

A significant amount of work had been performed to generate *isotropic* series coefficients for $\chi^{(n)}$ (by B. Nickel [6,7] up to order 116, then to order 257 by A.J. Guttmann and W. Orrick[4]). More recently, W. Orrick *et al.* [8], have generated coefficients[5] of $\chi(s)$ up to order 323 and 646 for high and low temperature series in s, using some *non-linear Painlevé difference equations* for the correlation functions [8–12]. As a consequence of this non-linear Painlevé difference equation, and the remarkable associated *quadratic double recursion* on the correlation functions, the computer algorithm had a $O(N^6)$ *polynomial growth* of the calculation of the series expansion instead of an exponential growth that one would expect at first sight. However, in such a *non-linear*, non-holonomic, *Painlevé-oriented* approach, one obtains results directly for the total susceptibility $\chi(s)$ which do not satisfy any

[3]The fact this natural boundary may be a "porous" natural frontier allowing some analytical continuation through it is not relevant here: one just need an infinite accumulation of singularities (not necessarily on a curve ...) to rule out the D-finite character of χ.

[4]A.J. Guttmann and W. Orrick private communication.

[5]The short-distance terms were shown to have the form $(T - T_c)^p \cdot (log|T - T_c|)^q$ with $p \ge q^2$.

linear differential equation, and thus prevents the easily disentangling of the contributions of the various holonomic $\chi^{(n)}$'s.

In contrast, we consider here, a *strictly holonomic approach.* This approach [13–15] enabled us to get 490 coefficients[6] of the series expansion of $\chi^{(3)}$ (resp. 390 coefficients for $\chi^{(4)}$), from which we have deduced [13–16] the *Fuchsian differential equation* of order *seven* (resp.*ten*) satisfied by $\chi^{(3)}$ (resp. $\chi^{(4)}$). We will focus, here, on the *differential Galois group* of these order seven and ten Fuchsian ODE's.

2. The Fuchsian differential equations satisfied by $\tilde{\chi}^{(3)}(w)$ and $\tilde{\chi}^{(4)}(w)$

Similarly to Nickel's papers [6,7], we start using the multiple integral form of the $\chi^{(n)}$'s, or more precisely of some normalized expressions $\tilde{\chi}^{(n)}$:

$$\begin{aligned} \chi^{(n)}(s) &= S_{\pm}\ \tilde{\chi}^{(n)}(s), && n = 3, 4, \cdots \\ S_+ &= \frac{(1-s^4)^{1/4}}{s}, && T > T_C \quad (n \quad odd) \\ S_- &= (1-s^{-4})^{1/4}, && T < T_C \quad (n \quad even) \end{aligned} \tag{2.1}$$

where:

$$\tilde{\chi}^{(n)}(w) = \int d^nV \quad \left(\prod_{i=1}^{n} \tilde{y}_i\right) \cdot R^{(n)} \cdot H^{(n)} \tag{2.2}$$

with (each angle ϕ_i varying from 0 to 2π):

$$\begin{aligned} d^nV &= \prod_{i=1}^{n-1} \frac{d\phi_i}{2\pi} \quad with \quad \sum_{i=1}^{n} \phi_i = 0, \qquad R^{(n)} = \frac{1+\prod_{i=1}^{n}\tilde{x}_i}{1-\prod_{i=1}^{n}\tilde{x}_i} \\ H^{(n)} &= \prod_{i<j} 4\frac{\tilde{x}_i\tilde{x}_j}{(1-\tilde{x}_i\tilde{x}_j)^2} \cdot \sin^2\left(\frac{\phi_i-\phi_j}{2}\right) \end{aligned} \tag{2.3}$$

Instead of the usual[6,7] variable s, we found it more suitable to use $w = \frac{1}{2}s/(1+s^2)$ which has, by construction, Kramers-Wannier duality invariance ($s \leftrightarrow 1/s$) and thus allows us to deal with both limits (high and low temperature, small and large s) on an equal footing [13–15]. The quantities

[6] We thank J. Dethridge for writing an optimized C++ program that confirmed the Fuchsian ODE we found for $\chi^{(3)}$, providing hundred more coefficients all in agreement with our Fuchsian ODE.

$\tilde{x}_j$ and $\tilde{y}_j$ can be written in the following form [13–15]:

$$\tilde{x}_j = \frac{2w}{1 - 2w\cos\phi_j + \sqrt{(1 - 2w\cos\phi_j)^2 - 4w^2}},$$
$$\tilde{y}_j = \frac{2w}{\sqrt{(1 - 2w\cos\phi_j)^2 - 4w^2}}. \qquad (2.4)$$

It is straightforward to see that $\tilde{\chi}^{(n)}$ is *only a function of the variable* w. From now on, we thus focus on $\tilde{\chi}^{(n)}$ seen as a function of the well-suited variable w instead of s [6,7]. One may expand the integrand in (2.2) in this variable w, and integrate the angular part.

We do not recall, here, the concepts, tricks and tools that have been necessary to generate very large series expansion for $\tilde{\chi}^{(3)}(w)$ and $\tilde{\chi}^{(4)}(w)$ with a *polynomial growth* of the calculations [13–15].

Given the expansion of $\tilde{\chi}^{(3)}(w)$ up to w^{490}, the next step amounts to encoding all the numbers in this long series into a *linear* differential equation. Note that such an equation should exist though its order is *unknown*[7]. Let us say that, using a dedicated program for searching[8] for such a finite order linear differential equation with polynomial coefficients in w, we succeeded finally in finding the following linear differential equation of order *seven* satisfied by the 490 terms we have calculated for $\tilde{\chi}^{(3)}$:

$$\sum_{n=0}^{7} a_n \cdot \frac{d^n}{dw^n} F(w) \; = \; 0 \qquad \text{with:} \qquad (2.5)$$

$$a_n \; = \; w^n \cdot (1 - 4\,w)^{\theta(n-2)}\, (1 + 4\,w)^{\theta(n-4)}\, P_n(w), \qquad n = 6, 5, \cdots, 0$$

$$\text{where:} \qquad \theta(m) \; = \; sup(m, 0), \qquad \text{and:} \qquad a_7 \; =$$

$$w^7 \cdot (1 - w)\, (1 + 2\,w)\, (1 - 4\,w)^5\, (1 + 4\,w)^3\, \left(1 + 3\,w + 4\,w^2\right)\, P_7(w)$$

where $P_7(w), P_6(w) \cdots, P_0(w)$ are polynomials of degree respectively 28, 34, 36, 38, 39, 40, 40 and 36 in w [13].

Furthermore, besides the known singularities (1.2) mentioned above, we remark the occurrence of the roots of the polynomial P_7 of degree 28 in w, and the two *quadratic numbers* roots of $1 + 3\,w + 4\,w^2 \; = \; 0$ which

---

[7] A lower bound for the order of this linear differential equation would be extremely useful : such a lower bound *does not* exist at the present moment.

[8] Note that we, first, actually found an order twelve Fuchsian ODE and, then, we reduced it (by factorization of the differential operator) to a seventh order operator. This order twelve differential equation requires much less coefficients in the series expansion to be guessed than the order seven Fuchsian ODE we describe here ! It is *actually easier to find the order twelve* differential equation than the order seven ODE !!

are not [17] Nickel singularities (they are not of the form (1.2)). The two quadratic numbers *are not* on the s-unit circle : $|s| = \sqrt{2}$ and $|s| = 1/\sqrt{2}$. These quadratic numbers do not occur in the "physical solution" $\chi^{(3)}$. For P_7, near any of its roots, all the local solutions carry *no logarithmic terms* and *are analytical* since the exponents *are all positive integers*. The roots of P_7 are thus *apparent singularities* [18,19] of the Fuchsian equation (2.5).

The order seven linear differential operator L_7 associated with the differential equation satisfied by $\tilde{\chi}^{(3)}$ has the following factorization properties [13,14,16]:

$$L_7 = L_1 \oplus L_6, \qquad L_6 = Y_3 \cdot Z_2 \cdot N_1 \tag{2.6}$$

where[9] L_1 is a first order differential operator which has the first contribution to the magnetic susceptibility, namely $\tilde{\chi}^{(1)} = 2w/(1-4w)$, as solution.

In the same way, we found that the order ten linear differential operator L_{10}, associated with the differential equation satisfied by $\tilde{\chi}^{(4)}$, has the following factorization properties [15,16]:

$$L_{10} = N_0 \oplus L_8, \qquad L_8 = M_2 \cdot G(L) \tag{2.7}$$

where N_0 is an order two differential operator which has the second contribution to the magnetic susceptibility, $\tilde{\chi}^{(2)}$ as solution and where $G(L)$ is an order four differential operator that can be factorized in a product of four order one differential operators [15].

3. Differential Galois group

A fundamental concept to understand (the symmetries, the solutions of) these exact Fuchsian differential equations is the so-called *differential Galois group* [20–24], which requires the computation of all the *monodromy matrices* associated with each (non apparent) regular singular point, these matrices being considered *in the same basis*[10]. Differential Galois groups have been calculated for simple enough second order, or even third order, ODE's. However, finding the differential Galois group of such *higher order* Fuchsian differential equations (order seven for $\chi^{(3)}$, order ten for $\chi^{(4)}$) with eight regular singular points (for $\chi^{(3)}$) is not an easy task. Along this side a first step amounts to seeing that the corresponding (order seven, ten)

[9]For the notations see [13,14,16] for $\tilde{\chi}^{(3)}$, and [15,16] for $\tilde{\chi}^{(4)}$.

[10]These monodromy matrices are the generators of the *monodromy group* which identifies with the differential Galois group when there are no irregular singularities, and, thus, no Stokes matrices [25]

differential operators *do factorize* in smaller order differential operators, as a consequence of some rational and algebraic solutions and other singled out solutions [16]. These factorizations yield a particular block-matrix form of the monodromy matrices [16]. The calculation of local monodromy matrices in some "well-suited" local (Frobenius series solution) bases is easy to perform, however the calculation of the so-called *connection matrices* corresponding to the "matching" of the various well-suited local bases associated with the various regular singularities is a hard *non-local* problem. Of course from the knowledge of all these connection matrices one can immediately write the monodromy matrices in a *unique* basis of solutions[16].

From exact Fuchsian ODE's one can calculate very large series expansions for these (well-suited local Frobenius) solutions, sufficiently large that the evaluation of these series far away from any regular singularity can be performed[11] with a very large accuracy (400, 800, 1000 digits ...). As far as $\chi^{(3)}$ is concerned one can reduce [16] the calculation of these connection and monodromy matrices, to the 6×6 matrices of an order six [16] differential operator L_6 appearing in the decomposition (2.6). Connecting various sets of Frobenius series-solutions well-suited to the various sets of regular singular points amounts to solving a linear system of 36 unknowns (the entries of the connection matrix). We have obtained these entries in floating point form with a very large number of digits (400, 800, 1000, ...). We have, then, been able to actually "recognize" these entries obtained in floating form with a large number of digits [16].

In particular it is shown in [16] that the *connection matrix* between the singularity points 0 and 1/4 (matching the well-suited local series-basis near $w = 0$ and the well-suited local series-basis near $w = 1/4$) is a matrix where the entries are expressions in terms of $\sqrt{3}$, π, $1/\pi$, $1/\pi^2$, ... and a (transcendental) constant I_3^+ introduced in equation (7.12) of [1]:

$$\begin{aligned}
&\frac{1}{2\,\pi^2} \cdot \int_1^\infty \int_1^\infty \int_1^\infty dy_1\, dy_2\, dy_3 \Big(\frac{y_2^2 - 1}{(y_1^2 - 1)\,(y_3^2 - 1)} \Big)^{1/2} \cdot Y^2 = \\
&= \; .000814462565662504439391217128562721997861158118508 \cdots \\
&Y \; = \; \frac{y_1 \; - y_3}{(y_1 \; + y_2)\,(y_2 \; + y_3)\,(y_1 \; + y_2 \; + y_3)}.
\end{aligned}$$

This transcendental constant can actually be written in term of the *Clausen function* Cl_2 :

$$I_3^+ \; = \; \frac{1}{2\pi^2} \cdot \Big(\frac{\pi^2}{3} + 2 - 3\sqrt{3} \cdot Cl_2(\frac{\pi}{3}) \Big) \tag{3.1}$$

[11] Within the radius of convergence of these series.

where Cl_2 denotes the Clausen function :

$$Cl_2(\theta) \,=\, \sum_{n=1}^{\infty} \frac{\sin(n\,\theta)}{n^2}$$

This constant I_3^+ *can also be written in terms of dilogarithms, polygamma functions or Barnes G-functions* :

$$I_3^+ \,-\, (1/6 + \pi^{-2}) \,=\, -\frac{3\sqrt{3}}{2\,\pi^2} \cdot Im\left(dilog\left(1/2 - 1/2\,i\sqrt{3}\right)\right)$$
$$= \frac{1}{16\,\pi^2} \cdot \Big(\Psi\,(1, 2/3) + \Psi\,(1, 5/6) - \Psi\,(1, 1/6) - \Psi\,(1, 1/3)\Big)$$
$$= -\frac{\sqrt{3}}{2\,\pi} \cdot \left(\ln\,(2) \,+ \ln\,(\pi) \,-\, 6\,\ln\left(\frac{G\,(7/6)}{G\,(5/6)}\right)\right)$$

The 6×6 connection matrix $C(0, 1/4)$ for the order six differential operator L_6 matching the Frobenius series-solutions around $w = 0$ and the ones around $w = 1/4$, reads:

$$C(0,\, 1/4) \,= \tag{3.2}$$
$$\begin{bmatrix} 1 & 0 & 0 & 0 & 0 & 0 \\ 1 & 0 & -\frac{9\sqrt{3}}{64\pi} & 0 & 0 & 0 \\ 0 & -\frac{3\pi\sqrt{3}}{32} & 0 & 0 & 0 & 0 \\ 5 & \frac{1}{3} - 2\cdot I_3^+ & \frac{3\sqrt{3}}{64\pi} & 0 & 0 & \frac{1}{16\pi^2} \\ -\frac{5}{4} & -\frac{3\pi\sqrt{3}}{32} & \frac{45\sqrt{3}}{256\pi} & 0 & \frac{1}{32} & 0 \\ \frac{29}{16} - \frac{2\pi^2}{3} & \frac{15\pi\sqrt{3}}{64} & -\frac{225\sqrt{3}}{1024\pi} - \frac{3\pi\sqrt{3}}{64} & \frac{\pi^2}{64} & 0 & 0 \end{bmatrix}$$

Not surprisingly[12] a lot of π's "pop out" in the entries of these connection matrices. We will keep track of the π's occurring in the entries of connection matrices through the introduction of the variable $\alpha \,=\, 2\,i\,\pi$.

The local monodromy matrices can easily be calculated [16] since they correspond, mostly, to "logarithmic monodromies" and will be deduced from simple calculations using the fact that each logarithm (or power of a logarithm) occurring in a (Frobenius series) solution, is simply changed as follows : $\ln(w) \,\rightarrow\, \ln(w) \,+\, \Omega$, where Ω will denote in the following $2\,i\,\pi$. From the local monodromy matrix $Loc(\Omega)$, expressed in the $w \,=\, 1/4$ well-suited local series-basis, and from the connection matrix (3.2), the

[12] One can expect the entries of the connection matrices to be *evaluations* of (generalizations of) hypergeometric functions, or solutions of Fuchsian differential equations.

monodromy matrix around $w = 1/4$, expressed in terms of the $(w = 0)$-well suited basis reads [16]:

$$24\,\alpha^4 \cdot M_{w=0}(1/4)(\alpha,\,\Omega) \;=\; \begin{bmatrix} \mathbf{A} & \mathbf{0} \\ \mathbf{B} & \mathbf{C} \end{bmatrix} \quad (3.3)$$

$$= \; C(0,\,1/4) \cdot Loc(\Omega) \cdot C(0,\,1/4)^{-1}$$

where $\begin{bmatrix} \mathbf{A} \\ \mathbf{B} \end{bmatrix}$ and $\begin{bmatrix} \mathbf{C} \end{bmatrix}$ read respectively:

$$\begin{bmatrix} -24\,\alpha^4 & 0 & 0 \\ -48\,\alpha^4 & 24\,\alpha^4 & -144\,\alpha^2\Omega \\ 0 & 0 & 24\,\alpha^4 \\ -48\,\rho_1 & 32\,\Omega\,\rho_2 & 48\,\Omega\,(9\,\alpha^2+80\,\Omega) \\ 12\,\alpha^2\rho_3 & 4\,(75-4\,\alpha^2)\,\alpha^2\Omega & -300\,\alpha^2\Omega \\ -(87+8\,\alpha^2)\alpha^4 & 0 & 3\,(4\,\alpha^2-75)\,\alpha^2\Omega \end{bmatrix},$$

with $\rho_1 = 5\,\alpha^4 + 8\,\Omega^2 + 8\,\Omega^2\alpha^2$, $\rho_2 = 4\,\Omega\,\alpha^2 - 75\,\Omega - 15\,\alpha^2$ and $\rho_3 = 5\,\alpha^2 + 4\,\Omega + 4\,\Omega\,\alpha^2$, and:

$$C = \begin{bmatrix} 24\,\alpha^4 & -384\,\alpha^2\Omega & 1536\,\Omega^2 \\ 0 & 24\,\alpha^4 & -192\,\alpha^2\Omega \\ 0 & 0 & 24\,\alpha^4 \end{bmatrix}$$

Note that the transcendental constant I_3^+ *has disappeared* in the final exact expression of (3.3) *which actually depends only on* α and Ω. This $(\alpha,\,\Omega)$ way of writing the monodromy matrix (3.3) enables to get straightforwardly the N–th power of (3.3):

$$M_{w=0}(1/4)(\alpha,\,\Omega)^N \;=\; M_{w=0}(1/4)(\alpha,\,N\cdot\Omega) \quad (3.4)$$

Let us introduce the following choice of ordering of the eight singularities, namely ∞, 1, $1/4$, w_1, $-1/2$, $-1/4$, 0, w_2 ($w_1 = (-3 + i\,\sqrt{(7)})/8$ and $w_2 = w_1^*$ are the two quadratic number roots of $1 + 3\,w + 4\,w^2 = 0$), the first monodromy matrix M_1 is, thus, the monodromy matrix $M_{w=0}(\infty)$ (see (3.3)) at infinity with $\alpha = \Omega = 2i\,\pi$, $\mathcal{M}(\infty)$, the second monodromy M_2 matrix being the monodromy matrix at $w = 1$, $\mathcal{M}(1)$, ... This is actually the particular choice of ordering of the eight singularities, such that a product of monodromy matrices *is equal to the identity matrix*[13]:

$$\begin{aligned} M_1 \cdot M_2 \cdot M_3 \cdot M_4 \cdot M_4 \cdot M_6 \cdot M_7 \cdot M_8 \;&=\; \mathbf{Id} \\ =\; \mathcal{M}(\infty) \cdot \mathcal{M}(1) \cdot \mathcal{M}(1/4) \cdot \mathcal{M}(w_1) & \\ \times \mathcal{M}(-1/2) \cdot \mathcal{M}(-1/4) \cdot \mathcal{M}(0) \cdot \mathcal{M}(w_2) & \end{aligned} \qquad (3.5)$$

It is important to note that relation (3.5) *is not verified* by the (α, Ω) extension (like (3.3)) of the monodromy matrices M_i. If one considers relation (3.5) for the (α, Ω) extensions of the M_i's, one will find that (3.5) is satisfied *only when* α *is equal to* Ω, but (of course[14]) this $\alpha = \Omega$ matrix identity *is verified for any value of* Ω, not necessarily equal to $2\,i\,\pi$.

3.1. *Mutatis mutandis for* $\chi^{(4)}$

Similarly to $\chi^{(3)}$ the differential operator for $\chi^{(4)}$ presents remarkable factorizations that yield a particular block-matrix form of the monodromy matrices [16]. Similarly, again, one can consider the (Frobenius series) solutions of the differential operator associated with $\chi^{(4)}$ around $x = 4\,w^2 = 0$ and around the ferromagnetic (and antiferromagnetic) critical point $x = 1$ respectively. Again the corresponding *connection matrix* (matching the solutions around the singularity points $x = 0$ and the ones around the singularity point $x = 4\,w^2 = 1$) have entries which are expressions in terms of π^2, rational numbers but also of constants like constant I_4^- introduced in [1] which can actually be written in term of the *Riemann zeta function*, as follows :

$$I_4^- \;=\; \frac{1}{16\pi^3} \cdot \Big(\frac{4\,\pi^2}{9} - \frac{1}{6} - \frac{7}{2} \cdot \zeta(3) \Big) \qquad (3.6)$$

The derivation of the two results (3.1), (3.6) for the two transcendental constants I_3^+ and I_4^- has never been published[15] but these results appeared in a conference proceedings [26]. We have actually checked that I_3^+ and I_4^- we got in our calculations of connection matrices displayed in [27] as floating numbers with respectively 421 digits and 431 digits accuracy, *are*

[13]Of course, from this relation, one also has seven other relations deduced by cyclical permutations.

[14]A matrix identity like (3.5) yields a set of polynomial (with integer coefficients) relations on $\Omega = 2\,i\,\pi$. The number π being transcendental it is not the solution of a polynomial with integer coefficients. These polynomial relations have, thus, to be *polynomial identities valid for any* Ω.

[15]We thank C. A. Tracy for pointing out the existence of these two results (3.1), (3.6) and reference [26].

actually in agreement with the previous two formula. These two results (3.1), (3.6) provide a clear answer to the question of how "complicated and transcendental" some of the constants occurring in the entries of the connection matrices can be. These two remarkable exact formulas (3.1), (3.6) are not totally surprising when one recalls the *deep link between zeta functions, polylogarithms and hypergeometric series* [28–30]. Along this line, and keeping in mind that we see all our Ising susceptibility calculations as a "laboratory" for other more general problems (Feynman diagrams, ...), we should also recall the various papers of D. J. Broadhurst [31] where $Cl_2(\frac{\pi}{3})$ and $\zeta(3)$ actually occur in a Feynman-diagram-hypergeometric-polylogarithm-zeta framework (see for instance equation (163) in [31]).

Similarly to the previous results for $\tilde{\chi}^{(3)}$ the monodromy matrices written in the *same* basis of solution, deduced from the connection matrices and the local monodromy matrices are such that a product in a certain order of them is the identity matrix. Denoting by $M_{x=0}(0)$, $M_{x=0}(1)$, $M_{x=0}(4)$ and $M_{x=0}(\infty)$ the monodromy matrices expressed in the same $x = 0$ well-suited basis, one obtains:

$$M_{x=0}(\infty) \cdot M_{x=0}(4) \cdot M_{x=0}(1) \cdot M_{x=0}(0) \; = \; \mathbf{Id} \qquad (3.7)$$

This matrix identity is valid irrespective of the "not yet guessed" constants [16].

4. Conclusion

The high order Fuchsian equations we have sketched here present many interesting mathematical properties close to the ones of the so-called *rigid local systems* [32], these rigid local systems exhibiting remarkable *geometrical interpretations* [33] as periods of some algebraic varieties. This "rigidity[16]" emerges through the log-singularities of the solutions of these Fuchsian ODE's: the powers of the logarithms of these solutions are "smaller" than one could expect at first sight. It is worth noting that almost all these mathematical structures, or singled-out properties, we sketched here, or in previous publications [13–16], are far from being specific of the two-dimensional Ising model : they also occur on many problems of lattice statistical mechanics or, even[17], as A. J. Guttmann and I. Jensen saw it recently, on

[16]Let us recall that hypergeometric functions are totally rigid.

[17]The wronskian of the corresponding differential equation in [34] is also rational, the associated differential operator factorizes in a way totally similar to the Fuchsian ODE's for $\chi^{(3)}$ and $\chi^{(4)}$, large polynomial corresponding to apparent singularities also occur, ...

enumerative combinatorics problems like, for instance, the generating function of the three-choice polygon [34].

We have also seen in some of our calculations [14,15] a clear occurence of hypergeometric functions, hypergeometric series and in some of our calculations (not displayed here) generalizations of hypergeometric functions to *several complex variables*: Appel functions [35], Kampé de Fériet, Lauricella-like functions, polylogarithms [31], Riemann zeta functions, multiple zeta values, ... The occurence of Riemann zeta function or dilogarithms in the two remarkable exact formulas (3.1), (3.6) is not totally surprising when one recalls the *deep link between zeta functions, polylogarithms and hypergeometric series* [28–30].

We think that such "collisions" of concepts and structures of different domains of mathematics (differential geometry, number theory, ...) are *not* a consequence of the free-fermion character of the Ising model, and that similar "convergence" should also be encountered on more complicated Yang-Baxter integrable models[18], the two-dimensional Ising model first "popping out" as a consequence of its simplicity. In a specific differential framework some of these interesting mathematical properties can clearly be seen in the analyzis of the differential Galois group of these Fuchsian equations.

We have underlined the fact that, beyond a general analyzis of the differential Galois group [20], one can actually find the exact expressions of the *non-local* connection matrices from very simple matching of series calculations, and deduce, even for such *high order* Fuchsian ODE's, *explicit representations* of all the monodromy matrices in the *same* (non-local) basis of solutions, *providing an effective way* of writing explicit representations of all the elements of the monodromy group. The remarkable form, structures and properties (see (3.2), (3.3), (3.4), (3.5)) of the monodromy matrices in the *same* (non-local) basis of solutions is something one could not suspect at first sight from the general description of the differential Galois group.

Acknowledgments. One of us (JMM) thanks Prof. Mo-Lin Ge for his kind hospitality in the Nankai University, Prof. C. Bai for help throughout his stay in Nankai University and Prof. F. Y. Wu for his interest in this work. One of us (JMM) thanks R.J. Baxter for kind hospitality and discussions. We thank A. J. Guttmann, I. Jensen and W. Orrick for a set of useful

[18]The comparison of the Riemann zeta functions equations obtained for the XXX quantum spin chain [36] with the evaluations of central binomial in [31] provides a strong indication in favor of similar structures on non-free-fermion Yang-Baxter integrable models.

comments. (S. B) and (S. H) acknowledge partial support from PNR3-Algeria.

References

1. T.T. Wu, B.M. McCoy, C.A. Tracy and E. Barouch, 1976 Phys. Rev. **B 13**, 316-374
2. C. R. Nappi, 1978 Nuovo Cim. A **44** 392
3. J. Palmer, C. Tracy, 1981, Adv. Appl. Math. **2** 329
4. K. Yamada, 1984 Prog. Theor. Phys. **71** 1416
5. K. Yamada, 1985 Phys. Lett. A **112** 456-458
6. B. Nickel, 1999, J. Phys. A: Math. Gen. **32** 3889-3906
7. B. Nickel, 2000, J. Phys. A: Math. Gen. **33** 1693-1711
8. W.P. Orrick, B.G. Nickel, A.J. Guttmann, J.H.H. Perk, 2001 J. Stat. Phys. **102** 795-841
9. W.P. Orrick, B.G. Nickel, A.J. Guttmann, J.H.H. Perk, 2001 Phys. Rev. Lett. **86** 4120-4123
10. B.M. McCoy, T.T. Wu, 1980 Phys. Rev. Lett. **45** 675-678
11. J.H.H. Perk, 1980 Phys. Lett. A**79** 3-5
12. M. Jimbo, T. Miwa, 1980 Proc. Japan Acad. A **56** 405; 1981 Proc. Japan Acad. A **57** 347
13. N. Zenine, S. Boukraa, S. Hassani, J.M. Maillard, 2004 J. Phys. A: Math. Gen. **37** 9651-9668 and arXiv:math-ph/0407060
14. N. Zenine, S. Boukraa, S. Hassani, J.M. Maillard, J. Phys. A: Math. Gen. **38** 1875-1899 and arXiv:hep-ph/0411051
15. N. Zenine, S. Boukraa, S. Hassani, J.M. Maillard, 2005, J. Phys. A: Math. Gen. **38** 4149-4173 and cond-mat/0502155
16. N. Zenine, S. Boukraa, S. Hassani and J-M. Maillard, J. Phys. A: Math. Gen. **42** (2005) and arXiv:hep-th/0506214
17. B. Nickel, 2005, J. Phys. A: Math. Gen. **38** 4517-4518
18. H.K. Ince, Ordinary differential equations, (Dover Pubs., NY, 1956)
19. A. R. Forsyth, Theory of Differential Equations, 6 vols.(N-Y: Dover, 1959)
20. V.I. Arnold and M. Monarstyrsky, *Hilbert's twenty-first problem for Fuchsian linear systems* in *Developments in Mathematics. The Moscow School.* V.I. Arnold and M. Monarstyrsky, editors,(1993) Chapman and Hall, London.
21. M. van der Put and M. F. Singer, *Galois theory of difference equations*, Springer-Verlag, Berlin, (1997), Lecture Notes in Mathematics.
22. P. Berman, M. Singer, J. Pure Appl. Alg. **139** (1999)3-24.
23. P. Berman *Computing Galois groups for certain classes of ordinary differential equations*, Thesis, North-Carolina State University, 2001.
24. J-A Weil, N. Zenine, S. Boukraa, S. Hassani, J.M. Maillard, in preparation.
25. *The Stokes phenomenon and Hilbert's 16th Problem*, (1995) World Scientific, B.L.J. Braaksma, G.K. Immink and M. van der Put editors
26. C.A. Tracy, *Painlevé transcendents and scaling functions of the two-dimensional Ising model*, Non-linear Equations in Physics and Mathematics, ed. A.O. Barut, pp. 221-237, D. Reidel Pub. Co. Holland, 1978

27. N. Zenine, S. Boukraa, S. Hassani and J-M. Maillard, 2005 in preparation
28. C. Krattenthaler, T. Rivoal, arXiv.org:math.NT/0311114, arXiv.org:math.CA/0312148
29. C. Krattenthaler, T. Rivoal, W. Zudilin, 2005, arXiv.org:math.NT/0311033, to be published in Journal de l'Institut de Mathematiques de Jussieu
30. T. Rivoal, J. Theorie des Nombres de Bordeaux, 15.1 (2003), 351-365
31. D. J. Broadhurst, hep-th/9803091, hep-ph/0211194, math.CA/9803067, hep-th/0004153 and 2001, Exp. Math. **10** 25-41
32. H. Darmon, 2000 Duke Math. Journ. **102**, 413-449
33. N. Katz, *Rigid local systems*, Annals of Math. Studies, Princeton University Press, Study 139.
34. A. J. Guttmann and I. Jensen, 2005, arXiv.org:math.CO/0506317
35. L.G Cabral-Rosetti amd M.A. Sanchis-Lozano, 2002, arXiv:hep-ph/0206081
36. H. E. Boos and V. E. Korepin, 2001, arXiv:hep-th/0104008

Conformal Triality of de Sitter, Minkowski and Anti-de Sitter Spaces *

Bin Zhou

Department of Physics, Beijing Normal University, Beijing 100875, China
Email: zhoub@bnu.edu.cn

Han-Ying Guo

Institute of Theoretical Physics, Chinese Academy of Sciences
P.O. Box 2735, Beijing 100080, China
Email: hyguo@itp.ac.cn

We describe how conformal Minkowski, dS- and AdS-spaces can be united into a single submanifold $[\mathcal{N}]$ of $\mathbb{R}P^5$. It is the set of generators of the null cone in $\mathcal{M}^{2,4}$. Conformal transformations on the Mink-, dS- and AdS-spaces are induced by $O(2,4)$ linear transformations on $\mathcal{M}^{2,4}$. We also describe how Weyl transformations and conformal transformations can be resulted in on $[\mathcal{N}]$. In such a picture we give a description of how the conformal Mink-, dS- and AdS-spaces as well as $[\mathcal{N}]$ are mapped from one to another by conformal maps. This implies that a CFT in one space can be translated into a CFT in another. As a consequence, the AdS/CFT-correspondence should be extended.

1. Introduction

In this talk we show how three kinds of spaces of constant curvatures are "unified" into a single space by conformal maps: the conformal Mink-, dS- and AdS-spaces are the same nature, resulted in from a hypersurface $[\mathcal{N}]$ of $\mathbb{R}P^5$. Here $[\mathcal{N}]$ is the quotient space from the null "cone" $\mathcal{N}$ of $\mathcal{M}^{2,4}$ with the vertex at the origin. Although no metric on $[\mathcal{N}]$ can be induced naturally from $\mathcal{M}^{2,4}$, a set of metrics can be obtained, differing from each other by a Weyl factor. For a given metric on $[\mathcal{N}]$, an $O(2,4)$ linear transformation on $\mathcal{M}^{2,4}$ induces a conformal transformation.

Starting from this picture, it is not astonishing that the conformal Mink-space, dS^4 and AdS^4 can be conformally mapped from one to another. This

*This work is supported partly by NSFC (10505004, 10375087, 90503002).

technique can be used to translate the content of a CFT from one space to another. Thus, if we have the AdS/CFT correspondence[1] between AdS^5 and the conformal Mink-space, then we also have various correspondences: AdS^5 between conformal dS^4, AdS^4 or $[\mathcal{N}]$.

2. The Hypersurface $[\mathcal{N}] \subset \mathbb{R}P^5$

2.1. *The $O(2,4)$-Invariant Hypersurface of $\mathbb{R}P^5$*

For the $(2+4)$-d Mink-space $\mathcal{M}^{2,4}$ endowed with the inner product

$$\zeta_1 \cdot \zeta_2 := \eta_{\hat{A}\hat{B}} \zeta_1^{\hat{A}} \zeta_2^{\hat{B}}, \qquad (\eta_{\hat{A}\hat{B}}) = \mathrm{diag}(1, -1, \ldots, -1, 1), \tag{2.1}$$

where $\hat{A}, \hat{B} = 0, 1, \ldots, 5$, we consider its null cone

$$\mathcal{N}: \quad \boldsymbol{\zeta} \cdot \boldsymbol{\zeta} = 0, \qquad (\boldsymbol{\zeta} \neq 0). \tag{2.2}$$

In $\mathcal{M}^{2,4}$ there is the standard equivalence relation $\sim$, defined by

$$\boldsymbol{\zeta}' \sim \boldsymbol{\zeta} \ \Leftrightarrow \ \boldsymbol{\zeta}' = c\boldsymbol{\zeta} \quad \text{for a nonzero } c \in \mathbb{R}, \tag{2.3}$$

which makes the quotient space $\mathcal{M}^{2,4} - \{0\}/\sim$ to be the projective space $\mathbb{R}P^5$. The equivalence class of a nonzero $\boldsymbol{\zeta} \in \mathcal{M}^{2,4}$ is denoted by $[\boldsymbol{\zeta}]$. Thus, $\mathcal{N}$ defines a quotient space $\mathcal{N}/\sim \subset \mathbb{R}P^5$, denoted by $[\mathcal{N}]$ for convenience. It is obvious that $[\mathcal{N}]$ is homeomorphic to $S^1 \times S^3$.

As well known, a general linear transformation on $\mathcal{M}^{2,4}$ induces a projective transformation on $\mathbb{R}P^5$. Since $\mathcal{N}$ is invariant under the $O(2,4)$ linear transformations†, they induce some transformations on $[\mathcal{N}]$. In §2.3 we shall show how these transformations on $[\mathcal{N}]$ can be made into a conformal transformation on $[\mathcal{N}]$. In §3 we shall show how these induced transformations on $[\mathcal{N}]$ can be viewed as "conformal transformations" on the Mink-space, dS^4 or AdS^4.

Before the topic of conformal transformations is concerned, we must investigate the problem of metric on $[\mathcal{N}]$. The metric $\boldsymbol{\eta} = \eta_{\hat{A}\hat{B}}\, d\zeta^{\hat{A}} \otimes d\zeta^{\hat{B}}$ on $\mathcal{M}^{2,4}$ cannot naturally induce a metric on $[\mathcal{N}]$. But it is not so bad.

A curve γ in $\mathcal{N}$ can be projected to be a curve $[\gamma]$ in $[\mathcal{N}]$. However, the projection from γ to $[\gamma]$ is not one-to-one. Another curve γ' in $\mathcal{N}$ can be projected to the same $[\gamma]$ in $[\mathcal{N}]$ iff their parameter equations differ from each other by a nonzero factor. We call such two curves in $\mathcal{N}$ are *equivalent*

---

†Strictly, $\mathcal{N}$ is invariant under the action of $O(2,4) \times \mathbb{R}$, where $r \in \mathbb{R}$ refers to a scale product on $\mathcal{M}^{2,4}$ by e^r. But the action of $\mathbb{R}$ induces the identity transformation on $\mathbb{R}P^5$. Thus it can be safely ruled out in our consideration.

to each other. Given two equivalent curves $\zeta^{\hat{A}} = \zeta^{\hat{A}}(t)$ and $\zeta^{\hat{A}} = c(t)\,\zeta^{\hat{A}}(t)$ in $\mathcal{N}$, their line elements, ds^2 and ds'^2, respectively, satisfy the relation

$$ds'^2 = c^2\, ds^2. \tag{2.4}$$

We can turn to the tangent spaces of $\mathcal{N}$ to formulate this result. We say that two tangent vectors, $\mathbf{X} \in T_\zeta\mathcal{N}$ and $\mathbf{X}' \in T_{\zeta'}\mathcal{N}$, are *equivalent* if $[\boldsymbol{\zeta}'] = [\boldsymbol{\zeta}]$ and $\pi_*\mathbf{X} = \pi_*\mathbf{X}'$ where π_* is the pull-back of the natural projection $\pi : \mathcal{N} \to [\mathcal{N}]$. Now suppose $\mathbf{X}', \mathbf{Y}' \in T_{\zeta'}\mathcal{N}$ are equivalent to $\mathbf{X}, \mathbf{Y} \in T_\zeta\mathcal{N}$, respectively. Then,

$$\boldsymbol{\eta}(\mathbf{X}', \mathbf{Y}') = c^2\, \boldsymbol{\eta}(\mathbf{X}, \mathbf{Y}), \tag{2.5}$$

where c is the number in $\boldsymbol{\zeta}' = c\boldsymbol{\zeta}$. This is the precise meaning that is implied, consciously or unconsciously, by eq. (2.4).

There are two important consequences of the above result. We describe them in §2.2 and §2.3, respectively.

2.2. *Induced Metric and Weyl Transformations on* $[\mathcal{N}]$

It is not only that $[\mathcal{N}]$ is a quotient manifold, but also that all its tangent vectors can be viewed as residue classes: each residue class is a set of equivalent tangent vectors of $\mathcal{N}$. The usual way to deal with $[\mathcal{N}]$ is select a representative from each point in $[\mathcal{N}]$. If all the representatives are selected perfectly, we obtain an embedding $\phi : [\mathcal{N}] \to \mathcal{N}$ satisfying

$$\pi \circ \phi = \mathrm{id}_{[\mathcal{N}]}, \tag{2.6}$$

where $\mathrm{id}_{[\mathcal{N}]}$ is the identity map on $[\mathcal{N}]$. Then the problem of selecting a representative for each tangent vector of $[\mathcal{N}]$ can be naturally solved by ϕ_*. In this way we obtain a metric

$$\mathbf{g} = \phi^* \boldsymbol{\eta} \tag{2.7}$$

on $[\mathcal{N}]$. It is easy to see that $\mathbf{g}$ is a Lorentzian metric on $[\mathcal{N}]$.

If $\phi' : [\mathcal{N}] \to \mathcal{N}$ is also an embedding satisfying $\pi \circ \phi' = \mathrm{id}_{[\mathcal{N}]}$, then for any $[\boldsymbol{\zeta}] \in [\mathcal{N}]$, we have $[\boldsymbol{\zeta}] = \pi(\phi([\boldsymbol{\zeta}])) = \pi(\phi'([\boldsymbol{\zeta}]))$. Thus there must be a nonzero real number $\Omega([\boldsymbol{\zeta}])$ so that

$$\phi'([\boldsymbol{\zeta}]) = \Omega([\boldsymbol{\zeta}])\, \phi([\boldsymbol{\zeta}]). \tag{2.8}$$

Therefore, the two embeddings ϕ and ϕ' define a nonzero function Ω on $[\mathcal{N}]$. It is obvious that Ω is smooth.

Let $\mathbf{g}' = \phi'^* \boldsymbol{\eta}$. Then it can be proved that

$$\mathbf{g}' = \Omega^2 \mathbf{g}. \tag{2.9}$$

That is, the consequence of the variation of embeddings is a Weyl transformation for the induced metric on $[\mathcal{N}]$.

2.3. *Conformal Transformations on* $[\mathcal{N}]$

In §2.1 we have pointed out that an $O(2,4)$ transformation O on $\mathcal{M}^{2,4}$ induces a transformation $[O]$ on $[\mathcal{N}]$, well defined by

$$[O]([\zeta]) := [O\zeta]. \tag{2.10}$$

For a given $O \in O(2,4)$, $[O]$ is a diffeomorphism on $[\mathcal{N}]$. Hence an action of $O(2,4)$ on $[\mathcal{N}]$ on the left is resulted in. However, such an action is not effective, because it can be easily verified that

$$[-E] = [E] = \mathrm{id}_{[\mathcal{N}]}, \qquad \text{or} \qquad [-O] = [O] \tag{2.11}$$

for arbitrary $O \in O(2,4)$, where E is the identity transformation on $\mathcal{M}^{2,4}$. It can be proved that, for an $O \in O(2,4)$, $[O] = \mathrm{id}_{[\mathcal{N}]}$ iff $O = \pm E$. It can be also proved that the action of $O(2,4)$ on $[\mathcal{N}]$ is transitive. So $[\mathcal{N}]$ is a homogeneous space of $O(2,4)$.

Let $\phi : [\mathcal{N}] \to \mathcal{N}$ be an embedding as described in §2.2, and O be an $O(2,4)$ linear transformation. For an arbitrary $[\zeta] \in [\mathcal{N}]$, we can set

$$\zeta := \phi([\zeta]), \qquad \zeta' := \phi([\zeta']) = \phi([O\zeta]), \tag{2.12}$$

which are contained in $\phi([\mathcal{N}]) \subset \mathcal{N}$ and can be treated as representatives of $[\zeta]$ and $[O][\zeta]$, respectively. On the other hand, since $[\zeta'] = \pi(\zeta') = (\pi \circ \phi)([O\zeta]) = \mathrm{id}_{[\mathcal{N}]}([O\zeta]) = [O\zeta]$, there must be a nonzero real number $\rho([\zeta])$, depending on $[\zeta]$, such that

$$\zeta' = \rho([\zeta])\, O\zeta. \tag{2.13}$$

In this way we obtain a nonzero function ρ on $[\mathcal{N}]$.

Now let $\mathbf{g}$ be the metric on $[\mathcal{N}]$ induced by the embedding $\phi : [\mathcal{N}] \to \mathcal{N}$, as shown in §2.2. It can be proved that $[O]$ is a conformal transformation:

$$[O]^*\mathbf{g} = \rho^2\, \mathbf{g}. \tag{2.14}$$

So, every $O(2,4)$ linear transformation on $\mathcal{M}^{2,4}$ induces a conformal transformation on $([\mathcal{N}], \mathbf{g})$. Due to eqs. (2.11), the conformal group of $([\mathcal{N}], \mathbf{g})$ is the quotient group $O(2,4)/\mathbb{Z}_2$.

3. Conformal Transformations on the Mink-Space, dS^4 and AdS^4

In §2.2 and §2.3 the representatives are selected in a perfect way that they form a submanifold $\phi([\mathcal{N}])$ diffeomorphic to $[\mathcal{N}]$. In this section we use a not so perfect method: only most of, but not all, points in $[\mathcal{N}]$ can find their respective representatives, located in a hyperplane $\mathcal{P}$ of $\mathcal{M}^{2,4}$ off $\boldsymbol{\zeta} = 0$. The resulted space $\mathcal{P} \cap \mathcal{N}$ are Mink, dS^4 or AdS^4 according to whether the normal vector $\mathbf{n}$ of $\mathcal{P}$ is null, timelike or spacelike. And "on" these spaces there are the "conformal transformations" which are of great interest in physics.

3.1. *The Minkowskian Case*

When the normal vector $\mathbf{n}$ is null, it can be extended to be a linear basis $\{\mathbf{e}_\mu, \mathbf{n}, \mathbf{l}\}$ of $\mathcal{M}^{2,4}$, with $\mathbf{e}_\mu$ for $\mu = 0, \dots, 3$ tangent to $\mathcal{N}$ and $\mathcal{P}$, satisfying

$$\mathbf{e}_\mu \cdot \mathbf{e}_\nu = \eta_{\mu\nu}, \quad \mathbf{e}_\mu \cdot \mathbf{n} = 0, \quad \mathbf{e}_\mu \cdot \mathbf{l} = 0, \quad \mathbf{l} \cdot \mathbf{l} = 0, \quad \mathbf{n} \cdot \mathbf{l} = 1. \qquad (3.1)$$

It is easy to see that a point $\boldsymbol{\zeta} \in \mathcal{P} \cap \mathcal{N}$ iff

$$\boldsymbol{\zeta} = x^\mu \, \mathbf{e}_\mu + x^+ \, \mathbf{n} + R\,\mathbf{l}, \qquad x^+ = -\eta_{\mu\nu} \, x^\mu x^\nu / (2R), \qquad (3.2)$$

with R a constant. And it is easy to verify that $\mathcal{N} \cap \mathcal{P}$ is a Mink-space because the induced metric on it is

$$ds_M^2 = \eta_{\mu\nu} \, dx^\mu \, dx^\nu . \qquad (3.3)$$

Now let us consider two equivalent curves with line elements $d\chi^2$ and ds_M^2, respectively. Assume that the former is just in $\mathcal{N}$, while the latter is in $\mathcal{P} \cap \mathcal{N}$. Then a relation similar to (2.4) can be obtained:

$$d\chi^2 = (\mathbf{n} \cdot \boldsymbol{\zeta} / R)^2 \, ds_M^2, \qquad (3.4)$$

where $\boldsymbol{\zeta}$ is the point along the former curve. From eq. (3.4) it can be derived that the $O(2,4)$ linear transformations induce the so-called "conformal transformations" on the Mink-space[2].

3.2. *The dS^4 and AdS^4 Cases*

When the normal vector $\mathbf{n}$ is timelike, the induced metric on $\mathcal{P}$ has a signature as diag$(1, -1, -1, -1, -1)$. Assume $\mathbf{n} \cdot \mathbf{n} = 1$ and extend it to be an orthonormal basis $\{\mathbf{e}_A, \mathbf{n} \,|\, A = 0, 1, \dots, 4\}$ of $\mathcal{M}^{2,4}$. Then $\boldsymbol{\zeta} \in \mathcal{P} \cap \mathcal{N}$ iff

$$\boldsymbol{\zeta} = \xi^A \, \mathbf{e}_A + R\,\mathbf{n}, \qquad \eta_{AB} \, \xi^A \xi^B = -R^2, \qquad (3.5)$$

where R is a positive constant and $(\eta_{AB}) = \mathrm{diag}(1,-1,-1,-1,-1)$. (We have carefully chosen $\mathbf{n}$ in order that $R > 0$.) Thus $\mathcal{N} \cap \mathcal{P}$ is a dS^4 of radius R.

Let $d\chi^2$ and ds_+^2 be the line elements of two equivalent curves γ and γ_+, respectively. Again the former is just in $\mathcal{N}$ and the latter is in $\mathcal{P} \cap \mathcal{N}$. Then, with $\boldsymbol{\zeta}$ the point along γ, there is similarly the relation

$$d\chi^2 = (\mathbf{n} \cdot \boldsymbol{\zeta}/R)^2\, ds_+^2. \tag{3.6}$$

Given an $O(2,4)$ linear transformation, γ can be transformed to be another curve γ', lying still in $\mathcal{N}$ and equivalent to a curve γ'_+ lying in $\mathcal{P} \cap \mathcal{N}$. Let their line elements be $d\chi'^2$ and ds'^2_+, respectively. Then a similar relation to (3.6) holds for $d\chi'^2$ and ds'^2_+. The $O(2,4)$ transformation preserves the line elements: $d\chi'^2 = d\chi^2$. Thus there will be

$$ds'^2_+ = \left(\frac{\mathbf{n} \cdot \boldsymbol{\zeta}}{\mathbf{n} \cdot \boldsymbol{\zeta}'}\right)^2 ds_+^2 \tag{3.7}$$

for γ_+ and γ'_+, where $\boldsymbol{\zeta}$ and $\boldsymbol{\zeta}'$ are a pair of equivalent points along γ and γ', respectively. Therefore, similar to the Minkowskian case, an $O(2,4)$ linear transformation induces a "conformal transformation" on dS^4.

In general a set of Beltrami coordinates [3–5] can be assigned to an equivalence class $[\boldsymbol{\zeta}]$. For $\boldsymbol{\zeta}$ as in eq. (3.5). The Beltrami coordinates for $[\boldsymbol{\zeta}]$ is

$$x^\mu := R\,\xi^\mu/\xi^4, \qquad (\mu = 0,1,2,3), \tag{3.8}$$

provided that $\xi^4 \neq 0$. In this coordinate system

$$ds_+^2 = \left[\frac{\eta_{\mu\nu}}{\sigma_+(x)} + \frac{\eta_{\mu\alpha}\eta_{\nu\beta}x^\alpha x^\beta}{R^2\sigma_+(x)^2}\right] dx^\mu\, dx^\nu, \quad \sigma_\pm(x) := 1 \mp R^{-2}\eta_{\mu\nu}x^\mu x^\nu. \tag{3.9}$$

Here $\sigma_-(x)$ is preserved for AdS^4. The Beltrami coordinates must satisfy $\sigma_+(x) > 0$ [4,5]. Isometries have the generic form as below[4,5]:

$$x'^\mu = \pm\frac{\sqrt{\sigma_+(a)}\, D^\mu{}_\nu\,(x^\nu - a^\nu)}{\sigma_+(a,x)}, \quad D^\mu{}_\nu := L^\mu{}_\nu + \frac{L^\mu{}_\alpha\,\eta_{\nu\beta}\,a^\alpha a^\beta}{R^2\sqrt{\sigma_+(a)}(1+\sqrt{\sigma_+(a)})}, \tag{3.10}$$

where $L = (L^\mu{}_\nu) \in O(1,3)$, $\pm 1 = \det L$ and the constants a^μ satisfy $\sigma_+(a) > 0$. In the above, $\sigma_\pm(a,x) := 1 \mp R^{-2}\eta_{\mu\nu}\,a^\mu x^\nu$, where $\sigma_-(a,x)$ is preserved for AdS^4. Other conformal transformations include

$$x'^\mu = \frac{x^\mu\sqrt{1-\beta^2}}{1 \pm \beta\sqrt{\sigma_+(x)}}, \qquad (|\beta| < 1) \tag{3.11}$$

and

$$x'^{\mu} = x^{\mu} - \frac{1 - \sigma_{+}(b, x)}{1 + \sqrt{\sigma_{+}(b)}} b^{\mu} \pm b^{\mu} \sqrt{\sigma_{+}(x)}, \tag{3.12}$$

where $\pm$ corresponds to the coordinate neighborhoods where $\xi^4 > 0$ or $\xi^4 < 0$.

When $\mathbf{n}$ is spacelike, it can be similarly proved that $\mathcal{N} \cap \mathcal{P}$ is AdS^4, Similarly, $O(2,4)$ transformations induce conformal transformations. Beltrami coordinates can be also introduced in the same way as on AdS^4, and the conformal transformations take a similar form as in the above.

4. The Extension of *AdS*/CFT Correspondence

4.1. *The Geometric Picture of AdS/CFT Correspondence*

The discussion in §2 and §3 reveals a wonderful geometric picture as follows. The 4-d space $[\mathcal{N}] \cong S^1 \times S^3$ is a hypersurface of $\mathbb{R}P^5$. Although no natural metric can be inherited from $\mathcal{M}^{2,4}$, $[\mathcal{N}]$ can be realized (by an embedding ϕ as in §2.2) as a hypersurface $\phi([\mathcal{N}])$ of $\mathcal{N}$, enabling it to receive a metric $\mathbf{g}$ from the realization. The variousness of realizations ends up with Weyl transformations for the metric. Thus, $[\mathcal{N}]$ is rather a Weyl space than a spacetime, having a vanishing Weyl tensor. Hence theory of physics in $[\mathcal{N}]$, if exists, should be Weyl-invariant — at least it should be conformally invariant.

If the infinity boundary is included in the Mink-space, dS^4 and AdS^4, they are also a realization of $[\mathcal{N}]$, as if the projective plane model for $\mathbb{R}P^2$.

What soever speaking, the Mink-space, dS^4 and AdS^4 can be embedded into $\mathcal{N}$, as shown in §3. These three kinds of spaces, together with the perfect realizations of $[\mathcal{N}]$ as in §2, can be related to each other by the projection. The maps from one to another are conformal maps, among which those from a Mink/dS^4/AdS^4 to a Mink/dS^4/AdS^4 are of special interest, which will be discussed elsewhere[6].

Using the above conformal maps, a CFT on the Mink-space can be transferred to be CFTs on both dS^4 and AdS^4, and vice versa. This fact can be summarized as the conformal triality of Mink-, dS- and AdS-spaces. In fact, a CFT on any of these spaces is a CFT on $([\mathcal{N}], \mathbf{g})$.

Topologically AdS^5 can be viewed as an open region in $\mathbb{R}P^5$, consisting of timelike 1-d linear subspaces of $\mathcal{M}^{2,4}$. In this sense $[\mathcal{N}] = \partial(AdS^5)$. If the AdS/CFT correspondence[1] is correct, then we can say that the corresponding CFT is on the Mink-space, on dS^4, on AdS^4, on $([\mathcal{N}], \mathbf{g})$. Thus we might have as many AdS/CFT correspondences as possible.

The AdS/CFT correspondence for higher dimensions can be also conjectured in the similar geometric picture.

4.2. *Null Geodesics*

As we know, up to re-parameterizations, null geodesics are invariant under conformal transformations and Weyl transformations. The null geodesics can also be illustrated in a geometric picture.

Suppose $[\zeta_0]$ and $[\zeta_1]$ are two distinct points in $[\mathcal{N}]$. Then ζ_0 and ζ_1 are two linearly independent null vectors, spanning a 2-d linear subspace (plane) Σ in $\mathcal{M}^{2,4}$. If, in addition,

$$\zeta_0 \cdot \zeta_1 = 0, \tag{4.1}$$

then the whole Σ except the origin 0 is contained in $\mathcal{N}$. Thus $\Sigma \cap \mathcal{P} \subset \mathcal{N} \cap \mathcal{P}$, no matter whether the latter is the Mink-, dS- or AdS-space. Obviously, $\Sigma \cap \mathcal{P}$ is a null straight line. If, in addition, we assume that ζ_0 and $\zeta_1 \in \mathcal{P}$, then the equation of $\Sigma \cap \mathcal{P}$ reads

$$\zeta(\lambda) = (1 - \lambda)\,\zeta_0 + \lambda\,\zeta_1. \tag{4.2}$$

For the 2-d linear subspace $\Sigma \subset \mathcal{M}^{2,4}$, an antisymmetric tensor

$$\omega := \zeta_0 \otimes \zeta_1 - \zeta_0 \otimes \zeta_1 \tag{4.3}$$

can be defined in terms of its linear basis $\{\zeta_0, \zeta_1\}$. If the linear basis of Σ is changed, then the antisymmetric tensor ω' corresponding to the new basis is proportional to ω. In fact, ω can be treated to be something like the area 2-form of Σ.

Meanwhile, for the straight line (4.2), a 6-d angular momentum tensor

$$\mathcal{L} := \zeta \otimes \frac{d\zeta}{d\lambda} - \zeta \otimes \frac{d\zeta}{d\lambda} \tag{4.4}$$

can be defined. Substituting eq. (4.2) into the above, we find that the angular momentum is conserved:

$$\mathcal{L} = \omega. \tag{4.5}$$

It is very intuitive and can be proved that, in the Mink, dS and AdS cases, $\Sigma \cap \mathcal{P}$ is a null geodesic. The 6-d angular momentum can be expressed in terms of the 4-d angular momentum and the 4-momentum of the massless particle.

In order to see what it looks like in the Minkowski or Beltrami coordinates, we consider two special cases. In the first case, $\mathcal{P}$ is a null hyperplane $\zeta^- = R$, where $\zeta^\pm = \frac{1}{\sqrt{2}}(\pm\zeta^4 + \zeta^5)$. Substitution of eq. (3.2) to (4.2) yields

$$x^\mu = (1 - \frac{\tau}{R})\, x_0^\mu + \frac{\tau}{R}\, x_1^\mu, \qquad \tau = R\, \frac{\lambda\, \zeta_1^-}{(1-\lambda)\,\zeta_0^- + \lambda\,\zeta_1^-}. \tag{4.6}$$

Let's write the energy-momentum and angular momentum as

$$P^\mu = m\, \frac{dx^\mu}{d\tau}, \quad L^{\mu\nu} = x^\mu P^\nu - x^\nu P^\mu, \tag{4.7}$$

respectively and formally introduce

$$P^+ = m\, \frac{dx^+}{d\tau}, \quad L^{\mu +} = -L^{+\mu} = x^\mu\, P^+ - x^+\, P^\mu, \tag{4.8}$$

with x^+ as shown in eqs. (3.2). Then it can be verified that

$$\mathcal{L}^{\mu\nu} = \frac{R}{m}\, L^{\mu\nu}, \quad \mathcal{L}^{\mu +} = \frac{R}{m}\, L^{\mu +}, \quad \mathcal{L}^{\mu -} = -\frac{R^2}{m}\, P^\mu, \quad \mathcal{L}^{+-} = -\frac{R^2}{m}\, P^+. \tag{4.9}$$

In the second case, we take $\mathcal{P}$ as $\zeta^5 = R$. In the corresponding Beltrami coordinate system, the equation of $\Sigma \cap \mathcal{P}$ is still of the form (4.6), only that τ is no longer an affine parameter. However, with the momentum and angular momentum still defined as in (4.7), they are conserved quantities, and there will be

$$\mathcal{L}^{\mu\nu} = \frac{\zeta_0^4 \zeta_1^4}{mR}\, L^{\mu\nu}, \quad \mathcal{L}^{\mu 4} = -\frac{\zeta_0^4 \zeta_1^4}{m}\, P^\mu, \tag{4.10}$$

$$\mathcal{L}^{\mu 5} = \mp \frac{\zeta_0^4 \zeta_1^4}{m}\, \eta^{\mu\nu}\, \sigma(x)^{\frac{3}{2}}\, g_{\nu\rho}(x)\, P^\rho, \tag{4.11}$$

$$\mathcal{L}^{45} = \mp \frac{\eta_{\mu\nu}\, x^\mu}{\sqrt{\sigma(x)}}\, \frac{\zeta_0^4 \zeta_1^4}{mR}\, P^\nu. \tag{4.12}$$

In the above, $\mp$ is opposite to the sign of ξ^4. If the normal vector of $\mathcal{P}$ is spacelike, the results are similar to the above.

5. Conclusion

From the null cone $\mathcal{N} \subset \mathcal{M}^{2,4}$, we can construct the Mink-space, dS^4 and AdS^4 on which the induced action of $O(2,4)$ is conformal. When $\mathcal{M}^{2,4}$ is viewed as the homogeneous space of $\mathbb{R}P^5$, $[\mathcal{N}] := \mathcal{N}/\sim$ is the conformal (extension of the) Mink-, dS- or AdS-space. Various metrics can be endowed on $[\mathcal{N}]$, differing from one another by Weyl transformations. For a given metric among them, the $O(2,4)$ transformations induce conformal transformations on $[\mathcal{N}]$. Since $[\mathcal{N}]$, the Mink-, dS- and AdS-spaces are related by conformal maps, a CFT in one space results in a CFT in each of other spaces. Therefore the AdS/CFT correspondence could be extended to all these spaces. The same idea can be generalized to higher dimensions.

We have shown some evidence that the role of Beltrami coordinates on dS/AdS-spaces is similar to that of the Mink- coordinates. In fact, in the study of kinematics and dynamics on dS/AdS-spaces[4,5], it is also revealed. The similarity is so strong that special relativity can be appealed for on dS/AdS-spaces[4,5].

Acknowledgments

Part of the contents of this talk is based on the co-operation[6] of Profs. Z. Xu, C. G. Huang and Dr. Y. Tian. Many thanks are expressed to them. We are also pleasant to thank Profs. Q. K. Lu, S. K. Wang, K. Wu and X. C. Song for their helpful discussions. B. Zhou would like to express his special thanks to Nankai Institute of Mathematics for the accommodation, especially to Profs. C. Bai and W. Zhang.

References

1. J. Maldacena, *Adv. Theor. Math. Phys.* **2** (1998) 231, hep-th/9711200.
2. R. Penrose and W. Rindler, *Spinors and Space-Time: Volume 2, Spinor and Twistor Methods in Space-Time Geometry*, Cambridge University Press, 1988.
3. E. Beltrami, *Opere Mat.*, **1** (1868) 374-405.
4. H.-Y. Guo, C.-G. Huang, Z. Xu and B. Zhou, *Mod. Phys. Lett.* **A19** (2004) 1701 , see also hep-th/0311156; H.-Y. Guo, C.-G. Huang, Z. Xu and B. Zhou, *Phys. Lett.* **A 331** (2004) 1–7, hep-th/0403171; H.-Y. Guo, C.-G. Huang, and B. Zhou, *Europhys. Lett.* **72**[6] (2005) 1-7, hep-th/0404010; H.-Y. Guo, C.-G. Huang, Y. Tian, Z. Xu and B. Zhou, hep-th/0405137.
5. The original idea is due to K. H. Look (Q. K. Lu), "Why the Minkowski metric must be used?" (1970) unpublished. For earlier works, see, K. H. Look, C. L. Tsou (Z.L. Zou) and H. Y. Kuo (H.-Y. Guo), *Acta Phys. Sinica* **23** (1974) 225; *Nature* (Shanghai, Suppl.), *Mod. Phys.* **1** (1980) 97; H. Y. Kuo, *Kexue Tongbao (Chinese Sci. Bull.)* **22** (1977) 487 (all in Chinese); H. Y. Kuo, *Proc. of the 2nd M. Grossmann Meeting on General Relativity*, ed. by R. Ruffini, (North-Holland Pub. Amsterdam, 1982) 801; H.-Y. Guo, *Nucl. Phys. B* (Proc. Suppl.) **6** (1989) 381.
6. H.-Y. Guo, B. Zhou, C.-G. Huang, Y. Tian and Z. Xu, in preparation; B. Zhou, H.-Y. Guo et al, in preparation.

Some Observations on Gopakumar-Vafa Invariants of Some Local Calabi-Yau Geometries

Jian Zhou

Department of Mathematical Sciences
Tsinghua University
Beijing, China
and Center of Mathematical Sciences
Zhejiang University
Hangzhou, China
Email: jzhou@math.tsinghua.edu.cn

We make some observations on the Gopakumar-Vafa invariants of the local Calabi-Yau geometries given by the canonical line bundles of the projective plane and the product of two projective lines. We conjecture some closed formulas.

Denote by F_X the generating series of Gromov-Witten invariants of a Calabi-Yau 3-fold X. In general such invariants are rational numbers. However, based on M-theory considerations, Gopakumar and Vafa [3] made a remarkable conjecture on the structure of F_X, in particular, its integral properties. More precisely, there are integers n^g_Σ such that

$$F_X = \sum_{\Sigma \in H_2(X)-\{0\}} \sum_{g\geq 0} \sum_{k \geq 1} \frac{1}{k} n^g_\Sigma (2 \sin \frac{k\lambda}{2})^{2g-2} Q^{k\Sigma}.$$

Recently there have been some progress in the calculations of F_X in the case of local Calabi-Yau geometries, both in the physics literature [2, 4, 1] and in the mathematics literature [7, 6]. In particular, the Gopakumar-Vafa invariants n^g_d for the local $\mathbb{P}^2$ geometry have been calculated in [2] for $0 \leq g \leq 55$, $1 \leq d \leq 12$. Calculations for the in [2] the Gopakumar-Vafa invariants $n^g_{(d_1,d_2)}$ for the local $\mathbb{P}^1 \times \mathbb{P}^1$ geometry have been calculated in [2] for $0 \leq g \leq 8$, $1 \leq d_1, d_2 \leq 6$.

For the reader's convenience, we reproduce their table of n^g_d for $0 \leq g \leq 28$ and $1 \leq d \leq 9$ below (Table 1). We will use their complete table for the results below.

As noted in [5, 2], for a given degree d, n^g_d vanishes for $g(d) > (d-1)(d-$

2)/2, and there are closed formulas for g close to $g(d)$. Indeed, $(d-1)(d-2)/2$ is the genus of a nondegenerate curve of degree d in $\mathbb{P}^2$. One has in this case

$$n_d^{g(d)} = \frac{(-1)^{d(d+3)/2}}{2}(d+1)(d+2). \tag{1}$$

For $d > 2$, we have contributions from curves with one node (therefore $g = g(d) - 1$):

$$n_d^{g(d)-1} = -(-1)^{d(d+3)/2}\binom{d}{2}(d^2 + d - 3). \tag{2}$$

Curves with two nodes start contributing at $d > 3$, and one finds:

$$n_d^{g(d)-2} = \frac{(-1)^{d(d+3)/2}}{4}(d-1)(d^5 - 2d^4 - 6d^3 + 9d^2 + 36). \tag{3}$$

For $d > 4$, curves with three nodes contribute to the Gopakumar-Vafa integral invariant:

$$\begin{aligned} n_d^{g(d)-3} = -\frac{(-1)^{d(d+3)/2}}{12}&(-96 + 222d - 323d^2 + 54d^3 \\ &- 34d^4 + 36d^5 + 2d^6 - 6d^7 + d^8). \end{aligned} \tag{4}$$

One expects closed formulas for $n_d^{g(d)-m}$ for $m > 3$ and large enough d. Such formulas will be very complicated because the degree in d increases very fast. It is not easy to guess their form.

In the original derivation of the Gopakumar-Vafa conjecture, n_Σ^g were obtained from some integers N_Σ^g as follows:

$$\sum_{g\geq 0} n_\Sigma^g(-1)^g(q^{\frac{1}{2}} - q^{-\frac{1}{2}})^{2g} = \sum_{g\geq 0} N_\Sigma^g R_g(q), \tag{5}$$

where $R_g(q) = q^g + q^{g-2} + \cdots + q^{-g}$. One can also obtain N_Σ^g from n_Σ^g from the following formula:

$$\begin{aligned} (q^{\frac{1}{2}} - q^{-\frac{1}{2}})^{2g} &= \sum_{k=0}^{g}(-2)^{g-k}\binom{g}{k}\sum_{i=0}^{[\frac{k}{2}]}\binom{k}{i}(R_{k-2i}(q) - R_{k-2i-2}(q)) \\ &= R_g(q) - 2gR_{g-1}(q) + (2g^2 - g - 1)R_{g-2}(q) + \cdots, \end{aligned}$$

where $R_{-1}(q) = 0$. Indeed, we have

$$\begin{aligned}(q^{\frac{1}{2}} - q^{-\frac{1}{2}})^{2g} &= (q + q^{-1} - 2)^g = \sum_{k=0}^{g} (-2)^{g-k} \binom{g}{k} (q + q^{-1})^k \\ &= \sum_{k=0}^{g} (-2)^{g-k} \binom{g}{k} \sum_{i=0}^{k} \binom{k}{i} q^{k-2i} \\ &= \sum_{k=0}^{g} (-2)^{g-k} \binom{g}{k} \sum_{i=0}^{[\frac{k}{2}]} \binom{k}{i} (R_{k-2i}(q) - R_{k-2i-2}(q)).\end{aligned}$$

In particular, if $g(\Sigma)$ is the largest g such that $n_\Sigma^{g(\Sigma)} \neq 0$, then we have

$$N_\Sigma^{g(\Sigma)} = (-1)^{g(\Sigma)} n_\Sigma^{g(\Sigma)}, \tag{6}$$

$$N_\Sigma^{g(\Sigma)-1} = (-1)^{g(\Sigma)-1} (2g(\Sigma) \cdot n_\Sigma^{g(\Sigma)} + n_\Sigma^{g(\Sigma)-1}), \tag{7}$$

$$\begin{aligned}N_\Sigma^{g(\Sigma)-2} &= (-1)^{g(\Sigma)} ((2g(\Sigma)^2 - g(\Sigma) - 1) n_\Sigma^{g(\Sigma)} \\ &\quad + 2(g(\Sigma) - 1) n_\Sigma^{g(\Sigma)-1} + n_\Sigma^{g(\Sigma)-2}).\end{aligned} \tag{8}$$

In this note we will show that for the local $\mathbb{P}^2$ case and the local $\mathbb{P}^1 \times \mathbb{P}^1$ case, N^g are much smaller and have some nice properties. For fixed m we observe some nice stable progression behavior of $N_d^{g(d)-m}$ for large enough d. This leads to some simple closed formulas for them.

Starting from the following table in [2], we get a table for N_d^g in the next two pages:

Table 1. Table for n_d^g in the local $\mathbb{P}^2$ case

g	d= 1	2	3	4	5	6	7	8	9
0	3	-6	27	-192	1695	17064	188454	-2228160	27748899
1	0	0	-10	231	-4452	80948	-1438086	25301295	-443384578
2	0	0	0	-102	5430	- 194022	5784837	-155322234	3894455457
3	0	0	0	15	-3672	290853	-15363990	649358826	-23769907110
4	0	0	0	0	1386	-290400	29056614	-2003386626	109496290149
5	0	0	0	0	-270	196857	-40492272	4741754985	-396521732268
6	0	0	0	0	21	-90390	42297741	-8802201084	1156156082181
7	0	0	0	0	0	27538	-33388020	12991744968	-2756768768616
8	0	0	0	0	0	-5310	19956294	-15382690248	5434042220973
9	0	0	0	0	0	585	- 9001908	14696175789	-8925467876838
10	0	0	0	0	0	-28	3035271	-11368277886	12289618988434
11	0	0	0	0	0	0	-751218	7130565654	-14251504205448
12	0	0	0	0	0	0	132201	-3624105918	13968129299517
13	0	0	0	0	0	0	-15636	1487970738	-11600960414160
14	0	0	0	0	0	0	1113	-490564242	8178041540439
15	0	0	0	0	0	0	-36	128595720	-4896802729542
16	0	0	0	0	0	0	0	-26398788	2489687953666
17	0	0	0	0	0	0	0	4146627	-1073258752968
18	0	0	0	0	0	0	0	-480636	391168899747
19	0	0	0	0	0	0	0	38703	-120003463932
20	0	0	0	0	0	0	0	-1932	30788199027
21	0	0	0	0	0	0	0	45	-6546191256
22	0	0	0	0	0	0	0	0	1138978170
23	0	0	0	0	0	0	0	0	-159318126
24	0	0	0	0	0	0	0	0	17465232
25	0	0	0	0	0	0	0	0	-1444132
26	0	0	0	0	0	0	0	0	84636
27	0	0	0	0	0	0	0	0	-3132
28	0	0	0	0	0	0	0	0	55

Table 2: Table of N_d^g in the local $\mathbb{P}^2$ case

g	d= 1	2	3	4	5	6	7	8	9	10
0	3	-6	7	-30	114	-550	3255	-22134	169750	-1431438
1	0	0	10	-33	150	-853	5466	-39372	311974	-2686038
2	0	0	0	-12	129	-900	6360	-49098	408924	-3639474
3	0	0	0	-15	96	-733	6168	-51882	456622	-4234206
4	0	0	0	0	51	-580	5268	-48846	461509	-4485894
5	0	0	0	0	18	-360	4200	-43116	434934	-448091
6	0	0	0	0	21	-258	3171	-35778	389385	-4201602
7	0	0	0	0	0	-138	2244	-28584	334482	-3815196
8	0	0	0	0	0	-72	1476	-21678	277812	-3360462
9	0	0	0	0	0	-25	996	-16278	224334	-2881647
10	0	0	0	0	0	-28	591	-11568	176964	-2418726
11	0	0	0	0	0	0	354	-8151	136008	-1987920
12	0	0	0	0	0	0	186	-5412	102545	-1607280
13	0	0	0	0	0	0	96	-3600	75536	-1275786
14	0	0	0	0	0	0	33	-2214	54666	-999060
15	0	0	0	0	0	0	36	-1440	38606	-769119
16	0	0	0	0	0	0	0	-780	26921	-584772
17	0	0	0	0	0	0	0	-462	18072	-436692
18	0	0	0	0	0	0	0	-240	12061	-322800
19	0	0	0	0	0	0	0	-123	7756	-233910
20	0	0	0	0	0	0	0	-42	4950	-168024
21	0	0	0	0	0	0	0	-45	2992	-118191
22	0	0	0	0	0	0	0	0	1840	-82302
23	0	0	0	0	0	0	0	0	990	-55851
24	0	0	0	0	0	0	0	0	582	-37836
25	0	0	0	0	0	0	0	0	300	-24648
26	0	0	0	0	0	0	0	0	153	-16110
27	0	0	0	0	0	0	0	0	52	-10110
28	0	0	0	0	0	0	0	0	55	-6372

Table 2 (continued): Table of N_d^g in the local $\mathbb{P}^2$ case

g	d= 10	11	12
0	-1431438	13025349	-126303034
1	-2686038	24811068	-243104587
2	-3639474	34387656	-342509526
3	-4234206	41210598	-419389857
4	-4485894	45176136	-471633948
5	-448091	46535634	-499923081
6	-4201602	45746832	-506850510
7	-3815196	43350270	-496239528
8	-3360462	39875763	-472228668
9	-2881647	35788092	-438935008
10	-2418726	31451790	-399896002
11	-1987920	27144348	-358135605
12	-1607280	23051883	-315914892
13	-1275786	19295790	-274989672
14	-999060	15936894	-236467548
15	-769119	13002726	-201130777
16	-584772	10481811	-201130777
17	-436692	8356338	-141171750
18	-322800	6587454	-116621626
19	-233910	5139042	-95518501
20	-168024	3965148	-77556222
21	-118191	3029214	-62468462
22	-82302	2287590	-49894916
23	-55851	1710720	-39545082
24	-37836	1264503	-31083092
25	-24648	925350	-24249023
26	-16110	668541	-18758088
27	-10110	478680	-14403727
28	-6372	337560	-10964512
29	-3729	235890	-8284122
30	-2280	235890	-6201922
31	-1221	110382	-4609204
32	- 714	73632	-3391380
33	-366	48894	-2477128
34	- 186	31449	-1789948
35	-63	20226	-1283697
36	- 66	12591	-909502

g	d= 11	12
37	7776	-639925
38	4533	-443502
39	2760	-305558
40	1473	-207008
41	858	-139293
42	858	-91742
43	222	-60365
44	75	-38430
45	78	-24571
46	0	-15100
47	0	-9297
48	0	-5404
49	0	-3280
50	0	-1746
51	0	-1014
52	0	-516
53	0	-261
54	0	-88
55	0	-91

Our first observation is that for fixed d, N_d^g have the same sign for $0 \leq g \leq g(d)$. This unexpected phenomenon might have an interpretation from the M-theory point of view. Secondly, for a positive integer d and a nonnegative integer m, set $M_d^m = (-1)^{d-1} N_d^{g(d)-m}$, then M_d^m is quadratic in d for large d. Indeed we have the following table for M_d^m:

Table 3: Table of M_d^m in the local $\mathbb{P}^2$ case

m	d= 1	2	3	4	5	6	7	8	9	10	11	12
0	3	6	10	15	21	28	36	45	55	66	78	91
1			7	12	18	25	33	42	52	63	75	88
2				33	51	72	96	123	153	186	222	261
3				30	96	138	186	240	300	366	438	516
4					129	258	354	462	582	714	858	1014
5					150	360	591	780	990	1221	1473	1746
6					114	580	996	1440	1840	2280	2760	3280
7						733	1476	2214	2992	3729	4533	5404
8						900	2244	3600	4950	6372	7776	9297

From this table one can verify the following formulas:

$$M_d^0 = \frac{(d+1)(d+2)}{2} = \frac{d^2+3d+2}{2}, \tag{9}$$

$$M_d^1 = \frac{(d+1)(d+2)}{2} - 3 = \frac{d^2+3d-4}{2}, \quad (d \geq 3) \tag{10}$$

$$M_d^2 = \frac{3(d^2+3d-6)}{2}, \quad (d \geq 4) \tag{11}$$

$$M_d^3 = 3(d^2+3d), \quad (d \geq 5) \tag{12}$$

$$M_d^4 = 6(d^2+3d-11), \quad (d \geq 6) \tag{13}$$

$$M_d^5 = \frac{21}{2}(d^2+3d) - 144, \quad (d \geq 7) \tag{14}$$

$$M_d^6 = 20(d^2+3d-16), \quad (d \geq 8) \tag{15}$$

$$M_d^7 = \frac{67}{2}(d^2+3d) - 626, \quad (d \geq 9) \tag{16}$$

$$M_d^8 = \frac{117}{2}(d^2+3d) - 1233, \quad (d \geq 10). \tag{17}$$

We expect in general M_d^m has the following form for $d \geq m+2$:

$$M_d^m = \frac{a(m)}{2}(d^2+3d) - b(m),$$

where $a(m), b(m)$ are positive integers (except for $b(5) = 0$). It is easy to see that (9) - (11) match with (1) - (3) by (6)-(8). One can also convert (12) -(17) to closed formulas for $n_d^{g(d)-m}$ for $4 \leq m \leq 8$. We leave that to the interested reader.

We observe similar behavior in the local $\mathbb{P}^1 \times \mathbb{P}^1$ case. We reformulate the table in [2] below.

Table 4: $n^g_{(d_1,d_2)}$ in the local $\mathbb{P}^1 \times \mathbb{P}^1$ case

g	$(d_1, d_2) = (1, 0)$	(2, 0)	(3, 0)	(4, 0)	(5, 0)	(6, 0)
0	- 2	0	0	0	0	0

g	$(d_1, d_2) = (1, 1)$	(2, 1)	(3, 1)	(4, 1)	(5, 1)	(6, 1)
0	-4	-6	-8	-10	-12	-14

g	$(d_1, d_2) = (2, 2)$	(3, 2)	(4, 2)	(5, 2)	(6, 2)
0	-32	-110	-288	-644	- 1280
1	9	68	300	988	2698
2	0	- 12	-116	-628	-2488
3	0	0	15	176	1130
4	0	0	0	-18	-248
5	0	0	0	0	21

g	$(d_1, d_2) = (3, 3)$	(4, 3)	(5, 3)	(6, 3)
0	-756	-3556	-13072	-40338
1	1016	7792	41376	172124
2	-580	-8042	-64624	-371980
3	156	4680	60840	501440
4	-16	-1560	-36408	-450438
5	0	276	13888	276144
6	0	-20	-3260	-115744
7	0	0	428	32568
8	0	0	-24	-5872

(In the table in [2], $n^4_{(5,3)} = 36048$. But this value does not seem to fit when one considers $N^g_{(d_1,d_2)}$.)

The following are noted in [2]. For a given bidegree (a, b), $n^g_{(a,b)}$ vanishes for $g > g(a, b) = (a - 1)(b - 1)$, which is indeed the arithmetic genus of a curve of bidegree (a, b) in $\mathbb{P}^1 \times \mathbb{P}^1$. One finds,

$$n^{g(a,b)}_{(a,b)} = -(-1)^{(a+1)(b+1)}(a + 1)(b + 1). \tag{18}$$

$$n^{g(a,b)-1}_{(a,b)} = 2(-1)^{(a+1)(b+1)}(a + b + ab - a^2 - b^2 + a^2b^2), \tag{19}$$

$$\begin{aligned} n^{g(a,b)-2}_{(a,b)} = & -(-1)^{(a+1)(b+1)}(-14 + 9(a + b) - 3ab - 3(a^2 + b^2) + 3a^2b^2 \\ & +2(a^3 + b^3 + a^2b + b^2a) - 2(a^3b + b^3a) - 2(a^3b^2 + b^3a^2) \\ & +2a^3b^3). \end{aligned} \tag{20}$$

For example,

Table 5

(a,b)	(6, 3)	(4, 4)	(5, 4)	(6, 4)	(7, 4)	(8, 4)	(9, 4)
$n^{g(a,b)}_{(a,b)}$	-28	25	-30	35	-40	45	- 50
$n^{g(a,b)-1}_{(a,b)}$	612	-496	776	-1116	1516	-1976	2496
$n^{g(a,b)-2}_{(a,b)}$	-5872	4266	-8982	16248	-26604	40590	-58746

We again expect closed formulas for $n^{g(a,b)-m}_{(a,b)}$ in the stable range, and since the degree in (a, b) increase very fast, such formulas are expected to be complicated. We again convert the above table to a table for N^g:

Table 6: Table of $N^g_{(d_1,d_2)}$ in the local $\mathbb{P}^1 \times \mathbb{P}^1$ case

g	(2, 2)	(3, 2)	(4, 2)	(5, 2)	(6, 2)	(3, 3)	(4, 3)	(5, 3)	(6, 3)
0	-14	42	-58	100	-148	112	330	788	1674
1	-9	20	-46	76	-127	112	364	984	2292
2	0	12	-26	58	-94	76	302	908	2262
3	0	0	-15	32	-70	28	184	656	1824
4	0	0	0	18	-38	16	100	428	1354
5	0	0	0	0	-21	0	36	232	860
6	0	0	0	0	0	0	20	124	524
7	0	0	0	0	0	0	0	44	280
8	0	0	0	0	0	0	0	24	148
9	0	0	0	0	0	0	0	0	52
10	0	0	0	0	0	0	0	0	28

We also have the following table from Table 5:

Table 7

(a,b)	(4, 4)	(5, 4)	(6, 4)	(7, 4)	(8, 4)	(9, 4)
$N^{g(a,b)}_{(a,b)}$	-25	30	-35	40	-45	50
$N^{g(a,b)-1}_{(a,b)}$	-46	56	-66	76	-86	96
$N^{g(a,b)-2}_{(a,b)}$	-130	160	-190	220	-250	280

We note for $g \leq g(a,b)$, $(-1)^{g(a,b)} N^g_{(a,b)}$ are all positive. Furthermore, if one set $M^m_{(a,b)} = (-1)^{g(a,b)} N^{g(a,b)-m}_{(a,b)}$, then one can verify that

$$M^0_{(a,b)} = (a+1)(b+1), \quad a \geq b \geq 1, \tag{21}$$

$$M^1_{(a,b)} = 2(a+1)(b+1) - 4, \quad a \geq b \geq 2, \tag{22}$$

$$M^2_{(a,b)} = 6(a+1)(b+1) - 20. \quad a \geq b \geq 3. \tag{23}$$

We also notice that

$$M^2_{(a,2)} = 12a - 2, \quad a \geq 4,$$
$$M^2_{(a,3)} = 24a + 4, \quad a \geq 3.$$

It is not hard to see that (21) - (23) are equivalent to (18)-(20). We conjecture that for $a \geq b \geq m + 1$, one has

$$M^m_{(a,b)} = x(m)(a+1)(b+1) - y(m)$$

for some positive integers $x(m)$ and $y(m)$ (except for $y(0) = 0$), and for any fixed m and b, when a is large enough,

$$M^m_{(a,b)} = u(m,b)a + v(m,b)$$

for some integers $u(m,b)$ and $v(m,b)$. We also conjecture that similar quadratic growth behavior of N^g hold for other local Calabi-Yau geometries. It is interesting to see if the arguments in [5] can be used to give an explanation of such phenomenon.

Acknowledgements. This research is partly supported by research grants from NSFC and Tsinghua University. The author thanks the organizer of the 23rd International Conference of Differential Geometric Methods in Theoretical Physics for the invitation.

References

1. M. Aganagic, A. Klemm, M. Marino, C. Vafa, *The topological vertex*, Commun.Math.Phys. 254 (2005) 425-478, hep-th/0305132.
2. M. Aganagic, M. Marino, C. Vafa, *All loop topological string amplitudes from Chern-Simons theory*, Commun.Math.Phys. 247 (2004) 467-512, hep-th/0206164.
3. R. Gopakumar, C. Vafa, *M-Theory and Topological Strings–II*, hep-th/9812127.
4. A. Iqbal, *All genus topological amplitudes and 5-brane webs as Feynman diagrams*, hep-th/0207114.
5. S. Katz, A. Klemm, C. Vafa, *M-theory, topological strings and spinning black holes*, Adv.Theor.Math.Phys. 3 (1999) 1445-1537, hep-th/9910181.
6. J. Li, C.-C. M. Liu, K. Liu, J. Zhou, *A mathematical theory of the topological vertex*, preprint, math.AG/0408426.
7. J. Zhou, *Localizations on moduli spaces and free field realizations of Feynman rules*, preprint, math.AG/0310283.